Ralph D'Agostino Sr. is Professor of Mathematics, Statistics, and Public Health at Boston University. A respected and widely published statistician, he has more than 30 years of experience in running clinical trials and epidemiological research. He is a Senior Editor of STATISTICS IN MEDICINE, Fellow of the American Statistical Association and Epidemiologic Section of the American Heart Association. He was Chair (2003) of the American Statistical Association Section of Statistics in Epidemiology. He serves as Executive Director of Biometrics and Data Management for the Harvard Clinical Research Institute. His interests are in biostatistics and robust procedures, longitudinal data analysis, and multivariate data. Dr. D'Agostinoís numerous awards include the Food & Drug Administration Commissioner's Special Citation in 1981 and 1995. He is Director of Data Management and Statistical Analysis for the Framingham Heart Study that for more than 50 years has searched for common factors that contribute to cardiovascular disease. He is co-author of five books in various fields of statistical methodology.

Lisa Sullivan is an Associate Professor of Biostatistics at the School of Public Health, Associate Professor of Mathematics and Statistics at Boston University, and Assistant Dean for Undergraduate Education in Public Health at Boston University, where she received her MA and her PhD. She has won numerous awards for excellence in teaching. Her research interests include applied biostatistics, longitudinal data analysis, design and analysis of clinical trials, and hierarchical modeling. Dr. Sullivan spends most of her time in the Boston University Statistics and Consulting Unit working on the Framingham Heart Study. Her recent research focuses on developing health risk appraisal functions to quantify individuals' risks of developing cardiovascular disease. Her dozens of articles are published in prestigious periodicals such as the JOURNAL OF THE AMERICAN MEDICAL ASSOCIATION, NEW ENGLAND JOURNAL OF MEDICINE, and STATISTICS IN MEDICINE. Away from work, she enjoys running and cooking.

Alexa Beiser is Professor of Biostatistics in the School of Public Health at Boston University. She received her MA from University of California at San Diego and her PhD from Boston University. Her research interests include clinical trials methodology, statistical computing, and survival analysis. Dr. Beiser joined the Framingham Study in 1994 after spending many years collaborating on pediatric research projects. She primarily investigates risk factors for stroke, dementia, and Alzheimer's disease using Framingham Study data. Her foremost methodological interest is in estimation of lifetime risk of disease. Dr. Beiser has published articles in the NEW ENGLAND JOURNAL OF MEDICINE, the JOURNAL OF THE AMERICAN MEDICAL ASSOCIATION, STATISTICS IN MEDICINE, STROKE, and NEUROLOGY. She enjoys reading, traveling, and spending time with her four children.

Introductory
Applied Biostatistics

Ralph B. D'Agostino, Sr.
Boston University

Lisa M. Sullivan
Boston University

Alexa S. Beiser
Boston University

THOMSON
™
BROOKS/COLE

Australia • Canada • Mexico • Singapore • Spain
United Kingdom • United States

THOMSON
BROOKS/COLE

Introductory Applied Biostatistics
Ralph B. D'Agostino, Sr., Lisa M. Sullivan, Alexa S. Beiser

Editor: Carolyn Crockett
Assistant Editor: Ann Day
Editorial Assistant: Daniel Geller
Technology Project Manager: Burke Taft
Marketing Manager: Stacy Best
Marketing Assistant: Jessica Bothwell
Executive Marketing Communications Manager: Nathaniel Bergson-Michelson
Project Manager, Editorial Production: Kelsey McGee
Art Director: Lee Friedman
Print Buyer: Lisa Claudeanos

Permissions Editor: Sarah Harkrader
Production Service: Matrix Productions/Merrill Peterson
Text Designer: Carolyn Deacy
Copy Editor: Pamela Rockwell
Cover Designer: Simple Design/Denise Davidson
Cover Image: PhotoDisc® Getty Images™
Compositor: Interactive Composition Corporation
Cover Printer: Phoenix Color Corp
Printing, Cover Printing, and Binding: R.R. Donnelley/Crawfordsville

Printed in the United States of America
1 2 3 4 5 6 7 09 08 07 06 05

For more information about our products, contact us at:
Thomson Learning Academic Resource Center
1-800-423-0563
For permission to use material from this text or product, submit a request online at **http://www.thomsonrights.com**. Any additional questions about permissions can be submitted by email to **thomsonrights@thomson.com**.

Library of Congress Control Number: 2004114720

ISBN 0-534-42399-X

Thomson Higher Education
10 Davis Drive
Belmont, CA 94002-3098
USA

Asia (including India)
Thomson Learning
5 Shenton Way
#01-01 UIC Building
Singapore 068808

Australia/New Zealand
Thomson Learning Australia
102 Dodds Street
Southbank, Victoria 3006
Australia

Canada
Thomson Nelson
1120 Birchmount Road
Toronto, Ontario M1K 5G4
Canada

UK/Europe/Middle East/Africa
Thomson Learning
High Holborn House
50-51 Bedford Row
London WC1R 4LR
United Kingdom

Latin America
Thomson Learning
Seneca, 53
Colonia Polanco
11560 Mexico
D.F. Mexico

Spain (including Portugal)
Thomson Paraninfo
Calle Magallanes, 25
28015 Madrid, Spain

Brief Contents

Contents

CHAPTER 6

Statistical Inference: Procedures for $(\mu_1 - \mu_2)$ 231

CHAPTER 7

Categorical Data 293

CHAPTER 8

Comparing Risks in Two Populations 359

CHAPTER 9

Analysis of Variance 407

CHAPTER 10

Correlation and Regression 465

CHAPTER 11

Logistic Regression Analysis 507

CHAPTER 12

Nonparametric Tests 545

CHAPTER 13

Introduction to Survival Analysis 585

APPENDIX A

Introduction to Statistical Computing Using SAS 591

APPENDIX B

Statistical Tables 609

APPENDIX C

Framingham Heart Study Longitudinal Data Documentation 631

Index 645

Preface

Biostatistics applies the principles of statistics to the medical or health fields. Although the techniques we discuss here can be applied broadly, our emphasis in the applications is generally medically oriented.

Why is statistics or biostatistics important? Consider an example. Most people believe that it is beneficial in terms of cardiovascular health to exercise regularly and to eat a low-fat diet. It is this underlying belief that often motivates us to exercise on a day when we otherwise might skip exercising or to feel guilty when we consume a high-fat meal or maybe two desserts. Should we be motivated by this belief or truly feel guilty? Maybe this idea of exercising and eating right is just a rumor that got out of control!

Why do we believe that exercising and eating healthy is associated with better cardiovascular health? Well, a number of research studies have shown that people who exercise regularly and eat low-fat diets are less likely to suffer from cardiovascular disease. An example of one such study is the Framingham Heart Study (Kannel, W. B.: "Habitual level of physical activity and risk of coronary heart disease: the Framingham Study." *Can. Med. Assoc. J.* 1967; 96: 811–812; Dawber, T. R., Kannel, W. B., Pearson, G., Shurtleff, D.: "Assessment of diet in the Framingham Study: methodology and preliminary observations. *Health News* 1961; 38: 4–6.). The Framingham Heart Study started in 1948 and involved over 5,000 men and women, who make up the Framingham cohort. At the beginning, each participant in the study had a complete physical examination and they came back for repeat examinations every 2 years. Along with monitoring such things as blood pressure, cholesterol, exercise, and nutrition, the study investigators also monitored whether participants had heart attacks or other cardiovascular disease over time. In the early 1970s, the offspring of the original cohort and their spouses were invited to join what is called the Framingham Offspring cohort. There are over 5,000 participants in the Offspring study and they have been examined approximately every 4 years. In 2002, a third generation was enrolled, which allows for more detailed investigations of genetic risk factors. This multigenerational study is still going on today, and much of what we understand about cardiovascular disease is derived from this very important study.

The details of investigations based on important studies often get reported in the newspaper or recapped on television newscasts. Should we believe that just because there was an association between exercise and cardiovascular disease among the participants in the Framingham Heart Study (or some other study) the same would hold for us? This idea of generalizing or inferring associations that are observed in a group of participants under study to the population at large is the crux of statistics.

Every day we are inundated with information. Information comes from newspapers, magazines, books, television newscasts, and the Internet. We need to distinguish important information from not-so-important information. We need to determine when the results of research studies apply to us and when they do not. It is unlikely that many of us will ever participate in a research study. However, we rely on well-conducted research studies to inform us of important associations, such as those between certain behaviors (e.g., exercise and diet) and better health. Statistics will provide us the tools to analyze and interpret studies.

Our primary goal in this book is to provide background for readers to apply and appropriately interpret statistical applications in the medical field. We have three specific aims:

- to provide an overview of *statistical vocabulary*
- to describe *statistical methodology and interpretation* and
- to introduce *statistical computing techniques*

Throughout the book, we present vocabulary associated with each application along with explicit definitions of terms and concepts. We provide readers with the tools to implement appropriate statistical techniques along with sufficient detail to understand each concept in a broad sense.

We cover a variety of methodological topics, from descriptive statistics and probability theory to statistical inference. The majority of the book is dedicated to methods for statistical inference, including one- and two-sample tests for means and proportions, analysis of variance techniques, and correlation and regression. In each topic area, we discuss the methodology, including assumptions and statistical formulas, along with appropriate interpretation of results. We introduce and discuss applications through real examples, most of which are taken from our work in applied biostatistics. We have purposely selected examples involving relatively few subjects to illustrate computations while minimizing the actual computation time. All of the techniques described can in practice be applied to larger problems.

Statisticians almost always use the computer and one or another statistical computing package to implement statistical techniques. There are a number of statistical computing packages available and most can be implemented on any computing platform (e.g., personal computer, Macintosh, UNIX, mainframe). In this book, we use SAS (SAS Institute Inc., SAS® User's Guide: Statistics, Version 8.02. Cary, NC: SAS Institute Inc., 1999–2001) to illustrate statistical computing techniques. In each chapter, we present SAS programming code and output for each technique. Our objectives are to familiarize the reader with both the format and content of computer output and to provide appropriate interpretation of results. In the statistical computing sections, we present programming techniques for interested readers. An introduction to the statistical computing package SAS is contained in Appendix A, which includes an overview of the SAS system, techniques for creating and executing SAS programs, and examples of various techniques for entering data. Throughout the statistical computing sections, we include advanced statistical computing techniques, in the sections marked *, for more sophisticated readers. These sections can be omitted, as appropriate; such omission will not disrupt the reader's understanding of the material.

However, even though statistical techniques are almost always implemented using statistical computing packages, we feel strongly that readers must master techniques by hand before moving to computer applications. We therefore focus on explicit computations throughout this book. Wherever possible, we illustrate techniques by hand and then perform the corresponding analysis using SAS.

We also provide examples of statistical analysis on data collected from the Framingham Heart Study Offspring cohort. The data set contains real data but is appropriately masked to protect participant confidentiality. The details of the data set are described in Appendix C.

Motivation

I

As an introduction to the topics covered in detail in subsequent chapters, we present here a simple example to illustrate the major concepts in applied biostatistics. We present, in overview fashion, a number of vocabulary terms along with brief definitions. General formulas are indicated in shaded boxes. Subsequent chapters contain more explicit and involved discussions of each term and concept. We recommend that readers refer back to the example presented here from time to time as a means of reminding themselves about the various concepts and their interrelationships.

In Section 1.1, we present some general vocabulary terms and the example. In Section 1.2, we discuss population parameters. In Section 1.3, we discuss sampling and sample statistics, and in Section 1.4, we describe statistical inference.

1.1 Vocabulary

Statistical analysis is the analysis of characteristics of subjects of interest. *Subjects* are the units on which these characteristics are measured. In most medical applications, subjects are human beings. However, subjects may be cells, blood samples, or animals used in research experiments. The characteristics are measurable properties, such as age, systolic blood pressure, outcome of surgery (e.g., success, failure), or total cholesterol level, and are called *variables*. It is important to define variables explicitly in each application along with appropriate measurement units. For example, age could be measured in years, weeks, days, hours, and so on. An investigation involving human subjects would most likely measure the age of each subject in years (except possibly an investigation of newborns, in which case weeks or days may be more appropriate). An investigation involving blood samples might measure the age of each sample in days or hours.

In statistical applications, we work with *data elements* or *data points*. Each data element is a representation of a particular measurable characteristic, or variable. Data elements may be subjects' ages measured in years (e.g., 51, 29, 36), systolic blood pressures measured in millimeters of mercury (mmHg) (e.g., 140, 160, 110), or classifications of disease stage (e.g., stage I, stage II, stage III).

In all statistical applications, it is important to define the *population* of interest explicitly. A population is simply the collection of *all* subjects of interest. For example, if we are interested in American males with heart disease, then the collection of *all* American males with heart disease would constitute the population. A *sample* is a subset of the population of interest. For example, American males aged 50 with heart disease would constitute one sample from the population of all American males with heart disease. One hundred randomly selected American males with heart disease would constitute another sample from the population of all American males with heart disease (assuming that there are many more than 100 American males with heart disease).

Many different samples can be selected from any given population. The number of distinct samples that can be taken depends on the numbers of subjects in both the population and in the sample. In subsequent chapters, we will outline explicit formulas to determine the number of possible samples from a particular population. We will also present some of the more popular methods used to select subjects from a population into a sample.

The number of subjects in a population, or the *population size,* is denoted N. The number of subjects in a sample, or the *sample size,* is denoted *n.* Any descriptive measure based on a population is called a *population parameter,* or simply a *parameter.* Any descriptive measure based on a sample is called a *sample statistic,* or simply a *statistic.* N is an example of a parameter, whereas *n* is an example of a statistic.

Data elements may have been measured on each member of a population or on each member of a sample. In most applications, populations are very large. There are exceptions, however, and an analyst cannot tell simply by looking at the data set or by its size alone (i.e., the number of data elements) whether the collection of data elements comprise a population or a sample. Someone involved in the design of the study, such as the statistician or an investigator in the particular substantive field, must convey such information to the data analyst.

Data elements (or observations) deriving from either a population or a sample are denoted X, where X is a variable name, or placeholder, representing the characteristic of interest. We will illustrate the use of this notation through examples in Chapter 2.

EXAMPLE 1.1

Relationship Between Population Parameters and Sample Statistics

Suppose we have a population consisting of 5 individuals who are 65 years of age or older (i.e., $N = 5$). In this example, our population is small and our interest lies solely in these 5 individuals. Suppose we are interested in analyzing the number of visits to primary-care physicians over a 3-year period in our population. We survey each member of our population and assess the number of visits to primary-care physicians over the previous 3 years. We exclude emergency room visits and hospitalizations from our assessment. The characteristic of interest, or the variable under investigation, is the number of visits to primary-care physicians in 3 years. The population data follow and are displayed in Figure 1.1 in a dotplot, where each observed data point is

Figure 1.1 *Number of Primary Care Visits in 3 Years (N = 5)*

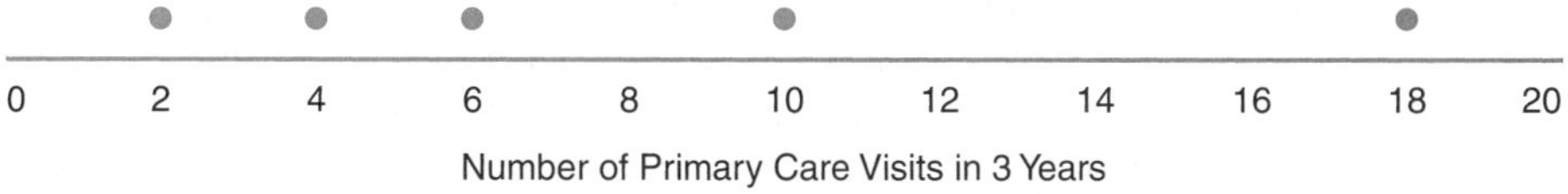

indicated by a dot. The horizontal scale represents the number of visits to primary-care physicians in 3 years.

Subject Number	Number of Visits to Primary-Care Physicians in 3 Years
1	2
2	4
3	6
4	10
5	18

■

In this population, 3 of 5 subjects reported 6 or fewer visits to primary-care physicians in 3 years. Two of the subjects reported many more visits (10 and 18). Recall that each subject is 65 years of age or older, which might explain the magnitude of reported numbers of visits. The 2 subjects reporting 10 and 18 visits to primary-care physicians in 3 years may have had chronic illnesses that required more frequent follow-up or acute illnesses or conditions that required frequent attention over shorter periods of time. As will become apparent in the remainder of this book, interpreting data (even a single characteristic measured on what seems to be a relatively homogenous collection of subjects—individuals 65 years of age or older) is often quite complicated.

Suppose that the population in Example 1.1 included $N = 500$ individuals 65 years of age or older or $N = 5{,}000$ individuals 65 years of age or older. In both cases, it would be impossible to understand the population with respect to the reported numbers of visits to primary-care physicians simply by inspecting the observed data elements. Even in the smaller population ($N = 500$), there would be too many data elements to draw conclusions simply by inspection. It is generally necessary to summarize characteristics measured on a population to understand the characteristic under investigation. Several summary measures are described next.

1.2 Population Parameters

As we have said, any measure computed on a population is called a *population parameter*, or simply a *parameter*. There are many parameters; we will present only a few here. The first parameter is the *population size*, denoted N. In Example 1.1, $N = 5$.

We are often interested in describing a population in terms of its average value on a particular characteristic. For Example 1.1, we may wonder, what is the typical number of visits that patients 65 years of age or older make to primary-care physicians in 3 years? The *population mean*, denoted μ ("mu"), addresses this question and is computed by summing the values and dividing by the population size.

$$\mu = \frac{\sum X}{N} \tag{1.1}$$

where $\sum$ (uppercase "sigma") denotes summation
X is a placeholder that represents the characteristic under consideration (e.g., number of primary-care visits in 3 years) and
N denotes the population size

In Example 1.1, the population mean is 8:

$$\mu = (2 + 4 + 6 + 10 + 18)/5 = 8$$

The mean is a very useful parameter. Since the population in Example 1.1 is so small, it is not necessary to summarize the data elements to understand the population with respect to the number of primary-care visits in 3 years. However, if the population size were $N = 500$ (or $N = 5,000$), the population mean would provide a very useful summary.

In addition to summarizing the population with respect to what a typical value looks like (i.e., $\mu = 8$), it is generally of interest to understand variability in the characteristic of interest. In Example 1.1, no two individuals reported the same number of visits to primary-care physicians in 3 years. If we take the population mean 8 as representative of a typical number of primary-care visits in 3 years, how close is each individual data element in the population to that typical value? In particular, are all of the reported numbers of visits close to 8 or do they vary widely above and below 8?

There are several measures of variability or dispersion. The first we discuss is the *population range,* which is defined as the difference between the largest, or maximum, score and the smallest, or minimum, score. In Example 1.1, the population range is $18 - 2 = 16$. Some statisticians would report the range as 2 to 18, whereas others would report the range as 16. Either is acceptable. A more sophisticated and intuitive measure of dispersion is the *population variance,* denoted σ^2 ("sigma squared"). The population variance is based on "deviations from the mean" or distances between each observation and the population mean. The following table displays the data elements from Example 1.1 along with deviations from the mean (i.e., distances from $\mu = 8$):

X	$(X - \mu)$
2	−6
4	−4
6	−2
10	2
18	10
	0

Notice that the sum of the deviations from the mean is zero, a property of the population mean. Recall that the goal is to generate an estimate of the dispersion in the population, in particular the dispersion in numbers of visits to primary-care physicians in 3 years relative to the population mean, $\mu = 8$. In the table, we computed the deviations from $\mu = 8$ for each subject in our population. Inspecting the deviations, we see a fair amount of variation in the reported numbers of visits. The first subject reported 6 fewer visits than the mean, the second subject reported 4 fewer visits than the mean, and the last subject reported 10 more visits than the mean number of visits. Since this population is small, we can evaluate the variability in the population by inspection. Again, imagine a population of size $N = 500$ (or $N = 5,000$), in which it would be impossible to inspect deviations from the mean and draw conclusions regarding variability.

Several techniques can be employed to summarize the magnitude of these deviations from the mean. Notice that it is not practical to take the mean deviation (i.e., sum the deviations and divide by N) since the sum of the deviations is always zero. The most popular method used to assess variation is the average squared deviation from the mean, or the *variance*. This measure proves to be the most straightforward mathematically. Another summary measure is the mean absolute deviation. However, this measure can be mathematically difficult, especially in mathematical proofs, which are beyond the scope of this discussion. The following table displays the numbers of primary-care visits in 3 years, denoted X, deviations from the mean, and squared deviations from the mean, respectively.

X	$(X - \mu)$	$(X - \mu)^2$
2	−6	36
4	−4	16
6	−2	4
10	2	4
18	10	100
	0	160

The sum of the squared deviations from the mean is 160. The population variance is the average of the squared deviations from the mean, defined as follows:

$$\sigma^2 = \frac{\sum (X - \mu)^2}{N} \tag{1.2}$$

In Example 1.1, the population variance is 32:

$$\sigma^2 = \frac{\sum (X - \mu)^2}{N} = \frac{160}{5} = 32$$

The population variance for Example 1.1 is interpreted as follows. The number of primary-care visits in 3 years is, on average, 32 visits squared from the mean of 8. In computing the variance, we squared each deviation from the mean to capture the magnitude of the deviations (since the negative deviations cancelled the positive deviations). To return to our original units, we take the square root, which produces the *population standard deviation,* denoted σ ("sigma"):

$$\sigma = \sqrt{\sigma^2} \qquad (1.3)$$

The standard deviation is $\sigma = \sqrt{32} = 5.7$. The population standard deviation in Example 1.1 is interpreted as follows: The typical deviation in number of primary-care visits in 3 years is about 5.7 visits from the mean of 8.

The population standard deviation is the parameter most widely used to describe the dispersion in a population and is interpreted as the typical deviation from the mean. It is difficult to quantify large and small values of the standard deviation, as the magnitude of the standard deviation depends on the measurement scale of the characteristic under investigation. The same is true for the mean value; however, the interpretation of the mean is more intuitive. The standard deviation is particularly useful in comparing populations with respect to a particular characteristic. For example, suppose we have two populations and wish to make comparisons with respect to numbers of primary-care visits in 3 years. Suppose the two populations have equal mean numbers of primary care visits (i.e., $\mu_1 = 8$ and $\mu_2 = 8$). Suppose that the standard deviations in numbers of visits are 5.7 and 2.3, respectively (i.e., $\sigma_1 = 5.7$ and $\sigma_2 = 2.3$). The difference in the standard deviations indicates that the numbers of visits are more dispersed in population 1 as compared to population 2, while both populations have the same means. In population 2, the reported numbers of visits are more tightly clustered about the mean of 8 (i.e., the reported numbers of visits deviate by about 2.3 visits from the mean of 8 visits).

1.3 Sampling and Sample Statistics

In Section 1.3 we outlined a number of parameters used to describe a population. In most statistical applications, we do not have the entire population available, but instead have only a subset or sample of individuals selected from the population of interest. In many situations, it is impossible and/or impractical to analyze the entire population. In some cases, the population may be so large that it is impossible to measure a specific characteristic or set of characteristics on each subject. In other cases, it may be too costly to measure certain characteristics on each subject. For example, suppose we wish to analyze a particular characteristic that is measured by a laboratory test costing more than $2,000 per subject. It may not be financially feasible to conduct tests on every member of the population. In some applications, it also may be too time-consuming to analyze the entire population. In such a case, it may take so long to collect and analyze the data that the results are not useful. We therefore rely on samples or subsets of the population for analysis. If the sample of subjects (i.e., the subset of the population) is representative of the population, then it is reasonable to assume that what we observe in the sample is similar to what we would observe in the population if the entire population were observed or analyzed. This notion is the basis for statistical inference, and we will provide a more complete justification in Chapters 3 and 4.

Many different samples can be drawn from any given population. The number of distinct samples that can be taken depends on the numbers of subjects in both the population and the sample. (Processes for selecting samples will be discussed in several subsequent chapters.) To illustrate the concept of sampling, suppose in Example 1.1 we enumerate all possible samples of size 3 from our population of 5 individuals. There are 10 samples of size $n = 3$ that do not contain any individual more than once (i.e., there are 10 simple random samples without replacement). In the following samples, X_1 denotes the number of visits to primary-care physicians reported by the first individual selected, X_2 denotes the number of visits reported by the second individual selected, and so on (see also Figure 1.2). For example, the first sample consists of subjects 1, 2, and 3 and the second sample consists of subjects 1, 2, and 4.

Sample	X_1	X_2	X_3
1	2	4	6
2	2	4	10
3	2	4	18
4	2	6	10
5	2	6	18
6	2	10	18
7	4	6	10
8	4	6	18
9	4	10	18
10	6	10	18

Notice how many distinct samples are possible from this very small population. Imagine how many samples of size $n = 10$ (or $n = 50$) are possible from a population of $N = 500$ (or from a population of size $N = 5,000$).

The samples shown are exhaustive (all-inclusive) under sampling without replacement. Each sample shown is equally likely to be the drawn sample. That is, the chance that any one sample is selected from our population is 1/10, or 10%.

Figure 1.2 *Simple Random Samples of Size n = 3 from a Population with N = 5*

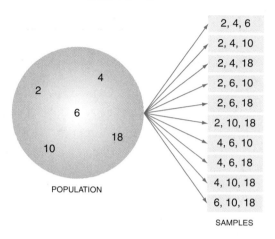

As we have mentioned, measures computed on a population are called *parameters;* measures computed on a sample are called *sample statistics,* or simply *statistics.* (In Chapter 2 we present and interpret a variety of statistics.) For illustration purposes, suppose we compute the *sample mean,* denoted $\overline{X}$ ("X bar") for each of the samples just shown. As in the case of the population mean, the sample mean is computed by summing the values and dividing by the *sample size* (i.e., $\overline{X} = \Sigma X/3$). Suppose we also compute the *sample range* (i.e., maximum–minimum) for each sample. The samples, their means, and ranges follow:

Sample				
X_1	X_2	X_3	Sample Mean $\overline{X} = \Sigma X/n$	Sample Range $= Max - Min$
2	4	6	4.0	4
2	4	10	5.3	8
2	4	18	8.0	16
2	6	10	6.0	8
2	6	18	8.7	16
2	10	18	10.0	16
4	6	10	6.7	6
4	6	18	9.3	14
4	10	18	10.7	14
6	10	18	11.3	12

Figure 1.3 *Sampling Distribution of the Sample Means (n = 3)*

0	2	4	6	8	10	12	14	16	18	20

Mean Number of Primary Care Visits in 3 Years

Notice that the sample means and sample ranges vary, depending on the individuals selected into each sample. The enumeration of all possible sample means is called the *sampling distribution of the sample means*. A sampling distribution is the listing of all values of a statistic (e.g., $\overline{X}$) based on all possible samples generated under a particular sampling strategy. The distribution of the sample means is shown in a dotplot in Figure 1.3.

To reiterate, we generally never have an entire population available; instead, we have a sample. In statistical applications, we attempt to draw inferences about a population based on that sample. The theory behind statistical inference is based on the relationship between the sampling distribution of the sample statistic and the population parameter (described in detail in Chapter 4). Here we introduced a population in Example 1.1 and enumerated all possible samples of size $n = 3$ without replacement. In the following section, we introduce the concept of statistical inference.

1.4 Statistical Inference

Suppose the administrators of a particular health maintenance organization (HMO) wish, in order to allocate resources, to estimate the number of visits to primary-care physicians that enrollees 65 years of age or older make in 3 years. Suppose that the HMO has $N = 5,489$ members who are 65 years of age or older (assume that some sociodemographic characteristics such as age and gender are available for each enrollee of the HMO in a centralized database but the number of primary-care visits is not). It would be very time-consuming and costly to survey each of the $N = 5,489$ enrollees with regard to the number of primary-care visits they made in the past 3 years. Statistical inference techniques can be used to estimate the mean number of visits among all members of the HMO who are 65 years of age or older based on a subset or sample of such enrollees of the HMO.

Figure 1.4 *Statistical Inference*

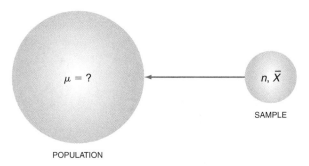

In statistical inference (discussed extensively in Chapters 5–13), we have a single sample and wish to make inferences about unknown population parameters based on sample statistics. Figure 1.4 illustrates a common situation in which the population mean of a particular characteristic, μ, is unknown. To generate an estimate of this population mean, we take a sample of subjects from the population and compute various statistics (e.g., the sample mean, $\overline{X}$). We use this information from the sample to make inferences about the unknown population mean. Inferences about a population are valid if the sample is a random sample (that is, all samples have an equal chance of being selected). It is intuitive, but worth noting, that the larger the sample size (larger n), the "better" the inference about the population parameter. We will define explicit criteria for determining how "good" estimates are in Chapter 5. Here, we use our simple Example 1.1 to illustrate the notion of statistical inference and to raise issues that will be addressed in complete detail in subsequent chapters.

In Example 1.1, we know the population and we computed the population mean, $\mu = 8$. Notice that only 1 sample out of 10 had a sample mean ($\overline{X}$) of 8. The sample means ranged from 4.0 to 11.3. In practice, we will not know the population and we will have only a single sample from the population of interest. Suppose that by chance we happened to select the first sample (which is as likely as any other sample to be selected). The reported numbers of primary-care visits among individuals in the first sample ($\overline{X} = 4.0$) are much lower than the numbers reported by other individuals in the population. If this were a real application, we would not know the population mean (or anything else about the population) and we would underestimate the mean number of primary-care visits in 3 years among enrollees 65 years of age or older based on this first sample. If we happened by chance to select the last sample ($\overline{X} = 11.3$), we would overestimate the mean number of visits in 3 years among enrollees 65 years of age or older.

Table 1.1 *Summary of Notation: Population Parameters*

Parameter	Notation/Formula	Description
Population size	N	Number of subjects in population
Population mean	$\mu = \dfrac{\sum X}{N}$	Typical or average value
Population variance	$\sigma^2 = \dfrac{\sum(X-\mu)^2}{N}$	Mean squared deviation from the mean
Population standard deviation	$\sigma = \sqrt{\dfrac{\sum(X-\mu)^2}{N}}$	Typical deviation from the mean

In practice, we have one sample and, in fact, never know for sure if we are underestimating population parameters, overestimating population para-meters, or are right on target. We will, however, be able to quantify how much "error" is in our estimate of the population parameter. We must be willing to accept some error due to the fact that our sample is only a subset of the population. In practice, we will also generally have samples of much larger size (larger n). Example 1.1 is a simple (somewhat unrealistic) exam-ple, used only to illustrate concepts.

If we compute the mean of the sample means (i.e., sum the 10 sample means—$\overline{X}$s—and divide by 10), we see that the mean of the sample means is equal to 8.0, the value of the population mean. Therefore, on the average, the sample mean $\overline{X}$ is equal to the population mean μ. Based on this property, we say that the sample mean is an *unbiased* estimator of the population mean.

We computed the population range as 16, and only 3 samples of 10 have a range of 16. The other samples all have ranges less than 16. On average, the sample range is not equal to the population range, but is less than the population range. The sample range is a biased estimator of the population range.

In statistical inference, discussed in detail in the majority of the chapters in this book, the first step in addressing a problem of estimating a population parameter is choosing a sample statistic with good properties, such as unbiasedness. Statistical inference is based on probability theory, reviewed in Chapter 3, and techniques for summarizing sample data are discussed in Chapter 2. Table 1.1 summarizes the notation and formulas for commonly used population parameters. The notation and formulas for sample statistics are presented in Chapter 2.

Descriptive Statistics
(Ch. 2)

Probability
(Ch. 3)

Sampling Distributions
(Ch. 4)

Statistical Inference
(Chapters 5–13)

OUTCOME VARIABLE	GROUPING VARIABLE(S)/ PREDICTOR(S)	ANALYSIS	CHAPTER(S)
Continuous	—	Estimate μ, compare μ to known, historical value	5, 12
Continuous	Dichotomous (2 groups)	Compare independent means (estimate/test $(\mu_1 - \mu_2)$) or the mean difference (μ_d)	6, 12
Continuous	Discrete ($>$ 2 groups)	Test the equality of K means using analysis of variance $(\mu_1 = \mu_2 = \cdots \mu_k)$	9, 12
Continuous	Continuous	Estimate correlation or determine regression equation	10, 12
Continuous	Several continuous or dichotomous	Multiple linear regression analysis	10
Dichotomous	—	Estimate p, compare p to known, historical value	7
Dichotomous	Dichotomous (2 groups)	Compare independent proportions (estimate/test $(p_1 - p_2)$)	7, 8
Dichotomous	Discrete ($>$ 2 groups)	Test the equality of k proportions (chi-square test)	7
Dichotomous	Several continuous or dichotomous	Multiple logistic regression analysis	11
Discrete	Discrete	Compare distributions among k populations (chi-square test)	7
Time to Event	Several continuous or dichotomous	Survival analysis	13

Summarizing Data

The first step in solving any problem is understanding it clearly. In statistics, the basis for any analysis is a clear understanding of the data. Describing, summarizing, and presenting data are central to all statistical applications. Before any statistical inferences about a population parameter (e.g., μ) are made based on a sample statistic (e.g., $\overline{X}$), the sample must be summarized appropriately. In Figure 2.1 we illustrate the concept of statistical inference. Specifically, we want to estimate the mean of a population based on an analysis of a sample. For example, suppose we want to estimate how frequently individuals in a population exercise; for this analysis we measure exercise as the number of hours of exercise per week. Suppose the population of interest includes all persons with adult-onset (or Type II) diabetes. These individuals are supposed to exercise (along with following a special diet) as part of their treatment plan. It would be impossible to study all persons with adult-onset diabetes, so instead we select a sample. We measure the number of hours of exercise per week among the members of our sample and use that information to make generalizations about exercise in the population of all persons with adult-onset diabetes. In statistical inference, we take into account the fact that we did not study the entire population by quantifying how much uncertainty exists in our analysis (described in detail in Chapter 5). Before we get to statistical inference, however, we must understand how to appropriately summarize our sample. In this chapter, we discuss both numerical and graphical descriptions, summaries, and presentations of sample data.

Section 2.1 provides background, including vocabulary and notation and also draws the distinction between continuous and discrete variables. Section 2.2 outlines an array of descriptive statistics and graphical methods and uses a number of examples to illustrate their applications. Sections 2.2.1 and 2.2.2, respectively, describe numerical and graphical summaries for continuous variables, and Sections 2.2.3 and 2.2.4, respectively, describe numerical and graphical summaries for discrete variables. We illustrate computations and graphical displays both by hand and by using SAS. Section 2.3 summarizes key formulas, and Section 2.4 provides SAS program code used to generate descriptive statistics and graphical displays. Section 2.5 uses data from the Framingham Heart Study to illustrate the concepts presented in the chapter.

Figure 2.1 *Statistical Inference*

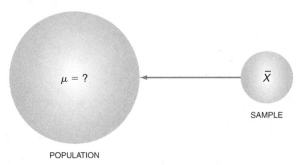

2.1 Background

2.1.1 Vocabulary

In statistics, we work with *data elements* or *data points*. Each data element is a representation of a particular measurable characteristic, or variable. For example, data elements may be individuals' ages measured in years (e.g., 51, 29, 36) or patients' systolic blood pressures measured in millimeters of mercury (mmHg) (e.g., 140, 160, 110). Data elements are measured on *subjects* or *units of measurement*. Although we assume subjects in most applications to be human beings, this is not always the case. Subjects might be animals involved in a research experiment where we might be interested in the time it takes each animal to complete a specific task. In this case, the time to complete the task is the variable of interest, and the observed times (in minutes) are the data elements. Subjects could be hospitals in a specific region of the country (e.g., the Northeast), and we might be recording the number of orthopedic procedures performed in each hospital in 1 year. In this case, the number of orthopedic procedures performed in 1 year is the variable of interest, and the observed number of procedures (measured in each hospital) are the data elements. The subjects or units of measurement are specific to each application and must be defined explicitly.

Subjects comprise either a *population* or a *sample*. A population is a collection of *all* subjects of interest. For example, if we are interested in American males with heart disease, then the collection of *all* American males with heart disease would constitute the population. A sample is a subset of the population of interest. For example, 100 randomly selected American males with heart disease would constitute one sample from the population of all American males with heart disease.

Many different samples can be taken from any given population. The number of distinct samples from a given population depends on the numbers of subjects in both the population and in the sample. In subsequent chapters, we will provide explicit formulas to determine the number of possible samples from a particular population. We will also present some of the more popular methods used to select subjects from a population into a sample. For now, we assume that the subjects in any sample are selected at random from the population of interest and that the sample, as a whole, is representative of the population of interest.

2.1.2 Classification of Variables

Each variable under investigation can be classified as either *continuous* or *discrete*. Continuous (or measurement) variables assume, in theory, any value between the minimum and maximum value on a particular measurement scale. Variables such as age, height, weight, cholesterol level, and systolic blood pressure are examples of continuous variables. Discrete variables take on a limited number of values, or categories. Discrete variables can be either *ordinal* or *categorical* (sometimes called *nominal*) variables. Ordinal variables take on a limited number of values or categories, but the categories are ordered. For example, symptom severity is an example of an ordinal variable with the ordered response options Minimal, Moderate, Severe. Self-reported health status is another example of an ordinal variable with ordered response options: Excellent, Very Good, Good, Fair, Poor. Categorical variables take on a limited number of categories and the categories are unordered. Gender is an example of an unordered categorical variable; its two response options are Male, Female. Marital status is another example of an unordered categorical variable, with the response options Married, Separated, Divorced, Widowed, Never Married. It is important to determine the nature of the variable under investigation, as the selection of the most appropriate technique to summarize it depends on whether the variable is continuous or discrete.

2.1.3 Notation

Any descriptive measure based on a population is called a *population parameter,* or just a *parameter.* Any descriptive measure based on a sample is called a *sample statistic,* or just a *statistic.* The number of subjects in a population, or the *population size,* is denoted N. The number of subjects in a sample, or the *sample size,* is denoted n. μ is an example of a parameter, and $\overline{X}$ is an example of a statistic.

The collection of data elements, called the *data set,* may have been measured on each member of a population or on each member of a sample. An analyst cannot tell simply by looking at the data set or by its size alone (i.e., the number of data elements) whether the collection of data elements comprise a population or a sample. Someone involved in the design of the study, such as the statistician or an investigator in the particular substantive field, must convey such information to the data analyst.

Data elements (i.e., observations) deriving either from a population or a sample are denoted X_i, where X is a variable name, or placeholder, representing the characteristic of interest and the subscript i refers to the subject number on which the measurement was taken. For example, suppose we have a sample consisting of 4 individuals or subjects (i.e., $n = 4$) and measure the age of each subject. Suppose the sample data set is as follows:

$$21 \quad 25 \quad 32 \quad 26$$

The first subject is 21 years of age, the second subject is 25 years of age, and so on. If we let $X = $ age, we can denote each data element as follows:

$$X_1 = 21 \qquad X_2 = 25 \qquad X_3 = 32 \qquad X_4 = 26$$

NOTE: In most applications the subscript, denoting the subject number, will be omitted for simplification.

This notation is employed to simplify computations and mathematical formulas, which are illustrated in the next section.

2.2 Descriptive Statistics and Graphical Methods

This section presents an array of descriptive statistics and graphical methods used to summarize sample data. In all cases, we assume that the subjects comprising the sample were selected at random from the population of interest (i.e., the sample is representative of the population). Each example illustrates different descriptive statistics; general formulas are shown in blue. In almost all cases, we consider small samples (i.e., small n) for ease of illustration, but in practice, these same descriptive statistics can be applied to larger samples (i.e., larger n). Section 2.2.1 presents numerical summaries for continuous variables, and Section 2.2.2 presents graphical summaries for continuous variables. Section 2.2.3 presents numerical summaries for discrete variables, and Section 2.2.4 presents graphical summaries for discrete variables. In Section 2.3 we summarize each of the descriptive statistics and provide a general rationale for their use. We encourage readers to refer often to Section 2.3 as they study the examples in this section.

2.2.1 Numerical Summaries for Continuous Variables

The following are examples of using descriptive statistics to summarize continuous or measurement variables.

EXAMPLE 2.1

Summary Statistics on Systolic Blood Pressures

Suppose we select a random sample of 7 subjects from a population of patients 50 years of age with diagnosed coronary artery disease. We measure systolic blood pressures, in millimeters of mercury (mmHg), on each subject. The sample data are

121   110   114   100   160   130   130

Because the sample size is small ($n = 7$), we can summarize the sample with respect to systolic blood pressure by inspection. The lowest blood pressure in the sample is 100 and the highest blood pressure is 160. A systolic blood pressure of 60 would be considered clinically problematic. The remaining blood pressures are between 110 and 130 and seem quite reasonable given that the subjects were sampled from the population of patients 50 years of age with diagnosed coronary artery disease. Had the patients been sampled from the population of patients 30 years of age free of cardiovascular disease, these observed systolic blood pressures might be slightly higher than expected.

If the sample included not 7 patients, but 700 instead, it would be impossible to summarize the sample with respect to systolic blood pressure by inspecting the values. In fact, the same would probably be true if the sample included 20 subjects. In most applications, it is necessary to use statistical techniques to summarize a sample. Several of these are described next. To simplify the computations, each data element (i.e., each observed systolic blood pressure) is represented by the variable X. Here X denotes systolic blood pressure, and the subscripts ($i = 1, 2, \ldots, 7$) denote the subject number in the sample:

$$X_1 = 121 \ \ X_2 = 110 \ \ X_3 = 114 \ \ X_4 = 100 \ \ X_5 = 160 \ \ X_6 = 130 \ \ X_7 = 130$$

It is generally of interest to summarize a continuous variable with respect to location. Location refers to the "center" of the data set and addresses the question, What is a typical systolic blood pressure? In the computations that follow, we can drop the subscripts, because the subject number (i.e., subscript) has no impact on them. To organize our calculations, we arrange the data elements in a column, as shown. Notice that the data elements are ordered from smallest to largest (this is not necessary but is sometimes convenient) and that the data element 130 is listed twice, as subjects 6 and 7 each have systolic blood pressures of 130.

X_i
100
110
114
121
130
130
160

The first descriptive statistic we consider is the *sample mean*, denoted $\overline{X}$ ("X bar"). The sample mean is one statistic that summarizes the average value of a sample; it gives a sense of what a typical value looks like. To compute the sample mean, we sum all of the observations and divide by the sample size. The sample mean of the systolic blood pressures is

$$\overline{X} = (100 + 110 + 114 + 121 + 130 + 130 + 160)/7$$

In mathematics, the symbol $\sum$ (uppercase "sigma") denotes summation. The sample mean of the systolic blood pressures can be represented as follows:

$$\overline{X} = \sum X_i / 7$$

where $\sum X_i = 100 + 110 + 114 + 121 + 130 + 130 + 160$

In general, the sample mean is denoted:

$$\overline{X} = \frac{\sum X_i}{n} = \frac{\sum X}{n} \tag{2.1}$$

NOTE: In the final expression of the formula, the subscript i is suppressed and the summation is understood to be over all subjects in the sample.

The mean systolic blood pressure is $\overline{X} = 865/7 = 123.6$. Reviewing the data elements, we see that some of the observed systolic blood pressures are above the mean of 123.6, and others are below the mean. The mean of 123.6 is interpreted as the average, or typical, systolic blood pressure in the sample. In journal articles and research reports, readers are generally not shown the actual data elements. Instead, summary statistics such as the sample size and sample mean are provided.

The sample mean is referred to as the balancing point, or pivot point, of the sample since the sum of the distances between observations below the mean and the sample mean is equal to the sum of the distances between observations above the mean and the sample mean (see dotplot in Figure 2.2).

These distances or "deviations from the mean" are denoted $(X - \overline{X})$. The following table displays the data elements along with their respective deviations from the mean (i.e., distance from $\overline{X} = 123.6$):

X	$(X - \overline{X})$
100	−23.6
110	−13.6
114	−9.6
121	−2.6
130	6.4
130	6.4
160	36.4
865	−0.2*

* This sum is theoretically zero; the difference here is due to rounding.

Figure 2.2 *Sample Mean as Balancing Point*

The sample mean measures the *location,* or central tendency of the sample. Location is very important in interpreting sample data. However, two very different samples might produce the same sample mean. Consider a second sample of 7 subjects from the population of patients 50 years of age with diagnosed coronary artery disease. Again we measure systolic blood pressures, in millimeters of mercury (mmHg), on each subject. The sample data are

120   121   122   124   125   126   127

The sample mean for this sample is $\overline{X} = (120 + 121 + 122 + 124 + 125 + 126 + 127)/7 = 865/7 = 123.6$. The sizes and the means are the same in the two samples, yet the samples are quite different.

For a more complete understanding of the data, we also need a measure of the *dispersion,* or variability, in the sample. Measures of dispersion address whether the data elements are tightly clustered together or widely spread. Specifically, we are interested in whether the data elements are tightly clustered about the mean or whether the data elements are widely spread above and below the mean. The goal is to generate an estimate of the dispersion in the sample, in particular the dispersion of the data elements about the sample mean. The deviations from the mean sum to zero, since the negative deviations "cancel out" the positive deviations (Figure 2.2). Of real interest is the magnitude of these deviations. Several techniques can be employed to summarize the magnitude of the deviations from the mean. One method is the mean absolute deviation (MAD), which is simply the mean of the absolute values of the deviations from the mean:

$$\text{MAD} = \frac{\sum |X - \overline{X}|}{n}$$

This technique is not generally used for mathematical reasons that are beyond the scope of this book. The more popular statistic, which proves to be the most straightforward mathematically, is based on *squared* deviations from the mean and is called the *sample variance,* defined as:

$$s^2 = \frac{\sum (X - \overline{X})^2}{n - 1} \tag{2.2}$$

NOTE: The denominator in the sample variance is $(n - 1)$, not n as was the case with the sample mean.

The following table organizes data for the computation of the sample variance. It displays the data elements, deviations from the mean, and squared deviations from the mean, respectively.

X	$(X - \bar{X})$	$(X - \bar{X})^2$
100	−23.6	556.96
110	−13.6	184.96
114	−9.6	92.16
121	−2.6	6.76
130	6.4	40.96
130	6.4	40.96
160	36.4	1,324.96
865	−0.2*	2,247.72

* The sum is not exactly zero due to rounding.

For Example 2.1, the sample variance is

$$s^2 = \frac{2,247.72}{6} = 374.6$$

The sample variance is interpreted as the average squared deviation from the mean. Therefore, on average, systolic blood pressures in our sample are 374.6 units, squared from the sample mean of 123.6. This information is important; however, in its present form it does not exactly achieve our original goal, which was to compute a measure of the typical deviation from the mean in the sample. Recall that we summed the square of each deviation from the mean since their sum was zero. Because of this step, the sample variance does not address our original objective directly. To return to our original units, we compute what is called the *sample standard deviation*, denoted *s*, defined as the square root of the sample variance.

$$s = \sqrt{s^2} \tag{2.3}$$

The sample standard deviation of the systolic blood pressures is

$$s = \sqrt{374.6} = 19.4$$

After taking the square root, we have a statistic which can be interpreted as the typical deviation from the mean. In this sample, systolic blood pressures are about 19.4 units from the sample mean. It is often difficult to interpret the value of a standard deviation (e.g., is 19.4 large, small, or appropriate?). The standard deviation, however, is very useful for comparing samples. Recall the second sample of $n = 7$ subjects we selected from the same population. The second sample had the same size ($n = 7$) and the same mean ($\bar{X} = 123.6$); however, the standard deviation for the second sample is $s = 2.6$. The standard deviation in the second sample is much smaller because all of the observations are tightly clustered around the sample mean of 123.6. There is much more variability in the systolic blood pressures measured among patients with diagnosed coronary artery disease in the first sample (sample 1: $s = 19.4$) as compared to the second sample ($s = 2.6$).

An alternative formula is available for computing the sample variance that is mathematically equivalent to the formulation provided in (2.2). This alternative formula is called the *computational formula for the sample variance* (Eq. 2.4). The formula provided in (2.2) is called the *definitional formula for the sample variance.*

$$s^2 = \frac{\sum X^2 - (\sum X)^2/n}{n - 1} \qquad (2.4)$$

where $\sum X^2$ = the sum of the squared observations, and
$(\sum X)^2$ = the square of the sum of the observations

The computational formula (2.4) can be easier to work with than the definitional formula given in (2.2), because the components in the computational formula are in most cases easier to compute (i.e., $\sum X^2$ and $(\sum X)^2$).

We will now illustrate the use of the computational formula for the sample variance using data from Example 2.1. The following table displays each data element, along with each data element squared.

X	X^2
100	10,000
110	12,100
114	12,996
121	14,641
130	16,900
130	16,900
160	25,600
865	109,137

Using the computational formula (2.4):

$$s^2 = \frac{109,137 - (865)^2/7}{7 - 1}$$

$$s^2 = \frac{109,137 - 106,889.3}{6} = \frac{2247.7}{6} = 374.6$$

As noted, the computations can be somewhat easier with the computational formula as compared to the definitional formula. To implement the computational formula, we need only to compute the sum of the data elements and the sum of the squared data elements, as opposed to deviations and deviations squared for each data element. The reduced number of calculations with the computational formula reduces the chance of error.

A standard data summary for a continuous variable in a sample consists of three statistics:

- sample size (n)
- sample mean ($\overline{X}$)
- sample standard deviation (s)

These three statistics provide information on the number of subjects in the sample, the location, and the dispersion of the sample, respectively.

We purposely chose a small data set to illustrate these statistics. It is easy to imagine applications with much larger sample sizes in which it would be impossible to view the entire sample. In such cases, the sample size, mean, and standard deviation provide a very informative and useful summary. Publications and reports almost always include these statistics.

As a general guideline, descriptive statistics should include no more than one decimal place beyond that observed in the original data elements. For example, the systolic blood pressures are recorded as whole numbers. Therefore, descriptive statistics are presented to the nearest tenths place (i.e., one decimal place).

The standard summary for Example 2.1 is $n = 7$, $\overline{X} = 123.6$, and $s = 19.4$. A number of other descriptive statistics beyond what we have called the standard summary statistics (i.e., n, $\overline{X}$, and s) are also widely used for continuous variables, including the median and quartiles. The *sample median* is defined as the middle value. It is the value that has as many values above it as below it. The median is computed by arranging the data elements from smallest to largest and successively counting from the right and left to arrive at the median, or middle, value. For example, if we arrange our 7 data elements from smallest to largest and count in from the right and left simultaneously, we arrive at the median or middle value after three steps:

Step 1:   ~~100~~   110   114   121   130   130   ~~160~~
Step 2:   ~~100~~ ~~110~~   114   121   130 ~~130~~ ~~160~~
Step 3:   ~~100~~ ~~110~~ ~~114~~   121 ~~130~~ ~~130~~ ~~160~~
                                  ⇑
                              Median

Because the number of observations in this sample is odd ($n = 7$), this procedure produces a single number, the median value, upon successively counting in from the right and left. In Example 2.2, we will illustrate the same procedure with an even number of data elements. The interpretation of the median in Example 2.1 is as follows: Half (50%) of the systolic blood pressures are greater than 121 and half (50%) are less than 121.

Both the mean and the median are statistics that measure the average or typical value of a particular characteristic. The median is particularly useful when there are extreme values (either very small or very large as compared to other values) in the sample. Suppose in Example 2.1 that the maximum value was not 160 but 260 instead. The sample mean would be $\overline{X} = (100 + 110 + 114 + 121 + 130 + 130 + 260)/7 = 137.9$, which does not look like a typical value (since 6 of 7 observations are below it). The sample mean is affected by extreme values. In this case, the value 260 inflates the mean value, and it is therefore no longer representative of a typical value. A better measure of location in this situation is the median, which is still 121 and is more representative of a typical systolic blood pressure in this sample. In the absence of extreme values, the sample mean and the sample median will be close in value and the sample mean is considered a better measure of location since all observations contribute to the sample mean. When the sample mean and the sample median are very different, it suggests that extreme values are affecting the mean and that the median might be a more appropriate measure of location. As we work through more examples, it will become clear which measure of location (the mean or the median) is more appropriate in specific applications.

Because the sample size in Example 2.1 is small ($n = 7$), the method of successively counting into the middle of the ordered data set to locate the median is easy to implement. When the sample size is larger, a more efficient method for computing the median involves two steps. In the first step, we compute the *position of the median in the ordered data set,* and in the second step, we locate the median value. When the number of observations is odd, the *position* of the median is computed as follows:

$$\frac{n+1}{2} \tag{2.5}$$

For Example 2.1, the median is in the fourth position ($(7 + 1)/2 = 4$) in the ordered data set and is equal to 121. The median represents the middle value; to further describe the sample, we now analyze the top and bottom halves.

The first and third *quartiles* are the values that separate, respectively, the bottom and top 25% of the data elements. The first quartile of the sample, denoted Q_1, is the sample value that holds approximately 25% of the data elements at or below it and approximately 75% above or equal to it. The third quartile, denoted Q_3, holds approximately 25% of the data elements at or above it and approximately 75% below or equal to it. The median is also referred to as the second quartile, Q_2. The best way to determine the quartiles is to follow the two-step procedure outlined for determining the median (i.e., first compute the positions of the quartiles in the ordered data set, and then locate the values). When the number of observations in a sample is odd, the positions of the quartiles are determined by the following formula:

$$\left[\frac{n+3}{4}\right] \tag{2.6}$$

where $[k]$ is the greatest integer less than k; for example, $[2.1] = 2$, $[2.9] = 2$, $[5.0] = 5$, $[10.8] = 10$, and so on

For Example 2.1,

$$\left[\frac{n+3}{4}\right] = \left[\frac{7+3}{4}\right] = \left[\frac{10}{4}\right] = [2.5] = 2$$

In Example 2.1, the quartiles are in the second positions from the top and bottom of the ordered data set. The first quartile is $Q_1 = 110$ and the third quartile is $Q_3 = 130$. Approximately 25% of the systolic blood pressures are 110 or lower and approximately 25% of the systolic blood pressures are 130 or higher. Again, because the sample in Example 2.1 is so small ($n = 7$), we do not need all of the statistics described to summarize and interpret these data. In larger samples, the quartiles are very informative statistics for understanding the distribution of a particular characteristic.

The *mode* of the data set is defined as the most frequent value. In Example 2.1, the mode is 130, since it appears twice and the remaining values appear only once. A sample can have one mode or several modes. A sample with no repeated values has no mode.

Other very informative descriptive statistics include the *minimum* and *maximum* values. In Example 2.1, the minimum is 100 and the maximum is 160. These values can be very useful, especially with regard to identifying *outliers*. Outliers are values that exceed the "normal" or expected range of values. For example, suppose ages are recorded on each of 20 individuals participating in an experimental study. Suppose the mean age for the sample is 83.5, with a standard deviation of 5.6. Suppose the minimum age is 70 and the highest five ages, in descending order, are 110, 90, 89, 89, and 87. Assuming that each age was recorded accurately, an age of 110 might be considered an outlier. It is not an incorrect value, just a value outside—in this case, above—the normal range. Outlying values can be determined by an expert in the particular substantive area or by using one of several statistical definitions (see Example 2.3). Assuming there are no errors in the data, the statistical analyst need not do anything in particular with respect to outliers, only be aware of their existence and their impact on certain descriptive statistics (e.g., the sample mean).

Another descriptive statistic that addresses dispersion in a data set is the *range*. The range is defined as the maximum value minus the minimum value. In Example 2.1 the range $= 160 - 100$, or 60. Some investigators report the range as "100 to 160"; others report the range as 60. Both reports are appropriate. As noted, the range addresses dispersion in the sample. In Example 2.1, the observed systolic blood pressures cover 60 units. The range is based on only two values in the sample, the maximum and minimum, and although it is a very useful statistic, it can be somewhat misleading, especially in the presence of outliers. For example, if the maximum value was 260 instead of 160 and all other data elements were unchanged, the range would be $260 - 100 = 160$. This would suggest much more dispersion in the sample than the range of 60 (based on the data presented in Example 2.1) when only a single observation changed. We suggest that the range be interpreted with caution and that the standard deviation be used to address dispersion in a sample. Consider the following samples, call them samples $A, B, C,$ and D. The samples are all of the same size ($n = 11$), have the same means (50) and ranges (100), yet the standard deviations are different. How are the samples different?

	Sample			
	A	B	C	D
Raw Data				
	0	0	0	0
	50	10	20	0
	50	20	20	0
	50	30	20	0
	50	40	20	0
	50	50	50	50
	50	60	80	100
	50	70	80	100
	50	80	80	100
	50	90	80	100
	100	100	100	100
Summary Statistics				
n	11	11	11	11
$\overline{X}$	50	50	50	50
Range	100	100	100	100
s	22	33	35	50

Based on the standard deviations, the first sample has the least variation among observations, and the last sample has the most. The range, in this example, does not differentiate the samples. ■

SAS EXAMPLE 2.1   **Summary Statistics on Systolic Blood Pressures Using SAS**

The following descriptive statistics were generated using SAS Proc Univariate (see the following interpretation and Section 2.4 for more details) and the data in Example 2.1. An interpretation of the relevant components appears after the output.

SAS Output for Example 2.1

```
                        Summary Statistics


                        Summary Statistics



                    The UNIVARIATE Procedure
            Variable: sbp (systolic blood pressure)


                             Moments
N                               7     Sum Weights                   7
Mean                   123.571429     Sum Observations            865
Std Deviation          19.3550781     Variance              374.619048
Skewness               1.04211435     Kurtosis              1.63467176
Uncorrected SS             109137     Corrected SS          2247.71429
Coeff Variation         15.663069     Std Error Mean        7.31553189


                  Basic Statistical Measures
            Location                      Variability
     Mean        123.5714      Std Deviation          19.35508
     Median      121.0000      Variance              374.61905
     Mode        130.0000      Range                  60.00000
                               Interquartile Range    20.00000


                  Tests for Location: Mu0=0
       Test           -Statistic-          -----p Value------
       Student's t    t   16.89165     Pr > |t|     <.0001
       Sign           M       3.5      Pr >= |M|     0.0156
       Signed Rank    S        14      Pr >= |S|     0.0156
```

```
                    Quantiles (Definition 5)
                    Quantile            Estimate
                    100% Max                 160
                    99%                      160
                    95%                      160
                    90%                      160
                    75% Q3                   130
                    50% Median               121
                    25% Q1                   110
                    10%                      100
                    5%                       100
                    1%                       100
                    0% Min                   100


                        Extreme Observations
         ----Lowest----                ----Highest----
         Value        Obs              Value        Obs
           100          4                114          3
           110          2                121          1
           114          3                130          6
           121          1                130          7
           130          7                160          5
```

Interpretation of SAS Output for SAS Example 2.1

The SAS Univariate Procedure is used to generate descriptive statistics on a continuous variable. SAS generates a number of descriptive statistics in the section labeled "Moments"; we will highlight only a few.

The sample size is 7 (notice that SAS uses uppercase N as opposed to lowercase n to denote sample size), the sample mean is 123.6, and the sample standard deviation is 19.4. The sum of the observations (i.e., $\sum X_i = \sum X$) is 865, and the sample variance, s^2, is 374.6. The skewness of a sample indicates the degree of asymmetry in the sample distribution. Values close to 0 are indicative of symmetry. The kurtosis of a sample indicates the thickness in the tails of the distribution (i.e., the degree of clustering of observations at the extremes). Again, values close to 0 indicate lack of clustering in the tails. Estimates of skewness and kurtosis are somewhat unreliable in small samples and should be interpreted with caution. The normal distribution (discussed extensively in Chapter 3) has skewness = 0 and kurtosis = 0.

The uncorrected sum of squares, "Uncorrected SS," is the sum of the observations squared (i.e., $\sum X^2 = 109{,}137$, which we used in the computational formula for the sample variance). The corrected sum of squares, "Corrected SS," is the numerator of the sample variance, $\sum(X - \bar{X})^2 = 2247.7$. The coefficient of variation, "Coeff Variation," is defined as the ratio of the sample standard deviation to the sample mean, expressed as a percentage (i.e., $CV = (s/\overline{X}) * 100$). The standard error of the mean, "Std Error Mean," is defined as $s/\sqrt{n}$.

The next part of the SAS output summarizes the most popular summary statistics for continuous variables in the section entitled "Basic Statistical Measures." Several measures of location are provided (the mean, median, and mode) as are several measures of dispersion (standard deviation, range, and interquartile range). The range is $160 - 100$, or 60; the interquartile range, the difference between the first and third quartiles ($Q_3 - Q_1$) is 20. The most appropriate measures of location and dispersion depend on whether there are outliers in the data set. If there are no outliers, the mean and standard deviation are the most appropriate measures of location and dispersion, respectively. If there are outliers, the median and interquartile deviation (defined as $(Q_3 - Q_1)/2)$) are the most appropriate measures of location and dispersion, respectively. The next part of the SAS output contains "Tests for Location," these will be discussed in detail in Chapter 5.

After "Tests for Location," SAS output displays the "Quantiles" (or percentiles) of the variable, where the kth quantile is defined as the score that holds $k\%$ of the data below it. For example, the maximum value is equivalent to the 100th quantile and equal to 160 in SAS Example 1, the 75th quantile is equivalent to the third quartile (130), and so on. SAS also presents the 99th quantile, which is equal to 160 in SAS Example 1. Since this is a small data set, these fine classifications are unnecessary and not meaningful.

SAS then prints the "Extreme Observations" in the data set. In particular, the five smallest and five largest values are printed. Next to each value, in parentheses, is the observation number (i.e., the position of the observation in the data set); for example, the smallest value is 100, the fourth observation among the seven. ■

EXAMPLE 2.2

Summary Statistics on Total Cholesterol Levels

Eight subjects are randomly selected from a population of patients with hypertension. Total serum cholesterol, in mg/100 ml, is measured on each subject and the sample data are

197 212 211 184 260 233 245 219

In the following, we summarize the cholesterol data using the statistics introduced in Example 2.1. Since total serum cholesterol is a continuous variable (as was systolic blood pressure), the same statistics will be computed. We will limit our discussion here to items and concepts that were addressed in detail in Example 2.1. Again, a small sample size is used to illustrate the calculation of descriptive statistics to keep actual computation time to a minimum. In practice, the same techniques can be applied to larger samples.

We will let the variable X denote total serum cholesterol. The table displays the data elements, which have been ordered from smallest to largest, along with the value of each data element squared (which is used in the computation of the sample variance):

X	X^2
184	33,856
197	38,809
211	44,521
212	44,944
219	47,961
233	54,289
245	60,025
260	67,600
1761	392,005

The number of patients, or sample size, is $n = 8$. The mean cholesterol level in this sample is

$$\overline{X} = \frac{\sum X}{n} = \frac{1,761}{8} = 220.1$$

Notice that some of the total cholesterol levels are above the mean and others are below. This will always be true, as displayed in Figure 2.2 using Example 2.1. The sample mean represents a typical cholesterol level in this sample.

To address dispersion in the sample, we use the computational formula for the sample variance (2.4):

$$s^2 = \frac{\sum X^2 - (\sum X)^2/n}{n-1} = \frac{392,005 - (1,761)^2/8}{7}$$
$$= \frac{392,005 - (3,101,121)/8}{7} = \frac{392,005 - 387,640.125}{7}$$
$$= \frac{4364.875}{7} = 623.6$$

Generally, the variance is not used to summarize dispersion. Instead, the sample standard deviation is computed:

$$s = \sqrt{623.6} = 25.0$$

The sample standard deviation represents how far each total cholesterol level is from the mean of 220.1. Again, by itself the standard deviation is often difficult to interpret. In particular, it is generally difficult to quantify what value of a standard deviation is considered large and what value is considered small. Individuals with substantive knowledge of the characteristic under investigation might have a feel for what is large and small with respect to the standard deviation.

Our standard summary of the cholesterol levels (a continuous variable) is $n = 8$, $\overline{X} = 220.1$, and $s = 25.0$. The maximum cholesterol level is 260 and the minimum is 184. The range is $260 - 184$, or 76. There is a substantial difference between the smallest and largest cholesterol levels, a difference of 76 units. Recall that the range is another measure of dispersion. In general, the range is less useful as a measure of dispersion than the sample standard deviation. The median value, or the value that holds 50% of the cholesterol levels above it and 50% of the cholesterol levels below it, can be computed by arranging the data from smallest to largest, and successively counting from the right and left to arrive at the median, or middle, value:

Step 1: ~~184~~ 197 211 212 219 233 245 ~~260~~

Step 2: ~~184~~ ~~197~~ 211 212 219 233 ~~245~~ ~~260~~

Step 3: ~~184~~ ~~197~~ ~~211~~ 212 219 ~~233~~ ~~245~~ ~~260~~

⇑   ⇑

Two Middle Values

Because the number of observations in this sample is even ($n = 8$), there are two middle values. *When the sample size is even, the median is defined as the mean of the two middle values:*

$$\text{Median} = \frac{212 + 219}{2} = 215.5$$

In Example 2.2, 50% of the cholesterol levels are above 215.5 and 50% of the cholesterol levels are below 215.5. We now have two statistics that represent a typical cholesterol level, the sample mean and the sample median. Although they convey different information, in general only one is necessary. Which is the best statistic to address location in this sample?

Reviewing the cholesterol levels in the sample, there do not appear to be outliers at either extreme, thus the mean is a better measure of location. An individual with clinical expertise would, however, be in a better position to make that assessment. In Example 2.3 we will present guidelines for assessing outliers based on statistical formulations. In this example, as in Example 2.1, the sample size is small and it is easy to order the data elements from smallest to largest and to count into the middle from right and left to determine the median. In applications where the sample size is larger, it may be more efficient to use the two-step method described in Example 2.1. First we compute the position(s) of the middle value(s) in the ordered data set and then locate those value(s). When the number of observations is even, there are two middle values, and their positions are computed as follows:

$$\frac{n}{2} \quad \text{and} \quad \left(\frac{n}{2}\right) + 1 \qquad (2.7)$$

For Example 2.2,

$$\frac{n}{2} = \frac{8}{2} = 4\text{th position}$$

$$\left(\frac{n}{2}\right) + 1 = \left(\frac{8}{2}\right) + 1 = 5\text{th position}$$

The median is the mean of the observations in the 4th and 5th positions in the ordered data set (i.e., $[212 + 219]/2 = 215.5$).

To further describe the sample, we now compute the quartiles. As noted in Example 2.1, the best way to determine the quartiles is to first compute the positions of the quartiles in the ordered data set and then to locate the values. When the number of observations in the sample is even, the positions of the quartiles are determined by the following formula:

$$\left[\frac{n+2}{4}\right] \tag{2.8}$$

For Example 2.2,

$$\left[\frac{n+2}{4}\right] = \left[\frac{8+2}{4}\right]\left[\frac{10}{4}\right] = [2.5] = 2$$

The quartiles are in the second positions from the top and bottom of the ordered data set. The first quartile is $Q_1 = 197$ and the third quartile is $Q_3 = 245$. ■

Summary Statistics on Total Cholesterol Levels Using SAS

The following descriptive statistics were generated using SAS. For Example 2.2, we produced the abbreviated summary statistics as opposed to the more extensive summary illustrated in Example 2.1. The following descriptive statistics were produced by SAS Proc Means (see the following interpretation and Section 2.4 for more details) using the data in Example 2.2. A brief interpretation appears after the output.

SAS Output for Example 2.2

```
                    Summary Statistics
                  The MEANS Procedure
          Analysis Variable : chol total serum cholesterol
```

N	Mean	Std Dev	Minimum	Maximum
8	220.1250000	24.9710547	184.0000000	260.0000000

Interpretation of SAS Output for Example 2.2

The SAS Means Procedure is used to generate descriptive statistics on a continuous variable. Many different statistics can be requested; the statistics shown are the default statistics. The sample size is 8 (notice that SAS uses uppercase "N" as opposed to lower case "n" to denote sample size), the sample mean is 220.1, and the sample standard deviation is 25.0. The minimum and maximum cholesterol levels are 184 and 260. SAS displays summary statistics to eight decimal places by default. Users should round appropriately to report summary statistics.

■

EXAMPLE 2.3

Summary Statistics on Ages

A sample of 51 individuals is selected for participation in a study of cardio-vascular risk factors. The following data represent the ages of enrolled individuals measured in years (continuous variable). Here, age is measured in the usual way with a person being recorded as 65, for example, until the day he/she turns 66. The data are as follows:

60	62	63	64	64	65	65	65	65	65	65	66	66
66	66	66	67	67	67	68	68	68	70	70	70	71
71	72	72	73	73	73	73	73	75	75	75	75	76
76	77	77	77	77	77	79	82	83	85	85	87

The number of subjects, or sample size, is $n = 51$, a much larger sample size than in previous examples. Here it is not possible to interpret the age data simply by inspecting the values. Instead, we need summaries of location and dispersion.

The mean age in this sample is

$$\overline{X} = \frac{\sum X}{n} = \frac{3,637}{51} = 71.3$$

In order to assess dispersion in the sample, we will compute the sample standard deviation. As a first step, we compute the sample variance using the computational formula presented in Example 2.1:

$$s^2 = \frac{261,439 - (3,637)^2/51}{(51 - 1)} = 41.4$$

The sample standard deviation is

$$s = \sqrt{41.4} = 6.4$$

In general, participants' ages deviate from the mean of 71.3 by 6.4 years. Notice that the magnitude of the standard deviation of the ages is smaller than the standard deviations we computed on the systolic blood pressures in Example 2.1

(s_{SBP} = 19.4) and on the cholesterol levels in Example 2.2 (s_{CHOL} = 25.0). Standard deviations are interpreted relative to their scale of measurement. In Example 2.3, the participants are very homogeneous with respect to age. It is possible that the study objectives were focused on individuals 60 years of age or older. Most, if not all, studies have very explicit inclusion and exclusion criteria, which must be recognized in order to appropriately interpret summary statistics.

As noted earlier, in many publications and research reports, investigators do not present raw data (i.e., observations measured on each member of a sample); instead, they present summary statistics. Suppose that we did not have access to the actual ages of each participant here, and that instead we had only the summary statistics: $n = 51$, $\overline{X} = 71.3$, and $s = 6.4$. The mean and standard deviation are used to understand where the data are located and how they are spread. The *Empirical Rule* can be used to learn more about a particular characteristic based on these commonly available statistics (discussed further in Chapter 4):

Empirical Rule

Approximately 68% of the observations fall between $\overline{X} - s$ and $\overline{X} + s$.
Approximately 95% of the observations fall between $\overline{X} - 2s$ and $\overline{X} + 2s$.
Approximately all of the observations fall between $\overline{X} - 3s$ and $\overline{X} + 3s$.

Using the data in Example 2.3 the Empirical Rule indicates that approximately 68% of the ages fall between $71.3 - 6.4 = 64.9$ and $71.3 + 6.4 = 77.7$, approximately 95% of the ages fall between 58.5 and 84.1, and almost all of the ages fall between 52.1 and 90.5. Because we have the actual observations here, we computed the percentages of 51 observations that actually fell into each range. The following table illustrates how closely the Empirical Rule approximates the distribution of ages in this sample:

Empirical Rule

Range	*Percent of Observations*	*Percent of Sample Data*
64.9–77.7	Approximately 68%	78.4%
58.5–84.1	Approximately 95%	94.1%
52.1–90.5	Almost all	100%

The Empirical Rule suggested that approximately 68% of the ages would fall between 64.9 and 77.7. In Example 2.3, 78.4% of the ages actually fell between 64.9 and 77.7. Similarly, the Empirical Rule suggested that approximately 95% of the ages would fall between 58.5 and 84.1. In Example 2.3, 94.1% of the ages actually fell between 58.5 and 84.1. Finally, the Empirical Rule suggested that almost all of the ages would fall between 52.1 and 90.5, and in fact they do. ■

The computation of the mean and standard deviation are cumbersome with a sample of 51 observations. We now use the SAS Proc Univariate procedure to generate descriptive statistics for the data in Example 2.3.

SAS EXAMPLE 2.3 **Summary Statistics on Ages Using SAS**

The following descriptive statistics were generated using SAS Proc Univariate and the data in Example 2.3. An interpretation of the relevant components appears after the output.

SAS Output for Example 2.3

Summary Statistics

The UNIVARIATE Procedure
Variable: age (age in years)

Moments

N	51	Sum Weights	51
Mean	71.3137255	Sum Observations	3637
Std Deviation	6.4358067	Variance	41.4196078
Skewness	0.57609039	Kurtosis	-0.2678579
Uncorrected SS	261439	Corrected SS	2070.98039
Coeff Variation	9.02463958	Std Error Mean	0.90119319

Basic Statistical Measures

Location		Variability	
Mean	71.31373	Std Deviation	6.43581
Median	71.00000	Variance	41.41961
Mode	65.00000	Range	27.00000
		Interquartile Range	10.00000

Tests for Location: Mu0=0

Test	-Statistic-		-----p Value------	
Student's t	t	79.13256	Pr > |t|	<.0001
Sign	M	25.5	Pr >= |M|	<.0001
Signed Rank	S	663	Pr >= |S|	<.0001

```
            Quantiles (Definition 5)
            Quantile          Estimate
            100% Max                87
            99%                     87
            95%                     85
            90%                     79
            75% Q3                  76
            50% Median              71
            25% Q1                  66
            10%                     65
            5%                      63
            1%                      60
            0% Min                  60


                  Extreme Observations
         ----Lowest----          ----Highest----
         Value      Obs          Value      Obs
            60        1             82       47
            62        2             83       48
            63        3             85       49
            64        5             85       50
            64        4             87       51
```

Interpretation of SAS Output for Example 2.3

The SAS Univariate Procedure generates a number of descriptive statistics in the section labeled "Moments"; we will highlight only a few. The sample size is 51, the sample mean is 71.3, and the sample standard deviation is 6.4. The sum of the observations (i.e., ΣX) is 3,637, and the sample variance, s^2, is 41.42. In the "Basic Statistical Measures" section, SAS also provides the median (71) and the mode (the mode is 65 which appears 6 times in the sample). The range in ages is $87 - 60$, or 27; the difference between the first and third quartiles (called the *Interquartile Range*, $Q_3 - Q_1$) is 10 years.

The oldest subject in the sample is 87, 75% of the subjects are 76 years of age or younger, 50% of the subjects are 71 or younger, 25% of the subjects are at or below age 66. The middle 90% of the sample are between 63 and 85 years of age (5% are above 85 and 5% are below 63).

If we compute statistics by hand, we determine the position of the median using (3.5), because n is odd:

$$\frac{n+1}{2} = \frac{51+1}{2} = 26\text{th position}$$

The median is in the 26th position (in the ordered data set) and equal to 71.

The positions of the quartiles are determined by (3.6):

$$\left[\frac{n+3}{4}\right] = \left[\frac{51+3}{4}\right] = \left[\frac{54}{4}\right] = [13.5] = 13$$

The quartiles are in the 13th positions from the top and bottom of the ordered data set: $Q_1 = 66$ and $Q_3 = 76$. ▪

Outliers in a sample can be determined using a number of different definitions. We present two of the more popular ones. The first, based on the Empirical Rule, is as follows:

Observations outside the range: $(\overline{X} \pm 3s)$

For Example 2.3, this range is (52.1 to 90.5). There are no values outside of this range and therefore no outliers according to the definition. A second definition is as follows:

Observations above $Q_3 + 1.5$ (IQR) or below $Q_1 - 1.5$ (IQR)

where IQR = Interquartile Range = $Q_3 - Q_1$

For Example 2.3, the upper limit is $76 + 1.5(10) = 91$ and the lower limit is $66 - 1.5(10) = 51$. There are no values exceeding the upper limit or falling below the lower limit, as defined. In Example 2.3 the most appropriate measure of location is the sample mean and the most appropriate measure of dispersion is the sample standard deviation.

If there are outliers in a data set, the most appropriate measure of location is the median and the most appropriate measure of dispersion is the *interquartile deviation* (IQR/2). If there are no outliers in the data, then the sample mean and standard deviation are the most appropriate measures of location and dispersion, respectively.

2.2.2 Graphical Summaries for Continuous Variables

Graphical presentations can also be very useful for summarizing data, although choosing the most appropriate presentation for any application can be difficult. One must always remember that the primary goal for generating any graphical presentation of data is to provide a simple, complete, and accurate representation of the data.

One very informative graphical display for continuous variables is the *box-and-whisker plot*. The box-and-whisker plot incorporates the minimum and maximum, the median, and the quartiles.

EXAMPLE 2.4   **Distribution of Systolic Blood Pressures**

The box-and-whisker plot for Example 2.1 is given in Figure 2.3. The minimum, first quartile (Q_1), median, third quartile (Q_3), and maximum are noted on the top line. These notations are not part of the plot, they are only noted here for reference.

The lengths of the four sections of the box-and-whisker plot, separated by the first quartile, the median, and the third quartile, reflect the clustering of observations within each section. The shorter the sections, relative to the others, the more observations are clustered. The box-and-whisker plot in Figure 2.3 indicates some clustering of observations between the median and third quartile (as indicated by the shorter section) and less clustering (more spread) among the observations between the third quartile and the maximum. With only $n = 7$ observations in Example 2.1, we are probably making too much of these relationships among observations. If the box-and-whisker plot was based on a larger sample, we could make more general statements about the distribution of observations based on the plot.

Box-and-whisker plots are extremely useful for comparing samples. For example, suppose we measure systolic blood pressures of a sample of patients free of coronary artery disease. We then generate summary statistics (i.e., minimum, Q_1, median, Q_3, maximum) on this sample. Figure 2.4 displays the box-and-whisker plots for the sample of patients with diagnosed coronary artery disease (Example 2.1) and for a sample of subjects free of the disease:

Figure 2.3 *Box-and-Whisker Plot for Example 2.1*

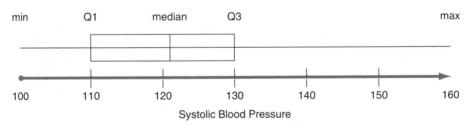

Figure 2.4 *Using Box-and-Whisker Plots to Compare Samples*

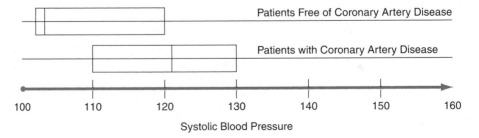

The ranges (i.e., the horizontal lines connecting minimum and maximum values) of observed systolic blood pressures are similar in the two samples. However, the majority of blood pressures in the sample of patients free of coronary artery disease are lower than those in the sample of patients with the disease. In fact, the middle 50% of the blood pressures (i.e., observations between Q_1 and Q_3) in the sample of patients free of disease are between 102 and 120, whereas the middle 50% of the blood pressures in the sample of patients with disease are between 110 and 130. ■

EXAMPLE 2.5

Distribution of Total Cholesterol Levels

The box-and-whisker plot for the cholesterol data in Example 2.2 is shown in Figure 2.5.

Based on the box-and-whisker plot, it appears that there is slight clustering at the lower end of the distribution as compared to the upper end (e.g., the lengths of the sections between the minimum and first quartile and between the first quartile and median are shorter than the length of the section between the median and the third quartile). Again, because the plot is based on such a small sample it is not appropriate to make much of these very slight indications.

Figure 2.5 *Box-and-Whisker Plot for Example 2.2*

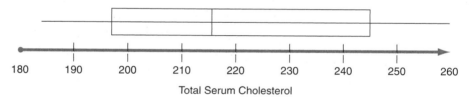

Total Serum Cholesterol

■

EXAMPLE 2.6

Distribution of Ages

The box-and-whisker plot for the data in Example 2.3 is a very informative display of the data. The distribution of ages is approximately symmetric, with a few values trailing off at the upper end. The middle 50% of the ages fall between 66 and 76.

Figure 2.6 *Box-and-Whisker Plot for Example 2.3*

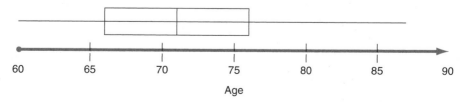

Age

Figure 2.7 *Stem-and-Leaf Plot*
        for Example 2.3

```
6 | 0 2 3 4 4
6 | 5 5 5 5 5 6 6 6 6 6 7 7 7 8 8 8
7 | 0 0 0 1 1 2 2 3 3 3 3 3 3
7 | 5 5 5 5 6 6 7 7 7 7 7 9
8 | 2 3
8 | 5 5 7
```

Another useful graphical presentation for continuous variables is the *stem-and-leaf plot*. In the stem-and-leaf plot, each data element (e.g., each systolic blood pressure, cholesterol level, age) is split into two pieces: the stem and the leaf. The leading digits (e.g., hundreds place or tens place) denote the stem, and the trailing digits (e.g., tens and units or just the units place) denote the leaf. It is up to the analyst to determine the appropriate point at which to split the data elements. *As a general guideline, best results are achieved when there are between 6 and 12 unique stems.*

In Example 2.3, ages are recorded in years and consist of two digits. We let the first digit (the tens place) denote the stem and the second digit (the units place) denote the leaf. (In other applications there can be more than one digit in the stem and/or the leaf.) Ages of individuals in the sample ranged from 60 to 87. Therefore, the stems include 6, 7, and 8 (the first digit, or tens place, of each age). Since our objective is to have at least 6 stems, we will split each decade (60s, 70s, 80s) into two parts (low 60s, high 60s; low 70s, high 70s; low 80s, high 80s) to generate the stem-and-leaf plot.

To construct the plot, we list the stems vertically, from smallest to largest, and draw a vertical line. The line separates the stems from the leaves, which appear to the right of their respective stems. The leaves consist of the second digit (or units place) of each age. For example, the first observation is 60 (stem = 6, leaf = 0), the second observation is 62 (stem = 6, leaf = 2), and so on. Leaves are displayed on the same horizontal line as their respective stems, separated by spaces. The stem-and-leaf plot for the age data is shown in Figure 2.7.

The stem-and-leaf plot is essentially a histogram (described in Section 2.2.4) with horizontal bars instead of vertical bars, but shows more detail in that the actual values of observations are displayed as opposed to only the number of observations in each category (or stem). ■

The "plot" option in the SAS Proc Univariate statement generates some graphical displays of continuous data. More elaborate graphical presentations are available in other SAS procedures.

SAS EXAMPLE 2.6 **Distribution of Ages Using SAS**

The following graphical displays were generated using SAS Proc Univariate and the data in Example 2.3. An interpretation appears after the output.

SAS Output for Example 2.3

```
                     Summary Statistics
                  The UNIVARIATE Procedure
                Variable: age (age in years)
      Stem    Leaf                    #           Boxplot
        86    0                       1              |
        84    00                      2              |
        82    00                      2              |
        80                                           |
        78    0                       1              |
        76    0000000                 7           +-----+
        74    0000                    4           |     |
        72    0000000                 7           |     |
        70    00000                   5           *--+--*
        68    000                     3           |     |
        66    00000000                8           +-----+
        64    00000000                8              |
        62    00                      2              |
        60    0                       1              |
              ----+----+----+----+
```

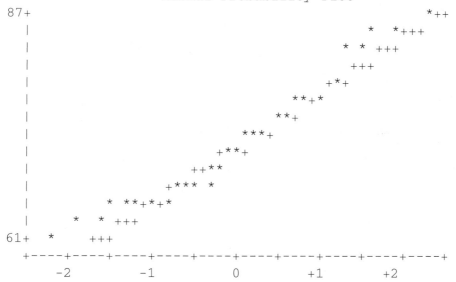

Normal Probability Plot

Interpretation of SAS Output for Example 2.3

The stem-and-leaf plot (labeled "Stem Leaf") for the age data is displayed on the left. Notice that the scale is displayed from largest (top) to smallest (bottom). SAS uses an internal algorithm to determine how to split data elements into stems and leaves. In this example, SAS elected to use the tens and units place as the stems and the first decimal place as the leaves. SAS shows only the even ages and uses—for example, 62 for ages 62 and 63; 64 for ages 64 and 65, and so on. Because each age was recorded to the nearest whole unit, the leaf for each value is 0 (e.g., 64 is shown as 64.0).

To the right of the stem-and-leaf plot is a vertical box-and-whisker plot. The top and bottom horizontal lines indicate the third and first quartiles, respectively. The center horizontal line indicates the median. The "+" sign indicates the value of the sample mean. In this example, the sample mean is very close to the sample median (71.3 and 71, respectively). SAS will indicate extreme values (or outliers) using asterisks (*) or zeros, as follows: Any value exceeding $Q_3 + 1.5(Q_3 - Q_1)$ or falling below $Q_1 - 1.5(Q_3 - Q_1)$ is denoted by 0. Any value exceeding $Q_3 + 3(Q_3 - Q_1)$ or falling below $Q_1 - 3(Q_3 - Q_1)$ is denoted by *. In this example there are no such values, thus there are no asterisks or zeros in the box-and-whisker plot. ■

2.2.3 Numerical Summaries for Discrete Variables

We now use examples to present descriptive statistics used to summarize discrete variables. Recall that discrete variables are variables which can take on only a limited number of response options. These response options can be ordered (ordinal variables) or unordered (categorical or nominal variables). It is not critical to draw the distinction between ordinal and categorical variables. However, it is critical to draw the distinction between continuous and discrete variables because the numerical summaries (and graphical summaries, presented in the next section) are quite different for continuous and discrete variables.

EXAMPLE 2.7

Frequency Distribution of Marital Status

Suppose we select a random sample of 50 subjects from the population of patients receiving care at a particular community health center. Each subject is given a self-administered survey containing a series of questions addressing medical history and personal or background characteristics. One of the questions in the background characteristics section of the survey asks patients their current marital status. The response options are: Married, Separated, Divorced, Widowed, Never Married. Each respondent is instructed to select the one category (of five possible) that reflects his or her current marital status. Marital status is an example of a categorical variable because the response options are unordered. The sample data are summarized as follows:

Marital Status	Number of Individuals (f)
Married	24
Separated	5
Divorced	8
Widowed	2
Never Married	11
Total	**50**

The number of times each data element (or response option) is observed is called its *frequency*, denoted *f*. A table that displays each response option and its frequency is called a *frequency distribution table*. A frequency distribution table is a useful summary for discrete data. Notice that the sum of the frequencies is equal to the sample size (i.e., $\sum f = n = 50$).

A more useful summary for these data is a frequency distribution table with one additional column (Table 2.1).

This frequency distribution table displays each response option (first column), along with the frequency each is observed (second column). The *relative frequency* (third column) incorporates the sample size and is computed as follows: Relative frequency = f/n, where n denotes the sample size ($n = 50$). The relative frequency can be presented as a proportion (i.e., a decimal value between 0 and 1), or as a percent (i.e., a value between 0 and 100). Relative frequencies are often used to summarize a discrete variable. In this sample, 48% of the patients are currently married,

Table 2.1 *Frequency Distribution Table for Example 2.7*

Marital Status	Number of Individuals (f)	Relative Frequency (f/n * 100)
Married	24	48%
Separated	5	10%
Divorced	8	16%
Widowed	2	4%
Never Married	11	22%
Total	**50**	**100%**

10% are separated, 16% are divorced, 4% are widowed, and 22% were never married.

SAS EXAMPLE 2.7 **Frequency Distribution of Marital Status Using SAS**

The following descriptive statistics were generated using SAS. For Example 2.7, we produced a frequency distribution table using SAS Proc Freq. A brief interpretation appears after the output.

SAS Output for Example 2.7

```
             Frequency Distribution Table
                  The FREQ Procedure


                                      Cumulative    Cumulative
marital         Frequency    Percent  Frequency     Percent
-------------------------------------------------------------
Divorced            8         16.00       8           16.00
Married            24         48.00      32           64.00
Never Married      11         22.00      43           86.00
Separated           5         10.00      48           96.00
Widowed             2          4.00      50          100.00
```

Interpretation of SAS Output for Example 2.7

The SAS Freq Procedure is used to generate a frequency distribution table for a discrete (either categorical or ordinal) variable. SAS lists the response options in the first column alphabetically. The most useful summary of this variable is given by the relative frequencies, which appear in the column labeled "Percent." SAS produces additional information in the frequency distribution table, namely, the cumulative frequency and the cumulative percent. These summaries are most appropriate for ordinal variables and we discuss their interpretation in Example 2.8. ■

EXAMPLE 2.8 **Frequency Distribution of Health Status**

Consider again the survey described in Example 2.7. Suppose that one of the questions in the medical history section of the survey asks patients to assess their current health status. The question in the survey reads, "How would you rate your current health?" The response options are: Excellent, Very Good, Good, Fair, Poor. Each respondent is instructed to select the one category (of five possible) that best describes his or her current health. Health status is an example of an ordinal variable because the response options are ordered— Excellent is better than Very Good, Very Good is better than Good, and so on.

The sample data are summarized in the following table:

Health Status	Number of Individuals (f)
Excellent	19
Very Good	12
Good	9
Fair	6
Poor	4
Total	50

For these data, we compute additional descriptive statistics (i.e., additional columns in the frequency distribution table), shown in Table 2.2.

This frequency distribution table displays each response option (first column), along with the frequency each is observed (second column). The *relative frequency* (third column) incorporates the sample size and is computed as follows: Relative frequency $= f/n$, where n denotes the sample size ($n = 50$). Here we report the relative frequency as a proportion (i.e., a decimal value between 0 and 1). Relative frequencies are often used to summarize a discrete variable. In this example, 38% of the patients in the sample reported that their health was excellent, 24% reported that their health was very good, 18% reported that their health was good, 12% that their health was fair, and 8% reported that their health was poor.

The *cumulative frequency* (fourth column) indicates the number of data elements at or better than each response option. For example, 19 patients reported that their health status was excellent. Thirty-one patients reported that their health was very good or better (very good or excellent). Forty patients reported that their health was good or better (good, very good, or excellent). Notice that the cumulative frequency for the last response option (poor) is equal to the sample size (i.e., $n = 50$). The *cumulative relative frequency* is shown in the fifth column and is computed by taking the ratio of the cumulative frequency

Table 2.2 *Frequency Distribution Table for Example 2.8*

Health Status	Frequency (f)	Relative Frequency (f/n)	Cumulative Frequency	Cumulative Relative Frequency
Excellent	19	0.38	19	0.38
Very Good	12	0.24	31	0.62
Good	9	0.18	40	0.80
Fair	6	0.12	46	0.92
Poor	4	0.08	50	1.00
Total	50	1.00		

to the sample size. (Again, the cumulative relative frequency can be presented as a proportion or a percent.) ■

Ordered scales, such as the one used in this example to measure health status are very popular, particularly in the social sciences. Sometimes, numeric values are assigned to each of the response options to produce an ordered scale. These numeric values are then used to generate summary statistics. For example, suppose we assign the following values to the response options:

1 = Poor   2 = Fair   3 = Good   4 = Very Good   5 = Excellent

According to this, higher values reflect better health status. Many assignment strategies are possible in applications such as this. For example, we could have assigned values in reverse order such that lower scores reflect better health. The exact strategy influences the interpretation of summary statistics.

Once the numerical values are assigned, we can then compute the mean and standard deviation using the same formulas used in Examples 2.1, 2.2, and 2.3. In applications in which numerical values are assigned to distinct response options, equally spaced assignments are usually made (e.g., 1, 2, 3, 4, 5; 0, 25, 50, 75, 100). Equally spaced numerical values imply that the distance between one response option and the next is the same across all possible responses. This may be true in many response scales, but not true in others. This issue as well as the exact numerical values assigned to each response option must be evaluated carefully when computing and interpreting numerical summary statistics such as the mean and standard deviation.

SAS EXAMPLE 2.8   Frequency Distribution of Health Status Using SAS

The following descriptive statistics were generated using SAS. For Example 2.8, we produced a frequency distribution table using SAS Proc Freq and generated descriptive statistics using the numerical assignments shown. A brief interpretation appears after the output.

SAS Output for Example 2.8

```
                  Frequency Distribution Table
                       The FREQ Procedure
                          health status
```

health	Frequency	Percent	Cumulative Frequency	Cumulative Percent
Poor	4	8.00	4	8.00
Fair	6	12.00	10	20.00
Good	9	18.00	19	38.00
Very Good	12	24.00	31	62.00
Excellent	19	38.00	50	100.00

```
                    Summary Statistics
                    The MEANS Procedure
             Analysis Variable : health health status


     N          Mean         Std Dev        Minimum         Maximum
 --------------------------------------------------------------------
    50       3.7200000      1.3099307      1.0000000       5.0000000
 --------------------------------------------------------------------
```

Interpretation of SAS Output for Example 2.8

The SAS Freq Procedure produced a frequency distribution table for the health status variable. In Section 2.4, we illustrate how to work with SAS to present the response options in order (as opposed to alphabetically) using formatting. With ordinal variables the ordering of responses is important; with categorical or nominal variables ordering is not important. The SAS Means Procedure is generally used to generate descriptive statistics on a continuous variable (e.g., Example 2.2). Here we assigned numerical values to each response option as described earlier and requested the default statistics. The sample size is 50, the sample mean is 3.7, and the sample standard deviation is 1.3. The minimum and maximum values are 1 and 5, respectively. All of these statistics must be interpreted with the specific numerical assignment strategy in mind. For example, the mean of 3.7 suggests that a typical health status assessment in this sample is Very Good (the response option assigned the numerical value 4). ■

EXAMPLE 2.9

Frequency Distribution of Age Classes

Recall the sample of 51 individuals selected for participation in a study of cardiovascular risk factors described in Example 2.3. Data were collected on each subject reflecting age measured in years (a continuous variable). In Example 2.3, we summarized the sample using an array of descriptive statistics appropriate for continuous variables (e.g., n, $\overline{X}$, s, median, quartiles). Sometimes, continuous data are organized into categories as a means of summarizing. The range of the ages was 60 to 87. Suppose we collapse the ages into 5-year categories: 60–64, 65–69, 70–74, 75–79, 80–84, and 85–89. In this example, these classifications are intuitive. In other examples, it is up to the investigator to define the most meaningful categories. *In all cases, the most useful presentations include between 6 and 12 distinct categories or classes.* In general, with fewer than 6 categories information may be lost, and with more than 12 categories data may be too sparse (and therefore not useful as a summary).

Collecting the numbers of subjects in each class (see the raw data in Example 2.3) results in the following frequency distribution table:

Age Class	Numbers of Individuals (f)
60–64	5
65–69	17
70–74	12
75–79	12
80–84	2
85–89	3
Total	51

We now add the relative frequencies, cumulative frequencies, and cumulative relative frequencies to the frequency distribution table (Table 2.3).

Table 2.3 *Frequency Distribution Table for Example 2.9*

Age Class	Frequency (f)	Relative Frequency (f/n)	Cumulative Frequency	Cumulative Relative Frequency
60–64	5	0.10	5	0.10
65–69	17	0.33	22	0.43
70–74	12	0.24	34	0.67
75–79	12	0.24	46	0.91
80–84	2	0.04	48	0.95
85–89	3	0.06	51	1.01
Total	51	1.00		

The expanded table includes each response option (age class) and the frequency or number of subjects in each class. The relative frequency is computed by dividing the frequencies by the sample size ($n = 51$). Here the relative frequencies are presented as proportions. The relative frequencies are the best summary of the age data: 10% of the subjects are between 60 and 64 inclusive, 33% are between 65 and 69, and so on. Almost half of the subjects are between 70 and 79, and 10% are 80 or older.

The cumulative frequency indicates the number of subjects in the particular age class or younger. For example, 5 subjects are in the 60 to 64 class, 22 are 69 or younger, 34 are 74 or younger. The cumulative relative frequency is shown in the fifth column and is computed by taking the ratio of the cumulative frequency to the sample size. In this example, 67% of the subjects are less than 75 years of age. ■

SAS EXAMPLE 2.9 **Frequency Distribution of Age Classes Using SAS**

The following frequency distribution table was generated using SAS Proc Freq. A brief interpretation appears after the output.

SAS Output for Example 2.9

```
            Frequency Distribution Table
                 The FREQ Procedure

                                 Cumulative   Cumulative
ageclass    Frequency    Percent  Frequency      Percent
----------------------------------------------------------
60-64               5       9.80          5         9.80
65-69              17      33.33         22        43.14
70-74              12      23.53         34        66.67
75-79              12      23.53         46        90.20
80-84               2       3.92         48        94.12
85-89               3       5.88         51       100.00
```

Interpretation of SAS Output for Example 2.9

The SAS Freq Procedure produced a frequency distribution table for the age class variable. The relative frequencies (labeled "Percent") and the cumulative relative frequencies are the best summary for this ordinal variable. ■

2.2.4 Graphical Summaries for Discrete Variables

Graphical presentations can also be very useful for summarizing discrete data. A popular graphical display for a categorical variable is the bar chart; a popular graphical display for an ordinal variable is the histogram. Bar charts and histograms are based on either the frequency or the relative frequency of responses in each category. We will illustrate bar charts and histograms using Examples 2.10, 2.11, and 2.12.

EXAMPLE 2.10

Distribution of Marital Status

Marital status is an example of a categorical variable because the response options are unordered: Married, Separated, Divorced, Widowed, Never Married. The sample data for Example 2.7 are displayed in a bar chart (produced in MS Word using the chart feature) in Figure 2.8.

Figure 2.8 *Bar Chart for Example 2.7 (Categorical Variable)*

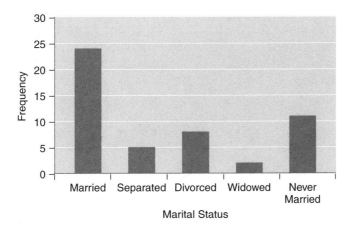

In the bar chart, each response option is listed along the horizontal axis. The vertical axis is scaled to accommodate either the frequencies (as in the example shown here) or relative frequencies. Rectangles are drawn above each response option to reflect the number of subjects (frequency) or proportion/percent of subjects (relative frequency) in each. In a bar chart, the breaks between rectangles indicate distinct, unordered response options. In histograms, (illustrated in Examples 2.11 and 2.12), there are no breaks between rectangles, suggesting order in the response options.

SAS EXAMPLE 2.10 Distribution of Marital Status Using SAS

The following bar chart was generated using SAS Proc Chart.

SAS Output for Example 2.7

```
                              Frequency Bar Chart

Frequency

         |                    * * * * *
         |                    * * * * *
         |                    * * * * *
         |                    * * * * *
   20  +                    * * * * *
         |                    * * * * *
         |                    * * * * *
         |                    * * * * *
         |                    * * * * *
   15  +                    * * * * *
         |                    * * * * *
         |                    * * * * *
         |                    * * * * *
         |                    * * * * *        * * * * *
   10  +                    * * * * *        * * * * *
         |                    * * * * *        * * * * *
         |        * * * * *    * * * * *        * * * * *
         |        * * * * *    * * * * *        * * * * *
         |        * * * * *    * * * * *        * * * * *
    5  +        * * * * *    * * * * *        * * * * *        * * * * *
         |        * * * * *    * * * * *        * * * * *        * * * * *
         |        * * * * *    * * * * *        * * * * *        * * * * *
         |        * * * * *    * * * * *        * * * * *        * * * * *      * * * * *
         |        * * * * *    * * * * *        * * * * *        * * * * *      * * * * *
         ---------------------------------------------------------------------------------
             Divorced      Married    Never Married    Separated    Widowed
                                        marital
```

Interpretation of SAS Output for Example 2.7

The SAS Chart Procedure is used to generate a bar chart for a categorical variable. The user can specify whether frequencies or relative frequencies are plotted on the vertical axis. Here we plotted frequencies. ■

EXAMPLE 2.11   **Distribution of Health Status**

Health status is an example of an ordinal variable because the response options are ordered: Excellent, Very Good, Good, Fair, Poor. Figure 2.9 displays the sample data in a relative frequency histogram.

Figure 2.9 *Relative Frequency Histogram for Example 2.8 (Ordinal Variable)*

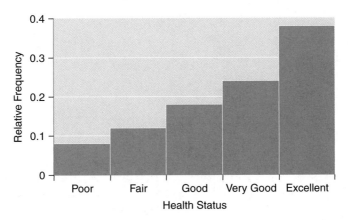

In this histogram, each response option is listed along the horizontal axis and the vertical axis is scaled to accommodate the relative frequencies. Rectangles are drawn above each response option to reflect the proportion of subjects in each. There are no breaks between rectangles, suggesting the ordering of the response options. It is clear from the histogram that most subjects reported excellent health status, while by comparison fewer reported fair or poor health status. ■

SAS EXAMPLE 2.11   **Distribution of Health Status Using SAS**

The following relative frequency histogram was generated using SAS Proc Chart.

SAS Output for Example 2.8

```
                    Relative Frequency Histogram

Percentage
       |                                        * * * * * * * * * *
       |                                        * * * * * * * * * *
       |                                        * * * * * * * * * *
  35 + |                                        * * * * * * * * * *
       |                                        * * * * * * * * * *
       |                                        * * * * * * * * * *
       |                                        * * * * * * * * * *
       |                                        * * * * * * * * * *
  30 + |                                        * * * * * * * * * *
       |                                        * * * * * * * * * *
       |                                        * * * * * * * * * *
       |                                        * * * * * * * * * *
       |                                        * * * * * * * * * *
  25 + |                                        * * * * * * * * * *
       |                          * * * * * * * * * * * * * * * * * *
       |                          * * * * * * * * * * * * * * * * * *
       |                          * * * * * * * * * * * * * * * * * *
       |                          * * * * * * * * * * * * * * * * * *
  20 + |                          * * * * * * * * * * * * * * * * * *
       |                          * * * * * * * * * * * * * * * * * *
       |              * * * * * * * * * * * * * * * * * * * * * * * * * *
       |              * * * * * * * * * * * * * * * * * * * * * * * * * *
       |              * * * * * * * * * * * * * * * * * * * * * * * * * *
  15 + |              * * * * * * * * * * * * * * * * * * * * * * * * * *
       |              * * * * * * * * * * * * * * * * * * * * * * * * * *
       |              * * * * * * * * * * * * * * * * * * * * * * * * * *
       |      * * * * * * * * * * * * * * * * * * * * * * * * * * * * * * * *
       |      * * * * * * * * * * * * * * * * * * * * * * * * * * * * * * * *
  10 + |      * * * * * * * * * * * * * * * * * * * * * * * * * * * * * * * *
       |      * * * * * * * * * * * * * * * * * * * * * * * * * * * * * * * *
       |  * * * * * * * * * * * * * * * * * * * * * * * * * * * * * * * * * * * *
       |  * * * * * * * * * * * * * * * * * * * * * * * * * * * * * * * * * * * *
       |  * * * * * * * * * * * * * * * * * * * * * * * * * * * * * * * * * * * *
   5 + |  * * * * * * * * * * * * * * * * * * * * * * * * * * * * * * * * * * * *
       |  * * * * * * * * * * * * * * * * * * * * * * * * * * * * * * * * * * * *
       |  * * * * * * * * * * * * * * * * * * * * * * * * * * * * * * * * * * * *
       |  * * * * * * * * * * * * * * * * * * * * * * * * * * * * * * * * * * * *
       |  * * * * * * * * * * * * * * * * * * * * * * * * * * * * * * * * * * * *
       ----------------------------------------------------------------
           Poor        Fair        Good    Very Good   Excellent
                            health status
```

Interpretation of SAS Output for Example 2.8

The SAS Chart Procedure can also be used to generate a histogram with certain options (e.g., no breaks between response options). In this type of histogram, the relative frequencies are expressed as percentages. The SAS code to generate the bar chart in SAS Example 2.10 and the relative frequency histogram in this example is in Section 2.4. ■

EXAMPLE 2.12 **Distribution of Age Classes**

The age class variable is an ordinal variable created in Example 2.9 from the continuous variable presented in Example 2.3. Figure 2.10 shows a frequency histogram for the age classes.

Figure 2.10 *Frequency Histogram for Example 2.9 (Ordinal Variable)*

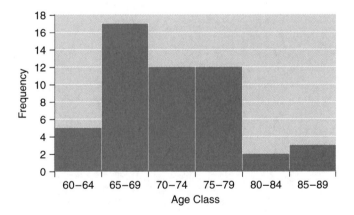

■

SAS EXAMPLE 2.12 **Distribution of Age Classes Using SAS**

The following frequency histogram was generated using SAS Proc Chart.

SAS Output for Example 2.10

```
                        Frequency Histogram

Frequency
  17 +                 * * * * * * * * * *
     |                 * * * * * * * * * *
  16 +                 * * * * * * * * * *
     |                 * * * * * * * * * *
  15 +                 * * * * * * * * * *
     |                 * * * * * * * * * *
  14 +                 * * * * * * * * * *
     |                 * * * * * * * * * *
  13 +                 * * * * * * * * * *
     |                 * * * * * * * * * *
  12 +                 * * * * * * * * * * * * * * * * * * * * * * * * * *
     |                 * * * * * * * * * * * * * * * * * * * * * * * * * *
  11 +                 * * * * * * * * * * * * * * * * * * * * * * * * * *
     |                 * * * * * * * * * * * * * * * * * * * * * * * * * *
  10 +                 * * * * * * * * * * * * * * * * * * * * * * * * * *
     |                 * * * * * * * * * * * * * * * * * * * * * * * * * *
   9 +                 * * * * * * * * * * * * * * * * * * * * * * * * * *
     |                 * * * * * * * * * * * * * * * * * * * * * * * * * *
   8 +                 * * * * * * * * * * * * * * * * * * * * * * * * * *
     |                 * * * * * * * * * * * * * * * * * * * * * * * * * *
   7 +                 * * * * * * * * * * * * * * * * * * * * * * * * * *
     |                 * * * * * * * * * * * * * * * * * * * * * * * * * *
   6 +                 * * * * * * * * * * * * * * * * * * * * * * * * * *
     |                 * * * * * * * * * * * * * * * * * * * * * * * * * *
   5 +* * * * * * * * * * * * * * * * * * * * * * * * * * * * * * * * * *
     |* * * * * * * * * * * * * * * * * * * * * * * * * * * * * * * * * *
   4 +* * * * * * * * * * * * * * * * * * * * * * * * * * * * * * * * * *
     |* * * * * * * * * * * * * * * * * * * * * * * * * * * * * * * * * *
   3 +* * * * * * * * * * * * * * * * * * * * * * * * * * * * * *         * * * * * * * * * *
     |* * * * * * * * * * * * * * * * * * * * * * * * * * * * * *         * * * * * * * * * *
   2 +* * * * * * * * * * * * * * * * * * * * * * * * * * * * * * * * * * * * * * * * * * * * *
     |* * * * * * * * * * * * * * * * * * * * * * * * * * * * * * * * * * * * * * * * * * * * *
   1 +* * * * * * * * * * * * * * * * * * * * * * * * * * * * * * * * * * * * * * * * * * * * *
     |* * * * * * * * * * * * * * * * * * * * * * * * * * * * * * * * * * * * * * * * * * * * *
     -----------------------------------------------------------------------
            60-64       65-69       70-74       75-79       80-84       85-89
                                     ageclass
```

Interpretation of SAS Output for Example 2.9

The SAS Chart Procedure was used to generate a frequency histogram. The SAS code used to generate the histogram is in Section 2.4. From the frequency histogram, we quickly see that the majority of the subjects in the sample are between 65 and 79 years of age. ▪

2.3 Key Formulas

Mathematical formulas and short descriptions of each of the statistics illustrated in Section 2.2 are outlined here. As a general guideline, descriptive statistics should include no more than one decimal place beyond that observed in the original data elements.

Numerical Methods: Location

Statistic	Notation/Formula	Description
Minimum	Min	Smallest value
Maximum	Max	Largest value
Mean	$\bar{X} = \dfrac{\sum X}{n}$	Average or typical value
Median	Order the data from smallest to largest Compute position(s) of middle value(s) n odd: $(n+1)/2$ n even: $(n/2)$ and $(n/2)+1$, compute mean	Middle value, holds 50% above it and 50% below it
Quartiles	Q_1	Holds 25% of the data at or below it
	Q_3	Holds 25% of the data at or above it

Order data from smallest to largest

Compute position(s) of quartiles

n odd: $\left[\dfrac{n+3}{4}\right]$

n even: $\left[\dfrac{n+2}{4}\right]$

Numerical Methods: Dispersion

Statistic	Notation/Formula	Description
Range	Maximum–Minimum	
Variance	$s^2 = \dfrac{\sum(X - \overline{X})^2}{(n - 1)}$ $= \dfrac{\sum X^2 - (\sum X)^2/n}{n - 1}$	Average squared deviation from the mean
Standard deviation	$s = \sqrt{s^2}$	Typical deviation from the mean
Empirical Rule	$\overline{X} \pm s$	Contains 68% of the distribution
	$\overline{X} \pm 2s$	Contains 95% of the distribution
	$\overline{X} \pm 3s$	Contains almost all of the distribution
Interquartile range	$Q_3 - Q_1$	Difference between first and third quartile (middle 50%)
Interquartile deviation	$\dfrac{Q_3 - Q_1}{2}$	Mean difference between quartiles and median

Numerical Methods: Distribution

Statistic	Notation/Formula	Description
Frequency	f	Number of subjects in each response category
Relative frequency	f/n	Proportion of subjects in each response category

Graphical Methods: Continuous Variables

Box-and-Whisker Plot

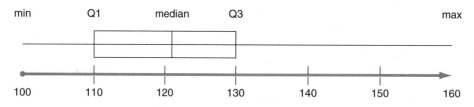

Stem-and-Leaf Plot

```
6 | 0 2 3 4 4
6 | 5 5 5 5 5 6 6 6 6 6 7 7 7 8 8 8
7 | 0 0 0 1 1 2 2 3 3 3 3 3
7 | 5 5 5 5 6 6 7 7 7 7 7 9
8 | 2 3
8 | 5 5 7
```

NOTE: Best results are achieved when there are between 6 and 12 stems.

Graphical Methods: Discrete Variables

Histogram (Frequency or Relative Frequency Histogram)

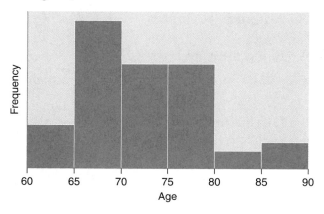

NOTE: Class limits, for classes of equal length, are displayed along the horizontal axis of the histogram. For ordinal variables, no breaks are left between the rectangles. For categorical variables, there are breaks between the rectangles, producing a bar chart (see below). Best results are achieved when there are between 6 and 12 distinct categories.

Relative Frequency Histogram

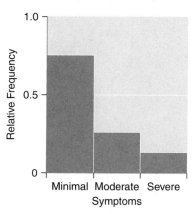

Bar Chart

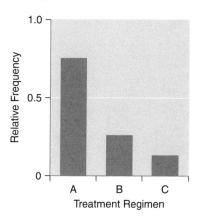

2.4 Statistical Computing

Following are the SAS programs which were used to generate the descriptive statistics and graphical summaries presented in Section 2.2 for Examples 2.1 through 2.12. The SAS procedures and brief descriptions are noted in the header to each example. Notes are provided to the right of the SAS programs (in blue) for orientation purposes and are not part of the programs. In addition, blank lines in the programs that follow are there solely to accommodate the notes. Blank lines and spaces can be used throughout SAS programs and are generally used to enhance readability. (More details are contained in Appendix A: "Introduction to Statistical Computing Using SAS.") A summary of the SAS procedures used in the examples is provided at the end of this section.

2.4.1 Continuous Variables

SAS CODE FOR

Systolic Blood Pressure Data

SAS EXAMPLE 2.1 *Proc Univariate (Summary Statistics for Continuous Variable)*

Seven subjects are randomly selected from a population of patients 50 years of age with diagnosed coronary artery disease. Systolic blood pressures, measured in millimeters of mercury (mmHg), are taken on each subject and are listed here. Summarize the sample data using SAS.

| 121 | 110 | 114 | 100 | 160 | 130 | 130 |

Program Code

```
options ps=64 ls=80;                    Formats the output page to 64 lines in length,
                                          80 columns in width
data in;                                Beginning of Data Step
  input sbp;                            Inputs variable sbp

label sbp='systolic blood pressure';   Attaches descriptive label to variable sbp

cards;                                  Beginning of Raw Data section
121                                        actual observations
110
114
100
160
130
130
run;                                    End of Raw Data section

title 'Summary Statistics';            Title for output
proc univariate;                       Procedure call. Proc Univariate generates
                                          summary statistics
  var sbp;                             Specification of variable sbp
run;                                   End of Procedure section            ■
```

SAS CODE FOR

Total Serum Cholesterol Data

SAS EXAMPLE 2.2 *Proc Means (Abbreviated Summary Statistics for Continuous Variable)*

Eight subjects are randomly selected from a population of patients with hypertension. Total serum cholesterol, in mg/100 ml, is a continuous variable measured on each subject. The sample data are given here. Summarize the sample data using SAS.

197   212   211   184   260   233   245   219

Program Code

```
options ps=64 ls=80;
```
Formats the output page to 64 lines in length, 80 columns in width

```
data in;
  input chol;
```
Beginning of Data Step

Inputs variable **chol**

```
label chol='total serum cholesterol';
```
Attaches descriptive label to variable **chol**

```
cards;
197
212
211
184
260
233
245
219
run;
```
Beginning of Raw Data section actual observations

End of Raw Data section

```
title 'Summary Statistics';
proc means;
```
Title for output

Procedure call. Proc Means generates abbreviated summary statistics

```
  var chol;
run;
```
Specification of variable **chol**

End of Procedure section ■

SAS CODE FOR
SAS EXAMPLE 2.3 AND 2.6

Age Data (continuous variable)

Proc Univariate (Summary Statistics for Continuous Variable)

A sample of 51 individuals are selected for participation in a study of cardiovascular risk factors. The following data represent the ages of enrolled individuals, measured in years. Summarize the age data using SAS.

60	62	63	64	64	65	65	65	65	65	65
66	66	66	66	66	67	67	67	68	68	68
70	70	70	71	71	72	72	73	73	73	73
73	75	75	75	75	76	76	77	77	77	77
77	79	82	83	85	85	87				

```
Program Code
options ps=64 ls=80;                 Formats output page
data in;                             Beginning of Data Step
  input age;                         Inputs variable age
label age='age in years';            Attaches label to variable age
cards;                               Beginning of Raw Data section
60                                       actual observations (one observation
62                                       per line, 51 lines total)
63
64
64
65
.
.
.
83
85
85
87
run;                                 End of Raw Data section

title 'Summary Statistics';          Title for output of Proc Univariate
proc univariate plot;                Procedure call (Proc Univariate); the plot option
                                         generates several graphical displays of the data
  var age;                           Specification of variable age
run;
```
■

2.4.2 Discrete Variables

SAS CODE FOR   **Marital Status Data**

SAS EXAMPLE 2.7   *Proc Freq (Frequency Distribution Table for Discrete Variable)*
AND 2.10   *Proc Chart (Bar Chart for Categorical Discrete Variable)*
A sample of 50 subjects respond to a survey. The following data reflect marital status of the subjects in the sample. Summarize the sample using SAS.

Marital Status	Number of Individuals (f)
Married	24
Separated	5
Divorced	8
Widowed	2
Never Married	11
Total	50

Program Code

```
options ps=64 ls=80;
data in;
  input marital $ 1-15 f;



cards;
Married               24
Separated              5
Divorced               8
Widowed                2
Never Married         11
run;


title 'Frequency Distribution Table';
proc freq;


  tables marital;
  weight f;


title 'Frequency Bar Chart';
proc chart;



  vbar marital/freq=f discrete;
run;
```

Formats output page
Beginning of Data Step
Inputs alphanumeric ($) variable **marital** from
    columns 1-15 and numeric variable *f*

Beginning of Raw Data section
    actual observations:
    two entries per line (**marital** and *f*)

Procedure call (Proc Freq to generate frequency
    distribution table)

Specification of variable **marital**
    weight *f* specifies variable containing frequencies (*f*)

Procedure call (Proc Chart to generate frequency
    bar chart)

Specification of variable **marital**; vbar option
    requests a vertical histogram, freq=*f* specifies
    variable containing frequencies (*f*), discrete
    indicates that variable is discrete in nature as
    opposed to continuous ■

SAS CODE FOR   **Health Status Data**

SAS EXAMPLE 2.8   *Proc Format (Attaches Descriptions to Discrete Response Options)*
AND 2.11   *Proc Freq (Frequency Distribution Table for Discrete Variable)*
    Proc Chart (Histogram for Ordinal Discrete Variable)
    Proc Means (Abbreviated Summary Statistics)

The sample of 50 subjects described in SAS Example 2.4 also report their
health status in the survey. The following data reflect the self-reported health
status of the subjects in the sample. Summarize the sample using SAS.

Health Status	Number of Individuals (f)
Excellent	19
Very Good	12
Good	9
Fair	6
Poor	4
Total	**50**

Program Code

```
options ps=64 ls=80;
```
Formats output page

```
proc format;
```
Procedure Call (Proc Format to attach descriptive labels to response options)

```
  value hlth   5='Excellent'
               4='Very Good'
               3='Good'
               2='Fair'
               1='Poor';
run;
```
Specification of a name for format (**hlth**)
Description for response options

```
data in;
  input health f;
format health hlth.;
label health='health status';
```
Beginning of Data Step
Inputs two numeric variables: **health** and *f*
Attaches description to each response option
Attaches label to variable **health**

```
cards;
5 19
4 12
3 9
2 6
1 4
run;
```
Beginning of Raw Data section
  actual observations:
  two entries per line (**health** and *f*)
The values 1–5 are used to represent the 5 distinct health status responses.

```
title 'Frequency Distribution Table';
proc freq;
```
Procedure call (Proc Freq to generate frequency distribution table)

```
  tables health;
  weight f;
run;
```
Specification of variable **health**
  weight *f* specifies variable containing frequencies (*f*)

```
title 'Summary Statistics';
proc means;
```
Procedure call (Proc Means to generate summary statistics)

```
  var health;
  freq f;
run;
```
Specification of variable **health**
Specification of variable containing frequencies (*f*)
End of procedure section

```
title 'Relative Frequency Histogram';
proc chart;
```
Procedure call (Proc Chart to generate relative frequency histogram)

```
  vbar health/freq=f type=pct levels=5
         space=0 width=10;
```
Specification of variable **health**; vbar option requests a vertical histogram, freq=*f* specifies variable containing frequencies (*f*), type=pct requests a relative frequency histogram, levels=5 indicates the number of response options, space=0 requests no space between response options (appropriate for a histogram), and width=10 requests that the rectangles above each response option have width 10

```
run;
```
■

SAS CODE FOR **Age Classes (ordinal variable)**

SAS EXAMPLE 2.9 *Proc Freq (Frequency Distribution Table for Discrete Variable)*
AND 2.12 *Proc Chart (Histogram for Ordinal Discrete Variable)*

A sample of 51 individuals are selected for participation in a study of cardio-vascular risk factors. The following data represent the ages of enrolled individuals, measured in years. Organize the subjects into age classes and summarize the age classes data using SAS.

60	62	63	64	64	65	65	65	65	65	65
66	66	66	66	66	67	67	67	68	68	68
70	70	70	71	71	72	72	73	73	73	73
73	75	75	75	75	76	76	77	77	77	77
77	79	82	83	85	85	87				

Program Code

```
options ps=64 ls=80;
data in;
  input age;
```
Formats output page
Beginning of Data Step
Inputs continuous variable **age**

```
if 60 le age le 64 then ageclass='60-64';
else if 65 le age le 69 then ageclass='65-69';
else if 70 le age le 74 then ageclass='70-74';
else if 75 le age le 79 then ageclass='75-79';
```
Creates **ageclass** variable from continuous variable **age** (measured in years)
(The SAS function le=less than or equal to)

```
else if 80 le age le 84 then ageclass='80-84';
else if 85 le age le 89 then ageclass='85-89';

cards;                                              Beginning of Raw Data section
60                                                     actual observations (one observation per line,
62                                                     51 lines total – See SAS Example 2.3)
63

85
87
run;                                                End of Raw Data section

title 'Frequency Distribution Table';    Title for output of Proc Freq
proc freq;                                           Procedure call (Proc Freq)
 tables ageclass;                                    Specification of variable ageclass
run;

title 'Frequency Histogram';
proc chart;                                          Procedure call (Proc Chart to generate frequency
                                                         histogram)
 vbar ageclass/levels=6              Specification of variable ageclass; vbar option
          space=0 width=10;              requests a vertical histogram, levels=6 indicates
                                                         the number of response options, space=0 requests
                                                         no space between response options (appropriate
                                                         for a histogram), and width=10 requests that the
                                                         rectangles above each response option have
                                                         width 10
run;                                                                                                    ■
```

2.4.3 Summary of SAS Procedures

The SAS procedures illustrated in this section are summarized in the following table. The SAS call statements for each procedure are shown, with optional statements in italics. Users should refer to the examples in this section for more description of each procedure and associated options. A general description of each procedure is provided in the table.

Procedure	Sample Procedure Call	Description
proc chart	proc chart; vbar x/*type* = *pct freq* = *f discrete*;	Generates (relative) frequency histogram for either continuous or discrete variables
proc freq	proc freq; tables x; *weight f*;	Generates frequency distribution table
proc means	proc means; var x1 x2; *freq f*;	Generates summary statistics (abbreviated version)
proc univariate	proc univariate *plot normal*; var x1; *freq f*;	Generates (extensive) summary statistics, stem-and-leaf plot, box-and-whisker plot

2.5 Analysis of Framingham Heart Study Data

The Framingham Heart Study is one of the most extensive studies of cardiovascular disease in the world. The study started in 1948 with the enrollment of over 5,000 participants from the town of Framingham, Massachusetts, located approximately 20 miles west of Boston. The study was designed to investigate, among other things, the risk factors for cardiovascular disease. A detailed description of the objectives of the study and its design can be found in "Epidemiological Background and Design: The Framingham Study," by Ralph B. D'Agostino, Sr., and William B. Kannel (*Proceedings of the American Statistical Association Sesquicentennial Meeting* 1989).

Members of the original cohort agreed to undergo extensive clinical examinations every 2 years. These exams included measurements like height, weight, systolic and diastolic blood pressure, laboratory and blood tests to measure cholesterol levels, echocardiography to measure heart function, and so on. In the early 1970s, their offspring and their spouses were enrolled in the Framingham Offspring Study, and they have participated in examinations approximately every 4 years. In 2002, a third generation was enrolled, which will allow for more investigation into genetic factors associated with cardiovascular and other diseases. In 1998, the Framingham Heart Study celebrated its 50th anniversary and the study is still continuing today.

Here we analyze data collected from the original cohort. A description of the data set is contained in Appendix C. The data set includes data collected from the original cohort from about 1956 to 1968. Participants contributed up to three examination cycles of data, approximately 6 years apart, and follow-up data on incident or newly developed cardiovascular disease, cerebrovascular disease, and death are available over a 24-year follow-up period. Here we focus on the first examination for each participant (called the period 1 examination). Using SAS, we generated descriptive statistics on an array of continuous and discrete variables using SAS Proc Means and SAS Proc Freq, respectively. The SAS code is outlined and we then provide some graphical displays of key variables.

FRAMINGHAM DATA ANALYSIS—SAMPLE SAS CODE

```
data in;
  set in.frmgham;
  if period=1;
run;

proc means; var age sysbp diabp totchol bmi cigpday; run;
proc freq; tables sex bpmeds cursmoke diabetes; run;
```

```
proc chart; vbar agegrp/levels=4 space=0 width=10; run;
proc sort; by bpmeds; run;
proc boxplot; plot sysbp*bpmeds; run;
```

FRAMINGHAM DATA ANALYSIS—SAS OUTPUT

The following summary statistics were generated using SAS Proc Means, which generates descriptive statistics on continuous variables.

Variable	Label	N	Mean	Std Dev
AGE	Age (years) at examination	4434	49.9258006	8.6769293
SYSBP	Systolic BP mmHg	4434	132.9077582	22.4215970
DIABP	Diastolic BP mmHg	4434	83.0835589	12.0559994
TOTCHOL	Serum Cholesterol mg/dL	4382	236.9842538	44.6510984
BMI	Body Mass Index (kr/(M*M)	4415	25.8461608	4.1018209
CIGPDAY	Cigarettes per day	4402	8.9663789	11.9317058

Variable	Label	Minimum	Maximum
AGE	Age (years) at examination	32.0000000	70.0000000
SYSBP	Systolic BP mmHg	83.5000000	295.0000000
DIABP	Diastolic BP mmHg	48.0000000	142.5000000
TOTCHOL	Serum Cholesterol mg/dL	107.0000000	696.0000000
CIGPDAY	Cigarettes per day	0	70.0000000

There are a total of $n = 4,434$ participants with an exam in period 1. Notice that the N's vary from one variable to the next. For example, for TOTCHOL there are $n = 4,382$ observations, because 52 participants have missing values. This is not unexpected in a study of this size and scope. It is, however, always important to assess the completeness of study variables.

These variables—AGE, SYSBP, DIABP, TOTCHOL, and BMI—could be well described by the Empirical Rule. Could we use the Empirical Rule to describe the number of cigarettes smoked per day (CIGPDAY)—why or why not?

The following frequency distribution tables were produced by SAS Proc Freq for discrete variables measured in period 1.

SEX

SEX	Frequency	Percent	Cumulative Frequency	Cumulative Percent
M	1944	43.84	1944	43.84
F	2490	56.16	4434	100.00

Anti-hypertensive meds Y/N

BPMEDS	Frequency	Percent	Cumulative Frequency	Cumulative Percent
N	4229	96.71	4229	96.71
Y	144	3.29	4373	100.00

Frequency Missing = 61

Diabetic Y/N

DIABETES	Frequency	Percent	Cumulative Frequency	Cumulative Percent
N	4313	97.27	4313	97.27
Y	121	2.73	4434	100.00

Current Cig Smoker Y/N

CURSMOKE	Frequency	Percent	Cumulative Frequency	Cumulative Percent
N	2253	50.81	2253	50.81
Y	2181	49.19	4434	100.00

In the Framingham cohort, there are 56% women, 3% of the participants are taking blood-pressure medications, 3% are diabetic, and 49% report being current smokers. How does the fact that just over half of the participants classify themselves as nonsmokers affect our interpretation of the mean and standard deviation of the number of cigarettes smoked per day? These data, particularly the percents of participants on blood-pressure medication and current smokers, must be interpreted with respect to the time at which they were collected (1950s–1960s). If the data were more contemporary, more participants would likely be on blood-pressure medication and fewer would be smokers.

The following figures describe the distributions of age, systolic blood pressure for the total sample, and then blood pressure for persons who are taking and not taking blood-pressure medications as measured in period 1.

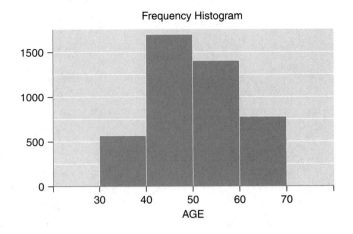

Frequency Histogram

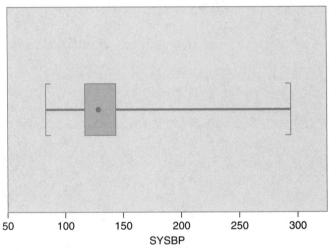

Box Plot

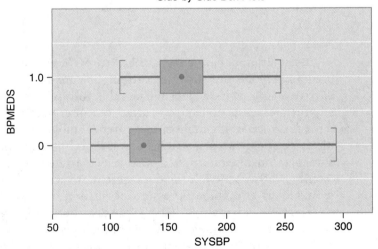

Side-by-Side Box Plots

How do the distributions of systolic blood pressure compare for persons taking and not taking blood-pressure medication?

The following figure shows the distribution of number of cigarettes smoked per day for the total sample. What is the value of the first quartile, the median, the third quartile?

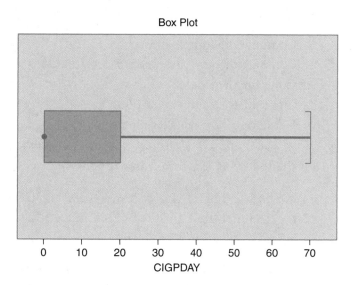

The next figure shows the number of cigarettes smoked per day classified by current smoking status.

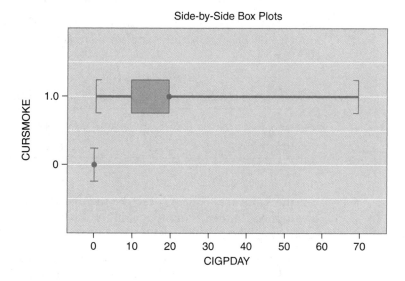

What is the value of the first quartile, the median, the third quartile among smokers?

2.6 Problems

1. Systolic blood pressures recorded on a random sample of six males are as follows:

   | 156 | 85 | 103 | 92 | 108 | 128 |

   a. Compute the sample mean.
   b. Compute the sample standard deviation.
   c. Compute the sample median.
   d. Compute the sample quartiles.
   e. Generate a box-whisker plot.

2. Twenty-five randomly selected, uncomplicated appendectomies lasted for the following lengths of time (minutes):

113	118	121	123	126	128	130	135	136	137
138	139	140	140	142	142	142	142	143	155
157	157	158	159	164

   a. Generate a frequency distribution table for the data using the following classes 110–119, 120–129, and so on.
   b. Generate a relative frequency histogram.
   c. Generate a stem-and-leaf plot for the data.

3. The following data represent summary statistics on selling prices of diabetes home-monitoring supplies from two different pharmaceutical companies:

	Mean Selling Price	Standard Deviation
Company #1	$75	$4
Company #2	$56	$16

   Use the Empirical Rule to describe the distributions of the selling prices of supplies from each company. How do the prices compare between companies (describe briefly)?

4. An investigator is interested in examining the effects of alcohol on individuals' abilities to perform physical activities. Twenty randomly selected individuals consumed a prescribed amount of alcohol, and

their time to complete a specific physical activity was recorded. The data are given here:

Subject ID Number	Time in Minutes
1	12.5
2	14.0
3	16.1
4	27.2
5	14.5
6	22.6
7	15.9
8	16.2
9	10.1
10	29.1
11	12.5
12	12.6
13	9.9
14	12.8
15	11.1
16	23.2
17	14.5
18	32.6
19	16.9
20	9.0

a. Compute the sample mean.

b. Compute the sample standard deviation.

c. Compute the sample median.

d. For this data, is the mean or the median a better measure of location? Why?

e. Generate a frequency histogram for the data.

f. The investigator suspects that there may be a difference for males versus females in times to completion of these tasks under the influence of alcohol. Subjects with identification numbers 1, 2, 5, 8, 9, 11, 13, 17, 19, and 20 are male; the remaining subjects are female. Generate and compare means, medians, and standard deviations for the males and females, considered separately.

5. The following data reflect participants' scores on a standardized exam designed to measure medical literacy. Scores range from 0 to 100, with higher scores indicating better performance.

   91   94   83   84   46   90   77   85

   a. Compute the sample mean.
   b. Compute the median.
   c. Compute the sample variance.
   d. Compute the sample standard deviation.
   e. Compute the quartiles.
   f. Display the box-whisker plot.

6. Cholesterol levels for males aged 50 have a mean of 210 and a standard deviation of 22.8. Use the Empirical Rule to describe the distribution of cholesterol levels.

7. A study was recently conducted in the Boston area to investigate various aspects of AIDS complications and care. The study involved nearly 300 AIDS patients. The following data represent CD4 counts for 10 randomly selected AIDS patients in the study.

   75   135   210   240   86   100   125   59   63   100

   a. Compute the mean CD4 count for the sample.
   b. Compute the median CD4 count.
   c. Compute the range of CD4 counts.
   d. Compute the standard deviation in the CD4 counts.
   e. Which is the best measure of location (e.g., mean or median)? Which is the best measure of dispersion (e.g., standard deviation or interquartile deviation)?

8. The following is a reproduction of a figure presented in a medical journal describing the distribution of ages (in years) of participants involved in a research study of cardiovascular risk factors:

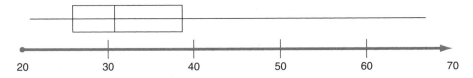

   a. What is the approximate value of the 75th percentile in the age distribution?
   b. Is there any evidence of outliers in the distribution? (*Hint:* Use the $Q \pm 1.5$ IQR rule.)

9. The following is a frequency distribution table of systolic blood pressures (SBP) measured on a sample of female patients who are 50 years of age:

SBP	f
105	6
108	2
110	3
120	4
124	8
128	6
130	7
136	8
140	2
170	3
190	2
204	1

   a. Compute the sample median.
   b. Compute the first and third quartiles.
   c. Construct a box-whisker plot of the SBPs.
   d. Are there outliers in the sample of SBPs? (Use the $Q \pm 1.5$ IQR rule.)

10. A figure similar to the following was shown in a paper comparing visits to primary physicians over 12 months between Canadian and U.S. patients (* = mean).

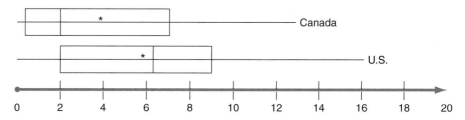

   Fill in the following statements:

   a. Half of the Canadian patients have fewer than _____ visits per year.
   b. Twenty-five percent of the U.S. patients had more than _____ visits per year.
   c. Briefly compare the numbers of primary-care visits per year between Canadian and U.S. patients.

11. The following reflect the times (in minutes) between prescription drop-off and pick-up at a local pharmacy:

    8   15   7   6   4   20   3   6   4   3

    a. Calculate the mean time between drop-off and pick-up.
    b. Calculate the standard deviation in time between drop-off and pick-up.
    c. Calculate the median time between drop-off and pick-up.

12. The times between successive routine primary-care visits (in months) in a sample of patients from a local health maintenance organization (HMO) are:

    6   18   12   9   10   14   6   5

    a. Compute the sample mean.
    b. Compute the sample median.
    c. Compute the quartiles.
    d. Compute the sample standard deviation.
    e. Construct a box-whisker plot.

13. The following data were taken from a paper from a study published in a medical journal investigating the effects of exercise in pregnant women.

    Characteristics of Women at Baseline*

	Exercise Group (n = 18)	*Control Group (n = 15)*
Age (yrs)	31.1 ± 5.4	29.7 ± 4.9
Education (yrs)	17.1 ± 2.2	15.9 ± 2.1
Race or ethnic group (Number of subjects)		
Non-Hispanic white	15	13
Hispanic	1	2
Asian	2	0

    * Table entries are numbers of subjects or means ± Standard Deviation (SD).

    a. Use the Empirical Rule to describe the distributions of ages of women in the exercise and control groups (separately).
    b. Is there a difference in the ages of women in the two groups? Answer briefly and base your answer only on part a.

14. A study is undertaken to investigate the self-reported health status among women with osteoporosis. Health status is measured on a scale from 0 to 100, with higher scores indicative of better health status. The following data are observed:

    35   48   70   42   80   57   74   39   40

    a. Compute the sample mean.
    b. Compute the sample standard deviation.
    c. Compute the median.
    d. Compute the quartiles.

15. The following figure represents the distribution of self-reported health status measured on a sample of women free of osteoporosis.

    a. Draw the box-whisker plot reflecting the self-reported health status among women with osteoporosis (using the data from Problem 14 below the plot).

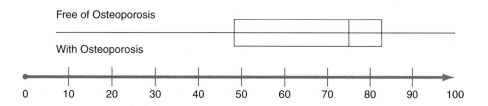

    b. How do the average self-reported health status scores compare between women with osteoporosis and those free of osteoporosis?
    c. How do the self-reported health status scores compare between women with osteoporosis and those free of osteoporosis with respect to variation?

16. A study is conducted to assess the extent to which patients who had coronary artery bypass surgery were maintaining their prescribed exercise programs. The following data reflect the numbers of times patients reported exercising over the previous month (4 weeks). For the purposes of this study, exercise was defined as moderate physical activity lasting at least 20 minutes.

    14   11   8   6   5   3
    6   13   12   8   1   4

    a. Compute the mean number of times patients exercised per month.
    b. Compute the median number of times patients exercised per month.

   c. Compute the standard deviation in the number of times patients exercised per month.

   d. If a patient reported exercising more than twice per week, that patient was considered adherent with respect to the exercise program. Assume that the reported number of times each participant exercised over the past month were evenly distributed over the 4-week reporting period. Classify each patient as adherent or not based on the number of times he or she reported exercising, and generate a relative frequency histogram displaying the distribution of adherent and nonadherent patients.

17. The following data were collected from a random sample of 10 students who started an MPH program in the fall of 1998. The data reflect GRE scores, which range from 200 to 800, with higher scores indicative of better achievement.

   520   680   470   560   510   610   670   560   525   475

   a. Compute the sample mean.
   b. Compute the sample standard deviation.
   c. Compute the sample median.
   d. Compute the sample range.

18. Twenty-five patients are randomly selected from the population of all patients with rheumatoid arthritis in a particular health maintenance organization (HMO). Each patient is surveyed and asked to rate the HMO with respect to satisfaction with the care received for the arthritis. The satisfaction scores range from 0 to 100, with higher scores indicative of more satisfaction. The data are given here:

   65   86   84   85   97   94   89   84   83   89
   88   78   77   76   82   72   92   99   94   83
   81   85   97   93   79

   a. Compute the sample mean.
   b. Compute the sample standard deviation.
   c. Compute the sample median.
   d. Compute the quartiles.
   e. Compute the sample range.
   f. Generate a frequency distribution table to summarize the data. A frequency distribution table is most informative when there are

between 6 and 12 classes or categories. For this example, suppose we wish to construct a frequency distribution table with 7 classes. In order to define the classes, the following formula is used to compute the width of each class (i.e., the range of values covered by each class): width = range/(number of intervals).

g. Construct a stem-and-leaf plot using stems defined by the classes given in the frequency distribution table in (f).

19. A study is conducted to assess the drinking behaviors of college seniors. A random sample of 10 seniors are selected and each is asked the number of alcoholic drinks consumed in a typical week. The data are

| 20 | 40 | 25 | 12 | 16 | 8 | 8 | 5 | 10 | 0 |

a. Compute the mean number of drinks per week.
b. Compute the standard deviation in the number of drinks per week.
c. Compute the range in the number of drinks per week.
d. Compute the median number of drinks per week.

20. The following box-whisker plots display the numbers of alcoholic drinks consumed by male and female students in a typical week.

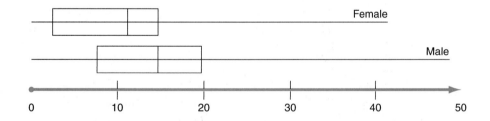

a. Estimate the median number of drinks per week for males.
b. Estimate the median number of drinks per week for females.
c. Estimate the quartiles for males.
d. Estimate the quartiles for females.
e. Are there any outliers among the male scores? (*Hint:* Use the quartile rule.)

21. The following displays the results of a standardized neuropsychological test administered to a sample of high school seniors. The test measures analytic skills and is scored on a scale of 0 to 100, with higher scores indicative of stronger analytic skills.

Males		Females
1	1	
8 8	2	
4 5 6	3	2 3 3 3 4 4 6 7 8
3 5 6 9	4	4 5 5 6 7 8 8 9 9
1 1 2 3	5	3 3 4 4 5 5 6 7
	6	1 1 1 2 2 3 3 4 4 4 5 5 5 7 7 8 9
3 4 5 5 5 7 8 9	7	2 2 3 3 3 4 5 6 6 7 8
4 5 6 6 6 8 9 9 9	8	2 2 3 6 7
1 2 2 2 3 3 3 3 4 4 4 4 4 4	9	1 3 4
0 0	10	

a. Compute the median scores for males and females.

b. Compute the quartiles for males and females.

c. Compute the median for the total sample (males and females combined).

d. Are there any outliers among the male scores? (*Hint:* Use the quartile rule.)

e. Based on your answer to (d), would the mean or median provide a better estimate of a typical result for males on the standardized exam?

22. A study is conducted among college freshman who smoke to assess the number of cigarettes smoked in a typical day. All freshman involved in the assessment report that they are regular smokers. The following reflects the reported numbers of cigarettes smoked per day:

    20   40   25   12   16   8   8   5   10   12

    a. Compute the mean number of cigarettes smoked.

    b. Compute the standard deviation in the number of cigarettes smoked.

    c. Compute the range in the number of cigarettes smoked.

    d. Compute the median number of cigarettes smoked.

23. A self-administered survey is fielded to measure the number of hours of exercise that patients with Type II diabetes get in a typical week. The following data are observed:

    5   3   2   0   2   7   4   3   4   6   0

    a. Compute the sample mean.

    b. Compute the sample median.

    c. Compute the sample standard deviation.

    d. Compute the sample range.

24. The following plots describe the lengths of stay following total knee replacement surgery for patients in three different hospitals (denoted A, B, and C).

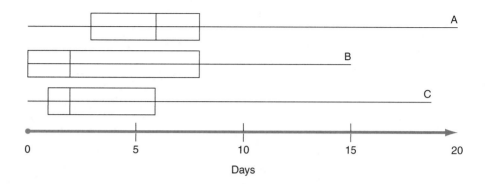

a. What is the median length of stay for patients in hospital B?
b. Which hospital has the shortest lengths of stay, on average, for patients undergoing knee replacement?
c. What is the range of lengths of stay in hospital C?
d. Are there any outliers in lengths of stay in hospital A? (*Hint:* Use the quartile rule to check for outliers.)

25. In the study described in Problem 24, suppose that patients in each hospital (A, B and C) were asked to rate their overall satisfaction with the total knee replacement surgery experience. The data are shown below:

	Hospital		
Satisfaction	*A*	*B*	*C*
Extremely Satisfied	62	50	20
Satisfied	20	30	16
Dissatisfied	14	12	17
Extremely Dissatisfied	9	8	42
Total number of patients	105	100	95

a. Generate frequency distribution tables for each hospital's satisfaction data, considered separately.
b. Generate a frequency distribution table for satisfaction data for all hospitals combined.

SAS Problems

Use SAS to solve each of the following problems.

1. Systolic blood pressures are recorded on a random sample of six males and are as follows:

   156   85   103   92   108   128

   Use SAS Proc Univariate to determine the following:
   a. The sample mean
   b. The sample standard deviation
   c. The sample median
   d. The sample quartiles

2. Twenty-five randomly selected, uncomplicated appendectomies lasted for the following lengths of time (minutes):

   113   118   121   123   126   128   130   135   136   137
   138   139   140   140   142   142   142   142   143   155
   157   157   158   159   164

   Use SAS Proc Chart to generate a frequency distribution table for the data.

3. The following data reflect participants' scores on a standardized exam designed to measure medical literacy. Scores range from 0 to 100, with higher scores indicating better performance.

   91   94   83   84   46   90   77   85

   Use SAS Proc Means to determine the following:
   a. The sample mean
   b. The sample variance
   c. The sample standard deviation
   d. The minimum and maximum

4. A study was recently conducted in the Boston area to investigate various aspects of AIDS complications and care. The study involved nearly 300 AIDS patients. The following data represent CD4 counts for 10 randomly selected AIDS patients involved in the study.

   75   135   210   240   86   100   125   59   63   100

Use SAS Proc Univariate to determine the following:

a. The sample mean and median.

b. The standard deviation and the interquartile range.

c. Which is the best measure of location (e.g., mean or median)? Which is the best measure of dispersion (e.g., standard deviation or interquartile deviation)?

5. The following is a frequency distribution table of systolic blood pressures (SBP) measured on a sample of female patients who are 50 years of age:

SBP	f
105	6
108	2
110	3
120	4
124	8
128	6
130	7
136	8
140	2
170	3
190	2
204	1

Use SAS Proc Univariate to generate a box-whisker plot of the SBPs. Are there outliers in the sample of SBPs?

6. The following reflect the times (in minutes) between prescription drop-off and pick-up at a local pharmacy:

8   15   7   6   4   20   3   6   4   3

Use SAS Proc Univariate to determine:

a. The mean time between drop-off and pick-up

b. The standard deviation in time between drop-off and pick-up

c. The median time between drop-off and pick-up

7. The time between successive routine primary-care visits (in months) in a sample of patients from a local health maintenance organization (HMO) are shown here:

6   18   12   9   10   14   6   5

Use SAS Proc Means to determine the following:
a. The sample mean
b. The minimum and maximum
c. The sample standard deviation

8. A study is conducted to assess the extent to which patients who had coronary artery bypass surgery were maintaining their prescribed exercise programs. The following data reflect the numbers of times patients reported exercising over the previous month (4 weeks). For the purposes of this study, exercise was defined as moderate physical activity lasting at least 20 minutes.

14   11   8   6   5   3
6   13   12   8   1   4

If a patient reported exercising more than twice per week, that patient was considered adherent with respect to the exercise program. Assume that the reported number of times each participant exercised over the past month were evenly distributed over the 4-week reporting period. Using SAS, classify each patient as adherent or not based on the number of times he or she reported exercising, and generate a relative frequency histogram (using SAS Proc Chart) displaying the distribution of adherent and nonadherent patients.

9. Twenty-five patients are randomly selected from the population of all patients with rheumatoid arthritis in a particular health maintenance organization (HMO). Each patient is surveyed and asked to rate the HMO with respect to satisfaction with the care received for arthritis. The satisfaction scores range from 0 to 100, with higher scores indicative of more satisfaction. The data are given here:

65   86   84   85   97   94   89   84   83   89
88   78   77   76   82   72   92   99   94   83
81   85   97   93   79

Use SAS Proc Univariate to generate a box-whisker plot and a stem-and-leaf plot for the satisfaction data.

10. The following data reflect satisfaction scores measured in patients undergoing total knee replacement surgery in three different hospitals (A, B, and C).

Satisfaction	Hospital		
	A	B	C
Extremely Satisfied	62	50	20
Satisfied	20	30	16
Dissatisfied	14	12	17
Extremely Dissatisfied	9	8	42
Total number of patients	105	100	95

Use SAS Proc Freq to generate a frequency distribution table for satisfaction data for all hospitals combined. Use SAS Proc Chart to generate a frequency histogram for the data (all hospitals combined).

Descriptive Statistics
(Ch. 2)

Probability
(Ch. 3)

Sampling Distributions
(Ch. 4)

Statistical Inference
(Chapters 5–13)

OUTCOME VARIABLE	GROUPING VARIABLE(S)/ PREDICTOR(S)	ANALYSIS	CHAPTER(S)
Continuous	—	Estimate μ; Compare μ to Known, Historical Value	5/12
Continuous	Dichotomous (2 groups)	Compare Independent Means (Estimate/Test $(\mu_1 - \mu_2)$) or the Mean Difference (μ_d)	6/12
Continuous	Discrete (>2 groups)	Test the Equality of k Means Using Analysis of Variance ($\mu_1 = \mu_2 = \cdots = \mu_k$)	9/12
Continuous	Continuous	Estimate Correlation or Determine Regression Equation	10/12
Continuous	Several Continuous or Dichotomous	Multiple Linear Regression Analysis	10
Dichotomous	—	Estimate p; Compare p to Known, Historical Value	7
Dichotomous	Dichotomous (2 groups)	Compare Independent Proportions (Estimate/Test $(p_1 - p_2)$)	7/8
Dichotomous	Discrete (>2 groups)	Test the Equality of k Proportions (Chi-Square Test)	7
Dichotomous	Several Continuous or Dichotomous	Multiple Logistic Regression Analysis	11
Discrete	Discrete	Compare Distributions Among k Populations (Chi-Square Test)	7
Time to Event	Several Continuous or Dichotomous	Survival Analysis	13

Probability

3

Figure 3.1 *Sampling*

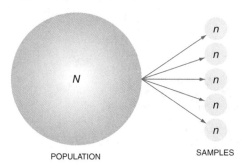

In Chapter 2 we discussed an array of statistics used to describe a sample. In this chapter we focus on the process of sampling from the population of interest. In particular, we address the experimental mechanisms and subsequent likelihood of sampling subjects from the population into a sample (Figure 3.1). The sampling process is a critical part of the theoretical basis for statistical inference.

In Section 3.1 we provide background and introduce vocabulary terms and definitions. In Section 3.2 we discuss the basic or first principles of probability. In Section 3.3 we address combinations and permutations. In Section 3.4 we discuss a popular discrete probability model, the binomial distribution model, and in Section 3.5 we discuss a popular continuous probability model, the normal distribution model. Throughout this chapter, we use simple examples to illustrate the concepts of probability theory. It is probability theory that provides the basis for statistical inference (where we make generalizations about unknown population parameters based on a single sample), which is discussed in subsequent chapters. We summarize key formulas and concepts in Section 3.6. In Section 3.7 we use SAS to simulate probability distributions, and in Section 3.8 we use data from the Framingham Heart Study to illustrate specific distributions.

3.1 Background

3.1.1 Vocabulary

An *experiment* is defined as a process by which a measurement is taken or observations are made or as a procedure that generates outcomes. An experiment may involve the selection of 3 individuals from a pool of 12 to participate in a particular activity, or the administration of a psychological test to a human subject to record a response (e.g., completion of the test—Yes/No—or the time to complete the test in minutes).

There are two types of experiments, *deterministic* and *random*. In deterministic experiments, the same outcome is observed each time the experiment is performed. In random experiments, one of several possible outcomes is observed each time the experiment is performed. For example, consider an experiment in which a coin is dropped from the roof of a 20-story building. Each time the experiment is performed (i.e., a coin is dropped) the same outcome is observed—the coin falls to the earth due to gravity. This is an example of a deterministic experiment. Consider a second experiment in which a single individual is selected from a pool of six individuals. Each time the experiment is performed (i.e., an individual is selected), only one of the six candidate individuals is selected. This is an example of a random experiment. In this experiment we know that there are six possible outcomes, but do not know with certainty which individual will be selected on a given performance of the experiment. Probability theory is concerned with random experiments.

The enumeration of all possible outcomes of an experiment is called the *sample space,* denoted S. Consider an experiment involving a population of six individuals ($N = 6$). Suppose each individual is assigned a unique identification number from 1 to 6. The sample space is the listing of all possible outcomes (or individuals) given by $S = \{1, 2, 3, 4, 5, 6\}$.

In probability we often look at certain characteristics or types of outcomes. For example, in an experiment that involves selecting individuals (or sampling) from the population, we might be interested in subjects with specific characteristics, such as female gender or ages over 65 years. Collections of outcomes are called *events* and are usually denoted with capital letters (e.g., A, B, C). Individual outcomes are referred to as *simple* events.

EXAMPLE 3.1 **Basic Probability**

Consider a population of $N = 6$ individuals in which each individual is assigned a unique identification number ranging from 1 to 6. Suppose we record gender and age on each member of the population. The population is described as follows:

Subject Number	Gender	Age
1	M	40
2	F	42
3	M	51
4	F	58
5	M	67
6	F	70

Suppose our experiment involves sampling or selecting one subject at random from the population. There are six possible outcomes of the experiment. Again, in most statistical applications we are generally not concerned with the probability that any particular individual is selected into a sample. Instead, we are concerned with selecting certain "types" of individuals. For example, if the population is comprised of 50% females, then a representative sample is one in which about half of the individuals are female. Consider the following events, denoted A, B, and C, and the outcomes or subjects that comprise each:

Event	Outcomes (i.e., Subject Numbers)
A = female subjects	{2, 4, 6}
B = male subjects	{1, 3, 5}
C = subjects over the age of 65	{5, 6}

In probability we assign a numeric value, a *probability,* to each outcome (i.e., simple event) and to each event to denote the likelihood that the outcome or event occurs. There are a number of strategies for assigning probabilities. We use a strategy in which each outcome is considered to have an equal chance of occurring. This strategy is appropriate when outcomes are assumed to occur at random. Probabilities of outcomes and of events (e.g., event A) are denoted as $P(\text{outcome})$ and $P(A)$, respectively. Probabilities are values between 0 and 1 (i.e., $0 \leq P(\text{outcome}) \leq 1$). A probability of 0 indicates that the outcome or event has no chance of occurring or is not possible, while a probability of 1 indicates that an outcome or event is certain to occur. The sum of the probabilities of the individual outcomes in any sample space is equal to 1 (i.e., $\sum_s P(\text{outcome}) = 1$). ■

3.2 First Principles

Assuming that each outcome in an experiment is equally likely, the probability of each outcome is given by

$$P(\text{outcome}) = 1/N \tag{3.1}$$

where N denotes the total number of outcomes in the experiment

Using the same logic, the probability of an event is given by

$$P(\text{event}) = (\#\,\text{outcomes in event})/N \tag{3.2}$$

In Example 3.1, we had a population of $N = 6$ individuals and each individual was assigned a unique identification number ranging from 1 to 6:

Subject Number	Gender	Age
1	M	40
2	F	42
3	M	51
4	F	58
5	M	67
6	F	70

The probability that any subject (e.g., subject assigned number 1 or 2) is selected is $1/6$ by (3.1): $P(1) = P(2) = P(3) = P(4) = P(5) = P(6) = 1/6$. Recall the events A, B, and C and the outcomes or subjects that comprise each:

Event	Outcomes (i.e., Subject Numbers)
A = female subjects	$\{2, 4, 6\}$
B = male subjects	$\{1, 3, 5\}$
C = subjects over the age of 65	$\{5, 6\}$

Using formula 3.2: $P(A) = 3/6$, $P(B) = 3/6$, and $P(C) = 2/6$. The probability of selecting a female is $3/6 = 50\%$, the probability of selecting a male is $3/6 = 50\%$, and the probability of selecting a subject over the age of 65 is $2/6 = 33\%$.

The *complement* of an event consists of all outcomes in the sample space that are not in the event. The complement of an event is denoted with a ′ following the event label (or sometimes with a bar over the event label). For example, the complement of event A in Example 3.1 is denoted A' (or $\bar{A}$) and consists of all outcomes in the sample space that are not in event A. Event A includes all female subjects. Therefore, the complement of event A includes all subjects who are not female (i.e., subjects who are male): $A' = \{1, 3, 5\}$. Similarly, $B' = \{2, 4, 6\}$ and $C' = \{1, 2, 3, 4\}$. The probabilities of these events can be found by applying formula 3.2: $P(A') = 3/6$, $P(B') = 3/6$, and $P(C') = 4/6$.

In probability theory there are a number of rules that can be applied in specific situations to generate probabilities. These rules can be extremely useful. However, we recommend using formula (3.2) wherever possible, as it is more intuitive in most cases. The Complement Rule is given next.

The *Complement Rule*: $P(A') = 1 - P(A)$

Figure 3.2 *Union of Events A and C:*
$$(A \cup C) = (A \text{ or } C)$$

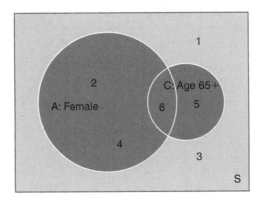

Using the Complement Rule, $P(A') = 1 - P(A) = 1 - 3/6 = 3/6$, $P(B') = 1 - P(B) = 1 - 3/6 = 3/6$, and $P(C') = 1 - P(C) = 1 - 2/6 = 4/6$.

The *union* of two events consists of outcomes in either event (or in both events). The union of two events is denoted by the symbol $\cup$ or with the word *or*. For example, the union of events A and C consists of outcomes in event A, in event C, or in both and is denoted: $(A \cup C) = (A \text{ or } C)$. Recall that event A consists of female subjects and event C consists of subjects over 65 years of age. The union of events A and C consists of individuals who are either female or over 65 years of age (or both): $(A \cup C) = (A \text{ or } C) = \{2, 4, 5, 6\}$. The union is displayed graphically in Figure 3.2 using a Venn diagram.

The probability of the union of two events can be computed using first principles, or formula (3.2): $P(A \cup C) = (A \text{ or } C) = 4/6$.

The *intersection* of two events consists of outcomes that are in both events. The intersection of two events is denoted by the symbol $\cap$ or with the word *and*. For example, the intersection of events A and C consists of all individuals that are both in event A and in event C and is denoted: $(A \cap C) = (A \text{ and } C)$. Event A consists of female subjects and event C consists of individuals over 65 years of age. The intersection of events A and C consists of the one female who is 70 years of age and assigned number 6: $(A \cap C) = (A \text{ and } C) = \{6\}$. The intersection is displayed graphically in Figure 3.3 using a Venn diagram.

The probability of the intersection of two events can be computed using first principles, or formula (3.2): $P(A \cap C) = (A \text{ and } C) = 1/6$.

A rule to compute the probability of the union of two events is called the *Addition Rule*:

The *Addition Rule*: $P(A \text{ or } B) = P(A) + P(B) - P(A \text{ and } B)$

Using the Addition Rule, $P(A \text{ or } C) = P(A) + P(C) - P(A \text{ and } C) = 3/6 + 2/6 - 1/6 = 4/6$.

Two events are said to be *mutually exclusive* if they have no outcomes in common. For example, events A and B in Example 3.1 are mutually exclusive

Figure 3.3 *Intersection of Events A and C: $(A \cap C) = (A \text{ and } C)$*

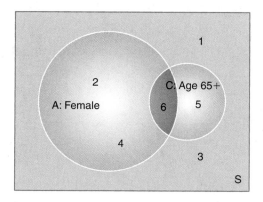

because event *A* consists of all female subjects while event *B* consists of all male subjects. Events *A* and *B* have no individuals in common and therefore are mutually exclusive. Events *A* and *C* are not mutually exclusive because both events include the individual assigned number {6}. Events *B* and *C* are not mutually exclusive because both events include the individual assigned number {5}. When two events are mutually exclusive, the probability of their intersection is zero (e.g., for Example 3.1, $P(A \cap B) = P(A \text{ and } B) = 0$). We now consider a more realistic example.

EXAMPLE 3.2A

Conditional Probability
A survey is conducted to assess the background characteristics of patients seeking care at an allergy clinic. All patients who visit the clinic over a 3-month period are asked to complete a self-administered survey that contains questions to measure age, gender, ethnicity, educational levels, and so on. Shown here are data collected on gender and educational level. The data are displayed in a cross-tabulation, or contingency table. The rows of the table indicate patients' gender and the columns display patients' highest educational level. Educational level is presented in categories—0–8 years, 9–12 years, 13–16 years, 17+ years—and reflect patients' highest level of formal education. The eight gender by educational level combinations (e.g., males with 0–8 years of education, males with 9–12 years of education) are called the *cells* of the table.

	Educational Level				
Gender	*0–8 years*	*9–12 years*	*13–16 years*	*17+ years*	*TOTAL*
Male	15	20	17	26	78
Female	30	42	31	27	130
TOTAL	**45**	**62**	**48**	**53**	**208**

Suppose a patient is selected at random, using formula (3.2); find the probability that

a. The patient is male. $P(male) = P(M) = 78/208$
b. The patient has 9–12 years of education. $P(9\text{–}12 \text{ years}) = 62/208$
c. The patient is female and has 9–12 years $P(F \text{ and } 9\text{–}12) = 42/208$
   of education.
d. The patient has 17+ years of education $P(17+ \text{ and } M) = 26/208$
   and is male.
e. The patient has at most 12 years $P(\leq 12 \text{ years})$
   of education. $= P(0\text{–}8 \text{ or } 9\text{–}12 \text{ years})$
                                                 $= 107/208$

From (a), the probability that a male is selected is $P(M) = 78/208 = 0.375$. The probability can be stated as a percent, for example there is a 37.5% chance that a male is selected at random from this population. Another interpretation is 37.5% of the population is male.   ■

Suppose we now want to compute the probability that a male is selected among those with 17+ years of education. That is, considering only those patients with the highest educational levels (17+ years), what is the probability of selecting a male? This question can be addressed with *conditional probability*. Specifically, we are interested in the probability of an event (i.e., that a male is selected at random) among a specific subgroup of the population (i.e., among those with 17+ years of education). In probabilistic terms, we "condition on the given event" (i.e., the event that a patient has 17+ years of education) and then compute the probability of the event of interest (i.e., that a male is selected). We write the conditional event as follows: $P(Male | 17+ \text{ years})$, where the | is read as "given." The probability that a male is selected "given" that a patient with 17+ years of education is selected is computed using first principles, or by formula (3.2). $P(M | 17+ \text{ years}) = 26/53 = 0.491$. Notice that the conditional probability is not equal to the unconditional probability: $P(M | 17+ \text{ years}) = 0.491$ and $P(M) = 0.375$, respectively. The proportion of males among patients with 17+ years of education is higher than the proportion of males in the population as a whole.

A rule for computing conditional probabilities that may be useful in certain applications follows:

The *Conditional Probability Rule*: $P(A|B) = P(A \text{ and } B)/P(B)$

We will illustrate the use of this rule using the data in Example 3.2A. We leave it to the reader to compute the following conditional probabilities using first principles or formula (3.2):

$$P(M \mid 9\text{--}12 \text{ years}) = P(M \text{ and } 9\text{--}12 \text{ years})/P(9\text{--}12 \text{ years})$$
$$= (20/208)/(62/208) = 20/62$$

$$P(13\text{--}16 \text{ years} \mid F) = P(13\text{--}16 \text{ years and } F)/P(F)$$
$$= (31/208)/(130/208) = 31/130$$

Two events are said to be *independent* if the probability of one is not "influenced" by the occurrence or nonoccurrence of the other.

EXAMPLE 3.2B

Independence

Suppose the data in Example 3.2A looked like the following:

Gender	0–8 years	Educational Level 9–12 years	13–16 years	17+ years	TOTAL
Male	25	30	25	25	105
Female	25	30	25	25	105
TOTAL	50	60	50	50	210

Using the data shown, $P(\text{Male}) = P(M) = 105/210 = 0.50$. The probability that a male is selected among those patients with 17+ years of education is: $P(M \mid 17+ \text{ years}) = 25/50 = 0.50$. In Example 3.2B, the unconditional probability (i.e., $P(M)$) is equivalent to the conditional probability (i.e., $P(M \mid 17+ \text{ years})$). Therefore, the two events (selecting a male and selecting a subject with 17+ years of education) are said to be independent because the occurrence of one event does not influence the probability of the other. ■

The definition of independence is:

Two events A and B are *independent* if: (3.3)

$$P(A) = P(A \mid B), \text{ or}$$
$$P(B) = P(B \mid A), \text{ or equivalently,}$$
$$P(A \cap B) = (A \text{ and } B) = P(A) \cdot P(B)$$

We illustrate the definition of independence by applying the last version of the definition. Using first principles (3.2) applied to the data in Example 3.2B, $P(M \text{ and } 17+ \text{ years}) = 25/210 = 0.119$, $P(M) = 105/210$, and $P(17+ \text{ years}) = 50/210$. To test if these two events (M and 17+ years of education) are independent, we must check whether the probability of their intersection, $P(M \text{ and } 17+ \text{ years})$, is equal to the product of their unconditional probabilities; that is, is $P(M \text{ and } 17+ \text{ years}) = P(M) \cdot P(17 + \text{years})$?

Substituting the probabilities,

$$P(M) \cdot P(17+ \text{ years}) = (105/210) \cdot (50/210) = 0.119$$

Thus, $P(M \text{ and } 17+ \text{ years}) = P(M) \cdot P(17+ \text{ years})$, and these events are independent by definition. The probability of selecting a male is the same, regardless of which educational subgroup we assess. This is only true for the data in Example 3.2B and is not the case for the data given in Example 3.2A. In Example 3.2A, the probability a male is selected at random is 0.375. The probability that a male is selected among the subgroup with 17 or more years of education is 0.491. There is a higher proportion of males among the patients with the highest educational levels, thus gender and educational level are not independent.

EXAMPLE 3.3

Assessing Independence of Depression Screen

Consider a screening tool comprised of a series of questions designed to assess whether patients exhibit symptoms of depression. Each participant completes the series of questions and the scores are summed to produce a total score. Patients are then classified as at low, moderate, or high risk for depression based on their total scores. In the present study, each patient completes the screening tool and, in addition, undergoes an extensive psychiatric examination. Based on the psychiatric examination, patients are classified as clinically depressed or not. The following table summarizes the results of the study:

| Clinical Assessment | Risk for Depression (Screening Tool) | | | |
	Low	Moderate	High	TOTAL
Not Depressed	522	127	39	688
Depressed	28	18	11	57
TOTAL	550	145	50	745

a. What is the probability that a patient who is at low risk for depression (based on the screening tool) is clinically depressed? $P(\text{depressed} | \text{low risk}) = 28/550 = 0.05$ using (3.2).

b. What is the probability that a patient who is at moderate risk for depression (based on the screening tool) is clinically depressed? $P(\text{depressed} | \text{moderate risk}) = 18/145 = 0.12$ using (3.2).

c. What is the probability that a patient who is at high risk for depression (based on the screening tool) is clinically depressed? $P(\text{depressed} | \text{high risk}) = 11/50 = 0.22$ using (3.2).

d. Is the clinical assessment of depression independent of the patient's risk for depression based on the screening tool? From (a)–(c) we see the likelihood that a patient is clinically depressed increases as a function of his/her risk. If we compare the unconditional and conditional probabilities for depression (3.3), we find the following: $P(\text{clinically depressed}) = 57/745 = 0.077$. The conditional probabilities of depression we computed in (a)–(c) are either below or above the unconditional probability of clinical depression for patients classified as low, moderate, or high risk. Thus, the diagnosis of depression is not independent of the patient's risk for depression based on the screening tool. The diagnosis of depression is related to the patient's risk; patients at higher risk are more likely to be clinically depressed. ■

EXAMPLE 3.3

Assessing Independence of Depression Screen (continued)

Suppose that 60% of the participants in the study described in Example 3.3 are male ($P(\text{male}) = 0.60$). In addition, suppose that 70% of the participants who are classified as clinically depressed are male (i.e., $P(\text{male}|\text{clinically depressed}) = 0.70$). Using the rules of probability, answer the following:

a. What proportion of participants is female?

   $P(\text{female}) = 1 - P(\text{male}) = 1 - 0.60 = 0.40$.

b. Are male gender and clinical depression mutually exclusive?

   If $P(\text{male and clinically depressed})$ is zero, then male gender and clinical depression are mutually exclusive. From the Conditional Probability Rule we have the following: $P(\text{male}|\text{clinically depressed}) = P(\text{male and clinically depressed})/P(\text{clinically depressed})$. From Example 3.3, $P(\text{clinically depressed}) = 0.077$. $P(\text{male and clinically depressed}) = P(\text{male}|\text{clinically depressed}) \cdot P(\text{clinically depressed}) = 0.70 \cdot 0.077 = 0.0539$. Because the probability is not zero, the events male and clinically depressed are not mutually exclusive.

c. Are gender and clinical depression independent?

   Because $P(\text{male}|\text{clinically depressed}) = 0.70$ and $P(\text{male}) = 0.60$ (the conditional and unconditional probabilities are not equal), these two events are not independent.

d. Who is more likely to be clinically depressed—a male or a female in this study?

   Find $P(\text{clinically depressed}|\text{male})$ and $P(\text{clinically depressed}|\text{female})$ and compare. $P(\text{clinically depressed}|\text{male}) = P(\text{male and clinically depressed})/P(\text{male}) = 0.0539/0.60 = 0.09$. $P(\text{clinically depressed}|\text{female}) = P(\text{female and clinically depressed})/P(\text{female}) = 0.02/0.40 = 0.05$. (Note that $P(\text{female and clinically depressed}) = P(\text{female}|\text{clinically depressed}) \cdot P(\text{clinically depressed})$.) To see these computations more clearly, try to construct a two-by-two table (with gender and clinical depression as the rows and columns) and fill in the information. In this study, men are almost twice as likely to be classified as clinically depressed. ■

3.3 Combinations and Permutations

Consider the population of six individuals ($N = 6$) in Example 3.1, where subjects are identified by unique numbers 1–6. There are several different strategies for selecting individuals from a population into a sample. It is important to know which sampling strategy is employed in any given application to ensure the validity of statistical inferences. Issues that must be considered in determining a sampling strategy include (1) whether individuals are sampled from a population with replacement or without replacement and (2) whether the order in which individuals are sampled is important or not.

We illustrate each of these considerations and their impact on the sampling process using our population of six individuals from Example 3.1. Suppose for the population in Example 3.1 we generate all possible samples of size $n = 2$. Suppose we sample with replacement (i.e., we select one individual at random from the population, record the unique subject number, and place the individual back into the population prior to the second selection) in a case where order is important (i.e., the samples (1, 2) and (2, 1) are considered two distinct samples). Thirty-six samples of size $n = 2$ are produced with this strategy:

Sampling with replacement, order important

(1, 1)	(1, 2)	(1, 3)	(1, 4)	(1, 5)	(1, 6)
(2, 1)	(2, 2)	(2, 3)	(2, 4)	(2, 5)	(2, 6)
(3, 1)	(3, 2)	(3, 3)	(3, 4)	(3, 5)	(3, 6)
(4, 1)	(4, 2)	(4, 3)	(4, 4)	(4, 5)	(4, 6)
(5, 1)	(5, 2)	(5, 3)	(5, 4)	(5, 5)	(5, 6)
(6, 1)	(6, 2)	(6, 3)	(6, 4)	(6, 5)	(6, 6)

Suppose we now sample with replacement in a case where order is not important (i.e., the samples (1, 2) and (2, 1) are considered the same sample). Twenty-one samples of size $n = 2$ are produced with this strategy:

Sampling with replacement, order not important

(1, 1)	(1, 2)	(1, 3)	(1, 4)	(1, 5)	(1, 6)
	(2, 2)	(2, 3)	(2, 4)	(2, 5)	(2, 6)
		(3, 3)	(3, 4)	(3, 5)	(3, 6)
			(4, 4)	(4, 5)	(4, 6)
				(5, 5)	(5, 6)
					(6, 6)

Usually we are interested in samples where individuals do not appear more than once. Suppose we sample without replacement (i.e., we select one individual at random, record the unique subject number, keep the individual aside, and make the second selection from the remaining five subjects) in a case where order is important. Thirty samples of size $n = 2$ are produced with this strategy:

Sampling without replacement, order important

	(1, 2)	(1, 3)	(1, 4)	(1, 5)	(1, 6)
(2, 1)		(2, 3)	(2, 4)	(2, 5)	(2, 6)
(3, 1)	(3, 2)		(3, 4)	(3, 5)	(3, 6)
(4, 1)	(4, 2)	(4, 3)		(4, 5)	(4, 6)
(5, 1)	(5, 2)	(5, 3)	(5, 4)		(5, 6)
(6, 1)	(6, 2)	(6, 3)	(6, 4)	(6, 5)	

Finally, suppose we sample without replacement in a case where order is not important. Fifteen samples of size $n = 2$ are produced with this strategy:

Sampling without replacement, order not important

(1, 2)	(1, 3)	(1, 4)	(1, 5)	(1, 6)
	(2, 3)	(2, 4)	(2, 5)	(2, 6)
		(3, 4)	(3, 5)	(3, 6)
			(4, 5)	(4, 6)
				(5, 6)

In most applications, we sample without replacement, thereby ensuring that the sample is comprised of different individuals rather than the same individual selected more than once. There are formulas that are useful to determine the numbers of distinct samples produced under specific sampling strategies. When sampling without replacement, the number of distinct arrangements (i.e., order important), called *permutations* of n individuals from a population of size N, is given by

$$_N P_n = \frac{N!}{(N - n)!} \tag{3.4}$$

where $N!$ (N "factorial") $= N \cdot (N - 1) \cdot (N - 2) \cdot \cdots \cdot 2\,(2) \cdot (1)$
(e.g., $5! = 5 \cdot 4 \cdot 3 \cdot 2 \cdot 1 = 120$, $3! = 3 \cdot 2 \cdot 1 = 6$, $1! = 1$, and $0!$ is defined as 1)

In Example 3.1, when sampling without replacement, the number of permutations (order important) of size 2 from the population of size 6 is given by

$$_6 P_2 = \frac{6!}{(6 - 2)!} = \frac{6!}{4!} = \frac{6 \cdot 5 \cdot 4 \cdot 3 \cdot 2 \cdot 1}{4 \cdot 3 \cdot 2 \cdot 1} = 30$$

When sampling without replacement, the number of samples in which order is not important, or *combinations*, of n individuals from a population of size N is given by

$$_NC_n = \begin{bmatrix} N \\ n \end{bmatrix} = \frac{N!}{n!(N-n)!} \tag{3.5}$$

In Example 3.1 when sampling without replacement, the number of combinations of size 2 from the population of size 6 is given by

$$_6C_2 = \begin{bmatrix} 6 \\ 2 \end{bmatrix} = \frac{6!}{2!(6-2)!} = \frac{6 \cdot 5 \cdot 4 \cdot 3 \cdot 2 \cdot 1}{2 \cdot 4 \cdot 3 \cdot 2 \cdot 1} = 15$$

From the previous illustration we see that specific sampling strategies produce different numbers of samples. In general, we are less interested in the number of distinct samples produced by a particular sampling strategy and more interested in the attributes of subjects selected into specific samples. For example, suppose we are interested in the number of females selected into each sample under the last sampling strategy (sampling without replacement, order not important).

Suppose we let the variable X denote the number of females selected into each sample. X is called a *random variable* and takes on the values 0, 1, or 2, depending on the number of females (subjects assigned even identification numbers) selected into each sample. The samples are shown next along with the value of $X =$ the number of females selected into each sample.

Sampling strategy: Sampling without replacement, order not important

Sample	$X =$ # females	Sample	$X =$ # females
(1, 2)	1	(2, 6)	2
(1, 3)	0	(3, 4)	1
(1, 4)	1	(3, 5)	0
(1, 5)	0	(3, 6)	1
(1, 6)	1	(4, 5)	1
(2, 3)	1	(4, 6)	2
(2, 4)	2	(5, 6)	1
(2, 5)	1		

In the population, three of six subjects, or 50%, are female. In the samples shown, females make up 0%, 50%, or 100% of each sample. Three of 15 samples (20%) have no females, 9 of 15 samples (60%) have exactly one female, and 3 of 15 samples (20%) have exactly two females. In statistical inference we will have only a *single* sample and will attempt to draw inferences about a population based on that one sample. Based on the preceding samples, we are most likely (probability = 0.60) to select a sample containing exactly one female. It is, however, possible that we select a sample containing no females or two females. To draw appropriate inferences, it is important to understand the relationship between characteristics of samples and characteristics of populations. We will elaborate on this notion in the remainder of this and subsequent chapters.

EXAMPLE 3.4

Sampling with and without Replacement

Consider a population consisting of four laboratory mice. Suppose that two are diseased and two are not. Let $D1$ and $D2$ denote the two diseased subjects and $N1$ and $N2$ denote the two nondiseased subjects. Suppose we take simple random samples of size $n = 2$ from the population.

The following table displays the samples generated under three different sampling strategies: sampling without replacement—order important, sampling without replacement—order not important, and sampling with replacement—order important. Next to each sample we display the probability that each sample is selected, assuming that all samples are equally likely.

Sampling without Replacement				Sampling with Replacement	
Ordered Samples	*Prob.*	*Unordered Samples*	*Prob.*	*Ordered Samples*	*Prob.*
N1, N2	1/12	N1, N2	1/6	N1, N1	1/16
N1, D1	1/12	N1, D1	1/6	N1, N2	1/16
N1, D2	1/12	N1, D2	1/6	N1, D1	1/16
N2, N1	1/12	N2, D1	1/6	N1, D2	1/16
N2, D1	1/12	N2, D2	1/6	N2, N1	1/16
N2, D2	1/12	D1, D2	1/6	N2, N2	1/16
D1, N1	1/12			N2, D1	1/16
D1, N2	1/12			N2, D2	1/16
D1, D2	1/12			D1, N1	1/16
D2, N1	1/12			D1, N2	1/16
D2, N2	1/12			D1, D1	1/16
D2, D1	1/12			D1, D2	1/16
				D2, N1	1/16
				D2, N2	1/16
				D2, D1	1/16
				D2, D2	1/16

Suppose we are interested in the likelihood that a diseased subject is selected into the sample. Let the *random variable X* denote the number of diseased subjects selected into each sample. In this example, X takes on the values 0, 1, or 2, depending on the number of diseased subjects selected into each sample. The following table displays the samples along with the corresponding values for the random variable X under each sampling strategy.

Sampling without Replacement				Sampling with Replacement	
Ordered Samples	*X*	*Unordered Samples*	*X*	*Ordered Samples*	*X*
N1, N2	0	N1, N2	0	N1, N1	0
N1, D1	1	N1, D1	1	N1, N2	0
N1, D2	1	N1, D2	1	N1, D1	1
N2, N1	0	N2, D1	1	N1, D2	1
N2, D1	1	N2, D2	1	N2, N1	0
N2, D2	1	D1, D2	2	N2, N2	0
D1, N1	1			N2, D1	1
D1, N2	1			N2, D2	1
D1, D2	2			D1, N1	1
D2, N1	1			D1, N2	1
D2, N2	1			D1, D1	2
D2, D1	2			D1, D2	2
				D2, N1	1
				D2, N2	1
				D2, D1	2
				D2, D2	2

The random variable X generates a *probability model* or *probability distribution*. The distribution has three possible values—0, 1, and 2—and corresponding probabilities under each different sampling strategy. The listing of the values of the random variable and corresponding probabilities constitutes a *probability distribution* or *model*. The probability distributions for the random variable X under each sampling strategy are as follows:

Sampling without Replacement				Sampling with Replacement	
Ordered Samples		*Unordered Sample*		*Ordered Samples*	
X	*P(X)*	*X*	*P(X)*	*X*	*P(X)*
0	2/12	0	1/6	0	4/16
1	8/12	1	4/6	1	8/16
2	2/12	2	1/6	2	4/16

Based on these probability distributions, we can now answer questions such as: What is the probability of selecting exactly one diseased subject into a sample of size 2 from a population in which 50% are diseased? The answer depends on the exact sampling strategy. Answers are provided here, relative to the three sampling strategies shown.

a. P(exactly one diseased subject is selected) $= P(X = 1)$

$P(X = 1) = 8/12$ $\qquad P(X = 1) = 4/6$ $\qquad P(X = 1) = 8/16$

b. P(no diseased subjects are selected) $= P(X = 0)$

$P(X = 0) = 2/12$ $\qquad P(X = 0) = 1/6$ $\qquad P(X = 0) = 4/16$

c. P(at least one diseased subject is selected) $= P(X = 1 \text{ or } X = 2)$

$P(X = 1 \text{ or } X = 2)$ $\qquad P(X = 1 \text{ or } X = 2)$ $\qquad P(X = 1 \text{ or } X = 2)$
$\quad = 10/12$ $\qquad\qquad\quad = 5/6$ $\qquad\qquad\quad = 12/16$ ■

Notice that the probabilities that exactly one diseased subject is selected, that no diseased subjects are selected, and that at least one diseased subject is selected differ depending on the sampling strategy employed. The order consideration does not influence probabilities (the first two sampling strategies produce equivalent probabilities); however, sampling with versus without replacement does influence results.

In order to construct the probability distributions, we enumerated every possible sample under a particular sampling strategy and computed the value of the random variable of interest (e.g., the number of diseased subjects selected into the sample). Example 3.4 involved a very small population ($N = 4$) and small samples ($n = 2$). When the population is large (as it is in most applications), it is practically impossible to enumerate every possible sample of a given size to develop a probability distribution.

To avoid having to enumerate every possible sample to develop a probability distribution, probability models based on specific attributes of the experiment under investigation are used. If a particular experiment satisfies the attributes of a specific probability model, that model can be used to compute probabilities. We present two of the most popular probability models in Sections 3.4 and 3.5. For each model, we specify the attributes that distinguish it from other models and illustrate the computation of probabilities using the models by way of examples.

3.4 The Binomial Distribution

We now focus on a probability distribution model, used in a variety of applications, called the binomial probability distribution model. The *binomial distribution model* has three specific attributes:

1. Each performance of the experiment, called a *trial,* results in only one of two possible outcomes, which we call either a success or a failure.

2. The probability of a success on each trial is constant, denoted p, with $0 \leq p \leq 1$.
3. The trials are independent.

Many applications satisfy these attributes. For example, suppose we are concerned with the risk of mortality (i.e., death) associated with a particular surgical procedure. The experiment involves performing the surgical procedure on different patients. Each patient who undergoes the surgical procedure either survives or does not survive the procedure (attribute 1). Suppose that for this procedure, the probability of surviving (or not surviving) is constant (attribute 2). (This may not be the case for some surgical procedures, as older patients or patients with comorbid conditions, for example, might be at higher risk for not surviving the procedure.) In this example, the trials (i.e., performances of the surgical procedure) are independent (attribute 3). The outcome observed in one patient (i.e., survives or does not survive the surgical procedure) has no effect on the outcome observed in another patient.

In applications with these attributes, the following mathematical formula (3.6), the binomial distribution formula, can be used to generate probabilities. The binomial distribution is defined for the discrete random variable X, which denotes the number of successes out of n trials.

$$P(X = x) = {_n}C_x p^x (1 - p)^{n-x} \qquad (3.6)$$

where x = # successes of interest $(x = 0, 1, \ldots, n)$
$p = P(\text{success})$ on any trial
n = # trials
${_n}C_x$ = the number of combinations of x successes in n trials

Recall $n!$ ("n factorial") $= n \cdot (n - 1) \cdots \cdot 2 \cdot 1$ (e.g., $5! = 5 \cdot 4 \cdot 3 \cdot 2 \cdot 1 = 120$, $3! = 3 \cdot 2 \cdot 1 = 6$, $1! = 1$, and $0! = 1$). The binomial distribution formula (3.6) could be used, for example, to compute the probability that a specified number of patients, x, do not survive a particular surgical procedure when it is applied to a collection of patients, n, with a known, constant risk of mortality (i.e., death), p. Model (3.6) is equivalent to

$$P(X = x) = \frac{n!}{x!(n - x)!} p^x (1 - p)^{n-x}$$

The binomial distribution is an appropriate model for the number of successes in a simple random sample of size n with replacement drawn from a population where there are two possible outcomes (success and failure). We illustrate the use of the binomial distribution in Examples 3.5 and 3.6 and revisit Example 3.4 from the previous section.

EXAMPLE 3.5

Binomial Probabilities: Antibiotic with $p = 0.7$

Suppose an antibiotic has been shown to be 70% effective against a common bacteria.

a. If the antibiotic is given to five unrelated individuals with the bacteria, what is the probability that it will be effective in exactly three?

In this experiment, we give the antibiotic to each of five unrelated patients affected with a common bacteria and monitor whether or not it is effective. Assume that it is possible to classify the outcome of a course of therapy as effective (success) or not effective (failure). Assuming that the probability of success is constant (i.e., the probability that the antibiotic is effective is the same for each patient) and the trials are independent (i.e., the outcome observed in one patient does not influence the outcome observed in another), then this experiment satisfies the attributes of the binomial probability model, and (3.6) can be used to compute the desired probability.

Using the binomial formula (3.6) with $n = 5$, $p = 0.70$, and $x = 3$, the probability is

$$P(X = 3) = \frac{5!}{3!(5 - 3)!} 0.70^3 (1 - 0.70)^{5-3} = 10(0.343)(0.09) = 0.3087$$

Thus, there is a 30.87% chance that the antibiotic is effective in exactly three patients out of five (when the probability of effectiveness in a single patient is 70%).

b. What is the probability that the antibiotic will be effective in all five? Using (3.6) with $n = 5$, $p = 0.70$, and $x = 5$, the probability is

$$P(X = 5) = \frac{5!}{5!(5 - 5)!} 0.70^5 (1 - 0.70)^{5-5} = 1(0.1681)(1) = 0.1681$$

c. What is the probability that the antibiotic will be effective in none of the five? Using (3.6) with $n = 5$, $p = 0.70$, and $x = 0$, the probability is

$$P(X = 0) = \frac{5!}{0!(5 - 0)!} 0.70^0 (1 - 0.70)^{5-0} = 1(1)(0.0024) = 0.0024$$

In this binomial experiment, the sample space consists of the following outcomes $S = \{0, 1, 2, 3, 4, 5\}$. Exactly one of these outcomes will be observed each time the experiment is run. The probability of a success on any one trial (i.e., the probability that the antibiotic is effective) is 70%. Because the probability of success is high, we are much more likely to observe three or five successes out of five trials than we are to observe none (probabilities 0.3087, 0.1681, and 0.0024, respectively). In fact, it is very unlikely that we will not observe any successes (less than a 1% chance) in five trials. ■

Table B.1 in Appendix B contains probabilities of the binomial distribution for specific values of n, p, and x computed using (3.6) for various combinations of n, p, and x. To use Table B.1, locate n along the left-hand margin, find x in the adjacent column, and locate p across the top row. Find the probability in the appropriate row and column of the table. The table provides probabilities of observing *exactly* x successes out of n trials for specific values of p. For example, if we locate $n = 5$ and $p = 0.70$ in Table B.1, we find the following probabilities (a probability distribution):

$X =$ *Number of Successes*	$P(X \mid n = 5, p = 0.70)$
0	0.0024
1	0.0284
2	0.1323
3	0.3087
4	0.3601
5	0.1681
	1.0

Formula (3.6) can be used to determine probabilities for the binomial distribution for those applications in which the table does not contain the desired probabilities. For example, the values of p in Table B.1 start at 0.10 and increase in increments of 0.10 to 0.90 (i.e., $p = 0.10, 0.20, 0.30, \ldots, 0.90$) and a specific application may involve $p = 0.27$. In that case, the binomial formula (3.6) can be used to compute the exact probability.

EXAMPLE 3.6

Binomial Probabilities: Antibiotic with $p = 0.5$

Consider an application similar to the one described in Example 3.5 involving a second antibiotic that has been shown to be only 50% effective against the same common bacteria. If 10 unrelated individuals with the bacteria take the antibiotic, what is the probability that it will be effective in more than 8?

Again, we let X denote the number of successes (i.e., the number of patients in whom the antibiotic is effective), and we wish to determine $P(X > 8)$. The binomial formula and Table B.1 produce the probability of observing *exactly* x successes out of n. In this example we are interested in the probability of observing more than 8 successes. Before using the binomial formula (or Table B.1), we must restate our problem in a format compatible with the binomial formula (and Table B.1):

$$P(X > 8) = P(X = 9 \text{ or } X = 10) = P(X = 9) + P(X = 10)$$

We can now compute $P(X = 9)$ and $P(X = 10)$ using the binomial formula (3.6), applied twice.

$$P(X = 9) = \frac{10!}{9!(10 - 9)!}0.50^9(1 - 0.50)^{10-9} = 10(0.0020)(0.5) = 0.0098$$

$$P(X = 10) = \frac{10!}{10!(10 - 10)!}0.50^{10}(1 - 0.50)^{10-10} = 1(0.0001)(1) = 0.0010$$

Thus, $P(X > 8) = 0.0098 + 0.0010 = 0.0108$ (see also Table B.1).

In this application, there is a 1.1% chance that the antibiotic will be effective in more than 8 patients when given to 10. The possible outcomes of this experiment are the values in the sample space, $S = \{0, 1, 2, 3, 4, 5, 6, 7, 8, 9, 10\}$. Because the antibiotic is 50% effective for any given patient, we are more likely to observe 4, 5, or 6 successes out of 10 than we are to observe few (0, 1, or 2) or many (8, 9, 10) successes (see Table B.1). ■

Recall Example 3.4 from Section 3.4 in which we had a population consisting of four laboratory mice, two diseased and two nondiseased. In the example we selected two subjects at random. Each selection resulted in one of two possible outcomes: selection of a diseased subject or selection of a nondiseased subject. The probability that we selected a diseased subject was 0.50 under the sampling with replacement strategy. Note that sampling with replacement ensures constant probability of success (in this example, probability of selecting a diseased subject is 0.5 on each selection as long as we sample with replacement). Suppose we call the selection of a diseased subject a success. (In general, we call the outcome of interest a success; often, particularly in medical applications, the outcome of interest is not the healthy outcome, e.g., disease, mortality). The random variable X under the sampling with replacement strategy illustrated in Example 3.4 is an example of a binomial random variable, whose probability distribution is given below. Recall we developed this probability distribution by enumerating every possible sample, assigning the value of X (the number of diseased subjects selected into each sample) and summarizing:

X	P(X)
0	4/16 = 0.25
1	8/16 = 0.50
2	4/16 = 0.25

This probability distribution is an example of a binomial probability distribution with $n = 2$ and $p = 0.5$. The binomial distribution formula and Table B.1 give the same probabilities; for example, with $n = 2$ and $p = 0.5$, the probability of selecting no diseased subjects $(X = 0)$ is

$$P(X = 0) = \frac{2!}{0!(2 - 0)!}0.50^0(1 - 0.50)^{2-0} = 1(1)(0.25) = 0.25$$

In the binomial distribution, it can be shown that the mean and variance of the random variable X (i.e., the mean and variance of the number of successes out of n binomial trials) are computed as follows:

$$\mu = np \tag{3.7}$$
$$\sigma^2 = np(1 - p)$$

In Example 3.5, we involved five patients in the experiment, and the probability of success was 70% for each patient. The possible outcomes of the experiment are given in the sample space $S = \{0, 1, 2, 3, 4, 5\}$. Each time the experiment is run, exactly one of these outcomes (number of successes) is observed. The mean (or expected) number of patients (out of five) in whom the antibiotic is effective is $\mu = np = 5 \cdot 0.7 = 3.5$. We could never observe 3.5 successes in any performance of the experiment; instead, we observe one of the values in the sample space (e.g., exactly 3 or 4 or 5 successes out of 5). The variance in the number of successes is $\sigma^2 = 5(0.7)(1 - 0.7) = 1.05$. The standard deviation is $\sigma = 1.02$. The mean number of successes represents the typical number of successes. In reviewing the probability distribution (shown in Example 3.5), we see that the most likely outcomes (those with the highest probabilities) of the experiment are 3 and 4 successes.

In Example 3.6, where the probability of success was 50% for each patient, the expected number of patients (out of 10) in whom the antibiotic is effective is $\mu = np = 10(0.5) = 5$. Again, any of the outcomes in the sample space can be observed on any performance of the experiment. We are most likely (or expect) to observe 5 successes out of 10 when $p = 0.50$.

3.5 The Normal Distribution

The normal distribution is our second probability model and is the most widely used probability distribution for continuous random variables. A characteristic (continuous variable) is said to follow a normal distribution if the distribution of values is bell- or mound-shaped (Figure 3.4).

The horizontal axis displays the values of the continuous normal random variable X. The vertical axis is scaled to accommodate the height of the curve at each value of the random variable that reflects the probability (or relative frequency) of observing that value. The total area under the normal curve

Figure 3.4 *The Normal Distribution*

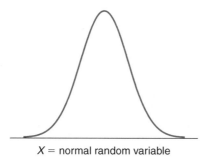

X = normal random variable

is 1.0, as it is a probability distribution. The normal distribution is one where values in the center of the distribution are more likely (have higher probabilities) than values at the extremes.

The mathematical formula for the normal probability distribution is

$$f(x) = \frac{1}{\sigma\sqrt{2\pi}} e^{-(1/2)[(x-\mu)/\sigma]^2} \tag{3.8}$$

where x is a continuous random variable $(-\infty < x < \infty)$
μ = mean of the random variable X
σ = standard deviation of the random variable X

The normal probability distribution has the following attributes, each of which will be illustrated through examples.

Attributes of normal distribution

1. The normal distribution is symmetric about the mean (i.e., $P(X > \mu) = P(X < \mu) = 0.5$, see Figure 3.5a.) A characteristic that follows a normal distribution is one in which there are as many values above the mean as below.

2. The mean = the median = the mode. This attribute follows directly from (1). If the distribution is symmetric at the mean, then half (50%) of the values are above the mean and half (50%) are below the mean. This is the definition of the median. Notice in Figure 3.5a that the peak of the distribution is exactly in the center of the distribution (at the mean = median). The height of the curve indicates the probability (or relative frequency) of observations at each point. The peak indicates the most frequent value, which is, by definition, the mode.

3. The mean and variance, μ and σ^2, completely characterize the normal distribution. If we know that a particular characteristic follows a normal distribution and we know μ and σ, then we know everything about that distribution. The mean and variance are the only parameters

Figure 3.5(a)—(c) *Properties of the Normal Distribution*

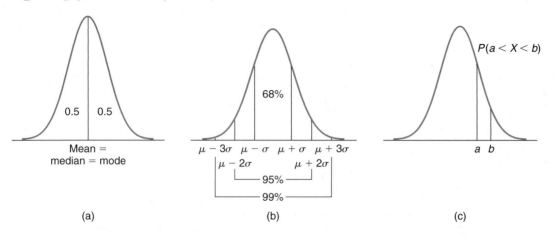

(a)   (b)   (c)

required to compute probabilities about a normal distribution (see the normal distribution function (3.8)).

4. $P(\mu - \sigma < X < \mu + \sigma) = 0.68$, $P(\mu - 2\sigma < X < \mu + 2\sigma) = 0.95$, $P(\mu - 3\sigma < X < \mu + 3\sigma) = 0.99$ (Figure 3.5b). These three properties of a normal distribution are the basis for the Empirical Rule presented in Chapter 2. For any normal distribution, 68% of the values fall between the mean minus 1 standard deviation and the mean plus 1 standard deviation, 95% of the values fall between the mean minus 2 standard deviations and the mean plus 2 standard deviations, 99% of the values fall between the mean minus 3 standard deviations and the mean plus 3 standard deviations. If we plot the mean of a normal distribution (in the center of the x axis) and count 3 standard deviations in either direction, the third standard deviation away from the mean should be very close to the end of the distribution.

5. $P(a < X < b) =$ the area under the normal curve from a to b (Figure 3.5c). In a normal distribution, the area under the curve reflects probability. We will compute probabilities for normal random variables by determining desired areas under the normal curve.

If a random variable X follows a normal distribution with mean μ and variance σ^2, we write $X \sim N(\mu, \sigma^2)$, where "~" is read "is distributed as."

EXAMPLE 3.7   **Normal Probabilities: Heights**

Suppose males aged 25 have an average height of 70" with a standard deviation of 2". Heights for a specific age and gender are approximately normally distributed. Let the random variable X denote height in inches for males aged 25, then $X \sim N(70, 2^2)$. Using the attributes for normal random variables, the distribution appears as shown in the following figure. Notice that the mean (70) is in the middle of the distribution. We also count 3 standard deviation units (units of 2") in either direction, and at the third standard deviations above and below the mean we are very close to the ends of the distribution.

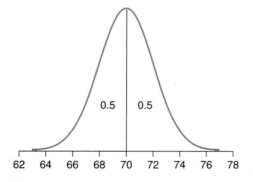

Suppose a male aged 25 is selected at random. Use the attributes of the normal distribution to solve the following.

a. What is the probability that his height is more than 70"? Because 70 is the mean of this normal distribution, we know from attribute (1) that the distribution is symmetric about the mean:

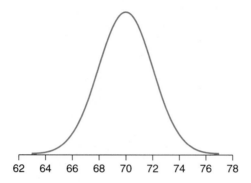

Therefore, $P(X > 70) = 0.5$.

b. What is the probability that his height is between 70" and 72"? From attribute (4) we know that the probability that his height is between 68 and 72 is 0.68 (the area under the curve between the mean minus 1 standard deviation and the mean plus 1 standard deviation). In addition, the normal distribution is symmetric about the mean (70), so the area to the right of 70 is identical to the area to the left of 70:

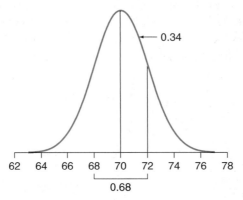

The desired probability is $P(70 < X < 72) = 0.34$.

c. What is the probability that his height exceeds 72"? From attribute (1) we know that the area under the curve above 70 (and the area under the curve below 70) is equal to 0.5. From part b we determined that the area under the curve between 70 and 72 was 0.34.

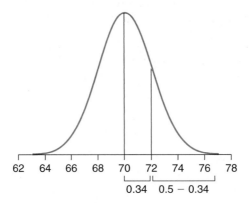

The area under the curve above 72 can be determined by subtraction:
$P(X > 72) = 0.5 - 0.34 = 0.16$ ■

The attributes are very useful so long as questions are focused on the mean and multiples of the standard deviation about the mean of a normal distribution. Of course, there will be other questions of interest. For example, suppose in Example 3.7 we wanted to know the probability that a male, aged 25, has a height exceeding 71"? From parts a and c, we know that this probability is between 0.5 and 0.16—but we cannot determine the exact probability using the attributes in the list.

To compute probabilities for a normal distribution, a table such as Table B.2 in Appendix B is needed. Table B.2 is a table of probabilities of the *standard normal distribution*. The standard normal distribution is the normal

Figure 3.6 *The Standard Normal Distribution*

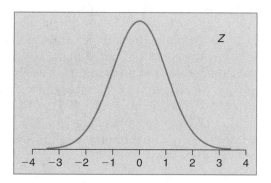

distribution with $\mu = 0$ and $\sigma = 1$ (Figure 3.6). The random variable Z is used to refer to the standard normal distribution (i.e., $Z \sim N(0, 1)$).

To calculate probabilities for any normal distribution X, we first standardize (i.e., subtract the mean and divide by the standard deviation) using (3.9), and then use Table B.2: Probabilities of the Standard Normal Distribution.

$$Z = \frac{X - \mu}{\sigma} \tag{3.9}$$

In Example 3.7, we posed the question, what is the probability that a male, age 25, has a height exceeding 71". To determine this probability, we convert the problem to a problem about Z, the standard normal distribution, using (3.9). Formula (3.9) converts a value from any normal distribution (X) to a value from the standard normal distribution (Z). The X value of interest is 71. It corresponds to a Z value of $Z = (71 - 70)/2 = 0.5$. The value 71 in a normal distribution with mean 70 and standard deviation 2 corresponds to $Z = 0.5$. The value 71 is one-half (0.5) a standard deviation above the mean. The area of interest is shown in the following figure (indicated by "?") both for the original distribution (X) and the standard normal distribution (Z).

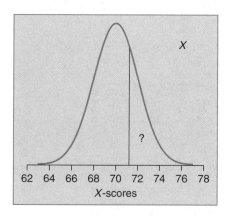 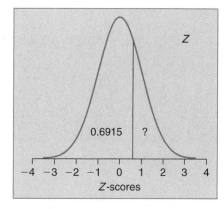

The areas under the two curves (above 71 in the normal distribution with mean 70, standard deviation 2 (X), and above 0.5 in the standard normal distribution (Z)) are identical; only the scales along the horizontal axes are different. The probability that Z exceeds 0.5 can be determined from Table B.2. Table B.2 contains areas (probabilities) under the standard normal curve below the desired Z value. If we locate $Z = 0.5$ in the table (find 0.5 down the left-hand margin, which contains the units and first decimal place of Z, and 0.00 across the top row, which contains the second decimal place of Z), we read $P(Z < 0.5) = 0.6915$. We want $P(Z > 0.5) = 1 - 0.6915 = 0.3085$. Thus, 31% of the males have heights exceeding 71". We illustrate the application of the normal probability distribution model, formula (3.9), and Table B.2 in the following example.

EXAMPLE 3.8

Normal Probabilities: Systolic Blood Pressures

Systolic blood pressures are assumed to follow a normal distribution with a mean of 108 and a standard deviation of 14 (this is denoted $X \sim N(108, 14^2)$). From the list of attributes, we know that the distribution of systolic blood pressures looks like that presented in Figure 3.7.

If an individual is selected at random from the population, find the following:

a. The probability that the systolic blood pressure is below 112 (i.e., $P(X < 112)$). This probability concerns the random variable X, which is approximately normally distributed with $\mu = 108$ and $\sigma = 14$. To find this probability, we must first standardize using (3.9) and then use Table B.2. We standardize the 112 as follows:

$$Z = (112 - 108)/14 = 0.29$$

The problem can be represented as follows:

$$P(X < 112) = P(Z < 0.29)$$

Table B.2 contains probabilities for the standard normal distribution Z. The probabilities contained in the table reflect the areas under the standard

Figure 3.7 *Distribution of Systolic Blood Pressures $\mu = 108, \sigma = 14$*

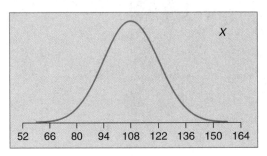

| 52 | 66 | 80 | 94 | 108 | 122 | 136 | 150 | 164 |

normal curve below each particular Z value. Here we wish to determine the area under the standard normal curve below 0.29. We locate the Z value 0.29 and read the probability. Using Table B.2, the desired probability is 0.6141. The probability that an individual's systolic blood pressure is below 112 is 61.41%. Another interpretation is that 61.41% of the population have systolic blood pressures below 112.

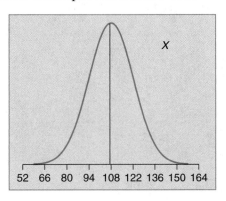 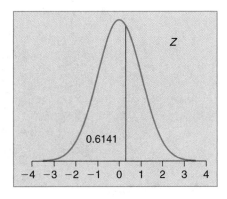

b. The probability that the systolic blood pressure is below 120 (i.e., $P(X < 120)$). To find this probability, we again standardize using (3.9) and then use Table B.2.

$$Z = (120 - 108)/14 = 0.86$$

The problem can be represented as follows:

$$P(X < 120) = P(Z < 0.86)$$

Now we wish to determine the area under the standard normal curve below 0.86. We locate the Z value 0.86 and read the probability. Using Table B.2, the desired probability is 0.8051. The probability that an individual's systolic blood pressure is below 120 is 80.51% (or 80.51% of the population have systolic blood pressures below 120).

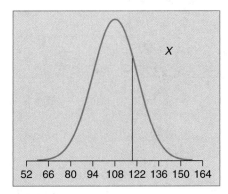 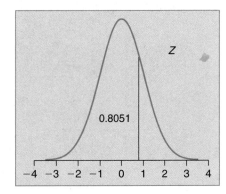

c. The probability that the systolic blood pressure is above 120 (i.e., $P(X > 120)$).

In part b, we standardized the value 120. The problem can be represented as follows: $P(X > 120) = P(Z > 0.86)$. The desired probability is computed by subtraction (from part b we determined $P(Z < 0.86) = 0.8051$): $P(X > 120) = P(Z > 0.86) = 1 - 0.8051 = 0.1949$.

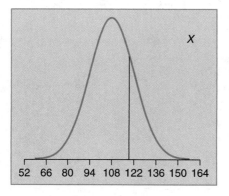
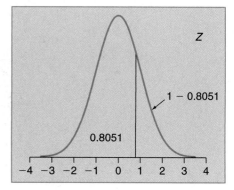

d. The probability that the systolic blood pressure is above 122 (i.e., $P(X > 122)$).

We first standardize using (3.9):

$$Z = (122 - 108)/14 = 1$$

The problem can be represented as follows: $P(X > 122) = P(Z > 1)$. Using Table B.2, we can determine $P(Z < 1) = 0.8413$. The desired probability is computed by subtraction: $P(X > 122) = P(Z > 1) = 1 - 0.8413 = 0.1587$.

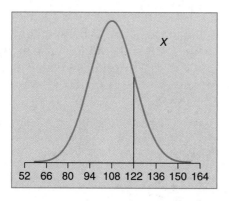
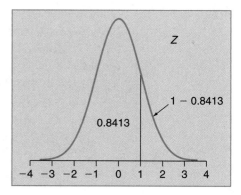

e. The probability that the systolic blood pressure is between 100 and 108 (i.e., $P(100 < X < 108)$).

We first standardize the two X values of interest: $Z = (100 - 108)/14 = -0.57$, $Z = (108 - 108)/14 = 0$.

The problem can be represented as follows: $P(100 < X < 108) = P(-0.57 < Z < 0)$. Recall that Table B.2 contains probabilities for the standard normal distribution, Z, and the probabilities reflect the areas under the standard normal curve below a particular Z value. If we locate the Z value -0.57 in Table B.2, we find the area under the curve below -0.57 is 0.2843. We want the area under the curve between -0.57 and 0. If we locate the Z value 0 in Table B.2, we find the area under the curve below 0 (because this is the mean, the area below 0 is 0.5). The desired probability is

$$P(100 < X < 108) = P(-0.57 < Z < 0) = 0.5 - 0.2843 = 0.2157$$

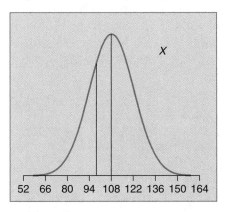
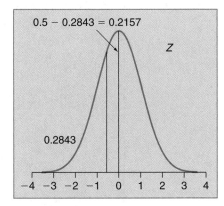

f. The probability that the systolic blood pressure is between 104 and 118 (i.e., $P(104 < X < 118)$).

   We first standardize the two X values of interest: $Z = (104 - 108)/14 = -0.29$, $Z = (118 - 108)/14 = 0.71$.

The problem can be represented as follows: $P(104 < X < 118) = P(-0.29 < Z < 0.71)$. Again, Table B.2 contains probabilities that reflect the areas under the standard normal curve below the selected Z value. To compute the desired probability, we enter Table B.2 twice, once with the value 0.71 (to determine the area below 0.71) and once with the value -0.29 (to determine the area below -0.29). The problem can be solved as follows:

$$P(104 < X < 118) = P(-0.29 < Z < 0.71) = 0.7611 - 0.3859 = 0.3752$$

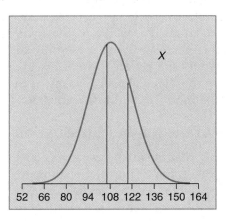
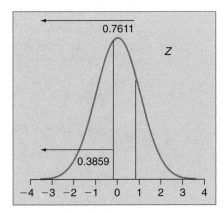

g. The probability that the systolic blood pressure is between 120 and 125 (i.e., $P(120 < X < 125)$).

We first standardize the two X values of interest: $Z = (120 - 108)/14 = 0.86$, $Z = (125 - 108)/14 = 1.21$.

The problem can be represented as follows: $P(120 < X < 125) = P(0.86 < Z < 1.21)$. Again, Table B.2 contains probabilities that reflect the areas under the standard normal curve below the selected Z value. To determine the desired probability, we need to enter Table B.2 twice, once with the value 1.21 (to determine the area below 1.21) and once with the value 0.86 (to determine the area below 0.86). The desired probability is then determined by subtraction:

$$P(120 < X < 125) = P(0.86 < Z < 1.21) = 0.8869 - 0.8051 = 0.0818$$

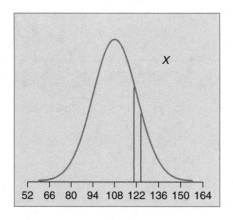
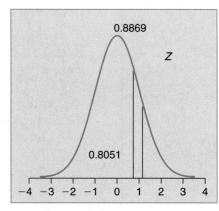

In determining probabilities for a normal distribution, it is always useful to draw a picture to display the desired probability (equivalent to the area under the normal curve). The picture is particularly useful as a check on the appropriateness of the solution. For example, in part g of Example 3.8, the desired probability is reflected by a narrow area under the normal curve. The solution was 0.0818, a small fraction of the total area under the normal curve. If an error was made in the computation and the solution was proposed as 0.818, it would not reflect this narrow area. It is not necessary, in general, to draw both the X and Z distributions, as the areas are identical. We presented both for Example 3.8 to illustrate this notion.

3.5.1 Percentiles of the Normal Distribution

Recall from Chapter 2, that the kth *percentile* is defined as the *score* that holds $k\%$ of the scores below it. For example, the 90th percentile in a distribution is the score that holds 90% of the scores below it. The first quartile, Q_1, is equivalent to the 25th percentile, the median is equivalent to the 50th percentile, and

Table 3.1 *Percentiles of the Standard Normal Distribution*

Percentile	Z
1	−2.326
2.5	−1.960
5	−1.645
10	−1.282
90	1.282
95	1.645
97.5	1.960
99	2.326

the third quartile, Q_3, is equivalent to the 75th percentile. For the normal distribution, the following formula (3.10) is used to compute percentiles.

$$X = \mu + Z\sigma \tag{3.10}$$

where μ = mean of the random variable X
σ = standard deviation of the random variable X
Z = value from the standard normal distribution for the desired percentile

Table 3.1 provides the values from the standard normal distribution (Z) corresponding to commonly used percentiles.

The percentiles of the standard normal distribution, Z, were computed using the following approach and Table B.2. The same approach could be used to determine any other percentile of the standard normal distribution.

Suppose we wish to determine the 95th percentile of Z. By definition, the 95th percentile is the Z score that holds 95% of the Z scores below it. The following figure displays the approximate (to be determined) value of the 95th percentile of Z.

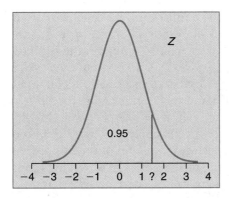

The probabilities in Table B.2 reflect the areas under the standard normal curve below the selected Z value. We need to find the probability 0.95 (the area below the Z value we wish to determine, denoted ? in the figure) in Table B.2 among the probabilities. The closest values are 0.9495 and 0.9505. The value we wish to locate (0.95) is in between. Using Table B.2, the Z value corresponding to the probability 0.9495 is 1.64 (i.e., the area under the curve below 1.64 is 0.9495). Similarly, the Z value corresponding to 0.9505 is 1.65 (i.e., the area under the curve below 1.65 is 0.9505). The 95th percentile is the Z value that holds 95% of the area below it. The 95th percentile of the standard normal distribution is 1.645. Other percentiles of the standard normal distribution can be determined in this same manner.

EXAMPLE 3.9

Percentiles of Systolic Blood Pressures

Consider the distribution of systolic blood pressures presented in Example 3.8. Systolic blood pressures were assumed to follow a normal distribution, with $\mu = 108$ and $\sigma = 14$ (i.e., $X \sim N(108, 14^2)$).

a. Find the 5th percentile in the systolic blood pressures. Using (3.10) and the appropriate value from Table 3.1:

$$X = 108 + (-1.645)(14) = 108 - 23.0 = 85.0$$

b. Find the 90th percentile in the systolic blood pressures. Using (3.10) and the appropriate value from Table 3.1:

$$X = 108 + (1.282)(14) = 108 + 17.9 = 125.9$$

Thus, 5% of the systolic blood pressures are at or below 85.0, and 90% of the systolic blood pressures are at or below 125.9 (i.e., 10% are above 125.9). ■

3.5.2 Normal Approximation to the Binomial

In Section 3.4 we discussed the binomial probability distribution model and presented two methods for computing probabilities: (1) the binomial distribution formula (3.6) and (2) Table B.1: Probabilities of the Binomial Distribution. Both the binomial distribution formula and the table produce the probability of observing exactly x successes out of n trials when the probability of success on a single trial is p. We used applications in which we determined the probability of observing a range of values. For example, suppose that 10 binomial trials are performed and it is of interest to compute the probability of observing more than 5 successes (i.e., $P(X > 5)$). To compute the probability, we first enumerate the values of x (the number of successes out of n) of interest and then proceed to the binomial distribution formula or to Table B.1. In this example, $P(X > 5) = P(X = 6) + P(X = 7) + P(X = 8) + P(X = 9) + P(X = 10)$. Each of these probabilities is computed with the binomial formula or using Table B.1. The final answer is the sum of the five probabilities. When the number of trials (n) is large, probabilities of observing ranges of values are increasingly cumbersome to compute. There are alternatives, however, which utilize the techniques described in the previous section.

Figure 3.8a *Binomial Distribution*
$n = 5, \; p = 0.7$

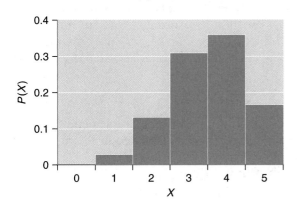

Figure 3.8b *Binomial Distribution*
$n = 10, \; p = 0.7$

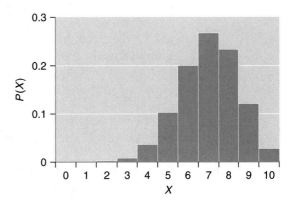

Figure 3.8c *Binomial Distribution*
$n = 15, \; p = 0.7$

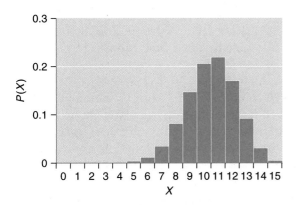

Figure 3.8d *Binomial Distribution*
$n = 25, \; p = 0.7$

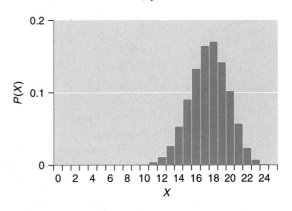

It can be shown that as the number of trials, n, in binomial applications becomes large, the distribution of the number of successes, x, is approximately normally distributed. Recall Example 3.5, which described an antibiotic that was 70% effective against a common bacteria. Figures 3.8a–d display the distributions of the number of successes, x, for $n = 5, 10, 15$, and 25, respectively. Notice that as n gets large (e.g., $n = 25$), the distribution of the number of successes looks approximately normally distributed.

If the distribution of the number of successes is approximately normal, which occurs when n is large and p is not close to 0 or 1, then the normal probability distribution can be used to approximate binomial probabilities. In general, the normal approximation is appropriate if

$$np \geq 5 \quad \text{and} \quad n(1 - p) \geq 5$$

The distribution displayed in Figure 3.8d is approximately normal and satisfies the preceding conditions ($n = 25, \; p = 0.7$: $np = 25(0.7) = 17.5$, and

$n(1 - p) = 25(0.3) = 7.5)$. For $n = 25$, $p = 0.7$, the normal distribution can be used to compute probabilities concerning the number of successes, X. Recall, for the binomial distribution, the mean number of successes is $\mu = np$ and the standard deviation is $\sigma = \sqrt{np(1 - p)}$. The following formula is used to convert values of the random variable X (reflecting the number of successes out of n binomial trials) into corresponding values of the standard normal distribution:

$$Z = \frac{X - np}{\sqrt{np(1 - p)}} \tag{3.11}$$

A continuity correction of 0.5 is generally applied prior to implementing (3.11). Formula (3.11) and the continuity correction are illustrated in Example 3.10.

EXAMPLE 3.10

Normal Approximation: Antibiotic with $p = 0.7$

Consider Example 3.5, which described an antibiotic that was 70% effective against a common bacteria. Suppose that the antibiotic is given to 25 patients with the bacteria.

a. What is the probability that the antibiotic is effective in more than 15 patients?

We wish to compute $P(X > 15)$. In order to solve the problem using the binomial distribution formula or Table B.1, we first enumerate the values of X of interest: $P(X > 15) = P(X = 16) + P(X = 17) + P(X = 18) + P(X = 19) + P(X = 20) + P(X = 21) + P(X = 22) + P(X = 23) + P(X = 24) + P(X = 25)$. Each of these probabilities could be computed by applying the binomial formula (3.9) or by using Table B.1. In either approach, particularly the former, the computations are cumbersome. In this application, we satisfy the criteria which suggest that the distribution of the number of successes, X, is approximately normally distributed (i.e., $np = 25(0.7) = 17.5 \geq 5$ and $n(1 - p) = 25(0.3) = 7.5 \geq 5$) with $\mu = 25(0.7) = 17.5$ and $\sigma = \sqrt{np(1 - p)} = \sqrt{25(0.7)(0.3)} = \sqrt{5.25} = 2.29$. The distribution of all possible outcomes for the binomial experiment with $n = 25$ and $p = 0.7$ was displayed in Figure 3.8d. The desired probability is the sum of the areas in the shaded bars, shown below.

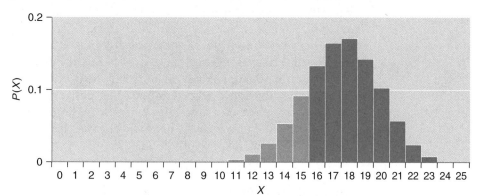

If we overlay a normal curve on this binomial histogram, we can see the area under the normal curve that approximates the desired probability:

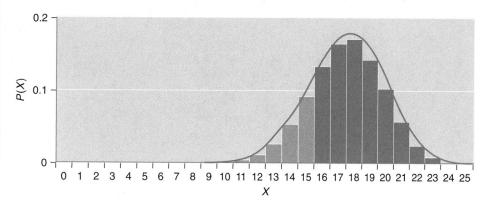

We wish to compute $P(X > 15)$. In the figure with the normal curve overlaid, this is approximated by the area under the normal curve to the right of the vertical bar (above the right-hand side of the bar over the value 15). Notice that each of the bars in the binomial histogram is centered above the respective X value. For example, the bar over the value $X = 15$ starts at 14.5 and ends at 15.5, the bar over the value $X = 16$ starts at 15.5 and ends at 16.5, and so on. The continuity correction essentially takes this into account. The binomial probability $P(X > 15)$ is approximated by the area under the normal curve above the value 15.5: $P(X_{\text{binomial}} > 15) \approx P(X_{\text{normal}} > 15.5)$. To compute the probability that a normal variable with mean 17.5 and standard deviation 2.29 exceeds 15.5, we convert to the standard normal distribution (Z):

$$Z = \frac{15.5 - 17.5}{2.29} = -0.87$$

$$P(X > 15.5) = P(Z > -0.87)$$

Now, using Table B.2,

$$P(Z > -0.87) = 1 - 0.1922 = 0.8078$$

There is a 80.78% probability that the antibiotic is effective in more than 15 patients. This is an approximation to the exact solution, which can be determined using Table B.1. The exact solution is 0.8105.

b. What is the probability that the antibiotic is effective in at most 12 patients?

We wish to compute $P(X \le 12)$. Again, if we overlay a normal curve on the binomial histogram, we can see the area under the normal curve that approximates the desired probability.

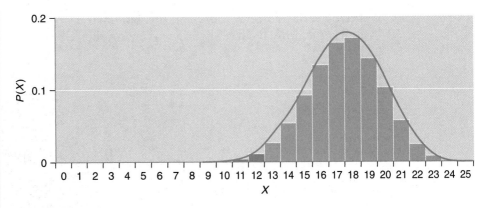

Adding the continuity correction:

$$P(X \le 12) \approx P(X < 12.5)$$

Now, converting to the standard normal distribution,

$$Z = \frac{12.5 - 17.5}{2.29} = -2.18$$

$$P(X < 12.5) = P(Z < -2.18)$$

Using Table B.2,

$$P(Z < -2.18) = 0.0146$$

There is a 1.46% probability that the antibiotic is effective in at most 12 patients. (Using Table B.1, the exact probability is 0.0175.) ▪

3.6 Key Formulas

Concept	Formula	Description
First principles	$P(A) = \dfrac{\# \text{ outcomes in } A}{\text{Total } \# \text{ outcomes}}$	where A is an event defined in the sample space S
Complement Rule	$P(A') = 1 - P(A)$	where A' is the complement of A (i.e., contains all outcomes in S that are not in A)
Addition Rule	$P(A \text{ or } B) = P(A) + P(B)$ $- P(A \text{ and } B)$	where A and B are events defined in the sample space S
Mutually exclusive events	$P(A \text{ and } B) = 0$	no outcomes in common
Conditional Probability Rule	$P(A|B) = \dfrac{P(A \text{ and } B)}{P(B)}$	probability of event A "given" event B
Independent events	$P(A|B) = P(A)$ $P(A \text{ and } B) = P(A) \cdot P(B)$	the probability of event A is not influenced by the occurrence or nonoccurrence of event B

Concept	Formula	Description
Permutations	$_NP_n = \dfrac{N!}{(N-n)!}$	number of distinct arrangements (order important) of size n from a population of size N without replacement
Combinations	$_NC_n = \dfrac{N!}{n!(N-n)!}$	number of samples (order not important) of size n from a population of size N without replacement
Binomial probability model	$P(X) = \dfrac{n!}{x!(n-x)!}p^x(1-p)^{n-x}$	probability of x successes out of n trials
Normal probability model	$f(x) = \dfrac{1}{\sigma\sqrt{2\pi}}e^{(-1/2)[(x-\mu)/\sigma]^2}$	normal probability model
Standard normal distribution	$Z = \dfrac{x-\mu}{\sigma}$ $Z \sim N(0,1)$	normal distribution with $\mu = 0, \sigma^2 = 1$, Table B.2
Percentiles of the normal distribution	$X = \mu + Z\sigma$	kth percentile holds $k\%$ of scores below it (Table 3.1)
Normal approximation to the binomial	$Z = \dfrac{(X \pm 0.5) - np}{\sqrt{np(1-p)}}$	standardizes binomial random variable X

3.7 Applications Using SAS

SAS has several functions that can generate random variables from specific probability distribution functions, such as the binomial and normal distributions. We illustrate the use of these functions as well as a function that can be used to determine specific values of the standard normal distribution, such as the 90th percentile.

A series of SAS programs and output are presented that illustrate several different SAS functions. The probability distribution to which the function relates is noted in the header to each example. Also noted is the specific SAS function illustrated and a brief description of its use. Notes are provided to the right of the SAS programs (in blue) for orientation purposes and are not part of the programs. The blank lines in the programs are solely to accommodate the notes. Blank lines and spaces can be used throughout SAS programs and are generally used to enhance readability.

The first example uses SAS to generate random variables from a binomial distribution with $n = 5$ and $p = 0.7$ (see Example 3.5). The second example uses SAS to generate random variables from a normal distribution with mean 108 and standard deviation 14 (see Example 3.8). The third example illustrates the computation of percentiles of the normal distribution using the data presented in Example 3.8.

For each example, we present three components:

1. the SAS program code
2. the computer output
3. a description of the relevant components of the computer output along with their interpretation

SAS EXAMPLE 3.5 **Generating Random Variables from the Binomial Distribution Effectiveness of Antibiotic (Example 3.5)**

Ranbin Function (Generates Random Variables from Binomial Distribution)

Suppose an antibiotic has been shown to be 70% effective against a common bacteria. If the antibiotic is given to five individuals with the bacteria, what is the probability that it will be effective on exactly three?

In this application, we will use SAS to generate random variables from the binomial distribution with $n = 5$ (number of trials) and $p = 0.70$ (probability of success on any trial). We will generate many such random variables, and in doing so estimate the theoretical probability distribution. The SAS program code follows.

Program Code

`options ps=64 ls=80;`	Formats the output page to 64 lines in length, 80 columns in width.
`data one;`	Beginning of Data Step (Data set name **one**).
`do i=1 to 5000;`	Beginning of a DO loop which will be repeated 5000 times. The index *i* counts the iterations.
`x=ranbin(21439,5,0.7);`	Generates a random variable, called *x*, from the binomial distribution with n=5, p=0.7. The first argument to the Ranbin function is the seed.[1]
`output;`	Writes the generated variable, *x*, to data set **one**.
`end;`	End of DO loop.
`run;`	End of Data Step.
`proc means;`	Procedure call (Proc Means to generate summary statistics).
`var x;`	Specification of variable *x*.
`run;`	End of Procedure section.
`proc freq;`	Procedure call (Proc Freq to generate frequency distribution=probability distribution).
`tables x;`	Specification of variable *x*.
`run;`	End of Procedure section.

(1) The seed is simply a random number used as a starting point for the random number generator (i.e., ranbin). When the seed is changed, different values are generated.

Computer Output

```
                        The MEANS Procedure
                        Analysis Variable : x
```

N	Mean	Std Dev	Minimum	Maximum
5000	3.4942000	1.0282888	0	5.0000000

```
                        The FREQ Procedure
```

x	Frequency	Percent	Cumulative Frequency	Cumulative Percent
0	15	0.30	15	0.30
1	135	2.70	150	3.00
2	686	13.72	836	16.72
3	1529	30.58	2365	47.30
4	1798	35.96	4163	83.26
5	837	16.74	5000	100.00

Interpretation

The first part of the output is from the Means procedure. Shown are the number of observations in the sample ($N = 5000$), the mean, standard deviation, minimum, and maximum. In this application, the summary statistics are computed on the random variable X, which reflects the numbers of patients in whom the antibiotic is effective. Recall for the binomial distribution, the mean is given by $\mu = np$. In Example 3.5, $n = 5$ and $p = 0.7$. The theoretical mean is $\mu = 5 \cdot 0.7 = 3.5$. In this data set in which we generated 5,000 random variables from a binomial distribution with $n = 5$ and $p = 0.7$, the observed mean is 3.494.

The next section of output displays a frequency distribution table for the random variable X. Notice that the values of X generated by SAS are between 0 and 5. The observed frequency distribution (based on 5000 observations) closely approximates the true binomial probability distribution (see Example 3.5). For example, from Table B.1 in Appendix B,

$P(X = 0) = 0.0024$, $P(X = 1) = 0.0284$, $P(X = 2) = 0.1323$, $P(X = 3) = 0.3087$, $P(X = 4) = 0.3601$, $P(X = 5) = 0.1681$. The observed distribution (shown in the SAS output) is 0.003, 0.027, 0.137, 0.306, 0.360, and 0.167. Using the SAS output, we can address the original question: If the antibiotic is given to five individuals, what is the probability that it will be effective in exactly three? From the SAS output, $P(X = 3) = 0.306$. ■

SAS Example 3.8 Generating Random Variables from the Normal Distribution Systolic Blood Pressures (Example 3.8)

Rannor Function (Generates Random Variables from the Standard Normal Distribution)

Systolic blood pressures are assumed to follow a normal distribution, with a mean of 108 and a standard deviation of 14.

In this application, we will use SAS to generate random variables from the standard normal distribution (Z) and then transform them into normal random variables with a mean of 108 and a standard deviation of 14.

Program Code

`options ps=64 ls=80;`	Formats the output page to 64 lines in length, 80 columns in width
`data one;`	Beginning of Data Step (Data set name **one**).
`mu=108;`	Create a variable called **mu**, assign the constant 108
`sigma=14;`	Create a variable called **sigma**, assign the constant 14
`do i=1 to 10000;`	Beginning of a DO loop which will be repeated 10,000 times. The index *i* counts the iterations.
`z=rannor(13755);`	Generates a random variable, called z, from the standard normal distribution. The only argument to the Rannor function is the seed.[1]
`x=mu+(z*sigma);`	Creates a new variable, **x**, which is a linear function of z ($x = \mu + z \cdot \sigma$).
`output;`	Writes the generated variables, **x** and **z**, to data set **one**.
`end;`	End of DO loop.
`run;`	End of Data Step.
`proc chart;`	Procedure call (Proc Chart to generate relative frequency histogram)
`vbar z/type=pct;`	Specification of variable z, type=pct displays relative frequencies
`vbar x/type=pct;`	Specification of variable x, type=pct displays relative frequencies
`run;`	End of Procedure section
`proc means;`	Procedure call (Proc Means to generate summary statistics)
`var z x;`	Specification of variables z and **x**.
`run;`	End of Procedure section

(1) The seed is simply a random number used as a starting point for the random number generator (i.e., rannor). When the seed is changed, different values are generated.

Computer Output

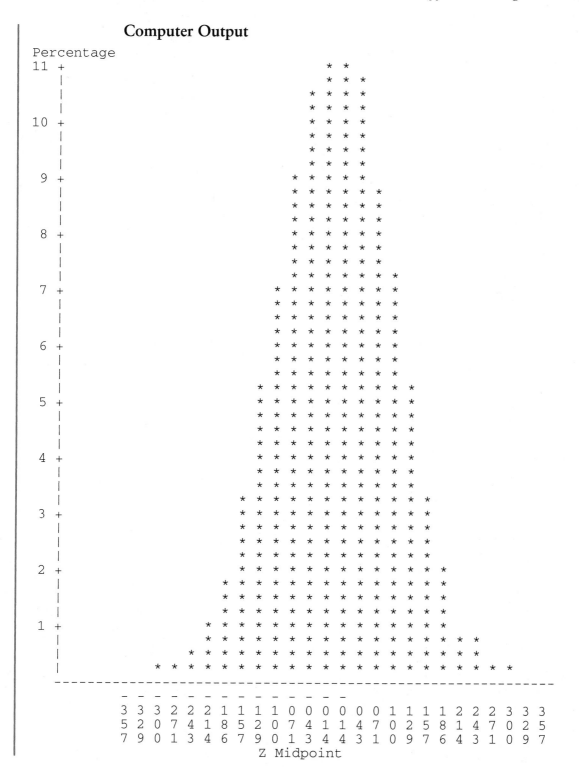

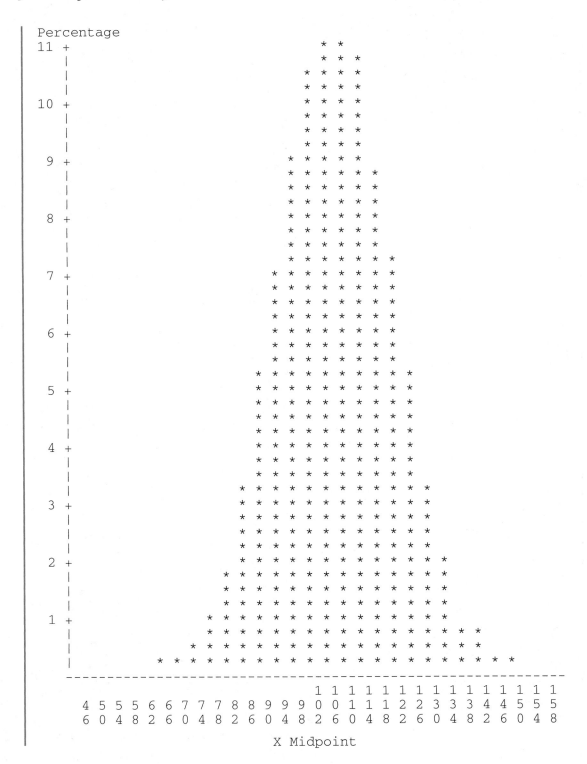

```
                          The MEANS Procedure
Variable       N           Mean        Std Dev       Minimum        Maximum
--------------------------------------------------------------------------
z           10000    0.000430808     0.9914063    -4.4742924      3.6632627
x           10000    108.0060313    13.8796876    45.3599064    159.2856784
--------------------------------------------------------------------------
```

Interpretation

The first part of the output is a relative frequency histogram for the random variable z. The random variable z follows the standard normal distribution, with a mean of 0 and a standard deviation of 1. The range of the observed values of z is listed along the horizontal axis and is approximately -3.0 to 3.0. Although the figure produced by SAS Proc Chart is somewhat crude, it does resemble the normal distribution curve.

The second figure displays the relative frequency histogram for the random variable x. The random variable x follows a normal distribution, with a mean of 108 and a standard deviation of 14. The range of the observed values of x are listed along the horizontal axis and is approximately 66 to 160. Again, the figure produced by SAS Proc Chart is crude, but it does resemble the normal distribution curve. In fact, it is identical to the relative frequency distribution of z, only the values along the horizontal axis are different.

The next section of output is from the Means procedure. Shown are the number of observations in the sample ($N = 10,000$), the mean, standard deviation, minimum, and maximum. Summary statistics are computed on both random variables z and x. Notice that the observed mean of z is 0.00043; the theoretical mean of z is 0. The observed standard deviation of z is 0.99141; the theoretical standard deviation is 1. The observed mean of x is 108.006; the theoretical mean of x is 108. The observed standard deviation of z is 13.8797; the theoretical standard deviation is 14. ▪

SAS EXAMPLE 3.9 **Computing Percentiles of the Normal Distribution Systolic Blood Pressures (Example 3.9)**

Probit Function (Returns Percentiles of the Standard Normal Distribution)

In Example 3.9 we computed the 5th and 90th percentiles of the systolic blood pressures illustrated in the previous example.

In this application, we will use SAS to determine the 5th and 90th percentiles of the systolic blood pressures, which are assumed to follow a normal distribution, with a mean of 108 and a standard deviation of 14.

Program Code

`options ps=64 ls=80;`	Formats the output page to 64 lines in length, 80 columns in width
`data one;`	Beginning of Data Step (Data set name **one**).
`mu=108;`	Create a variable called **mu**, assign the constant 108
`sigma=14;`	Create a variable called **sigma**, assign the constant 14
`z05=probit(0.05);`	Creates a variable called **z05**, which is assigned the 5th percentile of the standard normal distribution.
`z90=probit(0.90);`	Creates a variable called **z90**, which is assigned the 90th percentile of the standard normal distribution.
`x05=mu+(z05*sigma);`	Creates a new variable, **x05**, which is a linear function of **z05** $(x = \mu + z^*\sigma)$. **x05** is the 5th percentile of a normal distribution with a mean of 108 and a standard deviation of 14.
`x90=mu+(z90*sigma);`	Creates a new variable, **x90**, which is a linear function of **z90** $(x = \mu + z^*\sigma)$. **x90** is the 90th percentile of a normal distribution with a mean of 108 and a standard deviation of 14.
`run;`	End of Data Step.
`proc print;`	Procedure call (Proc Print to print raw data)
`run;`	End of Procedure section

Computer Output

```
OBS    MU    SIGMA       Z05         Z90         X05        X90
 1     108    14      -1.64485    1.28155    84.9720    125.942
```

Interpretation

In this application we computed several variables in the Data Step. Using Proc Print, we printed the values of each variable. Notice that no variables were specified in the Proc Print call statement (i.e., no var statement). In such a case, SAS simply prints the values for all variables. Shown are the values for each of the variables in the data set. SAS includes an observation number (labeled 'OBS'), which appears in the first column. In this example there is only one observation (with multiple variables). The values of the remaining variables are shown with the variable names on top, printed in all caps. The 5th percentile of the standard normal distribution is -1.64485 (i.e., 5% of the distribution is at or below -1.64485). The 90th percentile of the standard normal distribution is 1.28155. The 5th percentile of the systolic blood pressures, assumed to follow a normal distribution with a mean of 108 and a standard deviation of 14, is 84.9720, and the 90th percentile of the systolic blood pressures is 125.942. ■

3.7.1 **Summary of SAS Functions**

The SAS functions illustrated and a general description of each are summarized in the following table. SAS functions are used in the Data Step.

Function	Sample Call Statement	Description
ranbin	$x = \text{ranbin}(seed,n,p)$;	generates random variables from the binomial distribution with n trials and $P(\text{success}) = p$
rannor	$z = \text{rannor}(seed)$;	generates random variables from the standard normal distribution (z)
	$x = \mu + \text{rannorm}(seed)^*\sigma$;	transforms standard normal random variables to normal random variables with mean μ, standard deviation σ
probit	$z = \text{probit}(\alpha)$;	determines percentiles of the standard normal distribution with lower tail area α
	$x = \mu + \text{probit}(\alpha)^*\sigma$;	transforms percentile of the standard normal distribution to percentile of a normal distribution with mean μ and standard deviation σ

3.8 Analysis of Framingham Heart Study Data

The Framingham data set includes data collected from the original cohort. Participants contributed up to three examination cycles of data. Here we are analyzing data collected in the first examination cycle (called the period = 1 examination). Using SAS Proc Means, we generated descriptive statistics on key study variables and assessed whether these variables appear to follow the normal probability models discussed in this chapter. We also investigated graphical displays of the data for the same purpose.

Framingham Data Analysis—SAS Code

```
data in;
 set in.frmgham;
 if period=1;
run;


proc means n mean median std min max;
 var age sysbp diabp totchol bmi cigpday;
run;
```

Framingham Data Analysis—SAS Output

The MEANS Procedure

Variable	Label	N	Mean	Median
AGE	Age (years) at examination	4434	49.9258006	49.0000000
SYSBP	Systolic BP mmHg	4434	132.9077582	129.0000000
DIABP	Diastolic BP mmHg	4434	83.0835589	82.0000000
TOTCHOL	Serum Cholesterol mg/dL	4382	236.9842538	234.0000000
BMI	Body Mass Index (kr/(M*M)	4415	25.8461608	25.4500000
CIGPDAY	Cigarettes per day	4402	8.9663789	0

Variable	Label	Std Dev	Minimum	Maximum
AGE	Age (years) at examination	8.6769293	32.0000000	70.0000000
SYSBP	Systolic BP mmHg	22.4215970	83.5000000	295.0000000
DIABP	Diastolic BP mmHg	12.0559994	48.0000000	142.5000000
TOTCHOL	Serum Cholesterol mg/dL	44.6510984	107.0000000	696.0000000
BMI	Body Mass Index (kr/(M*M)	4.1018209	15.5400000	56.8000000
CIGPDAY	Cigarettes per day	11.9317058	0	70.0000000

Based on the summary statistics, which characteristics would likely follow a normal distribution? Recall the relationship between the mean and the median for normal distributions and between the mean ±3 standard deviations and the range. Use the following distributional plots in conjunction with the preceding analysis to determine which characteristics appear to follow a normal distribution in the Framingham cohort.

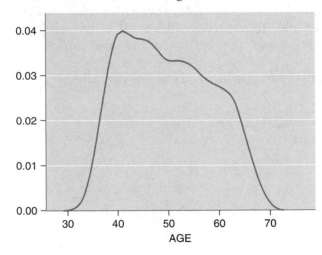

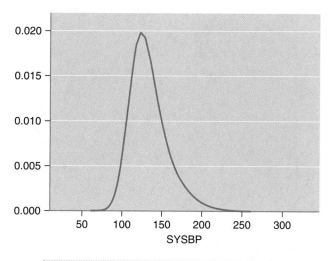

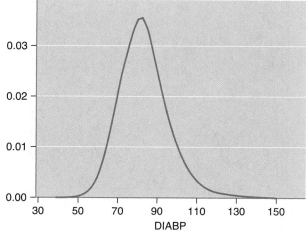

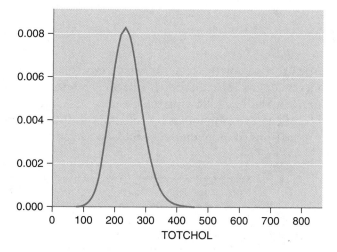

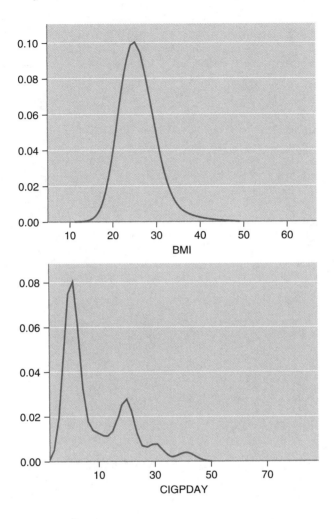

3.9 Problems

1. You are assisting in the design of a study to evaluate the effects of social factors on an individual's self-esteem. The study is to involve college-age individuals. The following table displays sociodemographic information on college-age individuals taken from the 2000 census. The table contains proportions of individuals in each age by gender category.

	Age						
	17	18	19	20	21	22	Total
Male	0.05	0.07	0.10	0.06	0.11	0.06	0.45
Female	0.09	0.08	0.11	0.07	0.12	0.08	0.55
Total	0.14	0.15	0.21	0.13	0.23	0.14	

If an individual is sampled from the population of interest (described above), calculate the probability that the sampled individual is

a. Female

b. At least 19 years of age

c. Female and age 18

d. Over 20 given that he is male

2. The following table displays the numbers of defective and nondefective medical devices produced by three plants working for the same manufacturer:

	Plant A	Plant B	Plant C
Nondefective	180	70	190
Defective	20	30	10

Suppose a medical device is selected at random, find the probability that

a. The device is defective.

b. The device is produced in Plant A.

c. The device is defective and produced in Plant A or in Plant C.

d. The device is defective given that it is produced in either Plant A or in Plant C.

3. The following table classifies institutions of higher education in the United States by region and type:

	Public	Private
Northeast	266	555
Midwest	359	504
South	533	502
West	313	242

Define the following events: P = Public Institution, N = Northeast, M = Midwest, S = South, W = West.

Suppose an institution is selected at random. Find the following:

a. $P(P)$

b. $P(W')$

c. $P(P' \text{ and } N)$

d. $P(P \text{ or } S)$

e. $P(P'|M)$

4. We wish to design a study to evaluate patients' satisfaction with the medical care received at a community health center. Before developing a sampling

plan, we will attempt to understand the patients in the population according to their insurance status (which might influence their satisfaction with medical care) and gender (investigators have shown that men tend to report more satisfaction with medical care than women). Some descriptive information is available on each patient who has been seen in the community health center within the last 3 years. Assume that patients who have been seen in the last 3 years constitute our population of interest.

Gender	Insurance Status		
	Private Health Insurance	Medicaid	No Insurance
Female	250	452	208
Male	128	680	157

a. What proportion of the population has no insurance?

b. What proportion of the population is female and has no insurance?

c. Among the patients with no insurance, what proportion is female?

d. Are the events "No Insurance" and "Female" independent? Justify your answer.

5. The probability that patients are treated (T) *given* they have a particular disease (D) is 0.8. Thirty percent of all patients have the disease. The proportion of all patients not treated (T') is 0.76.

a. What proportion of all patients has the disease and is treated?

b. What proportion of all patients is treated?

c. Are the events T' and D mutually exclusive?

d. Are the events T and D independent?

6. The following data were collected from the 2003 administrative records of a local community center:

Age	Primary Diagnosis			
	Diabetes	Asthma	Arthritis	Cardiac
30–39	27	56	20	30
40–49	32	32	25	24
50–59	30	14	43	43
60+	29	7	65	41

a. What proportion of patients had a primary diagnosis of asthma in the 2003 records?

b. What proportion of patients aged 40–49 had a primary diagnosis of asthma in the 2003 records?

c. What proportion of patients with a primary diagnosis of asthma in 2003 were 40–49 years of age?

d. What proportion of patients with a primary diagnosis of diabetes in 2003 was at least 50 years of age?

7. The following table summarizes ages (organized into groups) and disease stages (I–IV, with higher stages indicative of more advanced disease) of patients seen in the oncology clinic in the summer of 2002:

| | Disease Stage | | | |
Age	I	II	III	IV
18–39	70	55	40	32
40–59	10	31	45	51
60 years or older	5	12	48	84

a. What proportion of patients has stage III disease?

b. What proportion of patients has stage II disease and is 60 years of age or older?

c. What proportion of patients 60 years of age or older has stage III or IV disease?

d. Are the events A = stage IV disease and B = 60 years of age or older independent?

8. A ventilation perfusion scan is used to generate the probability of pulmonary embolism (PE) in patients. Based on the results of the scan, patients are classified as high probability for PE, moderate probability for PE, or low probability for PE. Consider the following data, classified by patient's age:

| | | Age | | |
		45–54	55–64	65–74
	High Prob. for PE	13	15	20
Scan	Moderate Prob. for PE	14	18	17
	Low Prob. for PE	25	20	31

If a patient is selected at random, find the probability that

a. The patient has a high probability for PE.

b. The patient is at least 55 years of age.

c. The patient has a high probability for PE given that the patient is at least 55 years of age.

d. Are the events high probability for PE and at least 55 years of age independent? Justify your answer.

9. The following table cross-classifies primary-care physicians by their gender and the size of their patient panel:

			Size of Patient Panel		
Gender	<100	100–299	300–499	500+	Total
Male	65	225	310	256	856
Female	78	120	650	308	1156
Total	143	345	960	564	2012

If a physician is selected at random, find the probability that

a. A female who has a panel of 500 or more patients is selected.

b. A male from among those with 500 or more patients is selected.

c. What proportion of the physicians has panel sizes of at least 300 patients?

d. What proportion of females has panel sizes of at least 300 patients?

e. What proportion of physicians with the largest panels is female?

10. Consider the following probabilities:

$P(A) = 0.5$, $P(B) = 0.3$, $P(C) = 0.1$, $P(A|C) = 1$, $P(A \text{ and } B) = 0.2$

a. Find $P(A \text{ or } B)$.

b. Suppose that B and C are mutually exclusive, find $P(B \text{ or } C)$.

c. Find $P(A \text{ and } C)$.

d. Are A and C independent? Justify your answer.

e. Are B and C independent? Justify your answer. (*Hint:* See part b.)

11. The following table classifies patients in a research study of cardiovascular risk factors according to their gender and whether they have a family history of cardiovascular disease:

		History of Cardiovascular Disease	
		No	Yes
Gender	Female	120	28
	Male	165	34

Compute the following probabilities:

a. $P(\text{family history of cardiovascular disease})$

b. $P(\text{family history of cardiovascular disease}|\text{male})$

c. $P(\text{family history of cardiovascular disease}|\text{female})$

d. Are family history of cardiovascular disease and gender independent? Justify your answer.

12. The following table displays medication adherence (defined as the percent of doses taken over the past month measured to the nearest whole number), classified by the time since initiation of therapy:

	Percent of Doses Taken		
Time Since Initiation of Therapy	*<70%*	*70%–89%*	*90%–100%*
Less than 3 months	18	58	24
Between 3 and 12 months	33	40	27
More than 12 months	21	27	52

a. What proportion of patients takes 70% or more of the prescribed doses?

b. What proportion of patients on medication therapy for more than 12 months takes 70% or more of the prescribed doses?

c. What proportion of patients on medication therapy for less than 3 months takes 70% or more of the prescribed doses?

d. What is the nature of the relationship between medication adherence and time since initiation of medication therapy? (Briefly, in words.)

13. A recent report described a clinical trial comparing an existing medication for asthma to a newly developed medication with respect to efficacy. In the trial, all patients who were seen in an asthma clinic at a particular hospital were approached and asked to participate. The following table classifies patients according to gender, eligibility, and enrollment status:

	Eligibility/Enrollment Status			
Gender	*Ineligible*	*Eligible–Enrolled*	*Eligible–Refused to Enroll*	*TOTAL*
Male	35	80	20	135
Female	60	70	35	165
TOTAL	95	150	55	300

a. What proportion of patients approached was eligible for the trial?

b. What proportion of patients approached was ineligible?

c. What proportion of male patients enrolled?

d. What proportion of ineligible patients was female?

e. Are gender and eligibility status independent?

14. Researchers claim that a new treatment for chronic bronchitis is 80% effective in reducing symptoms. If the new treatment is given to 12 patients suffering from bronchitis, what is the probability that

    a. The treatment is effective in exactly 10 patients?

    b. The treatment is effective in exactly 10 patients if the true effectiveness rate is 82%?

    c. If the true effectiveness rate is 80%, how many patients would you expect to have reduced symptoms?

15. A political pollster knows from a very elaborate poll that 60% of voters favors stricter regulations on Medicaid. If nine voters are randomly selected, find

    a. The probability that exactly three voters favor stricter regulations on Medicaid.

    b. The probability that at least three voters favor stricter regulations on Medicaid.

    c. The mean number of voters who favor stricter regulations on Medicaid.

    d. The probability that exactly three favor stricter regulations on Medicaid if the probability that voters favor stricter regulations on Medicaid is really 55%.

16. A longitudinal study is conducted requiring patients to follow up with research associates every month for assessments. The probability that a patient fails to follow up in a given month is 10%. A pilot study is conducted to assess feasibility involving 20 patients. What is the probability that at most 3 patients fail to follow up in the first month?

17. Data from a national survey conducted in 2002 reported that 10% of all women over age 50 changed their primary-care physicians at least once during a 3-year period.

    a. Assuming that this rate applies to all women over 50, what is the probability that less than 3 of 10 women over 50 years of age change their primary-care physicians within 3 years?

    b. If the true rate is higher than 10%, would the probability in (a) increase or decrease? Justify your answer. (Be brief, but complete.)

    c. If a primary-care physician's practice included 650 women over the age of 50, how many would be expected to leave the practice (i.e., change primary-care physicians) over 3 years?

18. An HMO is considering a marketing plan to increase the numbers of patients who get flu shots, particularly patients with chronic diseases. Before implementing such a plan, some preliminary analyses are conducted. Available data indicate that 40% of patients with chronic diseases get flu shots.

   a. If 10 patients with chronic diseases are sampled, what is the probability that at least 4 get flu shots regularly?

   b. What is the probability that at most 8 get flu shots regularly?

   c. If a particular clinic (which is part of the HMO) serves 5250 patients with chronic disease, how many would be expected to get flu shots ?

19. A article was published recently suggesting that persons who exercise regularly can reduce their risk of major clinical events (e.g., diabetes, cardiovascular disease) by up to 50% in some instances. It is believed that only 30% of adult Americans exercise on a regular basis.

   a. If eight adult Americans are analyzed, what is the probability that more than half of them exercise regularly?

   b. If eight adult Americans are analyzed, what is the probability that less than half of them exercise regularly?

   c. How many of the eight adults would you expect to exercise regularly?

20. Suppose we plan to conduct a clinical trial and will recruit patients from a hospital clinic. In this clinic, approximately 10 new patients are seen in each session. We expect 70% of patients approached to be eligible for the new trial.

   a. What is the probability that at least 7 patients will be eligible per clinic session?

   b. How many patients would you expect to be eligible each clinic session?

21. Data show that 20% of all patients who make appointments at a primary-care clinic never show up. If in a given clinic session, 15 patients make appointments,

   a. What is the probability that at most 4 patients do not show up?

   b. What is the probability that all patients show up?

   c. How many patients would be expected to not show up?

   d. How many patients would be expected to show up?

   e. What is the probability that all patients show up if the true "no-show" percentage is 18%?

22. A study is proposed to evaluate whether a new medication is effective in controlling the symptoms of asthma. Prior to mounting the full study, a pilot study is conducted to test specific aspects of the study protocol. A sample of 10 patients with asthma is enrolled in the pilot study. Each patient agrees to take the study medication and to report back to the study evaluation team in 30 days for a clinical evaluation. Similar studies report that 90% of all patients complete

follow-up visits when they are scheduled for 1 month from the initial contact.

   a. What is the probability that at least 8 patients show up for the clinical assessment?

   b. What is the probability that all patients show up for the clinical assessment?

   c. What is the probability that no more than 4 patients DO NOT show up for the clinical assessment?

   d. What is the probability that all show up if the true show rate is 87%?

23. Using the data in Problem 22, determine the expected (or mean) number of patients that would show up for clinical assessments if 10 were enrolled. What if 100 were enrolled?

24. The number of telephone calls to emergency phone numbers per hour is assumed to follow a normal distribution, with a mean of 7.5 calls per hour and a standard deviation of 2.1 calls per hour.

   a. What is the probability that more than 12 calls are received per hour?

   b. What is the probability that between 5 and 10 calls are received per hour?

   c. Complete the following statement: 90% of the time, the number of calls received per hour is less than _____.

25. Durations of clinical research studies awarded by a particular federal agency are assumed to follow a normal distribution, with a mean of 24 months and a standard deviation of 2.5 months.

   a. Find the probability that a study is awarded with a duration of more than 22 months.

   b. Find the probability that a study is awarded with a duration between 15 and 25 months.

   c. Fill in the following statement. 90% of studies awarded have durations between _____ and _____ months. (Consider the middle 90% of the distribution.)

26. Graduate Record Exam (GRE) scores are approximately normally distributed, with a mean of 500 and a standard deviation of 100.

   a. What proportion of individuals scores between 580 and 620?

   b. What proportion of individuals scores between 400 and 800?

   c. What is the 90th percentile score?

27. Weights, in pounds, for specific gender and age subgroups are assumed to follow a normal distribution. The mean weight for males aged 25 years is 160 pounds, with a standard deviation of 19.2 pounds.

What proportion of males aged 25 have weights

a. Exceeding 140 pounds?

b. Between 140 and 160 pounds?

c. Compute the weight that separates the top 10% of males aged 25 from the remaining males.

28. Total cholesterol levels are assumed to follow a normal distribution for males and females in specific age groups (e.g., 20–39, 40–59). The following parameters are available on total cholesterol levels for patients enrolled in a particular HMO:

Gender, Age Group	Mean	Standard Deviation
Female, 30–49	185	28
Male, 30–49	192	24

a. What proportion of females 30–49 years of age has total cholesterol levels exceeding 200?

b. What proportion of males 30–49 years of age has total cholesterol levels exceeding 200?

c. Compute the 90th percentile of total cholesterol levels among females 30–49 years of age.

d. Compute the 90th percentile of total cholesterol levels among males 30–49 years of age.

29. Prior to medical visits, patients are asked to complete a self-administered medical history form. The time it takes each patient to complete the form is approximately normally distributed, with a mean of 12.4 minutes and a standard deviation of 2.1 minutes.

a. What proportion of patients completes the medical history form in 10 minutes or less?

b. What proportion of patients takes 20 minutes or more to complete the medical history form?

c. What is the 90th percentile of the time to complete the medical history form?

30. Total serum cholesterol levels for individuals 65 years of age and older are assumed to follow a normal distribution, with a mean of 182 and a standard deviation of 14.7.

a. What proportion of individuals 65 years of age and older have cholesterol levels of 175 or more?

b. What proportion of individuals 65 years of age and older have cholesterol levels between 150 and 175?

    c. If the top 10% of the cholesterol levels are assumed to be abnormally high, what is the upper limit of the normal range?

31. Data are collected in a national survey and weighted to reflect the population of all U.S. adults. Suppose that the mean Quality of Life (QOL) score is 70, with a standard deviation of 10. The range of QOL scores is 0–100, with higher scores indicative of better QOL.

    a. What proportion of U.S. adults have QOL scores of 90 or higher?

    b. What is the 90th percentile of the QOL scores?

32. Heights of children are approximately normally distributed at each age and gender. If the mean height for 2-year-old males is 28 inches, with a standard deviation of 2.4 inches, find the following:

    a. The probability that a 2-year-old male is more than 26 inches.

    b. The probability that a 2-year-old male is between 30 and 35 inches.

    c. Suppose the smallest 10% and largest 10% of heights are considered "abnormal." What range of heights is "normal"?

33. A census of persons recovering from lower-extremity fractures finds they work a mean of 8 hours per week, with a standard deviation of 2 hours per week, while in the first 6 months of recovery. One year postinjury, these persons work a mean of 12 hours per week, with a standard deviation of 3.5 hours per week.

    a. What proportion of persons works at least 10 hours per week during the first 6 months of recovery?

    b. What proportion of persons works at least 10 hours per week one year postinjury?

    c. What is the median number of hours worked during the first 6 months of recovery?

    d. What is the 90th percentile in the number of hours worked per week for persons one year postinjury?

34. Body mass index (BMI) is computed as the ratio of weight in kilograms to height in meters squared. The distribution of BMI is approximately normal for specific gender and age groups. For females aged 30–39, the mean BMI is 24.5, with a standard deviation of 3.3.

    a. What proportion of females aged 30–39 has a BMI over 25?

    b. Persons with a BMI of 30 or greater are considered obese. What proportion of females aged 30–39 is obese?

    c. Suppose we classify females aged 30–39 in the top 10% of the BMI distribution as high risk. What is the threshold for classifying a female as high risk?

35. Based on a set of risk factors such as systolic blood pressure, total cholesterol level, and smoking status, risks for developing cardiovascular disease (CVD) are calculated. The risk scores range from 0 to 100, with higher scores indicative of increased risk. Suppose for men over age 50, the CVD risk scores are approximately normally distributed, with a mean of 30 and standard deviation of 8.5.

    a. What proportion of men over age 50 has CVD risk scores exceeding 25?
    b. What is the 90th percentile of the CVD risk scores?
    c. If a man aged 65 has a CVD risk score of 50, what percentile is he in?

SAS Problems

Use SAS to solve each of the following problems.

1. Use the SAS ranbin function to generate a probability distribution for a binomial random variable with $n = 10$ and $p = 0.25$. Run 5000 replications (see SAS Example 3.1), and use the probability distribution to find

    a. $P(X = 2)$
    b. $P(X < 2)$
    c. $P(X > 2)$

2. Use the SAS probit function to find the following percentiles for a normal random variable with mean 500 and standard deviation 100 (see SAS Example 3.3).

    a. 5th percentile
    b. 50th percentile
    c. 80th percentile
    d. 95th percentile

3. Use the SAS rannor function to generate a probability distribution for a normal random variable with mean 65 and standard deviation 8.6. Use Proc Chart to generate a relative frequency histogram for this variable (see SAS Example 3.3).

4. Use the SAS ranbin function to generate a probability distribution for a binomial random variable with $n = 50$ and $p = 0.25$. Run 5000 replications (see SAS Example 3.1), and use SAS Proc Chart to generate a relative frequency histogram for this variable. Does the variable look approximately normally distributed?

5. Use the SAS rannor function to generate a probability distribution for a normal random variable with mean 50 and standard deviation 2. Use the SAS probit function to find the 5th, 10th, 50th, 90th, and 95th percentiles. Also run a Proc Univariate, and compare results.

Descriptive Statistics
(Ch. 2)

Probability
(Ch. 3)

Sampling Distributions
(Ch. 4)

Statistical Inference
(Chapters 5–13)

OUTCOME VARIABLE	GROUPING VARIABLE(S)/ PREDICTOR(S)	ANALYSIS	CHAPTER(S)
Continuous	—	Estimate μ; Compare μ to Known, Historical Value	5/12
Continuous	Dichotomous (2 groups)	Compare Independent Means (Estimate/Test $(\mu_1 - \mu_2)$) or the Mean Difference (μ_d)	6/12
Continuous	Discrete (>2 groups)	Test the Equality of k Means Using Analysis of Variance ($\mu_1 = \mu_2 = \cdots = \mu_k$)	9/12
Continuous	Continuous	Estimate Correlation or Determine Regression Equation	10/12
Continuous	Several Continuous or Dichotomous	Multiple Linear Regression Analysis	10
Dichotomous	—	Estimate p; Compare p to Known, Historical Value	7
Dichotomous	Dichotomous (2 groups)	Compare Independent Proportions (Estimate/Test $(p_1 - p_2)$)	7/8
Dichotomous	Discrete (>2 groups)	Test the Equality of k Proportions (Chi-Square Test)	7
Dichotomous	Several Continuous or Dichotomous	Multiple Logistic Regression Analysis	11
Discrete	Discrete	Compare Distributions Among k Populations (Chi-Square Test)	7
Time to Event	Several Continuous or Dichotomous	Survival Analysis	13

Sampling Distributions

4

In the previous chapter, we introduced the concept of probability, presented rules for computing probabilities of outcomes and events, and discussed two probability models. The purpose of that discussion was to provide the mathematical background to understand the process of sampling subjects at random from a population. In statistical inference, which is addressed in the remaining chapters of this book, we will discuss the techniques of drawing inferences about unknown population parameters (e.g., the population mean μ) based on sample statistics (e.g., the sample mean $\overline{X}$). In this chapter, we discuss the relationships between sample statistics and population parameters. In particular, we focus on the relationship between the sample mean and the population mean. The statistical inference techniques described in subsequent chapters are based on the relationship between statistics and parameters.

In Section 4.1 we provide background, and in Section 4.2 we present the Central Limit Theorem, which is the mathematical theorem that formalizes the relationship between the sampling distribution of the sample mean and the population mean. In Section 4.3 we summarize key formulas, and in Section 4.4 we illustrate the relationship between the sampling distribution of the sample mean and the population mean using SAS.

4.1 Background

Suppose we have a population with mean μ and standard deviation σ. For now we do not specify any distributional form for the population, such as the normal distribution. Suppose we take simple random samples (s.r.s.) of size n from this population (we illustrated this process with our example in Chapter 1), and we compute the sample mean for each sample. The collection of all possible sample means is called the *sampling distribution of the sample means*. If we compute the mean and standard deviation over all possible sample means, the following results hold.

For simple random samples *with* replacement:

$$\mu_{\overline{X}} = \mu, \qquad \sigma_{\overline{X}} = \frac{\sigma}{\sqrt{n}} \tag{4.1}$$

where $\mu_{\overline{X}}$ and $\sigma_{\overline{X}}$ denote the mean and standard deviation of the sample means, respectively. The standard deviation of the sample means, $\sigma_{\overline{X}}$, is called the *standard error*.

For simple random samples *without* replacement:

$$\mu_{\overline{X}} = \mu, \qquad \sigma_{\overline{x}} = \frac{\sigma}{\sqrt{n}}\sqrt{\frac{N-n}{N-1}} \tag{4.2}$$

If either the population size (N) is large or the ratio of the sample size to the population size (n/N) is small (e.g., less than 0.10), the quantity

$\sqrt{(N-n)/(N-1)}$ will be close to 1 and the standard error in (4.2) is approximately equal to that given in (4.1). In most applications, this is the case, as the population is almost always very large. We illustrate these formulas in Examples 4.1 and 4.2.

EXAMPLE 4.1

Relationship Between Population Parameters and Sample Statistics

Suppose we have a small population of five patients with Type II (adult-onset) diabetes and measure each patient's glycosolated hemoglobin (blood sugar) level. The population data are

8.9   9.1   10.4   11.0   10.1

The population mean is $\mu = \sum X/N = 9.9$, and the population standard deviation is $\sigma = \sqrt{\sum(X-\mu)^2/N} = 0.79$. (These parameters were computed using formulas (1.1) and (1.3) presented in Chapter 1.)

Suppose we take simple random samples of size 3 from the population. For illustration purposes, consider *all* possible samples of size $n = 3$ from the population without replacement (i.e., once selected into the sample, an individual cannot be selected again). There are 10 samples of size $n = 3$ ($_5C_3 = 5 \cdot 4/2 = 10$) in which no individual appears more than once. The 10 samples and the sample means computed on each sample are shown in Table 4.1.

The right-hand column is the sampling distribution of the sample means. Notice that the sample means range from 9.37 to 10.50. The mean and standard deviation of this population (of all possible sample means based on simple random samples of size 3 without replacement) can be computed

Table 4.1 *Simple Random Samples of Size n = 3 and Sample Means*

Sample (X_1, X_2, X_3)	Sample Mean $\overline{X}$
8.9, 9.1, 10.4	9.47
8.9, 9.1, 11.0	9.67
8.9, 9.1, 10.1	9.37
8.9, 10.4, 11.0	10.10
8.9, 10.4, 10.1	9.80
8.9, 11.0, 10.1	10.00
9.1, 10.4, 11.0	10.17
9.1, 10.4, 10.1	9.87
9.1, 11.0, 10.1	10.07
10.4, 11.0, 10.1	10.50

directly using (1.1) and (1.3), respectively: $\mu_{\overline{X}} = 9.9$ and $\sigma_{\overline{X}} = 0.32$. By (4.2), the mean of the 10 sample means is $\mu_{\overline{X}} = \mu = 9.9$. The standard error, or standard deviation of the sample means, by (4.2) is

$$\sigma_{\overline{X}} = \frac{\sigma}{\sqrt{n}}\sqrt{\frac{N-n}{N-1}} = \frac{0.79}{\sqrt{3}}\sqrt{\frac{5-3}{5-1}} = 0.32$$

■

Results (4.1) and (4.2) are very useful for statistical inference, where we make inferences about a population parameter (e.g., μ) based on a sample statistic (e.g., $\overline{X}$). Results (4.1) and (4.2) indicate that the sample mean is an *unbiased* estimator of the population mean (i.e., on the average, the sample mean is equal to μ) regardless of whether samples are taken with or without replacement. The sampling strategy, however, does affect the variation in the sample means (i.e., the standard error $\sigma_{\overline{x}}$). As noted, when the population size (N) is large, this effect is negligible.

EXAMPLE 4.2   **Relationship Between Sample Size and the Standard Error**

Suppose we have a population with mean 100 and standard deviation 10. (Note that the form of the distribution of the population is not specified.) If we take simple random samples of size $n = 4$ with replacement, the following are true by (4.1):

$$\mu_{\overline{X}} = \mu = 100$$

$$\sigma_{\overline{X}} = \frac{\sigma}{\sqrt{n}} = \frac{10}{\sqrt{4}} = \frac{10}{2} = 5$$

Suppose we increase our sample size to $n = 25$:

$$\mu_{\overline{X}} = \mu = 100$$

$$\sigma_{\overline{X}} = \frac{\sigma}{\sqrt{n}} = \frac{10}{\sqrt{25}} = \frac{10}{5} = 2$$

Notice that under each sampling strategy the mean of the sample means is equal to the population mean. However, when the sample size is increased from 4 to 25 the standard error (or variation in the sample means) is reduced from 5 to 2. When the samples are larger in size, there is less variation in the sample means; the sample means are more tightly clustered about the population mean.

■

4.2 The Central Limit Theorem

The following mathematical theorem is perhaps the most important theorem in statistics.

Central Limit Theorem

Suppose we have a population with mean μ and standard deviation σ. If we take simple random samples of size n with replacement from the population, for large n, the sampling distribution of the sample means is approximately normally distributed with

$$\mu_{\overline{X}} = \mu, \qquad \sigma_{\overline{X}} = \frac{\sigma}{\sqrt{n}}$$

where, in general, $n \geq 30$ is sufficiently large.

The Central Limit Theorem (CLT) is important because in statistical inference we will make inferences about the population mean (μ) based on the value of a *single* sample mean. The CLT states that for large samples ($n \geq 30$), the distribution of the sample means is approximately normal. If the population is normal, then the results hold for any size n. If the population is binomial, then the following criteria are required, $np > 5$ and $n(1 - p) > 5$. In Chapter 3, we computed probabilities about normal random variables by standardizing (transforming to Z) and using Table B.2. The CLT tells us that we can use that same process to compute probabilities about the sample mean even if the population is not normal. This will be useful in statistical inference because when we make inferences about population parameters based on sample statistics, we will attach probability statements that quantify the precision in our inferences.

To reinforce the results of the Central Limit Theorem, we now present three illustrations. In each illustration we display the population distribution along with the sampling distributions of the sample means based on samples of size 5, 15, 30, and 50. We display the distributions graphically and present parameters associated with each in tabular form. The illustrations are presented in Figures 4.1–4.3. Figures 4.1 and 4.2 depict nonnormal populations; Figure 4.3 illustrates the normal population distribution. Figures 4.1a–4.1d display the sampling distributions of the sample mean based on s.r.s. of size $n = 5$, $n = 15$, $n = 30$, and $n = 50$, respectively, from the population displayed in Figure 4.1. Similar graphs are shown for sampling distributions from the populations displayed in Figures 4.2 and 4.3. Notice for the normal population distribution (Figure 4.3), the sampling distributions of the sample mean are normal for each sample size considered. In the nonnormal cases (Figures 4.1 and 4.2), the sampling distributions of the sample means approach normality as the sample size increases (e.g., $n \geq 30$).

The mean of the uniform population shown in Figure 4.1 is 100, with a standard deviation of 57.7. When simple random samples are drawn, the means of the sampling distributions for samples of each size are equal to 100. The standard deviations of the sample means, or the standard errors, decrease as the sample size increases (see Table 4.2 and Figures 4.1a–d). Notice how the shapes of the sampling distributions of the sample means start to look approximately normal for sample sizes of 30 or larger.

Figure 4.1 *Uniform Population*

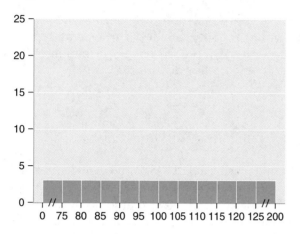

Table 4.2 *Uniform Population*

Population	Sample Size	Mean	SD
Population distribution	—	$\mu = 100$	$\sigma = 57.7$
Sampling distribution	5	$\mu_{\overline{X}} = 100$	$\sigma_{\overline{X}} = 25.9$
Sampling distribution	15	$\mu_{\overline{X}} = 100$	$\sigma_{\overline{X}} = 14.9$
Sampling distribution	30	$\mu_{\overline{X}} = 100$	$\sigma_{\overline{X}} = 10.5$
Sampling distribution	50	$\mu_{\overline{X}} = 100$	$\sigma_{\overline{X}} = 8.2$

Figure 4.1a *Sampling Distribution of $\overline{X}$, $n = 5$*

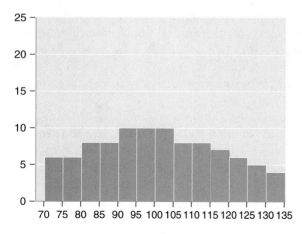

Figure 4.1b *Sampling Distribution of $\overline{X}$, $n = 15$*

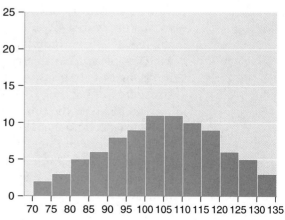

Figure 4.1c *Sampling Distribution of*
$\overline{X}, n = 30$

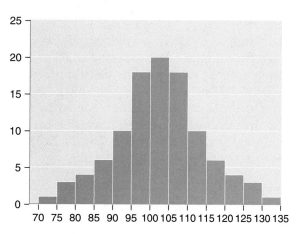

Figure 4.1d *Sampling Distribution of*
$\overline{X}, n = 50$

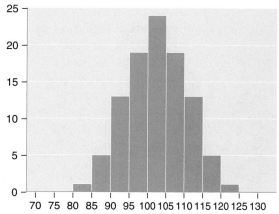

Figure 4.2 *Skewed Population*

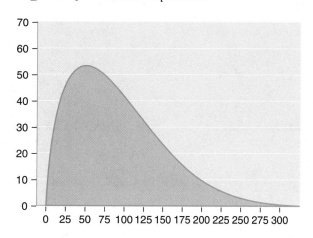

Table 4.3 *Skewed Population*

Population	Sample Size	Mean	SD
Population distribution	—	$\mu = 100$	$\sigma = 100$
Sampling distribution	5	$\mu_{\overline{X}} = 100$	$\sigma_{\overline{X}} = 44.8$
Sampling distribution	15	$\mu_{\overline{X}} = 100$	$\sigma_{\overline{X}} = 25.8$
Sampling distribution	30	$\mu_{\overline{X}} = 100$	$\sigma_{\overline{X}} = 18.2$
Sampling distribution	50	$\mu_{\overline{X}} = 100$	$\sigma_{\overline{X}} = 14.2$

We now consider a second nonnormal population, one which is skewed to the right, with most observations clustered at the low end of the distribution (Figure 4.2).

The mean of the skewed population shown in Figure 4.2 is 100, with a standard deviation of 100. When simple random samples are drawn, the means of the sampling distributions for samples of each size are equal to 100. The standard deviations of the sample means, or the standard errors, decrease as the sample size increases (see Table 4.3 and Figures 4.2a–d). Again, notice how the shapes of the sampling distributions of the sample means start to look approximately normal for sample sizes of 30 or larger.

The last example involves a normal population (Figure 4.3). Notice that the sampling distributions of the sample means are approximately normally

Figure 4.2a *Sampling Distribution of $\overline{X}$, $n = 5$*

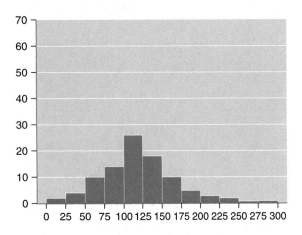

Figure 4.2b *Sampling Distribution of $\overline{X}$, $n = 15$*

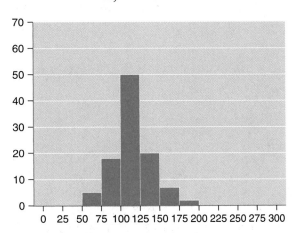

Figure 4.2c *Sampling Distribution of $\overline{X}$, $n = 30$*

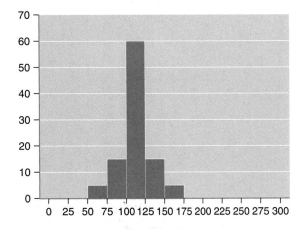

Figure 4.2d *Sampling Distribution of $\overline{X}$, $n = 50$*

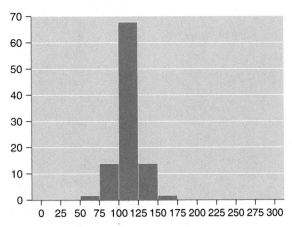

Figure 4.3 *Normal Population*

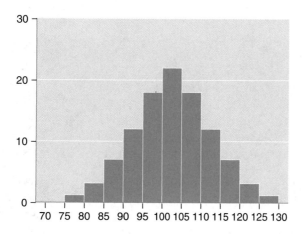

Table 4.4 *Normal Population*

Population	Sample Size	Mean	SD
Population distribution	—	$\mu = 100$	$\sigma = 10$
Sampling distribution	5	$\mu_{\overline{X}} = 100$	$\sigma_{\overline{X}} = 4.5$
Sampling distribution	15	$\mu_{\overline{X}} = 100$	$\sigma_{\overline{X}} = 2.6$
Sampling distribution	30	$\mu_{\overline{X}} = 100$	$\sigma_{\overline{X}} = 1.8$
Sampling distribution	50	$\mu_{\overline{X}} = 100$	$\sigma_{\overline{X}} = 1.4$

Figure 4.3a *Sampling Distribution of $\overline{X}$, $n = 5$*

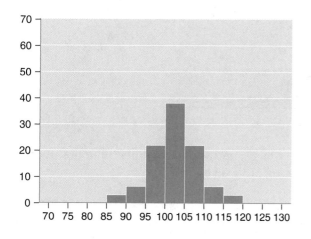

Figure 4.3b *Sampling Distribution of $\overline{X}$, $n = 15$*

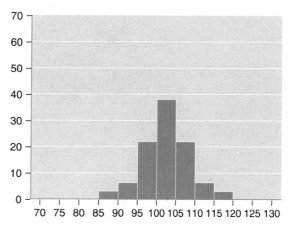

Figure 4.3c *Sampling Distribution of*
$\overline{X}, n = 30$

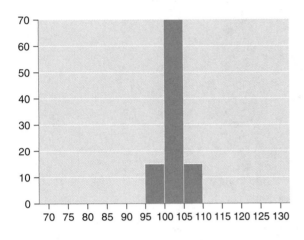

Figure 4.3d *Sampling Distribution of*
$\overline{X}, n = 50$

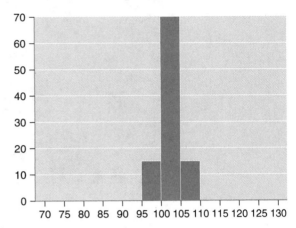

distributed for even small sample sizes (e.g., $n = 5$). In Figures 4.1 and 4.2, the sampling distributions of the sample mean looked normal only when the sample size was 30 or greater.

EXAMPLE 4.3

Applying the Central Limit Theorem: Non-Normal Population

Telephone calls placed to a drug hotline during the hours of 9–5 weekdays have a mean length of 5 minutes with a standard deviation of 5 minutes. The distribution of all calls (i.e., the population) is of the form:

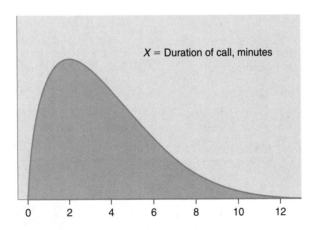

X = Duration of call, minutes

Suppose the population is very large, and we take simple random samples of three calls ($n = 3$). The following are true by (4.1):

$$\mu_{\overline{X}} = 5, \qquad \sigma_{\overline{X}} = \frac{5}{\sqrt{3}} = 2.89$$

Similarly, if we take s.r.s. of size $n = 100$, then by (4.1):

$$\mu_{\overline{X}} = 5, \qquad \sigma_{\overline{X}} = \frac{5}{\sqrt{100}} = 0.5$$

By the Central Limit Theorem the distribution of the sample means for samples of size $n = 100$ is approximately normal as shown here:

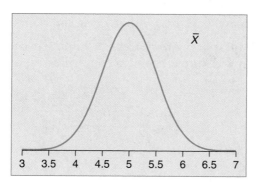

The distribution of the lengths of phone calls (X) to the drug hotline for all callers (the population) was not normally distributed. The population distribution was skewed to the right, with the majority of calls lasting a shorter time and fewer calls lasting a longer time. When we look at collections of 100 calls at a time (samples of $n = 100$), the distribution of the mean length of calls ($\overline{X}$) is normally distributed (as shown). ▪

Most of the topics addressed in subsequent chapters deal with statistical inference based on a single sample from the population of interest. We summarize the sample and then make inferences about unknown population parameters (e.g., μ) based on sample statistics (e.g., $\overline{X}$). Without knowing the population distribution, as long as the sample is sufficiently large (usually size 30 is sufficient), we can appeal to the Central Limit Theorem to make probabilistic statements about the relationship between the sample mean and the unknown population mean. For example, the Central Limit Theorem states that the sample mean is approximately normally distributed with mean and standard deviation $\mu_{\overline{X}}$ and $\sigma_{\overline{X}}$, respectively. We can make statements about the distribution of sample means of size n by transforming to the standard normal distribution using (4.3) and Table B.2 (see Section 3.6).

$$Z = \frac{\overline{X} - \mu_{\overline{X}}}{\sigma_{\overline{X}}} = \frac{\overline{X} - \mu}{\dfrac{\sigma}{\sqrt{n}}} \tag{4.3}$$

The following example illustrates the use of the CLT and formula (4.3).

EXAMPLE 4.4   **Applying the Central Limit Theorem: Normal Population**

Suppose we have a normal population with $\mu = 100$ and $\sigma = 10$. If we take a simple random sample of size 225, find the probability that the sample mean falls between 99 and 102.

The problem of interest is stated as follows:

$$P(99 < \overline{X} < 102)$$

By the Central Limit Theorem, we know that the distribution of the sample mean is approximately normal, with $\mu_{\overline{X}} = \mu$ and $\sigma_{\overline{X}} = \dfrac{\sigma}{\sqrt{n}}$. Because n is large ($n = 225$), we can appeal to the Central Limit Theorem and transform the problem concerning $\overline{X}$ into a problem concerning the standard normal distribution Z.

For $\overline{X} = 99$:

$$Z = \frac{99 - 100}{10/\sqrt{225}} = -1.5$$

For $\overline{X} = 102$:

$$Z = \frac{102 - 100}{10/\sqrt{225}} = 3$$

Thus, $P(99 < \overline{X} < 102) = P(-1.5 < Z < 3)$. Using Table B.2,

$$P(-1.5 < Z < 3) = 0.9987 - 0.0668 = 0.9319$$

Therefore, the probability that the sample mean based on 225 observations falls between 99 and 102 is 93.2%. If we were to use the sample mean (based on a sample size of 225) to estimate the unknown population mean, this result suggests that there is a high probability that the sample mean is close to the population mean.

Suppose instead that an *individual* is selected at random from the population. Find the probability that his/her score falls between 99 and 102 (i.e., find $P(99 < X < 102)$). Because X is distributed as a normal random variable with mean 100 and standard deviation 10, we can transform the problem into a problem concerning Z using (3.9) (i.e., $Z = (X - \mu)/\sigma$).

For $X = 99$:

$$Z = \frac{99 - 100}{10} = -0.1$$

For $X = 102$:

$$Z = \frac{102 - 100}{10} = 0.2$$

Thus, $P(99 < X < 102) = P(-0.1 < Z < 0.2)$. Using Table B.2,

$$P(-0.1 < Z < 0.2) = 0.5793 - 0.4602 = 0.1191$$

Therefore, the probability that a single observation falls between 99 and 102 is only 11.91%. ■

EXAMPLE 4.5

Applying the Central Limit Theorem

Suppose we have a population with $\mu = 50$, $\sigma = 10$. (Notice that no distributional form is specified for the population.) Taking a simple random sample of size 100, find the probability that the sample mean is between 48 and 51, using (4.1), the CLT, and Table B.2.

We want to compute $P(48 < \overline{X} < 51)$. We first standardize each value:

$$Z = \frac{48 - 50}{10/\sqrt{100}} = -2$$

$$Z = \frac{51 - 50}{10/\sqrt{100}} = 1$$

Thus, $P(48 < \overline{X} < 51) = P(-2 < Z < 1) = 0.8413 - 0.0228 = 0.8185$. ∎

The following example illustrates the application of the Central Limit Theorem in statistical inference.

EXAMPLE 4.6

Estimating the Mean of a Population

Suppose we wish to estimate the mean of a population (μ) whose standard deviation is known and equal to 12. Suppose a simple random sample of 100 individuals is selected from the population. Find the probability that the sample mean is no more than 2 units from the population mean.

In this example we do not know the population mean and wish to estimate it based on a single random sample. Here we ask the question, what is the likelihood that the sample mean is close to the true population mean (which is unknown)? We define close as within 2 units in either direction.

Because the sample size is large ($n = 100$), we can appeal to the Central Limit Theorem, which indicates that the distribution of the sample means is approximately normal, with $\mu_{\overline{X}} = \mu$ (as shown in the figure).

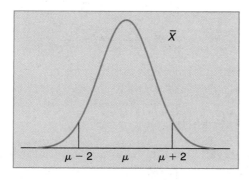

We wish to find the following:

$$P(\mu - 2 < \overline{X} < \mu + 2)$$

Standardizing using (4.3):

$$Z = \frac{(\mu - 2) - \mu}{12/\sqrt{100}} = \frac{-2}{1.2} = -1.67$$

$$Z = \frac{(\mu + 2) - \mu}{12/\sqrt{100}} = \frac{2}{1.2} = 1.67$$

This is equal to $P(-1.67 < Z < 1.67) = 0.9525 - 0.0475 = 0.9050$.

There is a 90.5% probability that the sample mean, based on 100 observations, will be within 2 units of the true mean, μ. There is a very high likelihood (exceeding 90%) that the sample mean will be close to the population mean (within 2 units in either direction). If the goal of the application was to generate an estimate of the unknown population mean, this example suggests that the value of the sample mean (based on a sample of size $n = 100$) would be quite close to the unknown population mean. ■

Would you feel comfortable using the value of the sample mean as an estimate of the unknown population mean? We will explore these types of applications in further detail in the next several chapters.

4.3 Key Formulas

Concept	Formula	Description
Simple random sampling with replacement	$\mu_{\overline{X}} = \mu, \quad \sigma_{\overline{X}} = \frac{\sigma}{\sqrt{n}}$	mean and standard error of sampling distribution
Simple random sampling without replacement	$\mu_{\overline{X}} = \mu, \quad \sigma_{\overline{X}} = \frac{\sigma}{\sqrt{n}}\sqrt{\frac{N-n}{N-1}}$	mean and standard error of sampling distribution
Standardize a sample mean (if CLT applies, e.g., $n \geq 30$)	$Z = \frac{\overline{X} - \mu_{\overline{X}}}{\sigma_{\overline{X}}} = \frac{\overline{X} - \mu}{\sigma/\sqrt{n}}$	standardizes sample mean

4.4 Applications Using SAS

In this section we review another SAS function, one which generates random variables from the uniform probability distribution. We take repeated samples from the uniform distribution, compute sample means, and generate the sampling distribution of the sample means. Specifically, using SAS we illustrate the results of the Central Limit Theorem presented in Section 4.3 for the uniform distribution.

A SAS program is presented illustrating the use of the SAS ranuni function, which is used to generate random variables from the uniform probability distribution. Notes are provided to the right of the SAS program (in blue) for orientation purposes and are not part of the program. The blank lines in the programs are solely to accommodate the notes. Blank lines and spaces can be used throughout SAS programs and are generally used to enhance readability.

We present a single example related to the example presented in Figure 4.1. For the example, we present three components:

1. The SAS program code
2. The computer output
3. A description of the relevant components of the computer output along with their interpretation

SAS EXAMPLE 4.1 **Uniform Population, Simple Random Samples of Size $n = 30$**
Ranuni Function (Generates Random Variables from the Uniform Distribution)

Consider a population that follows a uniform distribution, like the one illustrated in Figure 4.1. The uniform distribution is a continuous distribution. Suppose for this example that the population covers the interval 0 to 1. Using SAS, we illustrate the results of the Central Limit Theorem presented in Section 4.2.

In this application, we will generate 30 random variates from the uniform distribution, which is equivalent to drawing 30 observations at random from a uniform population. We will compute the sample mean for each set of 30 observations and will then summarize the distribution of the sample means. We will generate 10,000 samples of size $n = 30$. Due to the large number of samples, our observed probability distribution of the sample means will closely approximate the theoretical sampling distribution. The SAS program code is shown next.

Program Code

```
options ps=66 ls=80;
```
Formats the output page to 66 lines in length, 80 columns in width

```
data pop;
```
Beginning of Data Step (Data set name **pop**).

```
do i=1 to 10000;
```
Beginning of a DO loop which will be repeated 10,000 times. The index *i* counts the iterations.

```
 x=ranuni(13755);
```
Generates a random variable, called **x**, from the uniform distribution. The only argument to the ranuni function is the seed.[i]

`output;`	Writes the generated variable, x, to data set **pop.**
`end;`	End of DO loop.
`run;`	End of Data Step.
`title 'population distribution';`	Title for output
`proc chart;`	Procedure call (Proc Chart to generate relative frequency histogram)
`vbar x/type=pct;`	Specification of variable x, type=pct displays relative frequencies
`run;`	End of Procedure section
`proc means;`	Procedure call (Proc Means to generate summary statistics)
`var x;`	Specification of variable x.
`run;`	End of Procedure section
`data samples;`	Beginning of Data Step (Data set name **samples**).
`do i=1 to 10000;`	Beginning of a DO loop which will be repeated 10,000 times. The index i counts the iterations.
`sumx=0;`	Create a variable called *sumx*, initialized to 0.
`do n=1 to 30;`	Beginning of a DO loop which will be repeated 30 times (for each value of i). The index n counts the iterations.
`x=ranuni(13755);`	Generates a random variable, called x, from the uniform distribution.
`sumx+x;`	Increase the value of *sumx* by the value of x.
`end;`	End of DO loop indexed by n.
`xbar=sumx/30;`	Create a variable called **xbar**, computed by dividing *sumx* by 30.
`output;`	Writes the generated variable, **xbar**, to data set *samples*.
`end;`	End of DO loop indexed by i.
`run;`	End of Data Step.
`title 'sampling distribution of the sample means';`	Title for output.
`proc chart;`	Procedure call (Proc Chart to generate relative frequency histogram)
`vbar xbar/type=pct;`	Specification of variable **xbar**, type=pct displays relative frequencies

```
run;                              End of Procedure section
proc means;                       Procedure call (Proc Means to generate summary
                                      statistics)
 var xbar;                        Specification of variable xbar.
run;                              End of Procedure section
```

(1) The seed is simply a random number used as a starting point for the random number generator (i.e., ranuni). When the seed is changed, different values are generated.

Computer Output

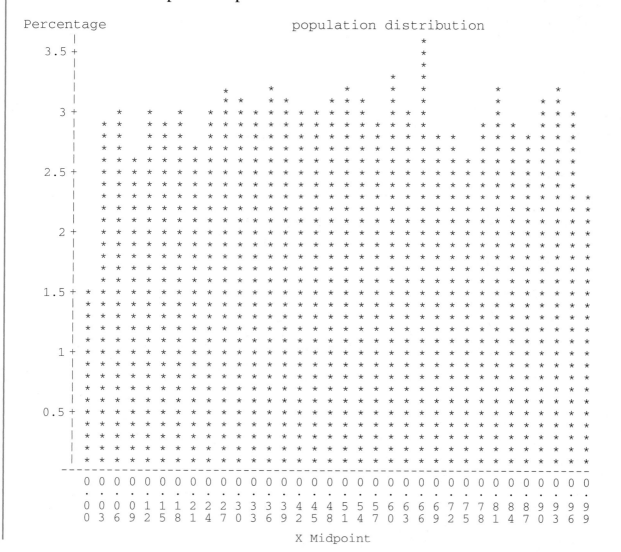

X Midpoint

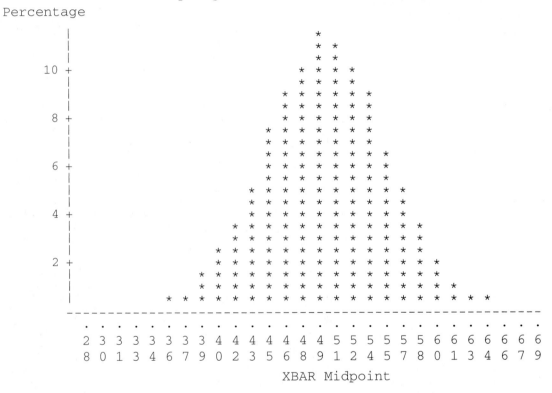

```
                        population distribution

Analysis Variable : X

    N          Mean       Std Dev       Minimum      Maximum
    ------------------------------------------------------------
  10000     0.5014338    0.2860532   0.000025837   0.9999899
    ------------------------------------------------------------



                sampling distribution of the sample means
Percentage
         |
         |                              *
         |                            * *
         |                            * *
    10 +                          * * * *
         |                          * * * *
         |                        * * * * * *
         |                        * * * * * *
     8 +                          * * * * * *
         |                      * * * * * * *
         |                      * * * * * * *
         |                      * * * * * * * *
     6 +                        * * * * * * * *
         |                      * * * * * * * *
         |                  * * * * * * * * * *
         |                    * * * * * * * * * *
     4 +                    * * * * * * * * * * *
         |                  * * * * * * * * * * *
         |                * * * * * * * * * * * *
         |                * * * * * * * * * * * * *
     2 +                * * * * * * * * * * * * * *
         |            * * * * * * * * * * * * * * *
         |          * * * * * * * * * * * * * * * *
         |    * * * * * * * * * * * * * * * * * * * *
         ------------------------------------------------------------
         .  .  .  .  .  .  .  .  .  .  .  .  .  .  .  .  .  .  .  .  .  .
         2  3  3  3  3  3  3  3  4  4  4  4  4  4  4  5  5  5  5  5  5  5  6  6  6  6  6  6
         8  0  1  3  4  6  7  9  0  2  3  5  6  8  9  1  2  4  5  7  8  0  1  3  4  6  7  9
                                 XBAR Midpoint

            sampling distribution of the sample means

            Analysis Variable : XBAR

    N          Mean       Std Dev      Minimum       Maximum
    ------------------------------------------------------------
  10000     0.5011598    0.0527851   0.2880476     0.7361218
    ------------------------------------------------------------
```

Interpretation

The first part of the output is a relative frequency histogram for the random variable X, which follows a uniform distribution over the interval 0 to 1. In theory, the distribution should be perfectly rectangular. Our simulated distribution has some spikes in various places. The summary statistics for the random variable X, the population variable, are given in the next section. The mean of X is 0.501, the standard deviation is 0.2861, and the range is from 0.000026 to 0.999989.

The next two sections of the output relate to the variable xbar, which is the sample mean based on 30 randomly selected observations from the uniform population. Notice that the distribution of xbar is approximately normal. Although the figure produced by SAS Proc Chart is somewhat crude, it does resemble the normal distribution. The mean of the variable xbar is 0.501, the standard deviation is 0.0528, and the range is from 0.2880 to 0.7361. Recall from the Central Limit Theorem that for sufficiently large samples (usually $n \geq 30$), that the following are true: $\mu_{\overline{X}} = \mu$ and $\sigma_{\overline{X}} = \dfrac{\sigma}{\sqrt{n}}$. The mean of xbar is identical to the mean of x (0.501). The theoretical standard deviation of xbar $= 0.2861/\sqrt{30} = 0.0522$ (assuming that $\sigma = 0.2861$). We observed a standard deviation of xbar equal to 0.0528.   ■

4.4.1 Summary of SAS Functions

The SAS functions illustrated and a general description of each are summarized in the following table. SAS functions are used in the Data Step.

Function	Sample Call Statement	Description
ranuni	$x = \text{ranuni(seed)};$	generates random variables from the uniform distribution with range 0–1

4.5 Problems

1. Scores on a standardized exam are assumed to follow a normal distribution, with a mean of 100 and standard deviation of 32.
   a. If a simple random sample of 5 exams is selected, what is the mean exam score? What is the standard error of the scores?
   b. If a simple random sample of 50 exams is selected, what is the mean exam score? What is the standard error of the scores?

    c. If a simple random sample of 500 exams is selected, what is the mean exam score? What is the standard error of the scores?

    d. Compare the results of (a)–(c).

2. The mean number of calories ingested per day for females aged 25 is 2750, with a standard deviation of 276. If a simple random sample of 100 females is selected, what is the probability that the mean number of calories ingested is more than 2800?

3. A dispensing machine is set to produce 1-pound lots of a particular compound. The machine is fairly accurate, producing mean weights of lots equal to 1.0 pound with a standard deviation of 0.12 pounds. Thirty-five lots are randomly selected.

    a. What is the expected weight of the sample (i.e., find $\mu_{\overline{X}}$)?

    b. What is the standard error in the weights (i.e., find $\sigma_{\overline{X}}$)?

    c. Find the probability that the mean weight is greater than 1 pound.

    d. Find the probability that the mean weight is less than 0.95 pound.

4. Following cardiac surgery, patients are encouraged to exercise regularly (assume that regular exercise is defined as exercising on 3 or more days per week). A physician suspects that patients exercise regularly immediately following cardiac surgery but tend to reduce, even stop exercising completely, over time. An investigation is planned to estimate the mean number of weeks that patients exercise regularly following cardiac surgery. Assume that the standard deviation in the number of weeks cardiac patients exercise regularly following surgery is 6.3 weeks.

    a. If a sample of 40 cardiac patients is followed, and the number of weeks in which each patient exercises regularly is recorded, what is the probability that the sample mean will be no more than 1 week higher than the true mean?

    b. If the sample is increased to 100 cardiac patients, what is the probability that the sample mean will be no more than 1 week higher than the true mean?

5. We wish to estimate the mean cholesterol level for Type II diabetic patients free of cardiovascular disease. Suppose the standard deviation is known and equal to 20 ($\sigma = 20$). A random sample of 50 Type II diabetic patients free of cardiovascular disease is selected. What is the probability that their mean cholesterol level will be no more than 4 units above the true mean?

6. We wish to estimate the mean gestational age (in days) in high-risk pregnancies. If we sample 50 high-risk pregnancies, and if the standard deviation is 4 days, what is the probability that the point estimate is within 1 day of the true mean gestational age?

7. Data are collected in a national survey and weighted to reflect the population of all U.S. adults. Suppose that the mean Quality of Life (QOL) score is 70, with a standard deviation of 10. The range of QOL scores is 0–100, with higher scores indicative of better QOL. What is the probability that the mean QOL score in a random sample of 40 adults is 72 or higher?

8. Total serum cholesterol levels for individuals 65 years of age and older are assumed to follow a normal distribution, with a mean of 182 and a standard deviation of 14.7.
   a. If a random sample of 20 individuals aged 65 years or older is selected, what is the probability that their mean total serum cholesterol level is between 180 and 185?
   b. Suppose that the mean total serum cholesterol level for individuals less than 65 years of age is 170 with a standard deviation of 26.8. If a random sample of 40 individuals less than 65 years of age is selected, what is the probability that their mean total serum cholesterol level is between 180 and 185?

Age Group	n	Mean
65 Years of age or older	20	182
Less than 65 years of age	40	170

9. An article reported that patients under care for HIV have CD4 tests every 3 months, on average, with a standard deviation of 1.5 months. What is the probability that in a sample of 40 patients, the mean time between tests exceeds 3.5 months?

10. We want to estimate the mean of a population. A random sample of subjects is selected and the sample mean is computed. What is the probability that the sample mean is within 3 units of the true mean if the standard error is 1.8?

11. Suppose we measure length of hospital stay (in days) for patients undergoing a particular surgical procedure. The mean is 3 days with a standard deviation of 3.2 days.
   a. Does it appear that the length-of-stay data follow a normal distribution? Justify your answer briefly.

    b. If a sample of 35 patients is selected, what is the probability that their mean length of stay is less than 2 days?

    c. If a sample of 35 patients is selected, what is the probability that their mean length of stay is more than 5 days?

SAS Problem

Use SAS to solve each of the following problems.

1. Use the SAS ranuni function to generate random variables from a uniform distribution over the range of 0–1. Take samples of size $n = 10$, $n = 20$, and $n = 30$, and plot the distributions of the sample means based on these different sample sizes. In addition, compute the mean and standard error of each distribution. Run 5000 replications (see SAS Example 4.1) for each sampling distribution.

2. Use the SAS rannor function to generate random variables from a normal distribution with mean 100 and standard deviation 8. Take samples of size $n = 5$, $n = 10$, $n = 20$, and $n = 30$, and plot the distributions of the sample means based on these different sample sizes. In addition, compute the mean and standard error of each distribution. Run 5000 replications for each sampling distribution. Do the sampling distributions look normally distributed?

3. Use the SAS probit function to find the following percentiles for the mean of the sampling distribution generated in Problem 1 for samples of size 30 (see SAS Example 3.3).

    a. 5th percentile

    b. 50th percentile

    c. 80th percentile

    d. 95th percentile

4. Use the SAS probit function to find the following percentiles for the mean of the sampling distribution generated in Problem 2 for samples of size 10.

    a. 5th percentile

    b. 50th percentile

    c. 80th percentile

    d. 95th percentile

5. Use the SAS ranexp function to generate random variables from an exponential distribution (the SAS ranexp function is similar to the SAS ranuni function in that the only input required is a seed, e.g.,

ranexp(15243).) Take samples of size $n = 10$, $n = 20$, and $n = 30$, and plot the distributions of the sample means based on these different sample sizes. Also plot the population distribution. What does the exponential distribution look like (i.e., the population)? What characteristics might follow an exponential distribution? Run 5000 replications (see SAS Example 4.1) for each sampling distribution.

Descriptive Statistics
(Ch. 2)

Probability
(Ch. 3)

Sampling Distributions
(Ch. 4)

Statistical Inference
(Chapters 5-13)

OUTCOME VARIABLE	GROUPING VARIABLE(S)/ PREDICTOR(S)	ANALYSIS	CHAPTER(S)
Continuous	—	Estimate μ; Compare μ to Known, Historical Value	5/12
Continuous	Dichotomous (2 groups)	Compare Independent Means (Estimate/Test $(\mu_1 - \mu_2)$) or the Mean Difference (μ_d)	6/12
Continuous	Discrete (>2 groups)	Test the Equality of k Means using Analysis of Variance $(\mu_1 = \mu_2 = \cdots = \mu_k)$	9/12
Continuous	Continuous	Estimate Correlation or Determine Regression Equation	10/12
Continuous	Several Continuous or Dichotomous	Multiple Linear Regression Analysis	10
Dichotomous	—	Estimate p; Compare p to Known, Historical Value	7
Dichotomous	Dichotomous (2 groups)	Compare Independent Proportions (Estimate/Test $(p_1 - p_2)$)	7/8
Dichotomous	Discrete (>2 groups)	Test the Equality of k Proportions (Chi-Square Test)	7
Dichotomous	Several Continuous or Dichotomous	Multiple Logistic Regression Analysis	11
Discrete	Discrete	Compare Distributions Among k Populations (Chi-Square Test)	7
Time to Event	Several Continuous or Dichotomous	Survival Analysis	13

5

Statistical Inference: Procedures for μ

In this chapter we introduce the techniques of statistical inference concerned with the mean (μ) of a population. Similar techniques applied to other parameters (e.g., the population proportion, p; the difference between two population means, $\mu_1 - \mu_2$; and so on) will be discussed in subsequent chapters.

There are two broad areas of statistical inference, *estimation* and *hypothesis testing*. In estimation, the population parameter (in this case, μ) is unknown. A random sample is drawn from the population of interest and sample statistics are used to generate estimates of the unknown parameter. In hypothesis testing, an explicit statement or hypothesis is generated about the population parameter. A random sample is drawn from the population of interest and sample statistics are analyzed to determine whether to either support or reject the hypothesis about the parameter.

In Section 5.1 we introduce the techniques of estimation applied to the population mean μ. We present vocabulary and notation (Section 5.1.1), followed by a series of examples (Section 5.1.2) and a discussion of issues (section 5.1.3) regarding precision in estimation and sample size requirements to achieve certain levels of precision. In Section 5.2 we introduce the techniques of hypothesis testing concerning the population mean μ. We present vocabulary and notation (Section 5.2.1), followed by a series of examples (Section 5.2.2), and introduce the concept of statistical power and sample size requirements to achieve certain levels of power (Section 5.2.3). In Section 5.3 we summarize key formulas, and in Section 5.4 we provide SAS program code used to perform statistical applications throughout this chapter. In Section 5.5 we use data from the Framingham Heart Study to illustrate the applications presented in this chapter.

5.1 Estimating μ

The goal in estimation is to make valid inferences about the population parameter based on a single random sample from the population. In the previous chapters, we discussed the theoretical basis for statistical inference; in particular, we described properties of sampling distributions of sample statistics. The relationship between sample statistics and population parameters is based on the sampling distribution of the sample statistics. With regard to estimating the population mean μ, we are particularly concerned with the relationship between the distribution of the sample mean $\overline{X}$ and μ.

5.1.1 Vocabulary and Notation

There are two types of estimates for population parameters: *point estimates* and *confidence interval estimates*. A point estimate for a population parameter is the "best" single-number estimate of that parameter. A confidence interval estimate is a range of values for the population parameter with a level of confidence attached (e.g., 95% confidence that the range or interval contains the unknown parameter).

The point estimate for the population mean μ is the "best" single-valued estimate of the population mean derived from the sample. The "best" single-number estimate for the population mean is the value of the sample mean. Estimates of a parameter are denoted by placing a ^ (read "hat") above the parameter notation. For example, the point estimate for the population mean is denoted as follows:

$$\hat{\mu} = \overline{X} \tag{5.1}$$

In Chapters 1 and 4 we learned that the sample mean is an unbiased estimator of the population mean (i.e., on average, the sample mean is equal to the population mean, $\mu_{\overline{X}} = \mu$). Unbiasedness is a desirable property in a point estimate. Consider the following example.

EXAMPLE 5.1

Estimating Mean Waiting Time in the Emergency Room

Suppose the administration of a particular suburban hospital wants to estimate the mean waiting time (in minutes) for patients in their emergency room (ER) during weekends. In order to estimate the unknown mean waiting time, we take a random sample of patients visiting this ER on weekends and record the time that each patient waits. Suppose that there are resources available for this study to sample 100 patients. For each patient selected into the sample, the waiting time is recorded, measured as the number of minutes between the time of entering the ER and the time when seen by a clinician in the ER. Once the data are collected on each of the 100 patients, summary statistics for the waiting times can be produced. The mean waiting time in our sample is 37.85 minutes. A point estimate of the mean waiting time in the ER during weekends is given by (5.1).

$$\hat{\mu} = \overline{X} = 37.85$$

where μ = mean waiting time in the ER during weekends

Suppose that the point estimate is reported to the administration, which is disappointed that the waiting time is so long and decides to take a second look. A second investigation is conducted in a similar fashion, involving a second (and distinct) sample of patients visiting the ER on weekends. The mean waiting time in the second sample is 33.75 minutes. A point estimate of the mean waiting time in the ER during weekends based on the second sample is 33.75 minutes.   ■

Which point estimate is more appropriate? From both the administration and the patient point of view, the second estimate is more appealing. Which one is better statistically? Both are point estimates (assuming that both samples are random samples from the population of interest), but they are not the same. The point estimate alone is usually not sufficient in an estimation problem. In particular, it is generally of interest to assess how close the point estimate is to the true (unknown) mean μ.

Suppose the second sample involves 35 patients. Intuitively, one would think that a sample involving more patients would produce a more precise estimate of the unknown parameter. In Chapter 4 we presented the Central Limit Theorem which states that if we take simple random samples (s.r.s.) with replacement from a population, then the distribution of the sample mean is approximately normal with $\mu_{\overline{X}} = \mu$ and $\sigma_{\overline{X}} = \dfrac{\sigma}{\sqrt{n}}$. (Note that if n is reasonably large, the same results hold for sampling without replacement.) Suppose that the standard deviation in waiting times on weekends is known to be 9.5 minutes (i.e., $\sigma = 9.5$). (In practice, the value of the population standard deviation is often unknown. We generally know nothing about the parameters of the population (e.g., μ, σ). For now assume that the standard deviation is known. Later we will present examples in which the population standard deviation is unknown.) From the Central Limit Theorem, we know that the distributions of sample means of size 100 (sample 1, $\overline{X} = 37.85$) and sample means of size 35 (sample 2, $\overline{X} = 33.75$) are approximately normal with

$$\mu_{\overline{X}} = \mu, \qquad \sigma_{\overline{X}} = \frac{\sigma}{\sqrt{n}} = \frac{9.5}{\sqrt{100}} = \frac{9.5}{10} = 0.95 \quad \text{and}$$

$$\mu_{\overline{X}} = \mu, \qquad \sigma_{\overline{X}} = \frac{\sigma}{\sqrt{n}} = \frac{9.5}{\sqrt{35}} = \frac{9.5}{5.9} = 1.61$$

respectively. The distributions of the sample means based on samples of size 100 and size 35 are displayed, respectively, in Figures 5.1a and 5.1b.

Notice that there is less variation in the distribution of sample means when the sample size is 100 as compared to the distribution when the sample size is 35. On average, the sample means based on $n = 100$ observations are 0.95 units from the population mean μ, whereas the sample means based on $n = 35$ observations are 1.61 units from the population mean. The variation in the sample means, denoted $\sigma_{\overline{X}}$, is called the *standard error*. The standard error is a measure of the variation in the sample means, specifically the extent to which the $\overline{X}$s vary around $\mu_{\overline{X}} = \mu$. As the sample size increases the standard error decreases, resulting in a more precise estimate of the population mean.

Figure 5.1a *Distribution of $\overline{X}$ for $n = 100$, $\sigma_{\overline{X}} = 0.95$*

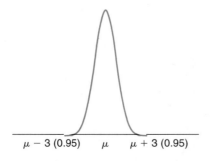

$\mu - 3\ (0.95)$ μ $\mu + 3\ (0.95)$

Figure 5.1b *Distribution of $\overline{X}$ for $n = 35$, $\sigma_{\overline{X}} = 1.61$*

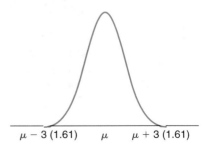

$\mu - 3\ (1.61)$ μ $\mu + 3\ (1.61)$

In Example 5.1 we described two samples and two point estimates for the unknown mean waiting time. In practice, a single sample would be selected and analyzed. We described two replications to illustrate the concept that the point estimate provides very valuable information, but alone is generally not sufficient in estimation problems. Once a point estimate is determined, the next issue of concern is how close that estimate is to the true, unknown value. The standard error addresses this issue, and the confidence interval incorporates both the point estimate and the standard error.

5.1.2 Confidence Intervals for μ

A confidence interval (CI) is a range of values that are likely to cover the true parameter (e.g., μ). CIs start with the point estimate, our best single-number estimate, and build in a component that addresses how close (or how far) the point estimate is from the true, unknown parameter. This component is called the *margin of error*.

In all statistical applications we must accept the fact that we will not be 100% accurate in estimating a population parameter (e.g., μ) based on a sample statistic (e.g., $\overline{X}$). In statistical inference we make statements about a population parameter based on analysis of only a subset (a random sample) of the population. Using the concepts of probability we discussed in Chapter 3 and the properties of sampling distributions we discussed in Chapter 4, we can quantify the margin of error in each application.

The general form of a confidence interval is: point estimate ± margin of error. In developing confidence intervals, we select a level of confidence that reflects the likelihood the confidence interval contains the true, unknown parameter. Usually, confidence levels of 90%, 95%, and 99% are chosen, although theoretically any likelihood could be selected. With a 95% confidence interval, for example, we are 95% confident that the range of values will contain or cover the true, unknown parameter. With a 99% confidence interval, we are 99% confident that the range of values will contain or cover the true population parameter. In the rest of this section, we derive the 95% confidence interval.

Recall the standard normal distribution Z presented in Chapter 3. The following statement is true for the standard normal distribution:

$$P(-1.96 < Z < 1.96) = 0.95 \qquad (5.2)$$

From the Central Limit Theorem (Chapter 4), we know that for large samples (usually $n \geq 30$ is sufficient) the sample means are approximately normally distributed, with $\mu_{\overline{X}} = \mu$ and $\sigma_{\overline{X}} = \dfrac{\sigma}{\sqrt{n}}$ (Figures 5.1a and 5.1b). Thus, for samples that are sufficiently large,

$$Z = \frac{\overline{X} - \mu_{\overline{X}}}{\sigma_{\overline{X}}} \qquad (5.3)$$

follows the standard normal distribution.

Substituting (5.3) into (5.2) gives the following:

$$P\left(-1.96 < \frac{\overline{X} - \mu_{\overline{X}}}{\sigma_{\overline{X}}} < 1.96\right) = 0.95 \tag{5.4}$$

Using algebra, (5.4) is equivalent to

$$P(\overline{X} - 1.96\sigma_{\overline{X}} < \mu_{\overline{X}} < \overline{X} + 1.96\sigma_{\overline{X}}) = 0.95 \tag{5.5}$$

Because $\mu_{\overline{X}} = \mu$, (5.5) indicates that the probability that the interval between $(\overline{X} - 1.96\sigma_{\overline{X}})$ and $(\overline{X} + 1.96\sigma_{\overline{X}})$ contains the mean μ is 0.95, or 95%. Confidence intervals are based on this result.

The *95% confidence interval* for μ is given by

$$\overline{X} \pm 1.96\sigma_{\overline{X}}$$

or, equivalently,

$$\overline{X} \pm 1.96\frac{\sigma}{\sqrt{n}}$$

where $\overline{X}$ is the point estimate and $1.96\frac{\sigma}{\sqrt{n}}$ is the margin of error (1.96 reflects the fact that we chose a 95% confidence level and $\frac{\sigma}{\sqrt{n}}$ is the standard error).

Using the data in Example 5.1 (sample 1), the 95% confidence interval for the mean waiting time in the ER during weekends is

$$37.85 \pm 1.96\frac{9.5}{\sqrt{100}}$$

$$37.85 \pm 1.96(0.95)$$

$$37.85 \pm 1.86$$

$$(35.99 \text{ to } 39.71)$$

Thus, we are 95% confident that the mean waiting time in the ER during weekends is between 35.99 and 39.71 minutes. The point estimate is 37.85 minutes and the margin of error is 1.86 minutes.

Following the same strategy, the 95% confidence interval based on the second sample of size $n = 35$ is 33.75 ± 3.15, or 30.60 to 36.90. Notice that this second interval is wider (30.60 to 36.90 as compared to 35.99 to 39.71) due to the smaller sample size (35 as compared to 100). Although the point estimate based on the second sample might make the administration happier, it is less precise due to the smaller sample size.

Suppose we construct 95% confidence intervals for the mean waiting time in the ER during weekends (μ) based on 40 different random samples of the same size (e.g., $n = 100$). The intervals change from sample to sample depending on the value of the observed sample mean. The intervals cover the population mean approximately 95% of the time (Figure 5.2).

Shown in the figure are 40 different 95% CIs for μ. In theory, 38 (95% of 40) 95% confidence intervals will cover the true mean μ. In practice we take

Figure 5.2 *Interpreting Confidence Intervals (95% CI for μ)*

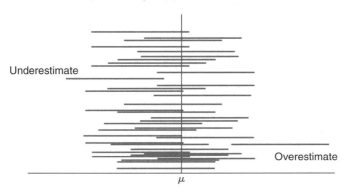

one random sample and develop a single confidence interval. We never know, with certainty, if our interval covers the true mean or not. It may be that our interval overestimates μ (i.e., the whole interval is above μ), or that the interval underestimates μ (i.e., the whole interval is below μ). Theoretically, using a 95% confidence interval, there is a 0.95 probability that any interval will cover the true mean.

Typically, confidence intervals are based on 80%, 90%, 95%, or 99% confidence levels, although it is possible to construct confidence intervals for any level (from 0% to 100%). The general form of the confidence interval for μ is

$$\overline{X} \pm Z_{1-(\alpha/2)}\frac{\sigma}{\sqrt{n}} \tag{5.6}$$

where $Z_{1-(\alpha/2)}$ is the value from the standard normal distribution with area in the lower tail (or area below it) equal to $1-\alpha/2$ corresponding to the $100(1-\alpha)\%$ confidence interval. Here, α refers to the total area in the tails of the standard normal distribution. Figure 5.3 illustrates the value(s) from the

Figure 5.3 $Z_{1-(\alpha/2)} = Z_{0.975}$ *for 95% Confidence Interval*

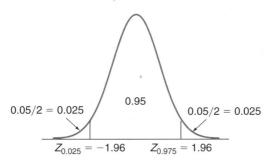

Table 5.1 *Values of* $Z_{1-(\alpha/2)}$ *for Confidence Intervals*

Confidence Level (%)	$Z_{1-(\alpha/2)}$	α^*
99.99	3.819	0.0001
99.9	3.291	0.001
99	2.576	0.01
95	1.960	0.05
90	1.645	0.10
80	1.282	0.20

* α refers to the total area in the tails of the standard normal distribution.

standard normal distribution corresponding to the 95% confidence interval. In the 95% confidence interval (i.e., total tail area = 0.05, $100(1-0.05)\% = 95\%$ confidence interval), $Z_{1-(\alpha/2)} = Z_{1-(0.05/2)} = Z_{0.975} = 1.96$.

Table 5.1 contains values from the standard normal distribution for commonly used confidence levels. The entries in the table are taken from the standard normal distribution table (Table B.2). If a confidence level and corresponding value of $Z_{1-(\alpha/2)}$ do not appear in Table 5.1, the standard normal distribution table can be used to extract the appropriate value of $Z_{1-(\alpha/2)}$ (see Figure 5.3).

Notice that as the confidence level increases, the value of $Z_{1-(\alpha/2)}$ increases and therefore so does the margin of error. In selecting the appropriate confidence level, investigators must evaluate the trade-off between higher levels of confidence and larger margins of error (i.e., wider confidence intervals).

EXAMPLE 5.2

Estimating Mean Age at Diagnosis of Hypertension

Suppose we wish to estimate, using a 95% confidence interval, the mean age at which patients with hypertension are diagnosed. We randomly select 12 subjects with diagnosed hypertension and record the age at which they were diagnosed. The following data are observed:

32.8 40.0 41.0 42.0 45.5 47.0 48.5 50.0 51.0 52.0 54.0 59.2

(Ages are computed based on dates of birth to the nearest tenth of a year.)

We assume that age at diagnosis is approximately normally distributed; that is, there is some mean value (unknown) and most patients are diagnosed around that mean value, whereas fewer are diagnosed at older or younger ages. The mean age at diagnosis in the sample is 47. A point estimate for the age at diagnosis among all hypertensives is 47. Suppose the standard deviation in the age at diagnosis among all hypertensives is known to be 7.2 (i.e., $\sigma = 7.2$). A 95% confidence interval (CI) for the true age at diagnosis is calculated as follows:

$$\overline{X} \pm Z_{1-(\alpha/2)} \frac{\sigma}{\sqrt{n}}$$

$$47 \pm 1.96 \frac{7.2}{\sqrt{12}}$$

$$47 \pm 4.1$$

$$(42.9, 51.1)$$

We are 95% confident that the mean age at diagnosis of hypertension is between 42.9 and 51.1. The point estimate is 47 years, with a margin of error of 4.1 years. ■

SAS EXAMPLE 5.2 **Estimating Mean Age at Diagnosis of Hypertension Using SAS**

The following output was generated using SAS Proc Means with an option to generate a 95% confidence interval. A brief interpretation appears after the output.

SAS Output for Example 5.2

```
              The MEANS Procedure
    Analysis Variable : age_dx Age at diagnosis of hypertension

N            Mean         Std Dev        Minimum          Maximum
-----------------------------------------------------------------
12      46.9166667      7.1598417      32.8000000       59.2000000
-----------------------------------------------------------------

    Analysis Variable : age_dx Age at diagnosis of hypertension
                    Lower 95%        Upper 95%
                    CL for Mean      CL for Mean
                    -----------------------------
                    42.3675203       51.4658131
                    -----------------------------
```

Interpretation of SAS Output for Example 5.2

The SAS Means Procedure is used to generate abbreviated descriptive statistics on a continuous variable. The default statistics are the sample size $n = 12$, the sample mean $\overline{X} = 46.9$, the sample standard deviation $s = 7.2$, the minimum 32.8, and the maximum 59.2. Here we requested that SAS generate a 95% confidence interval for the mean. The result is 42.4 to 51.5. The discrepancy between the interval we computed by hand in Example 5.2 and the interval computed by SAS in SAS Example 5.2 is due to rounding (SAS generally carries eight decimal places through computations). ■

EXAMPLE 5.3   **Estimating Mean Systolic Blood Pressure**

Suppose we wish to estimate, using a 90% confidence interval, the mean systolic blood pressure (SBP) of members of a health maintenance organization (HMO) who are 25 years of age. We select a random sample of size $n = 50$ members of the HMO who are 25 years of age. This sample has a mean SBP of 120 with a standard deviation of 14.2.

The population standard deviation (σ) is unknown. This is the case in most estimation problems (i.e., neither population parameter is known). If the sample size is large ($n \geq 30$ is usually sufficient), then the sample standard deviation (s) can be used to estimate σ (i.e., $\hat{\sigma} = s$) in the confidence interval. In this case, the confidence interval is given by

$$\overline{X} \pm Z_{1-(\alpha/2)} \frac{s}{\sqrt{n}} \tag{5.7}$$

Substituting our sample statistics, and using Table 5.1 to locate the appropriate value from the standard normal distribution, we have the following 90% confidence interval:

$$120 \pm 1.645 \frac{14.2}{\sqrt{50}}$$

$$120 \pm 3.30$$

$$(116.7, 123.3)$$

We are 90% confident that the mean SBP among 25-year-old members of the HMO is between 116.7 and 123.3.   ■

When the population standard deviation (σ) is unknown and the sample size is small ($n < 30$), the Central Limit Theorem no longer applies and we cannot assume that the distribution of the sample mean is approximately normal. In this case, (5.3) no longer holds and we cannot use $Z_{1-(\alpha/2)}$ in the formula for our confidence interval (5.6). If we can assume that the characteristic under investigation is approximately normally distributed in the population, then we can use a value from the t distribution in place of Z in (5.7). The t distribution is similar to the standard normal distribution (also bell-shaped and symmetric), except in the tail areas. The shape of the t distribution depends on the exact sample size (Figure 5.4).

t values for confidence intervals can be found in Table B.3 in the Appendix. Tabled entries are indexed by their degrees of freedom (df). In the case of the confidence interval for μ, the degrees of freedom (df) are computed by df $= n - 1$. As the degrees of freedom increase (with increasing sample size n), the t values approach the standard normal values, Z (Figure 5.4).

Table 5.2 summarizes three different formulas for confidence intervals for the population mean and the conditions under which each should be applied.

Figure 5.4 *t Distributions for n = 5, n = 10, and n = 20*

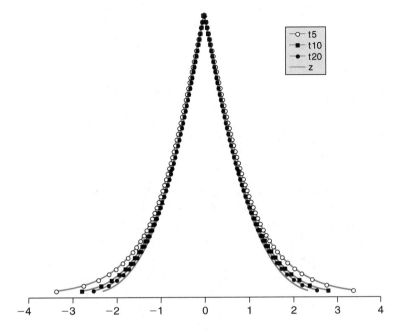

Table 5.2 *Confidence Intervals for* μ

Assumptions: Normal distribution or large samples (n ≥ 30); simple random samples

Case 1. σ known	$\overline{X} \pm Z_{1-(\alpha/2)} \dfrac{\sigma}{\sqrt{n}}$
Case 2. σ unknown and n ≥ 30	$\overline{X} \pm Z_{1-(\alpha/2)} \dfrac{s}{\sqrt{n}}$
Case 3. σ unknown and n < 30	$\overline{X} \pm t_{1-(\alpha/2)} \dfrac{s}{\sqrt{n}},\ \mathrm{df} = n - 1$

EXAMPLE 5.4

Estimating Mean IQ

Suppose we wish to estimate, using a 95% confidence interval, the mean IQ for all 12-year-olds. We select a random sample of sixteen 12-year-olds whose mean IQ score is 106 with a standard deviation of 12.4. Assuming that IQs are approximately normally distributed for a specific age group, because σ is unknown and the sample size is small (i.e., $n < 30$), the appropriate confidence interval for μ is the formula given in Case 3:

$$\overline{X} \pm t_{1-(\alpha/2)} \frac{s}{\sqrt{n}}, \quad \mathrm{df} = n - 1$$

We first need to find the appropriate value from Table B.3 for 95% confidence. In order to locate the appropriate value, we need to calculate the degrees of freedom: df $= n - 1 = 16 - 1 = 15$. The value from Table B.3 for 95% confidence is $t = 2.131$. Now substituting the sample data,

$$106 \pm 2.131 \frac{12.4}{\sqrt{16}}$$

Thus,

$$106 \pm 6.606$$

$$(99.4, 112.6)$$

We are 95% confident that the mean IQ score for all 12-year-olds is between 99.4 and 112.6. ■

EXAMPLE 5.5

Estimating Mean Number of Visits to Primary Care Physicians

We wish to estimate, using a 99% confidence interval, the mean number of visits to primary-care physicians over 3 years for all patients with Type II (adult-onset) diabetes. A sample of 65 patients with diabetes is followed over a 3-year period and the number of visits each makes to their primary-care physician is measured. The mean number of visits is 16 with a standard deviation of 1.4. In this example, σ is unknown and the sample size is large (i.e., $n \geq 30$); therefore, this is an example of Case 2.

$$\overline{X} \pm Z_{1-(\alpha/2)} \frac{s}{\sqrt{n}}$$

Substituting our sample statistics and the appropriate value from Table 5.1:

$$16 \pm 2.576 \frac{1.4}{\sqrt{65}}$$

$$16 \pm 0.45$$

$$(15.5, 16.5)$$

We are 99% confident that the mean number of visits to primary-care physicians over a 3-year period for diabetic patients is between 15.5 and 16.5. ■

In Example 5.5, the level of confidence is high, yet the width of the confidence interval is very small. This is primarily due to the magnitude of the standard error, or the variation in the sample means (i.e., $\sigma_{\overline{X}} = 1.4/\sqrt{65} = 0.174$). There is very little variation in the sample means, suggesting that the number of visits to primary-care physicians is very homogeneous among diabetic patients.

5.1.3 Precision and Sample Size Determination

The examples in the previous section illustrate estimation techniques for the population mean μ when presented with a random sample from the population of interest. In each example, sample statistics and sometimes population parameters (i.e., σ) were given. In order to construct confidence intervals for μ, we first reviewed the available information and then chose, based on that information, the appropriate formula from Table 5.2. Once the appropriate formula was selected, the computations were relatively straightforward. Looking back over Examples 5.1 through 5.5, we see that the margins of error in the confidence intervals vary widely from example to example, depending on both the variation in the characteristic under investigation (σ or s) and the sample size. The margin of error is small when the sample size is large and/or when the standard error is small.

In *experimental design,* statisticians are concerned with determining the numbers of subjects and the sampling strategy to be employed in a particular application so as to satisfy specific precision criteria in the statistical inference phase of the analysis. In some instances, the number of subjects selected depends solely on practical considerations (e.g., time and/or financial constraints). In other applications, there is more flexibility such that the sample size is determined based on statistical considerations. Consider the following scenario.

Suppose we wish to estimate the mean of a population μ, and it is important to produce an estimate that is within 5 units of the true mean with 95% confidence. The statistician concerned with experimental design determines the number of subjects required to produce such an estimate under these criteria (i.e., margin of error = 5 units with 95% confidence). Example 5.6 illustrates the experimental design aspect of an estimation problem.

EXAMPLE 5.6

Sample Size Determination to Estimate Mean Age at Diagnosis of Hypertension

In Example 5.2 we generated a 95% confidence interval estimate for the mean age at which patients with hypertension were diagnosed. Our interval was based on a sample of $n = 12$ subjects and had a margin of error of 4.1 years. Suppose that we wanted a more precise estimate, for example, an interval with a margin of error not exceeding 2 years. In applications in which one desires a confidence interval estimate for the mean of a population, formulas can be used to determine the necessary sample size. In order to design the experiment, specifically to determine the sample size, the following questions must be answered:

1. How much error can be tolerated in the estimate (i.e., how close must the estimate be to the true mean μ)?
2. What level of confidence is desired?

An investigator with substantive knowledge of the application at hand is in the best position to address question (1). With respect to question (2), the investigator must decide whether 90%, 95%, 99%, or some other level will satisfy the need.

Suppose the investigator decides that the estimate must be within 2 years of the true mean to be useful and that a 95% confidence level is sufficient. (This aspect of the design is relatively simple, as 95% confidence is the standard confidence level used in most applications.) The general form of the confidence interval for estimating μ is given by (5.6):

$$\overline{X} \pm Z_{1-(\alpha/2)} \frac{\sigma}{\sqrt{n}}$$

(In some applications we substitute s for σ; in other applications we use a value from the t distribution in place of Z. Here we use the general formula and make adjustments as necessary in the examples that follow.)

The general form of the confidence interval for estimating μ can be written as

$$\overline{X} \pm E$$

where E is the margin of error

$$E = Z_{1-(\alpha/2)} \frac{\sigma}{\sqrt{n}} \tag{5.8}$$

The primary issue in this design application is to determine the number of subjects, n, required to satisfy prespecified precision criteria. In particular, we want to determine the number of subjects required in order to generate an estimate for μ (the mean age at diagnosis) with a maximum margin of error of 2 years with 95% confidence.

Using algebra, we can solve for the sample size, n, in equation (5.8). We first multiply both sides by the square root of n:

$$\sqrt{n}E = Z_{1-(\alpha/2)}\sigma$$

Now divide both sides by the margin of error:

$$\sqrt{n} = Z_{1-(\alpha/2)} \frac{\sigma}{E}$$

Squaring both sides produces

$$n = \left(\frac{Z_{1-(\alpha/2)}\sigma}{E} \right)^2 \tag{5.9}$$

Equation (5.9) produces the *minimum* number of subjects required to ensure a margin of error equal to E in the confidence interval for μ with the specified level of confidence (reflected in $Z_{1-(\alpha/2)}$). Notice that equation (5.9) calls for the standard deviation (σ) of the population characteristic under investigation. We can use the value observed in Example 5.2, as it was based on a similar study. The number of subjects who must be sampled in order to produce an estimate for μ, the mean age at diagnosis, with $E = 2$ years and 95% confidence is given by (5.9):

$$n = \left(\frac{Z_{1-(\alpha/2)}\sigma}{E} \right)^2 = \left(\frac{1.96(7.2)}{2} \right)^2 = (7.056)^2 = 49.79$$

Because (5.9) produces the minimum number of subjects to satisfy the specified criteria, we always round up to the next integer. Therefore, 50 individuals with hypertension must be sampled and their age at diagnosis measured in order to generate a 95% confidence interval estimate for the mean age at diagnosis with a margin of error of no more than 2 years. With a sample size of 50, the resulting confidence interval for the mean age will be of the form: $\overline{X} \pm 2$ (where 2 is the margin of error). The exact value of the sample mean will be computed once the appropriate data are collected. ■

In design applications where the goal is to determine the sample size required to produce a confidence interval estimate for an unknown population mean, formula (5.9) can be applied. In order to apply (5.9), however, three inputs are required: the desired confidence level (reflected in $Z_{1-(\alpha/2)}$), the standard deviation of the characteristic under study (σ), and the margin of error (E). The confidence level is usually the easiest to determine because in most applications, the standard 95% confidence level is used ($Z_{1-(\alpha/2)} = 1.960$). The standard deviation of the characteristic under study can be difficult to determine, however. Recall that in these applications we are determining the number of subjects required so as to produce a confidence interval estimate with a specified level of precision for an unknown population mean μ. As we discussed in the previous section, it is often the case that the population standard deviation (σ) is unknown. If the value of σ is unknown, there are several options for generating a reasonable estimate: We can substitute a value derived from a similar application reported in the literature; we can generate an estimate based on a previous application (i.e., historical data); we can conduct a pilot study (e.g., select a nonrandom, convenience sample and compute the sample standard deviation s); or we can use an educated "guess." Because estimation of the sample size is not a problem of statistical inference, it is acceptable to generate a best guess. In doing so, however, one should always err on the conservative side. One method of generating a best guess is based on the Empirical Rule presented in Chapter 2. The Empirical Rule states that for unimodal distributions, approximately 68% of the observations fall between $\mu - \sigma$ and $\mu + \sigma$, approximately 95% of the observations fall between $\mu - 2\sigma$ and $\mu + 2\sigma$, and almost all of the observations fall between $\mu - 3\sigma$ and $\mu + 3\sigma$. Thus the range is equal to about 6 standard deviations. Based on the Empirical Rule, a *conservative* estimate of the standard deviation is given by the second part of the Empirical Rule:

$$\hat{\sigma} = \frac{\text{range}}{4} \tag{5.10}$$

To use (5.10) to estimate the standard deviation, we must specify the range. The range is the difference between the minimum and maximum values, which can often be approximated easily.

Consider the following example in which the population standard deviation is unknown and an estimate is used in (5.9) to determine the sample size requirements.

EXAMPLE 5.7   **Sample Size Determination to Estimate Mean Time to Travel Between Departments**

Suppose the administration of a hospital wants to estimate the mean time it takes for patients to get from one department to another on the hospital grounds (e.g., radiology to surgery, radiology to orthopedics) for appointment scheduling purposes. How large a sample would be required to ensure that the margin of error in the estimate of the mean time it takes for patients to get from one department to another is no more than 5 minutes with 95% confidence?

Suppose in this application we know nothing about the time it takes for patients to get from one department to another on the hospital grounds. The goal of the present analysis is to determine the number of subjects required to generate an estimate of the mean time it takes, and here we have no idea of what the variation in times might be (i.e., σ is unknown).

Suppose we cannot find any other, comparable studies and there are no historical data available. In order to generate an estimate of the variation in time it takes to get from one department to another, we conduct a pilot study. In the pilot study we survey a "convenience" (or nonrandom) sample of $n = 10$ patients traveling from department to department on hospital grounds. For each patient we record the time it takes to get from one department to another in minutes. Using the formulas provided in Chapter 2, we compute the sample standard deviation s based on the $n = 10$ observations. Suppose we find $s = 17$ minutes.

We can now implement formula (5.9) to determine the number of subjects required to estimate the mean time with a margin of error of 5 minutes with 95% confidence:

$$n = \left(\frac{Z_{1-(\alpha/2)}\sigma}{E} \right)^2 = \left(\frac{1.96(17)}{5} \right)^2 = (6.664)^2 = 44.4$$

A random sample of size 45 patients would ensure that the estimate of the mean time to travel between departments is within 5 minutes of the true mean with 95% confidence.

What would happen to the sample size estimate if we chose a smaller margin of error? Suppose in the previous application we wanted to estimate with 95% confidence the mean time it takes for patients to get from one department to another with a margin of error of 2 minutes:

$$n = \left(\frac{Z_{1-(\alpha/2)}\sigma}{E} \right)^2 = \left(\frac{1.96(17)}{2} \right)^2 = (16.667)^2 = 277.6$$

A random sample of size 278 patients would be required to ensure that the estimate of the mean time to travel between departments is within 2 minutes of the true mean with 95% confidence. To ensure a smaller margin of error, more subjects are needed in the sample.   ■

In design applications such as those described in Examples 5.6 and 5.7, investigators will often consider several scenarios. The different scenarios usually consider different margins of error. As we stated previously, there are often both practical and statistical considerations that must be evaluated in determining the sample size. In the previous application, it may be feasible (e.g., within budget constraints) to sample as many as 125 subjects. Statistical considerations suggest that a sample of size 45 would ensure a margin of error of no more than 5 minutes, but 278 subjects are required to ensure a margin of error of no more than 2 units. What level of precision could we achieve with 125 subjects?

SAS EXAMPLE 5.7 **Sample Size Determination to Estimate Mean Time to Travel Between Departments Using SAS**

The following output was generated using SAS to determine the sample size required to ensure different margins of error with 95% confidence. A brief interpretation appears after the output.

SAS Output for Example 5.7

Obs	c_level	z	sigma	e	n
1	0.95	1.95996	17	1	1111
2	0.95	1.95996	17	2	278
3	0.95	1.95996	17	3	124
4	0.95	1.95996	17	4	70
5	0.95	1.95996	17	5	45
6	0.95	1.95996	17	10	12

Interpretation of SAS Output for Example 5.7

There is no SAS procedure specifically designed to determine the number of subjects required to produce a confidence interval estimate with a specific margin of error and confidence level. However, SAS can be used to program the appropriate formula (5.9). Once the formula is programmed, different scenarios can be evaluated easily. In the example shown, six scenarios are considered (denoted Obs 1–6, respectively), related to SAS Example 5.7. In each scenario, the confidence level (c_level) is set at 0.95 (95%, which is reflected in $Z_{1-(\alpha/2)} = 1.96$) and the standard deviation (sigma) is set at 17. The scenarios differ in terms of the margins of error (e) specified. Here we considered margins of error 1, 2, 3, 4, 5, and 10 in scenarios 1–6, respectively. In scenario 1 (Obs = 1), the confidence level (c_level) is set at 0.95 (95%), the standard deviation (sigma) is set at 17, and the margin of error (e) is specified at 1. SAS produces a sample size of 1111, which is the required number of subjects to generate a 95% confidence interval for the mean time to get from one department to another with a margin of error equal to 1 minute. As the margin of error increases, fewer subjects are required. For example, when the margin of error is 2 (Obs = 2), 278 subjects are required to generate a 95% confidence interval for the mean time to get from one department to another, and when the margin of error is 10 (Obs = 6), only 12 subjects are required to generate a 95% confidence interval for the mean time to get from one department to another. This output would be weighed against practical constraints to determine the most appropriate sample size for the application. With 124 subjects, a margin of error of 3 minutes can be achieved. ■

5.2 Hypothesis Tests About μ

We now describe the second area of statistical inference, called *hypothesis testing*. In hypothesis testing, an explicit statement or hypothesis is generated about the population parameter. A random sample is drawn from the population of interest and sample statistics are analyzed and determined to either support or reject this hypothesis about the parameter. Based on the sample statistics, we determine the likelihood that the null hypothesis is true. As in the case of estimation, the theoretical basis for hypothesis testing is the relationship between the distribution of sample statistics and the population parameter.

5.2.1 Vocabulary and Notation

A number of new vocabulary terms are involved in hypothesis testing, and they will be used to discuss hypothesis tests about μ (presented in this chapter) and about other parameters (e.g., p, $\mu_1 - \mu_2$) presented in subsequent chapters. We use an example in this section to introduce these terms and their definitions. Section 5.2.2 contains a number of more concise examples that use these terms and concepts.

EXAMPLE 5.8

Testing Mean Systolic Blood Pressure Against Referent

A large, national study conducted in 2003 reported that the mean systolic blood pressure for males aged 50 was 130 with a standard deviation of 15. In 2004, an investigator hypothesized that due to increased stress in the workplace, faster-paced lifestyles, and poorer nutritional habits, systolic blood pressures have increased. In order to test this hypothesis (i.e., that the mean systolic blood pressure increased in 2004), we set up two competing hypotheses, called the *null and alternative hypotheses*, denoted H_0 and H_1, respectively.

$$H_0: \mu = 130$$
$$H_1: \mu > 130$$

(5.11)

The null hypothesis reflects the "no change" or "no effect" situation (i.e., systolic blood pressures are no different in 2004 and have a mean of 130); the alternative or research hypothesis reflects the investigator's claim (i.e., that the mean systolic blood pressure increased in 2004).

In order to test the hypotheses (5.11), a random sample is selected from the population of interest. Suppose we have resources to sample $n = 108$ males aged 50 in 2004. We record the systolic blood pressure on each male and generate summary statistics. Of primary interest is the mean systolic blood pressure in the sample $(\overline{X})$.

Consider the following scenarios and their interpretations with respect to the hypotheses (5.11):

■ Suppose the mean systolic blood pressure for the sample of $n = 108$ males selected in 2004 is $\overline{X} = 130$. The mean systolic blood pressure in the 2004 sample is identical to the population mean from 2003, which would lead us to believe that the null hypothesis, H_0, is most likely true (i.e., the population mean in 2004 is not changed and is equal to 130). Notice we cannot say with certainty that the null hypothesis is true because we have only a sample of males aged 50 in 2004. However, because the sample was selected at random, we would expect the mean systolic blood pressure among all males aged 50 in 2004 to be close to the observed 130.

■ Suppose $\overline{X} = 150$. The mean systolic blood pressure in the 2004 sample is substantially higher than the population mean from 2003, which would lead us to believe that the alternative hypothesis, H_1, is most likely true (i.e., the population mean in 2004 is greater than 130).

■ Suppose $\overline{X} = 135$. The mean systolic blood pressure in the 2004 sample is numerically higher ($135 > 130$) than the population mean from 2003, but is it solely due to chance fluctuation? Remember, we are evaluating statistics based on a random sample of 108 subjects from the population of interest, and sampling always involves some error. In order to evaluate whether the observed sample mean ($\overline{X} = 135$) provides evidence of a true increase in the underlying population mean, we must address the question: How likely are we to observe a sample mean of 135 from a population with $\mu = 130$? (This probability can be computed using the Central Limit Theorem and techniques we described in Chapter 4.)

The last scenario illustrates the crux of the hypothesis-testing problem. To conduct the test of hypothesis (i.e., to decide whether the null or alternative hypothesis is more likely true), we must determine a *critical value* such that if our sample mean is less than the critical value, we will conclude that H_0 is true (i.e., $\mu = 130$), and if our sample mean is greater than the critical value, we will conclude that H_1 is true (i.e., $\mu > 130$).

Instead of determining critical values for $\overline{X}$ that would be specific to each application (because $\overline{X}$ depends on the unit of measurement), we instead appeal to the Central Limit Theorem. In particular, for samples of sufficient size ($n \geq 30$), we know that the sample mean is approximately normally distributed with mean $\mu_{\overline{X}}$ and standard deviation $\sigma_{\overline{X}}$. Assuming that H_0 is true (or "under the null hypothesis"), we can standardize $\overline{X}$, producing a Z score:

$$Z = \frac{\overline{X} - \mu_{\overline{X}}}{\sigma_{\overline{X}}} = \frac{\overline{X} - \mu_0}{\sigma/\sqrt{n}} \tag{5.12}$$

where μ_0 is the mean specified in H_0 (i.e., $\mu_0 = 130$)

From Chapter 3, we know the properties of Z. If Z (5.12) is close to zero, which occurs when $\overline{X}$ is close to $\mu_0 = 130$, we suspect that H_0 is most likely true. When Z is large, which occurs when $\overline{X}$ is larger than $\mu_0 = 130$, we suspect that H_1 is most likely true. In hypothesis testing, we need to determine the point at which Z is "too large." That point is called the *critical value of Z*.

Here we know that $\sigma = 15$ and $n = 108$. Therefore, under the null hypothesis (i.e., when $\mu = 130$), the distribution of the sample means is approximately normal with mean $\mu_{\overline{X}} = \mu = 130$ and standard deviation $\sigma_{\overline{X}} = \sigma/\sqrt{n} = 15/\sqrt{108} = 1.44$ (Figure 5.5).

Under H_0: $\mu = 130$, it is possible to observe any value of $\overline{X}$ displayed in Figure 5.5. However, we know that it is unlikely (i.e., the probability is small) that $\overline{X}$ will take on a value in the tails of the distribution. For example, it is very unlikely to observe values of $\overline{X}$ exceeding 132.88 (which is 2 standard deviations above the mean: $\mu_{\overline{X}} + 2\sigma_{\overline{X}}$). Recall from Chapter 4, $P(\overline{X} > 132.88) = P(Z > 2) = 1 - 0.9772 = 0.0228$. Therefore, if we observe a sample mean that exceeds 132.88 and we reject the null hypothesis in favor of the alternative, the probability that we are making a mistake in rejecting is only 2.28%. However, if we reject the null hypothesis for values $\overline{X}$ that exceed 131.44, the probability that we are making a mistake in rejecting H_0 is $P(\overline{X} > 131.44) = P(Z > 1) = 1 - 0.8413 = 0.1587$. We must decide what level of error (specifically, error of this type where we incorrectly reject the null hypothesis) we can tolerate in the analysis. A 15.87% probability of making this type of mistake may be too high.

In hypothesis testing, investigators select a *level of significance*, denoted α, which is defined as the probability of rejecting H_0 when H_0 is true. The level of significance is generally in the range of 0.01 to 0.10, though any value from 0 to 1.0 can be selected. Because the level of significance reflects the likelihood of drawing an erroneous conclusion (i.e., $\alpha = P(\text{reject } H_0 | H_0 \text{ true})$), small levels of significance are purposely selected.

Once a level of significance is selected, a *decision rule* is formulated. The decision rule is a formal statement of the criteria used to draw a conclusion in the hypothesis test. For example, suppose we select a level of significance of 5% (i.e., $\alpha = 0.05$, we allow a 5% chance of rejecting H_0 when H_0 is true). The critical value and decision rule are displayed graphically in

Figure 5.5 *Distribution of $\overline{X}$ under*
H_0: $\mu = 130\,(\sigma_{\overline{X}} = 1.44)$

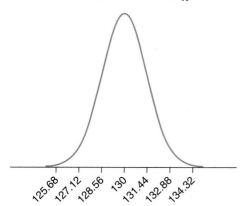

125.68 127.12 128.56 130 131.44 132.88 134.32

Figure 5.6 *Decision Rule for H_0: $\mu = 130$*
vs. H_1: $\mu > 130$, $\alpha = 0.05$

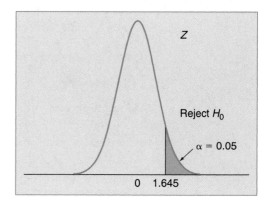

Figure 5.6. The region in the tail end of the curve (shaded) is called the *rejection region.*

The decision rule is given by

Reject H_0 if $Z \geq 1.645$

Do not reject H_0 if $Z < 1.645$

(5.13)

Once the decision rule is in place, we compute the value of the *test statistic*. The test statistic is given in (5.12) and is computed using the observed sample data. Suppose in Example 5.8, $\overline{X} = 135$. The test statistic is

$$Z = \frac{\overline{X} - \mu_0}{\sigma/\sqrt{n}} = \frac{135 - 130}{15/\sqrt{108}} = 3.46$$

The final step in the test of hypothesis is to compare the test statistic to the decision rule to draw a conclusion. In Example 5.8 the test statistic falls in the rejection region and therefore we reject H_0 because $3.46 \geq 1.645$. A final statement is made concerning our findings relative to the research or alternative hypothesis. Such a statement is as follows: We have significant evidence, $\alpha = 0.05$, that the mean systolic blood pressure for males aged 50 in 2004 has increased from 130. ■

Table 5.3 summarizes the steps involved in the test of hypothesis procedure. The steps are displayed with reference to a test about the population mean μ. The same steps will be used to test hypotheses concerning other parameters.

Table 5.4 contains critical values of Z for lower-, upper-, and two-tailed tests. The general form of the hypotheses are presented along with the form of the decision rule for each type of test.

Table 5.5 summarizes three different formulas for test statistics in tests concerning the population mean μ and the conditions under which each should be applied.

Table 5.3 *Tests of Hypothesis Concerning μ*

Step	Example
1. Set up hypotheses.	$H_0: \mu = \mu_0$
	$H_1: \mu > \mu_0$*
Select level of significance.	$\alpha = 0.05$
2. Select appropriate test statistic.	$Z = \dfrac{\overline{X} - \mu_0}{\sigma/\sqrt{n}}$
3. Generate decision rule.	Reject H_0 if $Z \geq Z_{1-\alpha}$*
	Do not reject H_0 if $Z < Z_{1-\alpha}$
4. Compute the value of the test statistic.	
5. Draw a conclusion about H_0 by comparing the test statistic (4) to the decision rule (3). Report findings relative to H_1. If we reject H_0, we compute the exact level of significance, called the p value.	

* *Note:* In hypothesis testing, there are three types of alternative or research hypotheses. The test of hypothesis presented in Example 5.8 is called an *upper-tailed test* due to the direction (i.e., >) specified in the alternative hypothesis and the location of the rejection region (see Figure 5.6). The other two types of tests are called *lower-tailed tests* and *two-tailed tests*. In lower-tailed tests, the alternative hypothesis is of the form: $H_1: \mu < \mu_0$; and in two-tailed tests, the alternative hypothesis is of the form: $H_1: \mu \neq \mu_0$. In lower-tailed tests, the research hypothesis indicates a decrease in the mean value; in two-tailed tests, the research hypothesis indicates a difference (either an increase or a decrease) in the mean value.

Notice in Case 3, where σ is unknown and the sample size is small ($n < 30$), the test statistic no longer follows a normal distribution (Z). Instead, the test statistic follows a t distribution. In this case, the critical value(s) must be selected from the t table (Table B.3). Use of this test statistic and selection of appropriate critical value(s) are illustrated in Example 5.9 in the next section.

In every test of hypothesis we draw a conclusion to either reject H_0 (in favor of H_1) or not to reject H_0. When the test indicates a rejection of H_0, a very strong statement is made. However, when the test does not indicate a rejection of H_0, a very weak statement is made, for the following reasons.

Suppose the true value of the population mean (or any parameter under investigation) is known (Table 5.6). If in reality H_0 is true (first row of Table 5.6) and we do not reject H_0 based on our test of hypothesis, a correct decision is made. However, if in reality H_0 is true and we reject H_0 based on our test of hypothesis, an error is committed. This particular type of error is called a Type I error, and its probability is given by $\alpha = P(\text{Type I error}) = P(\text{Reject } H_0 | H_0$ true$)$. If in reality H_0 is false (second row of Table 5.6) and we do not reject H_0 based on our test of hypothesis, an error is committed. This particular type of error is called a Type II error, and its probability is given by $\beta = P(\text{Type II error}) = P(\text{Do not reject } H_0 | H_0$ false$)$. If in reality H_0 is false and we reject H_0 based on our test of hypothesis, a correct decision is made.

Table 5.4 *Important Values for Statistical Inference*

Lower-Tailed Tests			
Hypotheses	α	Z_α	Decision Rule
$H_0: \mu = \mu_0$	.0001	-3.719	Reject H_0 if $Z \leq Z_\alpha$
$H_1: \mu < \mu_0$	.001	-3.090	
	.005	-2.576	
	.010	-2.326	
	.025	-1.960	
	.050	-1.645	
	.100	-1.282	

Upper-Tailed Tests			
Hypotheses	α	$Z_{1-\alpha}$	Decision Rule
$H_0: \mu = \mu_0$	.0001	3.719	Reject H_0 if $Z \geq Z_{1-\alpha}$
$H_1: \mu > \mu_0$	.001	3.090	
	.005	2.576	
	.010	2.326	
	.025	1.960	
	.050	1.645	
	.100	1.282	

Two-Tailed Tests			
Hypotheses	α	$Z_{1-\alpha/2}$	Decision Rule
$H_0: \mu = \mu_0$	.0001	3.819	Reject H_0 if $Z \leq -Z_{1-(\alpha/2)}$ or
$H_1: \mu \neq \mu_0$	.001	3.291	if $Z \geq Z_{1-(\alpha/2)}$
	.010	2.576	
	.050	1.960	
	.100	1.645	
	.200	1.282	

Because the true state of the parameter is unknown in practice, we never know which of the four cells of Table 5.6 we fall into in a given application. However, because we select a level of significance, α, a priori, we know that the probability of committing a Type I error is small. When our test of hypothesis indicates a rejection of H_0 (second column of Table 5.6), we can be confident that a correct decision has been made, since the chance of committing a Type I error is small. When our test of hypothesis does not indicate a

Table 5.5 *Test Statistics for Tests Concerning* μ

Assumptions: Normal distribution or large sample ($n \geq 30$); simple
random sample

Case 1. σ known	$Z = \dfrac{\overline{X} - \mu_0}{\sigma/\sqrt{n}}$
Case 2. σ unknown and n ≥ 30	$Z = \dfrac{\overline{X} - \mu_0}{s/\sqrt{n}}$
Case 3. σ unknown and n < 30	$t = \dfrac{\overline{X} - \mu_0}{s/\sqrt{n}}, \quad \mathrm{df} = n - 1$

Table 5.6 *Errors in Tests of Hypothesis*

	Statistical Test	
True State	Do Not Reject H_0	Reject H_0
H_0 true	Correct decision	Type I error
H_0 false	Type II error	Correct decision

rejection of H_0 (first column of Table 5.6), we cannot be so confident that a correct decision has been made, because the chance of committing a Type II error is unknown (see Section 5.2.3 for further discussion). In such a case, a weaker statement is made at the conclusion of the test (see Example 5.9).

5.2.2 Hypothesis Tests About μ

The following are examples of the test of hypothesis technique. We cover lower-tailed and two-tailed tests.

EXAMPLE 5.9

Testing Mean Cholesterol Level Against Referent

Suppose that the mean cholesterol level for males aged 50 is 241. An investigator wishes to examine whether cholesterol levels are significantly reduced by modifying diet only slightly. A random sample of 12 patients agrees to participate in the study and follow the modified diet for 3 months. After 3 months, their cholesterol levels are measured and summary statistics are produced on the $n = 12$ subjects. The mean cholesterol level in the sample is 235 with a standard deviation of 12.5. Based on the data, is there statistical evidence that the modified diet reduces cholesterol? Run the appropriate test using a 5% level of significance. We will now run the test of hypothesis using the five-step procedure outlined in Table 5.3.

1. Set up hypotheses.
    In this example, we wish to test whether the diet *reduces* cholesterol levels. The mean cholesterol level (without modifying diet) is taken to be

241 for males aged 50. The null hypothesis reflects the "no change" or "no effect" situation, and the alternative or research hypothesis reflects the situation where the diet is effective in reducing or lowering cholesterol levels. This is an example of a lower-tailed test.

$$H_0: \mu = 241$$
$$H_1: \mu < 241, \quad \alpha = 0.05$$

2. Select the appropriate test statistic.

   The appropriate test statistic is selected from Table 5.5 based on the information available. In this example, we have a sample of size $n = 12$ and we do not know the population standard deviation σ (i.e., the standard deviation in cholesterol levels among all males aged 50). Here we have the sample standard deviation $s = 12.5$ (computed on the data collected from the 12 participants).

   Because σ is unknown and the sample is small ($n < 30$), this is an example of Case 3, and the test statistic is

   $$t = \frac{\overline{X} - \mu_0}{s/\sqrt{n}}$$

3. Decision rule.

   The decision rule for any hypothesis testing application depends on three factors: (1) whether the test is an upper-, lower-, or two-tailed test; (2) the level of significance α; and (3) the form of the test statistic (e.g., Z or t as determined in step 2). Here we have a lower-tailed test, $\alpha = 0.05$, and we are using a t statistic. The critical value is found in Table B.3 (as opposed to Table 5.4, which contains critical values for Z). In order to determine the appropriate value, we first compute the degrees of freedom: $df = 12 - 1 = 11$. In Table B.3, we locate the 5% level of significance (across the top row labeled "one-sided α") and read down to 11 degrees of freedom. The appropriate critical value is $t = 1.796$. We can now formulate the decision rule:

   Reject H_0 if $t \leq -1.796$
   Do not reject H_0 if $t > -1.796$

4. Test statistic.

   In this step we substitute our sample data into the appropriate formula for the test statistic, selected in step 2. Recall that the value for μ_0 is 241 (as specified in H_0):

   $$t = \frac{\overline{X} - \mu_0}{s/\sqrt{n}} = \frac{235 - 241}{12.5/\sqrt{12}} = -1.66$$

5. Conclusion.

   In the final step, we draw a conclusion by comparing the test statistic (computed in step 4) to the decision rule (displayed in step 3). We do not reject H_0 because $-1.66 > -1.796$. We do not have significant evidence,

$\alpha = 0.05$, to show that the diet reduces cholesterol levels. Notice in our concluding statement that we do not state an "acceptance" of H_0 since it is possible that a Type II error has been committed. It is possible that the diet truly does lower cholesterol levels and it may be the case that our sample size is too small to detect the reduction. This issue will be discussed in the next section, where we consider experimental design issues related to tests of hypothesis. ■

EXAMPLE 5.10

Testing Mean Starting Salary Against Referent

In a managed care organization, the mean starting salary for males in entry-level, nonclinical positions is $29,500. We want to determine whether the mean starting salary for females in similar entry-level, nonclinical positions is different from $29,500. In order to make this assessment, we randomly select 10 females whose job titles and responsibilities fit our criteria, and we record their starting salaries (in $1000s). The data are

32	27	31	27	26
26	30	22	25	36

Is the mean starting salary for females significantly different from $29,500? Use $\alpha = 0.05$.

1. Set up hypotheses.

   In this example, we wish to test whether the mean starting salary for females is *different* from the reported starting salary for males in similar positions. This is an example of a two-sided test.

   $H_0: \mu = 29.5$
   $H_1: \mu \neq 29.5, \qquad \alpha = 0.05$

2. Select the appropriate test statistic.

   The appropriate test statistic is selected from Table 5.5 based on the information available. In this example, we have a sample of size $n = 10$ and we do not know the population standard deviation σ (i.e., the standard deviation in starting salaries for all females in entry-level, nonclinical positions).

   Because σ is unknown and the sample is small ($n < 30$), this is an example of Case 3, and the test statistic is

   $$t = \frac{\overline{X} - \mu_0}{s/\sqrt{n}}$$

3. Decision rule.

   Here we have a two-tailed test, $\alpha = 0.05$, and we are using a t statistic. The critical value is found in Table B.3 (as opposed to Table 5.4, which contains critical values for Z). In order to determine the appropriate value, we first compute the degrees of freedom: $df = 10 - 1 = 9$. In Table B.3, we locate the 5% level of significance (across the top row labeled "two-sided α") and read down to 9 degrees of freedom. The appropriate critical value is $t = 2.262$. We can now formulate the decision rule:

Reject H_0 if $t \leq -2.262$ or if $t \geq 2.262$
Do not reject H_0 if $-2.262 < t < 2.262$

4. Test statistic.

In this step we substitute our sample data into the appropriate formula for the test statistic, selected in step 2. In this example we are given raw scores as opposed to summary statistics, so we must first compute summary statistics using the formulas presented in Chapter 2.

X	X^2
32	1024
27	729
31	961
27	729
26	676
26	676
30	900
22	484
25	625
36	1296
$\sum X = 282$	$\sum X^2 = 8100$

$$\overline{X} = \frac{\sum X}{n} = \frac{282}{10} = 28.2$$

$$s^2 = \frac{\sum X^2 - \dfrac{(\sum X)^2}{n}}{n-1} = \frac{8100 - \dfrac{(282)^2}{10}}{9}$$

$$= 16.4, \qquad s = \sqrt{16.4} = 4.0$$

The summary statistics are: $n = 10$, $\overline{X} = 28.2$, $s = 4.0$.

We can now compute the test statistic:

$$t = \frac{\overline{X} - \mu_0}{s/\sqrt{n}} = \frac{28.2 - 29.5}{4.0/\sqrt{10}} = -1.02$$

5. Conclusion.

In the final step, we draw a conclusion by comparing the test statistic (computed in step 4) to the decision rule (displayed in step 3). Do not reject H_0, because $-2.262 < -1.02 < 2.262$. We do not have significant evidence, $\alpha = 0.05$, to show that the mean starting salary for females is significantly different from $29,500. Are the starting salaries the same? ■

SAS EXAMPLE 5.10 Testing Mean Starting Salary Against Referent Using SAS

The following output was generated using SAS Proc Means with an option to conduct a test of hypothesis. A brief interpretation appears after the output.

SAS will conduct a one-sample test of hypothesis in its Means procedure. However, in the one-sample test SAS assumes that the test of interest is H_0: $\mu = 0$ versus H_1: $\mu \neq 0$. In this application (and in most others), we are not interested in testing if the mean of the analytic variable is zero. We want to test if the mean starting salary among females is 29.5 (in $1000s, which is equal to $29,500). In order to use SAS to test the desired hypotheses, we create a new variable, which we call TESTSTAT; it is simply our original analytic variable (i.e., starting salary) minus 29.5. Using the variable TESTSTAT, we can use SAS to test the desired hypotheses. In particular, if the mean of TESTSTAT is significantly different from zero, then we can conclude that the mean salary is significantly different from 29.5 (since TESTSTAT is simply equivalent to salary -29.5). Conversely, if the mean of TESTSTAT is not significantly different from zero, then we can conclude that the mean salary is not significantly different from 29.5.

SAS Output for Example 5.10

```
                        The MEANS Procedure
Variable Label                     N         Mean      Std Dev   Std Error
---------------------------------------------------------------------------
salary    Annual Salary in $000s   10   28.2000000    4.0496913   1.2806248
teststat                           10   -1.3000000    4.0496913   1.2806248

          Variable    Label                        t Value    Pr > |t|
          -----------------------------------------------------------------
          salary      Annual Salary in $000s        22.02      <.0001
          teststat                                  -1.02      0.3366
          -----------------------------------------------------------------
```

Interpretation of SAS Output for Example 5.10

To illustrate the default procedure and the necessary modifications, we asked SAS to analyze both the original data (variable named salary) and our created variable (teststat, which is equal to salary -29.5). SAS first provides the number of observations ($n = 10$) and then summary statistics on each analysis variable. The mean salary is 28.2, and the mean of teststat is -1.3 (equal to $28.2 - 29.5$). The standard deviations of salary and teststat are identical, as subtracting a constant (e.g., 29.5) from a variable does not influence its standard deviation or variance. The standard error (i.e., $s/\sqrt{n}$) of both salary and teststat is 1.2806. The test statistic for salary (computed by taking the ratio of the mean—minus zero, as SAS assumes $\mu_0 = 0$—to the standard error) is 22.02. However, this is the test statistic for testing if the mean salary is zero. The test of interest is whether the mean salary is 29.5 or not. The test statistic for this test is $t = -1.0151$. If we were conducting this test of hypothesis by hand (as we did in Example 5.10), then we would compare the value of the test statistic to an appropriate critical value from the *t* distribution table (Table B.3) with 9 degrees of freedom to draw a conclusion. SAS instead produces a value, called a *p value,* that allows us to draw a conclusion. The *p* value for this test is 0.3366. The *p* value is defined as the exact level of significance of the data. In other words, the *p* value is the smallest level of significance that would lead to rejection of the null hypothesis. In this example, we could reject H_0 only if the level of significance was 0.3366 (or larger). Any level of significance smaller than 0.3366 would not lead to rejection. We generally use a 0.05 level of significance; therefore, we would not reject H_0. ■

INTERPRETING *p* VALUES FROM SAS

The following rule can be used to draw conclusions in tests of hypotheses based on *p* values:

$$\text{Reject } H_0 \text{ if } p \text{ value} \leq \alpha \qquad (5.14)$$

where α is the level of significance selected for the test (e.g., 0.05, 0.01).

NOTE: It is important that the level of significance, α, is selected prior to the implementation of the test. It is inappropriate to select a level of significance so as to ensure a particular conclusion in a test (i.e., based on the observed *p* value).

Consider the following example illustrating the computation of *p* values in tests of hypotheses.

Suppose we wish to perform the following two-sided test at $\alpha = 0.05$:

$$H_0: \mu = 30$$

$$H_1: \mu \neq 30$$

Suppose that $n = 25$, $\overline{X} = 35$, and $s = 10$. We perform the test using SAS, and SAS produces a test statistic $t = 2.50$ and *p* value $= 0.0194$. The graphical display of two-sided *p* value follows. The *p* value is the probability of observing a value as extreme or more extreme than the observed test statistic; that is, $P(t > 2.50 \text{ or } t < -2.50)$.

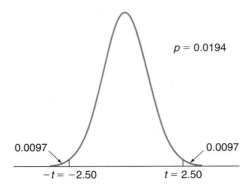

If this test were conducted by hand, critical values would be selected corresponding to the preselected level of significance α. The critical values for a two-sided test with $\alpha = 0.05$ are shown next. Because the test statistic $t = 2.50$ exceeds the critical value $t = 2.064$ (see figure), we would reject H_0 in favor of H_1 and conclude that the mean is significantly different from 30.

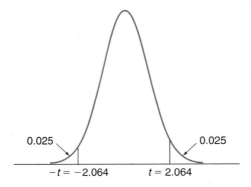

The same conclusion is reached by comparing the p value to the level of significance using rule (5.14). Because $p = 0.0194 \leq \alpha = 0.05$, we reject H_0. (Note: For the same test, if $\alpha = 0.01$, we would not reject H_0.)

Notice that when comparing the test statistic to a critical value, actual t statistics are compared, whereas when comparing the p value to the level of significance, areas in the tails of the distribution are compared.

For the one-sample tests of means (H_0: $\mu = \mu_0$ versus H_1: $\mu > \mu_0$, H_1: $\mu < \mu_0$, or H_1: $\mu \neq \mu_0$), SAS provides two-sided p values. If a two-sided test is desired, then the rule given in (5.14) is applied directly to draw a conclusion. However, if a one-sided test is desired, the rule must be modified as follows:

$$\text{Reject } H_0 \text{ if } (p \text{ value}/2) \leq \alpha \qquad (5.15)$$

where α is the one-sided level of significance selected by the investigator

For example, suppose we wish to perform the following upper-tailed test at $\alpha = 0.05$:

$$H_0: \mu = 30$$

$$H_1: \mu > 30$$

SAS produces $p = 0.0197$ (as shown). Using rule (5.15) (i.e., Reject H_0 if $(p \text{ value}/2) \leq \alpha$), we would reject H_0 since $0.0097 \leq 0.05$. This rule should only be applied if $\overline{X} > 30$.

COMPUTING p VALUES BY HAND

EXAMPLE 5.8 REVISITED: p-Value in Test for Mean Systolic Blood Pressure

In Example 5.8, we ran the following test at a 5% level of significance.

$$H_0: \mu = 130$$

$$H_1: \mu > 130$$

The decision rule we used was given by

Reject H_0 if $Z \geq 1.645$
Do not reject H_0 if $Z < 1.645$

We computed a test statistic of $Z = \dfrac{\overline{X} - \mu}{\sigma/\sqrt{n}} = \dfrac{135 - 130}{15/\sqrt{108}} = 3.46$ and rejected H_0. In this example we could have selected a smaller level of significance and still reached the same conclusion. The following table displays various levels of significance, α, and their associated critical values for upper-tailed Z tests. The boldface entry is the level of significance, 0.05, used in the test.

α	Z
0.0001	3.791
0.001	3.090
0.005	2.576
0.01	2.326
0.025	1.960
0.05	**1.645**

We ran the test at $\alpha = 0.05$ and rejected H_0 because $3.46 \geq 1.645$. The p value is defined as the smallest level of significance, α, where we still reject H_0. Using the table, we examine each level of significance smaller than 0.05 to determine if we could still reject H_0 at that level. For example, at $\alpha = 0.025$ we still reject H_0 because $3.46 \geq 1.960$. At $\alpha = 0.01$ we also reject H_0 because $3.46 \geq 2.326$, at $\alpha = 0.005$ we reject H_0 because $3.46 \geq 2.576$, at $\alpha = 0.001$ we reject H_0 because $3.46 \geq 3.090$, but at $\alpha = 0.0001$ we cannot reject H_0 because $3.46 < 3.791$. Therefore, the smallest level of significance where we still reject H_0 is 0.001. The significance of this data, or the p value, is 0.001. If we run this analysis using SAS, SAS produces an exact p value. Our hand computations produce only an approximate value (in fact, the exact p value is between 0.0001 and 0.001). To reflect the idea that this is an approximate p value, sometimes the value is reported as $p < 0.001$. ■

5.2.3 Power and Sample Size Determination

There are two types of errors that can be committed in hypothesis testing, a Type I error (i.e., reject H_0 when H_0 is true), or a Type II error (i.e., do not reject H_0 when H_0 is false). In Section 5.2.1 we introduced $\alpha = P(\text{Type I error}) = P(\text{Reject } H_0 | H_0 \text{ true})$ and $\beta = P(\text{Type II error}) = P(\text{Do not reject } H_0 | H_0 \text{ false})$. In each test of hypothesis, we specify α, purposely choosing small values (e.g., 0.01, 0.02, 0.05, or 0.10) so that the $P(\text{Type I error})$ is controlled. The probability of a Type II error, β, is difficult to control because it depends on several factors. In fact, one of the factors on which β depends is α: β decreases as α increases. Therefore, one must weigh the choice of a lower $\beta = P(\text{Type II error})$, which is desirable, against a higher level of significance, which is undesirable.

In hypothesis testing, we are concerned with the *power* of a test, defined as $1 - \beta$. The power of a test is defined as its ability to "detect" or reject a false null hypothesis. Specifically, power is defined as

$$\text{Power} = 1 - \beta = P(\text{Reject } H_0 | H_0 \text{ false}) \tag{5.16}$$

As power increases, β decreases, resulting in a better test. Power (and therefore β) is a complicated function of three components:

1. $n =$ sample size
2. $\alpha =$ level of significance $= P(\text{Type I error})$
3. ES $=$ the effect size $=$ the standardized difference in means specified under H_0 and H_1

The power of a particular test is higher (or better) with a larger sample size, a larger level of significance, and a larger effect size. In this section, we introduce the concept of statistical power as it applies to the one-sample test of hypothesis about μ and present a simple application.

Suppose we are interested in the following test.

$$H_0\text{: } \mu = 100$$
$$H_1\text{: } \mu > 100, \qquad \alpha = 0.05$$

To conduct the test, we select a random sample of subjects from the population of interest and analyze summary statistics, in particular $\overline{X}$. Under the null hypothesis (i.e., if $\mu = 100$), the distribution of sample means is as follows. Assume that $\sigma_{\overline{X}} = 6$.

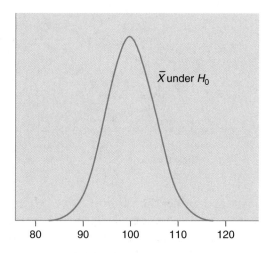

Suppose we want to determine the power of the test if the true mean is 110. The following displays the distributions of the sample mean under the null and alternative hypotheses.

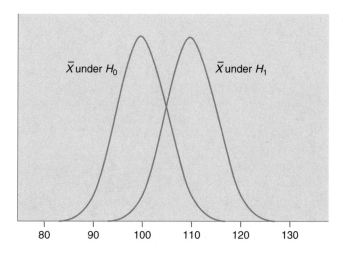

We now add $\alpha = P\,(\text{Type I error}) = P(\text{Reject } H_0 | H_0 \text{ true})$, the corresponding critical value, $\beta = P\,(\text{Type II error}) = P(\text{Do not reject } H_0 | H_0 \text{ false})$, and power $= 1 - \beta = P\,(\text{Reject } H_0 | H_0 \text{ false})$ to the figure, showing the distributions of the sample mean under $H_0\,(\mu = 100)$ and $H_1\,(\mu = 110)$.

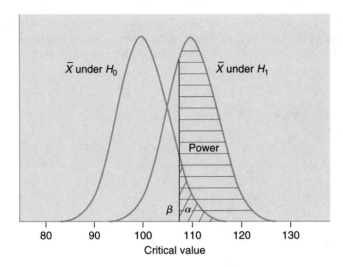

In the figure, α is the area under the leftmost curve (H_0 true) where we reject H_0, β is the area under the rightmost curve (H_1 true) where we do not reject H_0, and power is $1 - \beta$, or the remaining area under the rightmost curve (H_1 true).

Before we provide the formulas to compute the power of the test, we investigate the impact of each component on power. What happens to the power if we increase our level of significance? If $\alpha = 0.05$, what happens to the power if we increase α to 0.10?

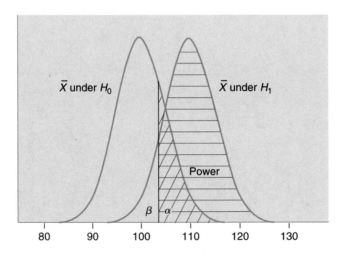

Notice that power increases when α increases. A better test is a more powerful test; however, it is unlikely that one would want to increase α so as to ensure a higher power in a test. Again, power is related to three components: sample size, α, and the effect size. The most appropriate component to modify is the sample size, if possible. A larger sample size ensures a more powerful test. We will illustrate this in the next example.

What is the impact of effect size on power? Suppose the mean under the alternative hypothesis is not 110 but instead 120. What happens to power? This situation is illustrated next, where we reset α to 0.05.

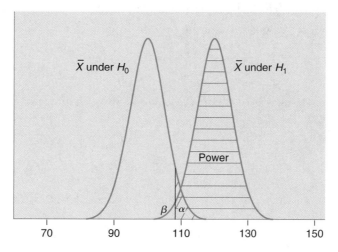

Notice that the power increases as the difference in means under H_0 and H_1 increases. This difference is called the *detectable difference*, and power increases as the alternative diverges from the null.

We now illustrate the calculation of power (or the determination of the area under the curve when H_1 is true). The calculations require the specification of the sample size, the level of significance, and the detectable difference and the use of the standard normal distribution table (Table B.2). The following formula is used to determine power of a two-sided test for μ.

$$\text{Power} = P\left(Z > Z_{1-(\alpha/2)} - \frac{|\mu_0 - \mu_1|}{\sigma/\sqrt{n}}\right) \qquad (5.17)$$

where μ_0 is the mean under H_0
μ_1 is the mean under H_1
σ is the standard deviation of the characteristic of interest
n is the sample size
$Z_{1-(\alpha/2)}$ is the Z value with lower-tail area $1 - \alpha/2$

The following formula is used to determine power of a one-sided test for μ.

$$\text{Power} = P\left(Z > Z_{1-\alpha} - \frac{|\mu_0 - \mu_1|}{\sigma/\sqrt{n}}\right) \qquad (5.18)$$

where μ_0 is the mean under H_0
μ_1 is the mean under H_1
σ is the standard deviation of the characteristic of interest
n is the sample size
$Z_{1-\alpha}$ is the Z value with lower-tail area $1 - \alpha$

EXAMPLE 5.11

Power in Test of Hypothesis for Mean

Suppose we wish to conduct the following test.

$$H_0: \mu = 80$$

$$H_1: \mu \neq 80, \qquad \alpha = 0.05$$

Find the power of the test if $\mu = 85$. Suppose that $n = 20$ and $\sigma = 9.5$. Using (5.17),

$$\text{Power} = P\left(Z > Z_{1-(\alpha/2)} - \frac{|\mu_0 - \mu_1|}{\sigma/\sqrt{n}}\right)$$

$$= P\left(Z > 1.96 - \frac{|80 - 85|}{9.5/\sqrt{20}}\right) = P(Z > 1.96 - 2.36)$$

$$= P(Z > -0.40) = 1 - 0.3446 = 0.6554$$

There is a 65% probability that this test will detect a difference of 5 units in means with $n = 20$ at $\alpha = 0.05$. The power is the same if $\mu = 75$.

What if the sample size is increased to $n = 50$?

$$\text{Power} = P\left(Z > Z_{1-(\alpha/2)} - \frac{|\mu_0 - \mu_1|}{\sigma/\sqrt{n}}\right)$$

$$= P\left(Z > 1.96 - \frac{|80 - 85|}{9.5/\sqrt{50}}\right) = P(Z > 1.96 - 3.72)$$

$$= P(Z > -1.76) = 1 - 0.0392 = 0.9608$$

There is a 96% probability that this test will detect a difference of 5 units in means (in either direction) with $n = 50$ at $\alpha = 0.05$.

Suppose we go back to $n = 20$ and consider a 3-point difference in means (i.e., $\mu_1 = 83$):

$$\text{Power} = P\left(Z > Z_{1-(\alpha/2)} - \frac{|\mu_0 - \mu_1|}{\sigma/\sqrt{n}}\right)$$

$$= P\left(Z > 1.96 - \frac{|80 - 83|}{9.5/\sqrt{20}}\right) = P(Z > 1.96 - 1.41)$$

$$= P(Z > 0.55) = 1 - 0.7088 = 0.2912$$

There is only a 29% probability that this test will detect a difference of 3 units in means (in either direction) with $n = 20$ at $\alpha = 0.05$. This is a small difference in means and with a small sample size we are not very likely to detect it. A larger sample would be required to ensure a higher probability of detecting such a difference. In the following we describe techniques for determining the sample size required to ensure a specified power. ■

In many applications, the number of subjects that can be sampled depends on financial and/or time constraints. In other cases, the investigators can choose a sample large enough to ensure a certain level of power. As described in Section 5.1.3, techniques from experimental design can be employed to determine the number of subjects required to achieve a certain level of power prior to mounting the study.

The sample size required to ensure a specific level of power in a *two-sided test* is

$$n = \left(\frac{Z_{1-(\alpha/2)} + Z_{1-\beta}}{ES} \right)^2 \tag{5.19}$$

where $Z_{1-(\alpha/2)}$ is the value from the standard normal distribution with
lower-tail area equal to $1 - \alpha/2$
$Z_{1-\beta}$ is the value from the standard normal distribution with
lower-tail area equal to $1 - \beta$
ES is the effect size, defined as the standardized difference in
means under the null and alternative hypotheses (5.20):

$$ES = \frac{|(\mu_1 - \mu_0)|}{\sigma} \tag{5.20}$$

where μ_0 is the mean specified in H_0
μ_1 is the mean specified in H_1
σ is the population standard deviation of the characteristic under
investigation.

The sample size required to ensure a specific level of power in a *one-sided test* is

$$n = \left(\frac{Z_{1-\alpha} + Z_{1-\beta}}{ES} \right)^2 \tag{5.21}$$

where $Z_{1-\alpha}$ is the value from the standard normal distribution with
lower-tail area equal to $1 - \alpha$
$Z_{1-\beta}$ is the value from the standard normal distribution with
lower-tail area equal to $1 - \beta$
ES is the effect size (5.20)

To implement formulas (5.19) and (5.21) to compute the sample size required to ensure a certain level of power to detect a specified difference in means, several inputs are required. First, we must specify the level of significance, α. This is usually straightforward because $\alpha = 0.05$ is considered standard. Second, we must specify $\beta = P(\text{Type II error})$. In many experimental design applications, β is set to 0.20, which reflects 80% power. With $\beta = 0.20$, there is an 80% chance of rejecting a false null hypothesis relative to a specific effect size. (In some instances, β is set to 0.10, which reflects 90% power.) The third input, the effect size, is the most difficult to specify. The effect size reflects the magnitude of a clinically important difference in means. The effect size is best determined by an expert in the substantive area under investigation. In order to compute the effect size, we also need to quantify the variation in the characteristic under investigation (σ). If no such value exists, the same options outlined in Section 5.1.3 can be used to determine a reasonable approximation. The following example illustrates the computations.

EXAMPLE 5.14   **Power in Test of Hypothesis for Mean**

Suppose we wish to conduct the following two-sided test at a 5% level of significance:

$$H_0: \mu = 100$$

$$H_1: \mu \neq 100$$

Suppose that a difference of 5 units in the mean score is considered a clinically meaningful difference. If the true mean is less than 95 or greater than 105, we do not want to fail to reject the null hypothesis. How many subjects would be required to ensure that the probability of detecting a 5-unit difference is 80% (i.e., power = 0.80)? For this example, suppose we know that $\sigma = 9.5$.

Because we wish to conduct a two-sided test, formula (5.19) is appropriate. First, we compute the effect size (5.20) by substituting the mean specified in H_0 ($\mu_0 = 100$), the mean we wish to detect when H_0 is false (here we could use either $\mu_1 = 95$ or $\mu_1 = 105$, both produce the same result), and the standard deviation $\sigma = 9.5$:

$$ES = \frac{|(105 - 100)|}{9.5} = 0.526$$

We now substitute the effect size into formula (5.19):

$$n = \left(\frac{Z_{0.975} + Z_{0.80}}{0.526} \right)^2$$

We use the standard normal distribution table (Table B.2) to determine $Z_{0.975}$ and $Z_{0.80}$. By definition, $Z_{0.975}$ is the Z value that holds 0.975 below it (or 0.025 above it, in the upper tail) and $Z_{0.80}$ is the Z value that holds 0.80 below it in the standard normal distribution, shown in the next figure:

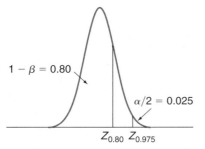

Using Table B.2 and the techniques we described in Chapter 3, we determine $Z_{0.975} = 1.96$ and $Z_{0.80} = 0.84$. We now substitute these values:

$$n = \left(\frac{1.96 + 0.84}{0.526}\right)^2 = (5.323)^2 = 28.33$$

A sample of size 29 (again, we always round up to the next integer) will ensure that power = 0.80 (or 80%, $\beta = 0.20$) to detect a 5-point difference in means. If the true mean is at least 5 units different from 100, there is an 80% chance that the test will lead to rejection of the null hypothesis H_0: $\mu = 100$. ■

How many subjects would be required to ensure a power of 80% to detect a difference of 3 units? This and other scenarios can be investigated by substituting into the formulas just shown or by using SAS.

SAS EXAMPLE 5.12 Power in Test of Hypothesis for Mean Using SAS

The following output was generated using SAS to determine the sample size required to ensure a specified power (we considered scenarios with 80% and 90% power) and differences in means of 5 and 3 units. A brief interpretation appears after the output.

SAS Output for Example 5.14

Obs	alpha	beta	mu0	mu1	sigma	power	es	n_2	n_1
1	0.05	0.2	100	105	9.5	0.8	0.52632	29	23
2	0.05	0.1	100	105	9.5	0.9	0.52632	38	31
3	0.05	0.2	100	103	9.5	0.8	0.31579	79	62
4	0.05	0.1	100	103	9.5	0.9	0.31579	106	86

Interpretation of SAS Output for Example 5.12

There is no SAS procedure specifically designed to determine the number of subjects required to detect a specific effect in the mean of a population.

However, similar to the approach taken in SAS Example 5.7, SAS can be used to program appropriate formulas. Once the formulas are implemented, users can evaluate different scenarios easily. In the example shown here, four scenarios are considered (denoted Obs 1–4, respectively). Scenario 1 corresponds to the situation presented in Example 5.12. Five variables are input into the program; the level of significance (alpha), the power (power), the mean under the null hypothesis (mu0), the mean under the alternative hypothesis (mu1), and the standard deviation (sigma). Several variables are created in the program and the values of all variables are printed in the output. A description of the variables and an interpretation of results follows.

In scenario 1 (Obs = 1), the level of significance (alpha) is set at 0.05 (5%) and the power was specified at 0.80 (80%); the probability of Type II error, β, is computed to be 0.20 (20%); the mean under the null hypothesis (mu0) was specified as 100; the mean under the alternative hypothesis was specified as 105 (mu1); and the standard deviation (sigma) was specified at 9.5. The effect size (es) was computed by dividing the absolute value of the difference in means under the null and alternative hypothesis by the standard deviation. In scenario 1 (Obs = 1), the effect size is 0.52632. Twenty-nine subjects (n_2) are required to ensure that the probability of detecting a 5-unit difference in means is 80% (i.e., power = 0.80), with a *two-sided* level of significance of 5%. Twenty-three subjects (n_1) are required to ensure that the probability of detecting a 5-unit difference in means is 80%, with a *one-sided* level of significance of 5%.

In scenario 2 (Obs = 2), we increase the power to 0.90 (90%). Thirty-eight subjects (n_2) are required to ensure that the probability of detecting a 5-unit difference in means is 90%, with a *two-sided* level of significance of 5%. Thirty-one subjects (n_1) are required to ensure that the probability of detecting a 5-unit difference in means is 90%, with a *one-sided* level of significance of 5%.

In scenario 3 (Obs = 3), we input a power of 0.80 (80%) but decrease the mean under the alternative hypothesis (mu1) to 103, which decreases the effect size (es) to 0.31579. Seventy-nine subjects (n_2) are required to ensure that the probability of detecting a 3-unit difference in means is 80%, with a *two-sided* test and 5% level of significance. Sixty-two subjects (n_1) are required to ensure that the probability of detecting a 3-unit difference in means is 80%, with a *one-sided* test and 5% level of significance.

In scenario 4 (Obs = 4) we input a power of 0.90 (90%) and consider the mean under the alternative hypothesis (mu1) as 83. One hundred six subjects (n_2) are required to ensure that the probability of detecting a 3-unit difference in means is 90%, with a *two-sided* test and 5% level of significance. Eighty-six subjects (n_1) are required to ensure that the probability of detecting a 3-unit difference in means is 90%, with a *one-sided* test and 5% level of significance.

Notice that more subjects are required to ensure a higher power. More subjects are also required to detect a smaller effect size. These results would be weighed against practical constraints to determine the most appropriate sample size for the application. ■

5.3 Key Formulas

Application	Notation/Formula	Description
Confidence interval estimate for μ	$\overline{X} \pm Z_{1-(\alpha/2)} \dfrac{s}{\sqrt{n}}$	See Table 5.2 for alternate formulas (find Z in Table 5.1)
Find n to estimate μ	$n = \left(\dfrac{Z_{1-(\alpha/2)}\sigma}{E} \right)^2$	Sample size to ensure margin of error E with confidence level reflected in Z
Test statistic for H_0: $\mu = \mu_0$	$Z = \dfrac{\overline{X} - \mu_0}{s/\sqrt{n}}$	See Table 5.3 for hypothesis testing procedure, Table 5.4 for critical values of Z, and Table 5.5 for alternate formulas
Find power for test of H_0: $\mu = \mu_0$	Power $=$ $P\left(Z > Z_{1-(\alpha/2)} - \dfrac{\lvert \mu_0 - \mu_1 \rvert}{\sigma/\sqrt{n}} \right)$	Power of two-sided test for mean
Find n to test H_0: $\mu = \mu_0$	$n = \left(\dfrac{Z_{1-(\alpha/2)}\, Z_{1-\beta}}{ES} \right)^2$, where $ES = \dfrac{\lvert \mu_1 - \mu_0 \rvert}{\sigma}$	Sample size to detect ES with power $= 1 - \beta$
p value	Reject H_0 if $p \leq \alpha$	Exact significance of the data

5.4 Statistical Computing

Following are the SAS programs used to generate the confidence interval, perform the test of hypothesis, and compute required sample sizes in the examples in Sections 5.2 and 5.3. The SAS procedures used and brief descriptions are noted in the header to each example. Notes are provided to the right of the SAS programs (in blue) for orientation purposes and are not part of the programs. The blank lines in the programs are solely to accommodate the notes. Blank lines and spaces can be used throughout SAS programs to enhance readability. A summary of the SAS procedures used in the examples is provided at the end of this section.

SAS Example 5.2 **Confidence Intervals for μ (Example 5.2)**

Age at Diagnosis of Hypertension

We randomly select 12 subjects with diagnosed hypertension and record the age at which they were diagnosed. Ages, measured in years, are recorded on each subject and are listed here. Generate a 95% confidence interval using SAS for the mean at which patients with hypertension are diagnosed.

32.8	40.0	41.0	42.0	45.5	47.0	48.5	50.0	51.0
52.0	54.0	59.2						

Program Code

`options ps=62 ls=80;`	Formats the output page to 62 lines in length and 80 columns in width
`data in;`	Beginning of Data Step
`  input age_dx;`	Inputs variable **age_dx**
`label age_dx='Age at diagnosis of hypertension';`	Attaches descriptive label to variable **age_dx**
`cards;`	Beginning of Raw Data section.
`32.8`	actual observations
`40.0`	
`41.0`	
`42.0`	
`45.5`	
`47.0`	
`48.5`	
`50.0`	
`51.0`	
`52.0`	
`54.0`	
`59.2`	
`run;`	
`proc means n mean std min max alpha=0.05 clm;`	Procedure call. **Proc Means generates summary** statistics for continuous variables. Certain statistics are requested along with a 95% confidence interval by the 'clm' option (see (iii) Interpretation)
`  var age_dx;`	Specification of variable **age_dx**
`run;`	End of procedure section. ■

SAS Example 5.7 **Determine the Number of Subjects Required to Generate a Confidence Interval for μ (Example 5.7)**

Time to Travel Between Hospital Departments

Compute the number of subjects required to generate a 95% confidence interval estimate of the mean time it takes patients to travel from one department to another on hospital grounds. We wish to estimate the mean within 5 minutes of the true value with 95% confidence. Assume $\sigma = 17$ (based on a pilot study).

Program Code

```
options ps=62 ls=80;
```
Formats the output page to 62 lines in length and 80 columns in width

```
data in;
  input c_level e sigma;
z=-probit((1-c_level)/2);
```
Beginning of Data Step
Inputs 3 variables **c_level**, *e* and **sigma**.
Determines the (positive) value from the standard normal distribution (z) with lower tail area (1-**c_level**)/2

```
tempn=(z*sigma/e)**2;
```
Creates a temporary variable, called **tempn**, determined by formula (5.8).

```
n=ceil(tempn);
```
Computes a variable *n* using the ceil function which computes the smallest integer greater than **tempn** (i.e., rounds up).

```
/* Input the following information (required)
  c_level: Confidence Level: range 0.0 to 1.0
          (e.g., 0.95),
  e: Margin of Error, and
  sigma: Standard Deviation */
cards;
0.95 1 17
0.95 2 17
0.95 3 17
0.95 4 17
0.95 5 17
0.95 10 17
run;
proc print;
  var c_level z sigma e n;
run;
```
Beginning of comment section

End of comment
Beginning of Raw Data section
  actual observations
  values of **c_level**, *e* and **sigma**
  on each line

Procedure call. Print to display input and computed variables.
End of procedure section. ■

SAS EXAMPLE 5.10 **Tests of Hypothesis About μ (Example 5.10)**

Starting Salaries

In a managed care organization, the mean starting salary for males in entry-level, nonclinical positions is $29,500. Starting salaries for a random sample of 10 females in similar positions (in $1000s) follow:

| 32 | 27 | 31 | 27 | 26 | 26 | 30 | 22 | 25 | 36 |

Is the mean starting salary for females significantly different from $29,500? Use a 5% level of significance.

Program Code

```
options ps=62 ls=80;
```
Formats the output page to 62 lines in length and 80 columns in width

```
data in;
  input salary;
```
Beginning of Data Step.

Inputs variable **salary.**

```
teststat=salary-29.5;
```
Creates a new variable, called **teststat,** by subtracting 29.5 (μ_0) from each **salary.**

```
label salary='Annual Salary in $000s';
```
Attaches a descriptive label to **salary.**

```
cards;
32
27
31
27
26
26
30
22
25
36
run;
```
Beginning of Raw Data section.

actual observations

```
proc means n mean std stderr t prt;
```
Procedure call. Proc Means generates summary statistics for continuous variables. Certain statistics are requested- *t* statistics and *p* values for conducting the tests of hypothesis (see (iii) Interpretation)

```
  var salary teststat;
```
Specification of variables.

```
run;
```
End of procedure section.   ■

SAS Example 5.12 Determine the Number of Subjects Required to Detect a Specific Effect Size in a Test of Hypothesis About μ (Example 5.12)

Sample Size Requirements

We wish to conduct the following test at the 5% level of significance:

$$H_0: \mu = 100 \quad \text{vs.} \quad H_1: \mu \neq 100$$

How many subjects would be required to ensure that the probability of detecting a 5- (or a 3-) unit difference is 80% (i.e., power = 0.80)? Also consider scenarios with 90% power and assume that $\sigma = 9.5$.

Program Code

```
options ps=62 ls=80;
```
Formats the output page to 62 lines in length and 80 columns in width

```
data in;
  input alpha power mu0 mu1 sigma;
```
Beginning of Data Step.

Inputs 5 variables **alpha, power, mu0, mu1** and **sigma.**

```
z_alpha2=probit(1-alpha/2);
```
Determines the value from the standard normal distribution with lower-tail area 1-alpha/2 (see $Z_{\alpha/2}$ above)

```
z_alpha1=probit(1-alpha);
```
Determines the value from the standard normal distribution with lower-tail area 1-alpha (see Z_α above)

```
beta=1-power;
```
Computes **beta.**

```
z_beta=probit(1-beta);
```
Determines the value from the standard normal distribution with lower-tail area I-**beta** (see Z_b above)

```
es=abs(mu1-mu0)/sigma;
```
Computes the effect size (**es**)

```
tempn_2=((z_alpha2+z_beta)/es)**2;
```
Creates a temporary variable, called **tempn_2**, determined by formula (5.14).

```
tempn_1=((z_alpha1+z_beta)/es)**2;
```
Creates a temporary variable, called **tempn_1**, determined by formula (5.16).

```
n_2=ceil(tempn_2);
```
Computes a variable **n_2** using the ceil function, which computes the smallest integer greater than **tempn_2** (i.e., rounds up).

```
n_1=ceil(tempn_1);
```
Computes a variable **n_1** using the ceil function, which computes the smallest integer greater than **tempn_1**.

```
/*
```
Beginning of comment section

```
 Input the following information (required)
 alpha: Level of Significance: range 0.0 to 1.0 (e.g., 0.05),
 power: Power: range 0.0 to 1.0 (e.g., 0.80), and
 mu0: Mean Under the Null Hypothesis,
 mu1: Mean Under the Alternative Hypothesis, and
 sigma: Standard Deviation
*
/
```
End of comment section

```
cards;
```
Beginning of Raw Data section

```
0.05 0.80 100 105 9.5
0.05 0.90 100 105 9.5
0.05 0.80 100 103 9.5
0.05 0.90 100 103 9.5
```
actual observations

```
run;
```

```
proc print;
 var alpha beta mu0 mu1 sigma power es
   n_2 n_1;
```
Procedure call. Print to display computed variables.

```
run;
```
End of procedure section. ■

5.4.1 Summary of SAS Procedures and Functions

The SAS Means procedure is used to generate confidence intervals for μ and to run test of hypothesis about μ. The specific options to generate a confidence interval and to perform a test of hypothesis are shown in italics in the following table. Users should refer to the examples in this section for complete descriptions of the procedure and specific options.

Procedure	Sample Procedure Call	Description
proc means	proc means *n mean std min max alpha = 0.05 clm*; var x;	generates a $100(1 - alpha)\%$ confidence interval for μ
proc means	proc means *n mean std stderr t prt*; var teststat;	conducts a one-sample test of hypothesis (H_0: $\mu = 0$ vs. H_1: $\mu \neq 0$)

The following SAS function was also used in the programs in this section to determine sample size requirements.

Function	Sample Call Statement	Description
probit	Z_alpha = probit(*alpha*);	determines the value from the standard normal distribution with lower-tail area equal to *alpha*

5.5 Analysis of Framingham Heart Study Data

The Framingham data set includes data collected from the original cohort. Participants contributed up to three examination cycles of data. Here we are analyzing data collected in the first examination cycle (called the period = 1 examination). Using SAS Proc Means, we generated 95% confidence intervals for key study variables.

Framingham Data Analysis—SAS Code

```
data in;
 set in.frmgham;
 if period=1;
run;

proc means n mean std stderr clm;
 var age sysbp diabp totchol bmi cigpday;
run;

proc means n mean std stderr clm;
 class bpmeds;
 var age sysbp diabp;
run;
```

Framingham Data Analysis—SAS Output

The MEANS Procedure

Variable	Label	N	Mean	Std Dev	Std Error
AGE	Age (years) at examination	4434	49.9258006	8.6769293	0.1303071
SYSBP	Systolic BP mmHg	4434	132.9077582	22.4215970	0.3367198
DIABP	Diastolic BP mmHg	4434	83.0835589	12.0559994	0.1810529
TOTCHOL	Serum Cholesterol mg/dL	4382	236.9842538	44.6510984	0.6745218
BMI	Body Mass Index (kr/(M*M)	4415	25.8461608	4.1018209	0.0617321
CIGPDAY	Cigarettes per day	4402	8.9663789	11.9317058	0.1798364

Variable	Label	Lower 95% CL for Mean	Upper 95% CL for Mean
AGE	Age (years) at examination	49.6703336	50.1812677
SYSBP	Systolic BP mmHg	132.2476192	133.5678972
DIABP	Diastolic BP mmHg	82.7286049	83.4385128
TOTCHOL	Serum Cholesterol mg/dL	235.6618501	238.3066575
BMI	Body Mass Index (kr/(M*M)	25.7251349	25.9671868
CIGPDAY	Cigarettes per day	8.6138092	9.3189487

In the Proc Means statement we requested certain statistics, specifically the sample size, the mean, the standard deviation, the standard error, and the 95% confidence interval. Notice that for more variable characteristics (e.g., total cholesterol) there is a larger standard error and consequently a wider confidence interval. In fact, none of the intervals are very wide here due to the large sample size.

In the following analysis, we requested the same statistics for age and for systolic and diastolic blood pressure, but we stratified (or produced separate analyses) for participants taking and not taking antihypertensive medications. Notice the difference in these characteristics for the distinct subgroups.

The MEANS Procedure

Anti-hypertensive meds Y/N	N Obs	Variable	Label	N	Mean
N	4229	AGE	Age (years) at examination	4229	49.6578387
		SYSBP	Systolic BP mmHg	4229	131.7563254
		DIABP	Diastolic BP mmHg	4229	82.6188224
Y	144	AGE	Age (years) at examination	144	56.2152778
		SYSBP	Systolic BP mmHg	144	165.1354167
		DIABP	Diastolic BP mmHg	144	96.6458333

Anti-hypertensive meds Y/N	N Obs	Variable	Label	Std Dev	Std Error
N	4229	AGE	Age (years) at examination	8.6137540	0.1324566
		SYSBP	Systolic BP mmHg	21.3693122	0.3286031
		DIABP	Diastolic BP mmHg	11.7098268	0.1800660
Y	144	AGE	Age (years) at examination	7.7542348	0.6461862
		SYSBP	Systolic BP mmHg	27.2035335	2.2669611
		DIABP	Diastolic BP mmHg	13.4437282	1.1203107

```
Anti-hypertensive    N                                              Lower 95%     Upper 95%
meds Y/N            Obs    Variable   Label                       CL for Mean   CL for Mean
-----------------------------------------------------------------------------------------------
N                  4229    AGE        Age (years) at examination   49.3981542    49.9175232
                           SYSBP      Systolic BP mmHg            131.1120906   132.4005601
                           DIABP      Diastolic BP mmHg            82.2657986    82.9718463

Y                   144    AGE        Age (years) at examination   54.9379665    57.4925891
                           SYSBP      Systolic BP mmHg            160.6543323   169.6165010
                           DIABP      Diastolic BP mmHg            94.4313239    98.8603427
-----------------------------------------------------------------------------------------------
```

Are the participants taking antihypertensive medications different from those not taking antihypertensive medications? In the next chapter, we will discuss methods for comparing means between groups.

5.6 Problems

1. A hospital researcher wishes to estimate the mean weight of full-term newborns. A random sample of 13 full-term newborns had a mean birthweight of 7.3 pounds with a standard deviation of 2.2 pounds. Compute a 95% confidence interval for the mean weight of all full-term newborns.

2. A newspaper article estimated that the mean cost for preventive dental care was $165 per year. They indicated that the estimate was based on a sample of 100 people and that the margin of error was 2.8. Assuming a 95% confidence level was used, calculate the value of the standard error.

3. A manufacturer of tobacco products plans to market a new brand of cigarette. The regulatory commission needs to know the mean tar content for the new brand. Analysis of a random sample of 25 of the new brand of cigarettes gives a mean tar content of 10.98 milligrams with a standard deviation of 0.604 milligram. Construct a 95% confidence interval for the mean tar content of the new brand.

4. Suppose we wish to design a study to investigate the effects of loud music on teenagers' ability to concentrate. We know from previous studies that the standard deviation of time to complete this task is 3.4 minutes. How many subjects would be required to ensure with 90% confidence that the generated estimate is within 1 minute of the true mean time required?

5. The administrators of an Emergency Room (ER) at a local hospital wish to estimate the mean number of overtime hours worked by employees during December, typically a very busy month. A random sample of 15 employees worked an average of 27 overtime hours with a standard deviation of 4.2 hours. Estimate the mean number of overtime hours

among all employees of the ER during December using a 90% confidence interval.

6. A marketing research firm has put together a health survey that it would like to administer to individuals in malls across the state. They need to estimate the mean time it takes to complete the survey so that recruiters can give potential participants an idea of their time commitments. The survey is administered to a random sample of 40 individuals. The mean time to complete the survey is 14.6 minutes with a standard deviation of 2.8 minutes. Estimate the mean time to complete the survey using a 95% confidence interval.

7. Suppose we need to estimate the mean blood glucose levels of diabetic patients following a specific treatment regimen. To be useful to clinicians, the estimate must be within 5 units with 95% confidence. If the standard deviation in blood glucose levels is known to be $\sigma = 15.6$, how many subjects will be required to produce such an estimate?

8. A local health maintenance organization (HMO) wishes to estimate the mean age of its members for marketing purposes.

   a. How large a sample would be required to generate an estimate that is within 5 years of the true mean with 90% confidence? Assume $\sigma = 8.6$.

   b. Suppose the budget allows the HMO to randomly sample 50 individuals for their marketing survey. Suppose the mean age in the sample is 58.2 with a standard deviation of 7.4. Estimate the mean age of all members using a 90% confidence interval.

9. Adherence to antiretroviral therapy is critical in patients with HIV disease. Suppose we wish to estimate the mean number of doses missed per month among HIV patients currently on antiretroviral therapy. A random sample of 25 patients on antiretroviral therapy is selected, and each patient reports the number of doses missed over the previous month. The mean number of doses missed is 4.7 with a standard deviation of 1.2. Estimate the mean number of doses missed per month among all HIV patients using a 95% confidence interval.

10. The following table appeared in a recent article summarizing a study investigating the relationship between body mass index (defined as the ratio of weight in kilograms to height in meters squared) and dietary habits:

Characteristics of Study Sample ($n = 100$)	Mean $\pm$ SE
Body mass index	25 ± 0.67
Total calories per day	1875 ± 36.4
Total fat grams per day	21 ± 0.31

The study involved subjects 18 years of age or older. Suppose we wish to design a study to investigate the relationship between body mass index and dietary habits among individuals aged 15–17 years.

a. How many subjects would be required to ensure that the estimate of the mean body mass index among 15–17-year-olds is within 2 units of the true mean with 95% confidence?

b. How many subjects would be required to ensure that the estimate of the mean number of total calories per day among 15–17-year-olds is within 50 units of the true mean with 95% confidence?

c. What is the smallest sample size that will ensure that both criteria (a) and (b) are satisfied?

11. A study is conducted to assess the extent to which patients who had coronary artery bypass surgery were maintaining their prescribed exercise programs. The following data reflect the numbers of times patients reported exercising over the previous month (4 weeks). For the purposes of this study, exercise was defined as moderate physical activity lasting at least 20 minutes in duration.

14	11	8	6	5	3
6	13	12	8	1	4

a. Construct a 95% confidence interval for the mean number of times patients exercise following surgery.

b. How many subjects would have been required in (a) to ensure that the margin of error was no more than 4 with 95% confidence?

12. A community health center wants to assess the alcohol consumption of its patients. A random sample of 100 patients is selected to participate in the assessment. One survey item measures the number of alcoholic drinks consumed per drinking occasion, and the mean number of alcoholic drinks consumed per occasion is 3.2 with a standard deviation of 2.6. Construct a 95% confidence interval for the mean number of alcoholic drinks consumed per occasion among all patients of the community health center.

13. The following data were presented summarizing the background characteristics of 50 participants in a research study evaluating HIV medications:

Patient Characteristics	Mean (SD)/%
Age (years)	37.6 (6.8)
% male	75.1%
Education (years)	13.6 (2.4)
Annual income ($000s)	31.4 (5.2)
CD4 cell count (cells/μl)	376 (94)

    a. Construct a 95% confidence interval estimate of the mean CD4 cell count.

    b. How many subjects would be required to estimate the mean CD4 cell count within 15 cells of the true value with 95% confidence?

14. A clinical study is conducted to estimate the proportion of patients who relapse following treatment for drug addiction. Twenty of 85 patients relapsed in the study. The mean number of days until relapse was 31 with a standard deviation of 3.8. Estimate the mean number of days to relapse among patients who do relapse using a 95% confidence interval.

15. We wish to design a study to estimate the mean weight loss (in pounds) among participants in a study of a newly developed weight-loss drug. How many subjects would be required to ensure that the estimate is within 2 pounds of the true mean with 95% confidence?

A published study of a similar drug reported the following:

	Number of Subjects	Mean	Variance	Median	Range
Weight loss (pounds)	125	53	14.4	6.1	0–15

16. A recent article appeared in a journal describing the health status of HIV-infected patients. Based on the Centers for Disease Control classification algorithm, patients were classified into one of three HIV disease stages: Asymptomatic, Symptomatic, or AIDS. Demographic characteristics of the study sample were provided in the article and are summarized here:

	Demographic Characteristics by HIV Disease Stage		
	Asymptomatic	*Symptomatic*	*AIDS*
n	40	49	71
Mean age (± SE)	37.7 ± 1.3	36.1 ± 1.2	36.6 ± 1.0
Male (%)	70	61	68

    a. Is there a relationship between disease stage and age? If yes, what is the nature of the relationship? Explain (briefly).

    b. Compute a 90% confidence interval for the mean age of Asymptomatic patients.

17. The following data were collected from a random sample of 10 asthmatic children enrolled in a research study and reflect the number of days each child missed school during the past 3 months:

    6   12   14   3   2   4   7   8   10   6

a. Compute the sample mean.

b. Compute the sample standard deviation.

c. Construct a 95% confidence interval for the mean number of days asthmatic children miss school during a 3-month period.

d. How many children would be required in (c) to ensure that the margin of error in the estimate is 1 day with 95% confidence? (Use part b to estimate σ.)

18. A dentist has recently started a new private practice and wants to estimate how long patients are waiting, on average, in the waiting room before their appointments. He randomly selects 50 patients for observation and records how many minutes each waits in minutes. The mean waiting time is 18 minutes with a standard deviation of 3.6 minutes.

a. Estimate the mean waiting time among all patients using a 95% confidence interval.

b. Based on the interval, could you conclude that the mean waiting time is significantly different from 20 minutes? Justify briefly.

19. Consider the following table:

Table 1 *Description of Study Sample* ($n = 85$)

Characteristic	Mean (SD) or %
Age (years)	38 (5.7)
Gender: % female	46%
Race: % white	86%
Education (years)	11.5 (4.3)

a. What is the standard error in the ages?

b. Construct a 90% confidence interval estimate for the mean age in the study population.

c. How many subjects would be required to ensure that the margin of error in the confidence interval of (b) is no more than 2 units with 90% confidence?

d. Suppose we wish to estimate the mean educational level using a 95% confidence interval. How many subjects would be required to ensure that the margin of error is no more than 1 year?

e. How many subjects should be enrolled in a study to satisfy the requirements of *both* (c) and (d)?

20. Currently, there are highly effective medications available for patients infected with HIV. However, in order to achieve the maximum benefits of therapy, patients must be extremely adherent with respect to pill taking. Suppose we wish to estimate medication adherence among HIV-infected persons new to medication therapy (initiated within the past 3 months). Adherence is measured as the percent of prescribed doses taken over the past month (e.g., 100% = perfect adherence—all doses taken; 50% = half of prescribed doses taken). A random sample of 75 HIV-infected patients new to medication therapy agree to participate in the study. Each reports their medication regimen and the doses they took over the past month. The mean percent adherence is 78% with a standard deviation of 7.2%. Construct a 95% confidence interval estimate of the mean percent adherence for all HIV-infected patients new to medication therapy.

21. The mean lung capacity for nonsmoking males aged 50 is 2 liters. An investigator wants to examine if the mean lung capacities are significantly lower among former smokers of similar backgrounds (i.e., males aged 50 who smoked in the past and are not currently smokers). A random sample of 60 former smokers is selected. Their lung capacities have a mean of 1.8 liters with a standard deviation of 0.27 liter. Run the appropriate test at a 5% level of significance.

22. Among private universities in the United States, the mean ratio of students to professors is 35.2 (i.e., 35.2 students for each professor) with a standard deviation of 8.8.
    a. What is the probability that in a random sample of 50 private universities that the mean student-to-professor ratio exceeds 38?
    b. Suppose a random sample of 50 universities is selected and the observed mean student-to-professor ratio is 38. Is there evidence that the reported mean ratio actually exceeds 35.2? Use $\alpha = 0.05$.

23. The recommended daily allowance (RDA) of iron for adult females under the age of 51 is 18 mg. We wish to test if females under age 51 are, on average, getting less than 18 mg. A random sample of 48 females between the ages of 18 and 50 is selected. The average iron intake is 16.4 mg with a standard deviation of 4.1 mg. Run the appropriate test at the 5% level of significance.

24. If a statistical test is performed and H_0 is rejected at $\alpha = 0.01$, will it also be rejected at $\alpha = 0.05$?

25. A journal article reported that the mean hospital stay following a particular surgical procedure in 2001 was 7.1 days. A researcher feels

that the mean hospital stay in 2002 should be less due to initiatives aimed at reducing health care costs. A random sample of 40 patients undergoing the same surgical procedure in 2002 had a mean length of stay of 6.85 days with a standard deviation of 7.01 days. Run the appropriate statistical test at $\alpha = 0.05$.

26. A consumer group is investigating a producer of diet meals to examine if its prepackaged meals actually contain the advertised 6 ounces of protein in each package. Based on the following data, is there any evidence that the meals do not contain the advertised amount of protein? Run the appropriate test at a 5% level of significance.

| 5.1 | 4.9 | 6.0 | 5.1 | 5.7 | 5.5 | 4.9 | 6.1 | 6.0 | 5.8 |
| 5.2 | 4.8 | 4.7 | 4.2 | 4.9 | 5.5 | 5.6 | 5.8 | 6.0 | 6.1 |

27. An article reported that patients under care for HIV have CD4 tests every 3 months, on average. A concern at Boston Medical Center is that there is a longer lag between tests. To test the concern, a random sample of 15 patients currently under care for HIV is selected and the time between their two most recent CD4 tests is recorded. The mean time between tests is 3.9 months with a standard deviation of 0.4 month. Run the appropriate test at a 5% level of significance.

28. A study reports that the mean systolic blood pressure for patients with a history of cardiovascular disease is 125 with a standard deviation of 15. We wish to design a study to evaluate an experimental medication for reducing blood pressure. How many subjects would be required to detect a 10-unit reduction in systolic blood pressure with 80% power? Assume that a two-sided test will be run at a 5% significance level.

29. We wish to test the hypothesis that the mean weight for females who are 5'8" is 140 pounds. Assuming $\sigma = 15$, using a 5% level of significance and with $n = 36$, find the power of the test if $\mu = 150$. (Use a two-sided test of hypothesis.)

30. We wish to run the following test: H_0: $\mu = 100$ versus H_1: $\mu \neq 100$ at $\alpha = 0.05$. If $\sigma = 10$, how large a sample would be required so that $\beta = 0.04$ if $\mu = 110$?

31. In a normal population with $\sigma = 5$, we wish to test H_0: $\mu = 12$ versus H_1: $\mu \neq 12$ at $\alpha = 0.05$. With a sample of 64 subjects, what is the probability of rejecting H_0 if $\mu = 14$? If $\mu = 9$?

32. Results of an industry survey in the computer software field find that the mean number of sick days taken by employees is 9.4 per year with a standard deviation of 2.7 per year. A local computer software company feels its employees take significantly fewer sick days per year. A random sample of 15 employees is selected from the local company

and attendance records are reviewed. The following data represent the numbers of sick days taken by these employees over the past year:

8   10   5   0   6   9   5   15
5   4   3   2   0   4   15

Run the appropriate test at a 5% level of significance.

33. An analysis is conducted to compare the mean GRE scores among seniors in a local university to the national average of 500. Use the SAS output shown to address the following questions.

Variable	N	Mean	Std Dev	Std Error	T	Prob>|T|
GRE	250	512.0463595	86.2844894	5.4571103	93.835	0.0001
TESTSTAT	250	12.0463595	86.2844894	5.4571103	2.207	0.0282

a. Is the mean GRE score among the local seniors significantly different from the national average? Justify briefly (show all parts of the test).

b. Can we say that the mean GRE score among the local seniors is significantly higher than the national average? Support your conclusion with data (show all parts of the test).

34. An academic medical center surveyed all of its patients in 2002 to assess their satisfaction with medical care. Satisfaction was measured on a scale of 0 to 100, with higher scores indicative of more satisfaction. The mean satisfaction score in 2002 was 84.5. Several quality-improvement initiatives were implemented in 2003 and the medical center is wondering whether the initiatives increased patient satisfaction. A random sample of 125 patients seeking medical care in 2003 was surveyed using the same satisfaction measure. Their mean satisfaction score was 89.2 with a standard deviation of 17.4. Is there evidence of a significant improvement in satisfaction? Run the appropriate test at the 5% level of significance.

SAS Problems

Use SAS to solve each of the following problems.

1. A study is conducted to assess the extent to which patients who had coronary artery bypass surgery were maintaining their prescribed exercise programs. The following data reflect the numbers of times patients reported exercising over the previous month (4 weeks). For the purposes of this study, exercise was defined as moderate physical activity lasting at least 20 minutes in duration.

14   11   8   6   5   3
6   13   12   8   1   4

Use SAS Proc Means to generate summary statistics on the numbers of times patients exercised over the previous month and a 95% confidence interval estimate for the mean number of times patients exercised following surgery.

2. We wish to design a study to estimate the mean of a population. We wish to consider several scenarios and to estimate the required sample size for each. Use SAS to determine the sample sizes required for each scenario. Consider margins of error of 5, 10, and 20; confidence levels of 90% and 95%; and standard deviations of 55 and 65 (for a total of $3 \times 2 \times 2 = 12$ scenarios).

3. The following data were collected from a random sample of 10 asthmatic children enrolled in a research study and reflect the number of days each child missed school during the past 3 months:

6   12   14   3   2   4   7   8   10   6

Use SAS Proc Means to generate summary statistics and request a 95% confidence interval for the mean.

4. A consumer group is investigating a producer of diet meals to examine if its prepackaged meals actually contain the advertised 6 ounces of protein in each package. The group collected the following data:

5.1   4.9   6.0   5.1   5.7   5.5   4.9   6.1   6.0   5.8
5.2   4.8   4.7   4.2   4.9   5.5   5.6   5.8   6.0   6.1

Use SAS Proc Means to generate summary statistics on the ounces of protein contained in the packaged meals. In addition, run a test to determine if there is any evidence that the meals do not contain the advertised amount of protein. Run the appropriate test at a 5% level of significance.

5. We wish to design a study to test the following hypotheses: H_0: $\mu = 100$ versus H_1: $\mu \neq 100$. We wish to consider several scenarios and to estimate the required sample size for each. Use SAS to determine the sample sizes required for each scenario to ensure power = 80%. Consider means under the alternative hypothesis of 90, 95, and 120; levels of significance of 0.05 and 0.01; and standard deviations of 7 and 10 (for a total of $3 \times 2 \times 2 = 12$ scenarios).

6. Results of an industry survey in the computer software field finds that the mean number of sick days taken by employees is 9.4 per year with a standard deviation of 2.7 per year. A local computer software company feels its employees take significantly fewer sick days per year. A random sample of 15 employees is selected from the local company and attendance records are reviewed. The following data represent the

numbers of sick days taken by these employees over the past year:

| 8 | 10 | 5 | 0 | 6 | 9 | 5 | 15 |
| 5 | 4 | 3 | 2 | 0 | 4 | 15 |

Use SAS Proc Means to generate summary statistics on the numbers of sick days taken by employees. In addition, run a test to determine if there is any evidence that the employees take fewer than 9.4 sick days per year. Run the appropriate test at a 5% level of significance. (*Note:* SAS performs a two-sided test; make the adjustment to draw your conclusion.)

Descriptive Statistics
(Ch. 2)

Probability
(Ch. 3)

Sampling Distributions
(Ch. 4)

Statistical Inference
(Chapters 5–13)

OUTCOME VARIABLE	GROUPING VARIABLE(S)/ PREDICTOR(S)	ANALYSIS	CHAPTER(S)
Continuous	—	Estimate μ; Compare μ to Known, Historical Value	5/12
Continuous	Dichotomous (2 groups)	Compare Independent Means (Estimate/Test $(\mu_1 - \mu_2)$) or the Mean Difference(μ_d)	6/12
Continuous	Discrete (> 2 groups)	Test the Equality of k Means using Analysis of Variance ($\mu_1 = \mu_2 = \cdots = \mu_k$)	9/12
Continuous	Continuous	Estimate Correlation or Determine Regression Equation	10/12
Continuous	Several Continuous or Dichotomous	Multiple Linear Regression Analysis	10
Dichotomous	—	Estimate p; Compare p to Known, Historical Value	7
Dichotomous	Dichotomous (2 groups)	Compare Independent Proportions (Estimate/Test $(p_1 - p_2)$)	7/8
Dichotomous	Discrete (> 2 groups)	Test the Equality of k Proportions (Chi-Square Test)	7
Dichotomous	Several Continuous or Dichotomous	Multiple Logistic Regression Analysis	11
Discrete	Discrete	Compare Distributions Among k Populations (Chi-Square Test)	7
Time to Event	Several Continuous or Dichotomous	Survival Analysis	13

6

Statistical Inference: Procedures for $(\mu_1 - \mu_2)$

We now describe statistical inference procedures when there are two comparison groups and the outcome of interest is a continuous variable. In such cases, we compare means between groups. In Chapter 5 we described statistical inference procedures for a single sample (estimation of an unknown mean and tests of hypothesis about the mean of the population). Two sample applications are extremely common. For example, recall Example 5.9 in which a diet was evaluated for its ability to reduce cholesterol levels. In the example, a single sample of 12 individuals followed the diet for 3 months. At the end of the 3-month observation period, cholesterol levels were measured and compared against a known (or historical) value. In that example, we did not find statistically significant evidence of a reduction in cholesterol attributable to the diet. We made the assumption that the mean cholesterol level for males aged 50 not following the diet was 241, and we observed a mean cholesterol level in our sample of 235. The assumption that the mean cholesterol level would be 241 in persons not following the diet may or may not have been a valid assumption.

We could have used different study designs that might have given a better assessment of the impact of the diet on cholesterol. For example, we could have used a concurrent comparison group (instead of a historical comparison). This type of study involves selecting a group of individuals appropriate for the study (e.g., males aged 50) and randomly assigning them to one of two groups. (Later we will describe in detail the procedures for assigning individuals at random to comparison groups.) One group of individuals follows the diet while the comparison group does not. At the end of 3 months, we compare the mean cholesterol levels between comparison groups. If the groups are similar (and this is related to random assignment), except that one group followed the diet and the other did not, then differences in cholesterol can be attributed to the diet. This is an example of what we call a *two independent samples procedure,* in which the two comparison groups are physically distinct (i.e., they are comprised of different individuals). Another study design for the assessment of the effect of the diet on cholesterol involves the 12 subjects, but before starting the diet, we measure their initial (sometimes called baseline) cholesterol levels. After each individual follows the diet for 3 months, we then measure a final cholesterol level. The focus in this design is how much each individual changes over time. If individuals' cholesterol levels drop from where they started, we conclude that the diet is effective in reducing cholesterol. This is an example of what we call a *two dependent samples procedure,* in which each individual serves as his or her own control.

In this chapter, we will describe two independent and two dependent samples procedures in detail. The most appropriate design for a specific application depends on a variety of factors, including the treatment and outcome under investigation and characteristics of the study subjects. These details will be discussed in subsequent chapters.

The techniques described here are concerned with the difference between two means. The techniques for estimating the difference between two means as well as the techniques for testing if two means are significantly different (or if one is larger than the other) are identical in principle to the techniques described in Chapter 5, which were concerned with the mean of a single population μ. The assumptions necessary for valid applications of the techniques and formulas that follow are

1. random samples from the populations under consideration
2. large samples ($n_i \geq 30$, where $i = 1, 2$) or normal populations

In two independent samples procedures, the parameter of interest is the difference in population means: $(\mu_1 - \mu_2)$. Confidence intervals in two independent samples applications are concerned with estimating $(\mu_1 - \mu_2)$, the *difference in means*, as opposed to the *value* of either mean as was the case in the one-sample estimation problems. The same is true in tests of hypotheses in the two-sample case. Both the null and research or alternative hypothesis are concerned with the difference in means, for example, H_0: $\mu_1 - \mu_2 = 0$ (no difference in means) versus H_1: $\mu_1 - \mu_2 \neq 0$ (means are different). In two dependent samples procedures, the parameter of interest is the mean difference: μ_d. Confidence intervals in two dependent samples applications are concerned with estimating μ_d, the *mean difference*.

Techniques and formulas for estimation and test of hypotheses concerning the difference between two independent means and the mean difference depend on the specific attributes of the application. We classify these attributes into four cases (Table 6.1).

Cases 1, 2, and 3 apply when the two populations are independent. Going back to the example in which we evaluate the effect of the diet on cholesterol levels, two independent samples procedures would be appropriate if individuals were assigned to either the diet group or a comparison group. The outcome is cholesterol level, a continuous variable, measured after 3 months on the assigned treatment (diet or not). Interest lies in comparing the mean cholesterol level for persons on the diet with that of persons who are not. The treatment group defines two distinct (nonoverlapping) populations.

A second example might involve a comparison of the mean cholesterol levels for persons following a special diet as compared to persons taking

Table 6.1 *Procedures Concerning* $(\mu_1 - \mu_2)$

Case 1	Two independent populations—population variances known (i.e., σ_1^2 and σ_2^2 known)
Case 2	Two independent populations—population variances unknown but assumed to be equal
Case 3	Two independent populations—population variances unknown and possibly unequal
Case 4	Two dependent populations—the data are matched or paired

medication to reduce cholesterol. Again, the outcome variable measured on each subject is his or her cholesterol level. The treatment or grouping variable in this example is the assigned treatment (diet or medication). Assuming that patients are assigned to follow the diet OR to receive medication, the treatment defines two distinct (nonoverlapping) populations.

Case 4 applies when the two populations are dependent (sometimes referred to as matched or paired). For example, suppose we again wish to assess the effect of the diet on cholesterol levels. In order to assess whether the diet is effective in reducing cholesterol, we recruit a sample of subjects into the study. At the outset, we record each participant's cholesterol level (a continuous variable). Each then participates in the diet program, and after 3 months we again measure cholesterol (a continuous variable). We now have two samples of data: a sample of values reflecting baseline or initial cholesterol levels and a sample of values reflecting cholesterol levels after 3 months on the diet. In this application, different from the two just described, two measurements are taken on each individual. We again have two samples of data; however, the samples are matched by individual. This is an example of a Case 4 application.

It is extremely important to recognize the difference between a two independent samples application and a two dependent samples application in order to apply the appropriate statistical analysis. Incorrectly classifying an application could result in incorrect inferences. Using a series of examples, we will illustrate the difference between these types of applications in the sections that follow.

Examples of confidence intervals and the test of hypothesis for each case are provided in Section 6.1. Section 6.1.1 contains examples of Case 1, Section 6.1.2 contains examples of Case 2, Sections 6.1.3 and 6.1.4 contain examples of Cases 3 and 4, respectively. Power and sample size determination are discussed in Section 6.2. Key formulas are summarized in Section 6.3, and in Section 6.4 we provide SAS program code used to perform statistical computing applications presented in this chapter. In Section 6.5 we use data from the Framingham Heart Study to illustrate the applications presented here.

6.1 Statistical Inference Concerning $(\mu_1 - \mu_2)$

Similar to the applications described in Chapter 5 involving the mean of a single population, statistical inference procedures for the difference in two means can be classified as either estimation or hypothesis-testing applications. (Although we present these as distinct applications, there is a clear relationship between them, which we will discuss through examples in this section.)

Estimation in two-sample applications is concerned with estimating $(\mu_1 - \mu_2)$, the *difference* in means between groups. Tests of hypotheses in the two-sample case are also concerned with the difference in means, for example, $H_0: \mu_1 - \mu_2 = 0$ (no difference in means) versus either $H_1: \mu_1 - \mu_2 \neq 0$ (means

Table 6.2 *Case 1: Two Independent Populations—Population Variances Known*

Attributes	Test Statistic	Confidence Interval
σ_1^2 and σ_2^2 known	$Z = \dfrac{\overline{X}_1 - \overline{X}_2}{\sqrt{\dfrac{\sigma_1^2}{n_1} + \dfrac{\sigma_2^2}{n_2}}}$	$(\overline{X}_1 - \overline{X}_2) \pm Z_{1-(\alpha/2)} \sqrt{\dfrac{\sigma_1^2}{n_1} + \dfrac{\sigma_2^2}{n_2}}$

are different) or $H_1: \mu_1 > \mu_2$ (the mean of population 1 is larger than the mean of population 2), or $H_1: \mu_1 < \mu_2$ (the mean of population 1 is smaller than the mean of population 2). Following is a series of examples to illustrate estimation and hypothesis testing techniques for applications classified as Cases 1, 2, 3, or 4. In all cases, we assume that the outcome variable is approximately normally distributed or that we have sufficiently large samples.

6.1.1 Case 1: Two Independent Populations—Population Variances Known

Case 1 applies when there are two independent populations being compared with respect to their means and the population variances (σ_1^2 and σ_2^2) are known. The two independent populations can be defined on the basis of a characteristic inherent to the subjects under study (e.g., male gender versus female gender, age < 30 years versus age ≥ 30 years) or by design (e.g., medication A versus medication B, treatment plan 1 versus treatment plan 2). When an application satisfies the attributes of Case 1, the test statistic for testing $H_0: \mu_1 - \mu_2 = 0$ (equivalent to $H_0: \mu_1 = \mu_2$) and the confidence interval for estimating $(\mu_1 - \mu_2)$ are as given in Table 6.2.

We now illustrate the use of these formulas through examples.

EXAMPLE 6.1

Estimating Difference in Mean Walking Distance Between Two Physical Therapy Programs

Following total knee replacement (TKR) surgery, physical therapy is initiated almost immediately. Depending on a variety of factors (e.g., patients' physical abilities, age, health insurance coverage), patients receive varying numbers of physical therapy sessions. As a means of assessing the effectiveness of physical therapy, patients are given periodic walking tests. One particular test involves measuring the distance (in feet) that a patient can walk independently (i.e., unassisted). We wish to compare two physical therapy programs with respect to how far patients can walk independently. The first program involves four physical therapy sessions that take place 1 hour per day over 4 days following surgery. The second program involves two physical therapy sessions that take place 2 hours per day on the first and second days following surgery. The outcome of interest, the number of feet patients can walk independently, is measured on the fifth day following surgery for all patients.

To compare the programs, 34 patients are randomly assigned to either the 4-day or the 2-day program. After completing the assigned program, the number of feet each can walk independently is measured on day 5. The 18 patients undergoing physical therapy for 4 days following TKR walked a mean of 47.2 feet independently, whereas the 16 patients undergoing physical therapy for 2 days following TKR walked a mean of 24.6 feet independently. We will compare the programs by constructing a 95% confidence interval for the difference in mean walking distances between the two groups of patients. Suppose in this application that the variations in walking distances are known: $\sigma_1^2 = 6.4$ feet and $\sigma_2^2 = 4.2$ feet. (It is more common in practice to encounter applications in which the population variances are unknown. In the next two sections, we will illustrate the procedures to use when the population variances are unknown.) Because we have two independent populations and the population variances are known, this application is an example of Case 1. The appropriate formula for the confidence interval for the difference in population means is

$$(\overline{X}_1 - \overline{X}_2) \pm Z_{1-(\alpha/2)} \sqrt{\frac{\sigma_1^2}{n_1} + \frac{\sigma_2^2}{n_2}}$$

The data layout can be represented as follows:

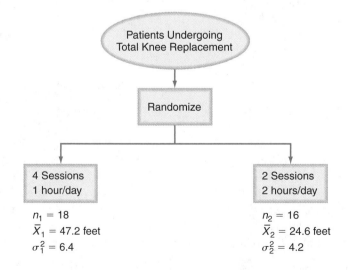

We now substitute the available data and the appropriate Z value for 95% confidence from Table B.2A in the Appendix (Z Values for Confidence Intervals; this table was also Table 5.1 in Chapter 5):

$$(47.2 - 24.6) \pm 1.96\sqrt{\frac{6.4}{18} + \frac{4.2}{16}}$$

$$22.6 \pm 1.96(0.786)$$

$$22.6 \pm 1.54$$

$$(21.06, 24.14)$$

We are 95% confident that the difference in mean walking distances between patients undergoing physical therapy for 4 versus 2 days is between 21.06 and 24.14 feet. ■

Again, in two-sample procedures, the confidence interval estimates the *difference in means,* as opposed to the value of either mean (as was the case in the one-sample applications). In this example, a point estimate for the difference in mean walking distances between patients undergoing physical therapy for 4 versus 2 days is 22.6 feet, and the 95% confidence interval for the difference in means is 21.06 to 24.14 feet. If you could choose either the 4-day or the 2-day program following TKR, which physical therapy program would you select? Why?

There is a relationship between confidence interval estimates and tests of hypothesis. In Example 6.1 we generated a 95% confidence interval estimate for $(\mu_1 - \mu_2)$. If we were to run a test of hypothesis, the hypotheses would be of the form: H_0: $\mu_1 = \mu_2$ versus H_1: $\mu_1 \neq \mu_2$. An alternate representation is H_0: $\mu_1 - \mu_2 = 0$ versus H_1: $\mu_1 - \mu_2 \neq 0$. If the confidence interval estimate contains the value specified in H_0, then we do not reject H_0 in a test of hypothesis. In Example 6.1 we can conclude that there is a statistically significant difference in mean walking distances between patients undergoing physical therapy for 4 versus 2 days (because the confidence interval estimate does not include 0).

EXAMPLE 6.2

Testing Difference in Mean Number of Sick Days Between Supervisors and Non-Supervisors

Hospital management hypothesizes that employees in nonsupervisory positions take significantly more sick days per calendar year than employees in supervisory positions. To assess this claim, a test of hypothesis is planned. The two groups under investigation are defined by job classification (non-supervisory versus supervisory). We take random samples of employees from each job classification and record the number of sick days that each took in the last calendar year. A random sample of 100 employees in nonsupervisory positions took a mean of 6.7 sick days in the last calendar year, and a random sample of 100 employees in supervisory positions took a mean of 4.8 sick days in the last calendar year. The variance in the number of sick days among nonsupervisory employees is known to be 2.2 days, and the variance in the number of sick days among supervisors is known to be 3.1 days. Notice that the mean number of sick days reported by employees in nonsupervisory positions is larger than the mean number reported by employees in supervisory positions (6.7 versus 4.8). The question is, "Is there evidence of a *significantly higher* mean in the population of all non-supervisory employees as compared to supervisors?" Run the appropriate test at a 5% level of significance.

NOTE: The same five steps used in testing hypotheses concerning μ (outlined in Table 5.3) are used in the tests of hypotheses concerning $(\mu_1 - \mu_2)$.

1. Set up hypotheses.

The null hypothesis in the two-sample problem again reflects the "no difference" or "no effect" situation. The alternative hypothesis here reflects the claim that the mean number of sick days taken by nonsupervisory employees is *higher* than the mean number of sick days taken by supervisors.

$$H_0: \mu_1 = \mu_2$$

$$H_1: \mu_1 > \mu_2, \quad \alpha = 0.05$$

where $\mu_1 =$ the mean number of sick days per calendar year among employees in nonsupervisory positions
$\mu_2 =$ the mean number of sick days per calendar year among employees in supervisory positions

2. Select the appropriate test statistic.

Because we have two independent populations and the population variances are known, this is an example of Case 1 (see Table 6.2), so the test statistic is

$$Z = \frac{\overline{X}_1 - \overline{X}_2}{\sqrt{\dfrac{\sigma_1^2}{n_1} + \dfrac{\sigma_2^2}{n_2}}}$$

3. Decision rule.

The decision rule in two-sample tests of hypothesis depends on the same three factors: (1) whether the test is upper-, lower-, or two-tailed, (2) the level of significance, and (3) the form of the test statistic. Here we have an upper-tailed test, $\alpha = 0.05$, and we are using a Z statistic. The critical value of Z is found in Table B.2B in the Appendix (Z Values for

Tests of Hypothesis; this table was also Table 5.4 in Chapter 5):

$$\text{Reject } H_0 \text{ if } Z \geq 1.645$$
$$\text{Do not reject } H_0 \text{ if } Z < 1.645$$

4. Test statistic.

We now substitute the available data:

$$Z = \frac{\overline{X}_1 - \overline{X}_2}{\sqrt{\dfrac{\sigma_1^2}{n_1} + \dfrac{\sigma_2^2}{n_2}}} = \frac{6.7 - 4.8}{\sqrt{\dfrac{2.2}{100} + \dfrac{3.1}{100}}} = \frac{1.9}{0.23} = 8.26$$

5. Conclusion.

Reject H_0 because $8.26 \geq 1.645$. We have significant evidence, $\alpha = 0.05$, to show that employees in nonsupervisory positions take more sick days per calendar year than employees in supervisory positions (i.e., $\mu_1 > \mu_2$).

What is the p value for this application? Use Table B.2B to determine the smallest level of significance α we could have selected and still rejected H_0. Because the test statistic is so large, we could have selected $\alpha = 0.0001$ and still rejected H_0. Thus, $p < 0.0001$. ■

6.1.2 Case 2: Two Independent Populations—Population Variances Unknown but Assumed Equal

Case 2 applies when there are two independent populations and the population variances (σ_1^2 and σ_2^2) are unknown but assumed to be equal. The sample variances (s_1^2 and s_2^2) are used as estimates and the formulas for test statistics and confidence intervals concerning $(\mu_1 - \mu_2)$ are given in Table 6.3.

Table 6.3 *Case 2: Population Variances Unknown but Assumed to Be Equal*

Attributes	Test Statistic	Confidence Interval
σ_1^2 and σ_2^2 unknown but assumed to be equal, $n_1 \geq 30$ and $n_2 \geq 30$	$Z = \dfrac{\overline{X}_1 - \overline{X}_2}{S_p\sqrt{\dfrac{1}{n_1} + \dfrac{1}{n_2}}}$	$(\overline{X}_1 - \overline{X}_2) \pm Z_{1-(\alpha/2)} S_p \sqrt{\dfrac{1}{n_1} + \dfrac{1}{n_2}}$
σ_1^2 and σ_2^2 unknown but assumed to be equal, $n_1 < 30$ or $n_2 < 30$	$t = \dfrac{\overline{X}_1 - \overline{X}_2}{S_p\sqrt{\dfrac{1}{n_1} + \dfrac{1}{n_2}}}$	$(\overline{X}_1 - \overline{X}_2) \pm t_{1-(\alpha/2)} S_p \sqrt{\dfrac{1}{n_1} + \dfrac{1}{n_2}}$
	$\text{df} = n_1 + n_2 - 2$	

where S_p is the pooled estimate of the common standard deviation:

$$S_p = \sqrt{\frac{(n_1 - 1)s_1^2 + (n_2 - 1)s_2^2}{n_1 + n_2 - 2}}$$

NOTE: S_p^2 is the pooled estimate of the common variance, defined as a weighted average of the sample variances (s_1^2 and s_2^2). The weights are determined by the sample sizes (n_1 and n_2). *If the sample sizes are equal (i.e., $n_1 = n_2$), then S_p reduces to* $S_p = \sqrt{\frac{s_1^2 + s_2^2}{2}}$.

In Case 2 we assume that the population variances are equal in the two comparison groups. This is a reasonable assumption in many applications. In this case, we estimate the unknown population variance by pooling data from both samples. In order to count more heavily the sample variance derived from the larger group, a weighted mean is used.

EXAMPLE 6.3

Testing Difference in Mean Number of Problems Solved Between Males and Females

A new curriculum has been implemented across medical schools that is designed to improve medical students' analytic skills. An evaluation committee is concerned that the new curriculum may be differentially effective among male and female students. To evaluate the new curriculum, random samples of male and female students who completed the new curriculum are selected and given a test to assess their analytic skills. The following data, which denote the numbers of analytic problems correctly solved by the students, are observed:

Statistic	Males	Females
Sample size	15	12
Mean	15.8	12.4
Standard deviation	4.2	3.6

Use the data to test if there is a significant *difference* in the mean numbers of analytic problems correctly solved by male and female medical students using a 5% level of significance.

1. Set up hypotheses.

   Here we wish to test whether there is a significant *difference* in means; therefore, a two-sided alternative is used.

   $$H_0: \mu_1 = \mu_2$$

   $$H_1: \mu_1 \neq \mu_2, \quad \alpha = 0.05$$

where $\mu_1 = $ mean number of analytic problems correctly solved by male students

$\mu_2 = $ mean number of analytic problems correctly solved by female students

2. Select the appropriate test statistic.

We assume that the variation in the numbers of analytic problems correctly solved is the same for male and female students (i.e., assume $\sigma_1^2 = \sigma_2^2$). Notice that the sample standard deviations ($s_1 = 4.2$ and $s_2 = 3.6$) are similar in magnitude, evidence that the assumption regarding the equality of population variances is appropriate. Because of this assumption and due to the small sample sizes ($n_1 < 30$ and $n_2 < 30$), the test statistic is

$$t = \frac{\overline{X}_1 - \overline{X}_2}{S_p\sqrt{\dfrac{1}{n_1} + \dfrac{1}{n_2}}}$$

where S_p is the pooled estimate of the common standard deviation (i.e., $\sigma_1 = \sigma_2 = \sigma$)

3. Decision rule.

Here we have a two-sided test, $\alpha = 0.05$, and we are using a t statistic. The critical value of t is found in Table B.3 in the Appendix. In order to determine the appropriate critical value, we first compute degrees of freedom: df $= n_1 + n_2 - 2 = 15 + 12 - 2 = 25$. The critical value is $t = 2.060$.

Reject H_0 if $t \geq 2.060$ or if $t \leq -2.060$

Do not reject H_0 if $-2.060 < t < 2.060$

4. Test statistic.

First, we compute S_p:

$$S_p = \sqrt{\frac{(n_1 - 1)s_1^2 + (n_2 - 1)s_2^2}{n_1 + n_2 - 2}}$$

$$= \sqrt{\frac{14(4.2)^2 + 11(3.6)^2}{15 + 12 - 2}} = \sqrt{15.581} = 3.95$$

(Notice that S_p, the pooled estimate of the common standard deviation, falls in between the values of the two sample standard deviations.)

Now, the test statistic:

$$t = \frac{15.8 - 12.4}{3.95\sqrt{\dfrac{1}{15} + \dfrac{1}{12}}} = \frac{3.4}{1.53} = 2.22$$

5. Conclusion.

Reject H_0 because $2.22 \geq 2.060$. We have significant evidence, $\alpha = 0.05$, to show that there is a significant difference in the mean number of analytic problems correctly solved by male and female medical students. The p value for this test is determined using Table B.3. With 25 degrees of freedom, the critical value for $\alpha = 0.05$ is 2.060 (see step 3). The next smallest level of significance in Table B.3 (for df = 25) is 0.02, which has a corresponding critical value of 2.485. Because we would not reject H_0 at $\alpha = 0.02$, $p = 0.05$. If this test were run on SAS, an exact p value would be computed. The exact p value is between 0.02 and 0.05 (thus we would report $p < 0.05$). ■

NOTE: The preceding test is appropriate for testing the equality of two independent population means when variances are assumed to be equal. The assumption regarding the equality of population variances must be evaluated carefully. If it is not reasonable to assume that the population variances are equal between comparison groups, then the techniques described in the next section must be employed. However, if the sample sizes are equal (i.e., $n_1 = n_2$), the formulas presented in this section are *robust*. Formulas and techniques are robust if they maintain their statistical properties under violations of assumptions (e.g., if the population variances are not equal). As a rule of thumb, if the sample sizes are equal or if the sample sizes are unequal but the sample variances are close in value, defined as $0.5 \leq s_1^2/s_2^2 \leq 2$, then the formulas presented in this section (Case 2) can be used.

EXAMPLE 6.4

Testing Difference in Mean Time to Pain Relief Between Treatments

An investigation is undertaken to examine the average times to relief from headache pain under two entirely different treatments: Medication versus Relaxation. Patients suffering from chronic headaches are enrolled in a study and randomly assigned to one of the two treatments under investigation. Patients are instructed to either take the assigned medication or to perform the relaxation exercises at the onset of their next headache. They are also instructed to record the time, in minutes, until the headache pain is resolved. Fifteen subjects are assigned to the medication treatment and report a mean time to relief of 33.8 minutes with a variance of 2.85 minutes. A second random sample of 15 subjects is assigned to the relaxation treatment and report a mean time to relief of 22.4 minutes with a variance of 3.07 minutes.

The data layout is as follows:

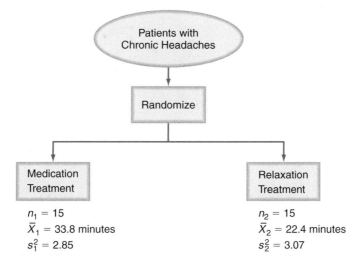

Do the observed sample means provide evidence of a statistically significant difference in means in the populations? Run the appropriate test to assess whether there is a significant *difference* in the mean time to relief under the two different treatments using a 5% level of significance.

1. Set up hypotheses.

$$H_0\colon \mu_1 = \mu_2$$
$$H_1\colon \mu_1 \neq \mu_2, \quad \alpha = 0.05$$

where μ_1 = mean time to relief with medication treatment
μ_2 = mean time to relief with relaxation treatment

2. Select the appropriate test statistic.
   Assuming that population variances are equal (since $n_1 = n_2$, we need not be concerned with violating this assumption), the test statistic is

$$t = \frac{\overline{X}_1 - \overline{X}_2}{S_p\sqrt{\dfrac{1}{n_1} + \dfrac{1}{n_2}}}$$

3. Decision rule.
   Here we have a two-sided test, $\alpha = 0.05$, and are using a t statistic. The appropriate critical value is found in Table B.3. We first compute degrees of freedom: df $= n_1 + n_2 - 2 = 15 + 15 - 2 = 28$.

Reject H_0 if $t \geq 2.048$ or if $t \leq -2.048$

Do not reject H_0 if $-2.048 < t < 2.048$

4. Test statistic.

First, we compute S_p:

$$S_p = \sqrt{\frac{(n_1 - 1)s_1^2 + (n_2 - 1)s_2^2}{n_1 + n_2 - 2}}$$

$$= \sqrt{\frac{14(2.85) + 14(3.07)}{15 + 15 - 2}} = \sqrt{2.96} = 1.72$$

Notice that S_p^2 falls in between the values of the two sample variances. Because the sample sizes are equal here, the following can also be used:

$$S_p = \sqrt{\frac{s_1^2 + s_2^2}{2}} = \sqrt{\frac{2.85 + 3.07}{2}} = 1.72$$

Now, the test statistic:

$$t = \frac{33.8 - 22.4}{1.72\sqrt{\frac{1}{15} + \frac{1}{15}}} = \frac{11.4}{0.63} = 18.10$$

5. Conclusion.

Reject H_0 since $18.10 \geq 2.048$. We have significant evidence, $\alpha = 0.05$, to show that the mean time to relief from headache pain is different under medication as compared to relaxation treatment. This test statistic is so large that $p < 0.0001$. ■

6.1.3 Case 3: Two Independent Populations— Population Variances Possibly Unequal

In Case 3, we are again concerned with estimating the difference between two independent population means or conducting a test concerning the equality of two independent population means. Population variances, however, are not known and cannot be assumed to be equal (as in Case 2). In order to determine if the population variances are equal, we conduct a *preliminary test* of H_0: $\sigma_1^2 = \sigma_2^2$ against H_1: $\sigma_1^2 \neq \sigma_2^2$. It is necessary to conduct this preliminary test in order to determine the appropriate formula for estimating the difference in means or for the test statistic in the two independent samples test of means. If, based on the preliminary test, we determine that the population variances are equal, then the formulas given in the previous section (Case 2) can be used. If, however, we find that the population variances are not equal, then the formulas given in this section (Case 3) are used.

The preliminary test is conducted in the same fashion as the tests for means, that is, by following the same five steps used in tests of hypotheses concerning μ and tests of hypotheses concerning $(\mu_1 - \mu_2)$. The preliminary test for equality (or homogeneity) of variances is outlined next and is summarized in Table 6.4. The hypotheses to be tested in the preliminary test are

$$H_0: \sigma_1^2 = \sigma_2^2$$

$$H_1: \sigma_1^2 \neq \sigma_2^2$$

The preliminary test is always a two-sided test, as we are solely interested in determining whether the population variances are equal or not for the purposes of selecting the appropriate test statistic (or confidence interval formula) for evaluating the difference in means $(\mu_1 - \mu_2)$.

> NOTE: In some applications, a larger level of significance is chosen for the preliminary test (e.g., $\alpha = 0.10$ or 0.15).

The test statistic is determined by the ratio of the sample variances, which follows an F distribution:

$$F = \frac{s_1^2}{s_2^2} \tag{6.1}$$

The F distribution has two degrees of freedom, denoted df_1 and df_2, called the *numerator* and *denominator* degrees of freedom, respectively:

$$df_1 = n_1 - 1 \quad \text{(numerator degrees of freedom)}$$

$$df_2 = n_2 - 1 \quad \text{(denominator degrees of freedom)}$$

If the test statistic, F, is close to 1 (which occurs when s_1^2 and s_2^2 are approximately equal in value), then H_0 is most likely true (i.e., $\sigma_1^2 = \sigma_2^2$). However, if F is significantly smaller or larger than unity, then H_1 is most likely true. In order to make a decision about H_0 or H_1, critical values from the F distribution must be determined to formulate the decision rule.

The F distribution is an asymmetric distribution (Figure 6.1) that takes on values greater than or equal to zero.

The decision rule is given by

Reject H_0 if $F \leq 1/F_{1-(\alpha/2)}(df_2, df_1)$ or if $F \geq F_{1-(\alpha/2)}(df_1, df_2)$

Do not reject H_0 if $1/F_{1-(\alpha/2)}(df_2, df_1) < F < F_{1-(\alpha/2)}(df_1, df_2)$

Figure 6.2 displays the critical values in the preliminary test for equality of variances (always a two-sided test).

For example, suppose we conduct a preliminary test of the equality of population variances (i.e., $H_0: \sigma_1^2 = \sigma_2^2$ versus $H_1: \sigma_1^2 \neq \sigma_2^2$) using a 5% level of significance based on samples of size $n_1 = 20$ and $n_2 = 20$. The critical

Figure 6.1 *F Distribution*

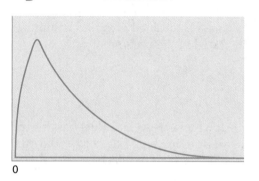

0

Figure 6.2 *Critical Values in the Preliminary Test*

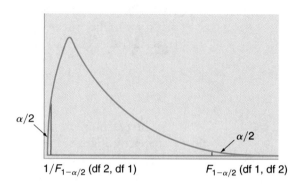

$\alpha/2$

$\alpha/2$

$1/F_{1-\alpha/2}$ (df 2, df 1) $F_{1-\alpha/2}$ (df 1, df 2)

values from the F distribution are found in Table B.4 and are given here for $\mathrm{df}_1 = n_1 - 1 = 20 - 1 = 19$ and $\mathrm{df}_2 = n_2 - 1 = 20 - 1 = 19$:

$$F_{0.975}(19, 19) = 2.51 \qquad 1/F_{0.975}(19, 19) = 1/2.51 = 0.40$$

The critical values from the F distribution table (Table B.4) with numerator degrees of freedom equal to 19 and denominator degrees of freedom equal to 19 are 0.40 and 2.51.

The steps involved in the preliminary test of variances are summarized in Table 6.4.

If the null hypothesis in the preliminary test is not rejected (i.e., $\sigma_1^2 = \sigma_2^2$), then the formulas given in Table 6.3 (Case 2) are used for the *main test* of

Table 6.4 *Preliminary Test for Homogeneity of Variances*

Step	Example
1. Set up hypotheses.	$H_0: \sigma_1^2 = \sigma_2^2$ $H_1: \sigma_1^2 \neq \sigma_2^2$ *
Select level of significance.[†]	$\alpha = 0.05$
2. Select appropriate test statistic.	$F = \dfrac{s_1^2}{s_2^2}$
3. Generate decision rule.	Reject H_0 if $F \leq 1/F_{1-(\alpha/2)}(\mathrm{df}_2, \mathrm{df}_1)$ or if $F \geq F_{1-(\alpha/2)}(\mathrm{df}_1, \mathrm{df}_2)$ Do not reject H_0 if $1/F_{1-(\alpha/2)}(\mathrm{df}_2, \mathrm{df}_1) < F < F_{1-(\alpha/2)}(\mathrm{df}_1, \mathrm{df}_2)$
4. Compute the value of the test statistic.	
5. Draw a conclusion by comparing the test statistic (4) to the decision rule (3). Report findings relative to H_1.	

* In the preliminary test, we are not concerned with upper- or lower-tailed tests. The preliminary test is conducted solely to determine if the population variances are or are not equal, for the purposes of selecting the correct formula for statistical inference concerning the difference in population means.

[†] Larger levels of significance (e.g., $\alpha = 0.10, 0.15$) are sometimes used in the preliminary test.

Table 6.5 *Case 3: Two Independent Populations—Population Variances Unequal*

Attributes	Test Statistic	Confidence Interval
σ_1^2 and σ_2^2 unknown but assumed to be unequal, $n_1 \geq 30$ and $n_2 \geq 30$	$Z = \dfrac{\overline{X}_1 - \overline{X}_2}{\sqrt{\dfrac{s_1^2}{n_1} + \dfrac{s_2^2}{n_2}}}$	$(\overline{X}_1 - \overline{X}_2) \pm Z_{1-(\alpha/2)}\sqrt{\dfrac{s_1^2}{n_1} + \dfrac{s_2^2}{n_2}}$
σ_1^2 and σ_2^2 unknown but assumed to be unequal, $n_1 < 30$ or $n_2 < 30$	$t = \dfrac{\overline{X}_1 - \overline{X}_2}{\sqrt{\dfrac{s_1^2}{n_1} + \dfrac{s_2^2}{n_2}}}$	$(\overline{X}_1 - \overline{X}_2) \pm t_{1-(\alpha/2)}\sqrt{\dfrac{s_1^2}{n_1} + \dfrac{s_2^2}{n_2}}$

$$\mathrm{df} = \frac{\left(\dfrac{s_1^2}{n_1} + \dfrac{s_2^2}{n_2}\right)^2}{\dfrac{\left(s_1^2/n_1\right)^2}{n_1 - 1} + \dfrac{\left(s_2^2/n_2\right)^2}{n_2 - 1}}$$

H_0: $\mu_1 = \mu_2$ or for the confidence interval for $(\mu_1 - \mu_2)$. If the null hypothesis in the preliminary test is rejected (i.e., $\sigma_1^2 \neq \sigma_2^2$), then the formulas given in Table 6.5 (Case 3) are used for the *main test* of H_0: $\mu_1 = \mu_2$ or for the confidence interval for $(\mu_1 - \mu_2)$.

The formula for the degrees of freedom (df) to determine both the critical value(s) in the test of hypothesis and the appropriate value for the confidence interval is based on the Welch–Satterthwaite solution [Welch, B. L. (1938). The significance of the difference between two means when the population variances are unequal. *Biometrika*, 29, 350–362; Satterthwaite, F. E. (1946). An approximate distribution of estimates of variance components. *Biometrics Bulletin*, 2, 110–114]. It allows for the use of the t distribution table (Table B.3) in applications involving these formulas (Case 3).

To simplify the computations of the degrees of freedom, consider the following:

$$c_i = \frac{s_i^2}{n_i} \tag{6.2}$$

where $i = 1, 2$

Then,

$$\mathrm{df} = \frac{(c_1 + c_2)^2}{\dfrac{c_1^2}{n_1 - 1} + \dfrac{c_2^2}{n_2 - 1}} \tag{6.3}$$

We illustrate the use of this formula in the following examples. In Example 6.5, we use the Case 3 formulas; we illustrate the complete procedure (i.e., the preliminary test followed by the main test or estimation) in the subsequent examples.

EXAMPLE 6.5

Testing Difference in Mean Heart Rates Between Females 20–24 Years and 25–30 Years of Age

Consider an experiment involving the comparison of the mean heart rates following 30 minutes of aerobic exercise in females aged 20 to 24 as compared to females aged 25–30. For this experiment, 10-second heart rates are recorded on each participant following 30 minutes of intense aerobic exercise and converted to beats per minute (i.e., heart rate per 60 seconds). The sample data are

Statistic	Age: 20–24	Age: 25–30
Sample size	15	10
Mean heart rate	146.22	141.10
Variance in heart rates	40.0	10.0

Use the data to test if there is a significant *difference* in the mean heart rates following 30 minutes of aerobic exercise between the two groups, using a 5% level of significance. Assume that the variances in heart rates are not equal between comparison groups. (Notice that the sample variances are quite different—40.0 versus 10.0, $s_1^2/s_2^2 = 4$, which is outside of the range 0.5–2—suggesting a true difference in population variances.)

1. Set up hypotheses.

$$H_0: \mu_1 = \mu_2$$
$$H_1: \mu_1 \neq \mu_2, \quad \alpha = 0.05$$

where μ_1 = mean heart rate for females 20–24 years of age
μ_2 = mean heart rate for females 25–30 years of age

2. Select the appropriate test statistic.
    (Because population variances are assumed to be unequal, this is an example of Case 3.)

$$t = \frac{\overline{X}_1 - \overline{X}_2}{\sqrt{\dfrac{s_1^2}{n_1} + \dfrac{s_2^2}{n_2}}}$$

3. Decision rule.

   To determine the appropriate critical value, we first compute the degrees of freedom:

$$
df = \frac{\left(\dfrac{s_1^2}{n_1} + \dfrac{s_2^2}{n_2} \right)^2}{\dfrac{\left(s_1^2/n_1\right)^2}{n_1 - 1} + \dfrac{\left(s_2^2/n_2\right)^2}{n_2 - 1}}
$$

To simplify the computations, using (6.2) we first compute

$$
c_i = \frac{s_i^2}{n_i}, \qquad c_1 = 40/15 = 2.67, \qquad c_2 = 10/10 = 1
$$

Then,

$$
df = \frac{(c_1 + c_2)^2}{\dfrac{c_1^2}{n_1 - 1} + \dfrac{c_2^2}{n_2 - 1}} = \frac{(2.67 + 1)^2}{\dfrac{2.67^2}{14} + \dfrac{1^2}{9}} = \frac{13.47}{0.62} = 21.73 = 22
$$

Using the t distribution table (Table B.3), the two-sided critical value with 22 degrees of freedom is $t = 2.074$. The decision rule is

$$\text{Reject } H_0 \text{ if } t \geq 2.074 \text{ or if } t \leq -2.074$$

$$\text{Do not reject } H_0 \text{ if } -2.074 < t < 2.074$$

4. Test statistic.

$$
t = \frac{\overline{X}_1 - \overline{X}_2}{\sqrt{\dfrac{s_1^2}{n_1} + \dfrac{s_2^2}{n_2}}} = \frac{146.22 - 141.10}{\sqrt{\dfrac{40}{15} + \dfrac{10}{10}}} = \frac{5.12}{1.92} = 2.67
$$

5. Conclusion.

   Reject H_0 since $2.67 \geq 2.074$. We have significant evidence, $\alpha = 0.05$, to show that there is a difference in the mean heart rates following 30 minutes of aerobic exercise for females aged 20–24 years as compared to females aged 25–30 years. For this example, $p = 0.02$. ■

   We now illustrate the preliminary test for homogeneity of variances followed by a test of hypothesis concerning means of two independent populations.

EXAMPLE 6.6

Testing Difference in Mean Public Health Awareness Scores Between Males and Females

Random samples of 11 male high school students and 12 female high school students are selected within a particular school district for an investigation. Students' scores on a public health (PH) awareness test are recorded; the descriptive statistics follow. The test is scored on a scale of 0–1000, with higher scores indicative of more awareness.

Statistic	Males	Females
Sample size	11	12
Mean PH awareness score	560.0	554.2
Standard deviation in PH awareness scores	133.1	129.4

Test if the male students score significantly *higher* than the female students on the public health awareness test within this school district using a 5% level of significance.

1. Set up hypotheses.

$$H_0: \mu_1 = \mu_2$$
$$H_1: \mu_1 > \mu_2, \quad \alpha = 0.05$$

where μ_1 = mean score for males and μ_2 = mean score for females

2. Select the appropriate test statistic.
   In order to determine if this application falls into Case 2 or Case 3, a preliminary test of the equality of population variances must be conducted.

   1. Set up hypotheses.

   $$H_0: \sigma_1^2 = \sigma_2^2$$
   $$H_1: \sigma_1^2 \neq \sigma_2^2, \quad \alpha = 0.05$$

   2. Select the appropriate test statistic.

   $$F = \frac{s_1^2}{s_2^2}$$

   3. Decision rule.

   $$df_1 = n_1 - 1 = 11 - 1 = 10 \quad \text{(numerator degrees of freedom)}$$
   $$df_2 = n_2 - 1 = 12 - 1 = 11 \quad \text{(denominator degrees of freedom)}$$
   $$F_{0.975}(10,11) = 3.53 \text{ and } F_{0.975}(11,10) = 3.72$$

   Reject H_0 if $F \leq 1/3.72 = 0.269$ or if $F \geq 3.53$
   Do not reject H_0 if $0.269 < F < 3.53$

4. Test statistic.

$$F = \frac{s_1^2}{s_2^2} = \frac{(133.1)^2}{(129.4)^2} = 1.06$$

5. Conclusion.

Do not reject H_0 since $0.269 < 1.06 < 3.53$. We do not have significant evidence to show that $\sigma_1^2 \neq \sigma_2^2$. Therefore, for the purposes of this test of means, we assume that the population variances are equal (i.e., $\sigma_1^2 = \sigma_2^2$) and apply the test statistic given under Case 2:

$$t = \frac{\overline{X}_1 - \overline{X}_2}{S_p \sqrt{\dfrac{1}{n_1} + \dfrac{1}{n_2}}}$$

$$df = n_1 + n_2 - 2 = 11 + 12 - 2 = 21$$

3. Decision rule.

$$\text{Reject } H_0 \text{ if } t \geq 1.721$$

$$\text{Do not reject } H_0 \text{ if } t < 1.721$$

4. Test statistic.

We first compute S_p:

$$S_p = \sqrt{\frac{(n_1 - 1)s_1^2 + (n_2 - 1)s_2^2}{n_1 + n_2 - 2}}$$

$$= \sqrt{\frac{10(133.1)^2 + 11(129.4)^2}{11 + 12 - 2}} = \sqrt{17253.11} = 131.4$$

Now the test statistic:

$$t = \frac{560.0 - 554.2}{131.4\sqrt{\dfrac{1}{11} + \dfrac{1}{12}}} = \frac{5.8}{54.8} = 0.11$$

5. Conclusion.

Do not reject H_0 since $0.11 < 1.721$. We do not have significant evidence, $\alpha = 0.05$, to show that the male students score significantly higher than the female students on the public health awareness test within this school district. ▪

NOTE: The two sample t tests (Case 2 and Case 3) concerning $(\mu_1 - \mu_2)$ are generally robust (i.e., insensitive to violations in assumptions such as normality and/or equality of variances) when the sample sizes are equal (i.e., $n_1 = n_2$). However, the F test for homogeneity of variances is sensitive to violations in the normality assumption. For example, if the analytic variable is not normally distributed and the F test is applied, the actual level of significance, α, may exceed the specified level (e.g., 0.05).

SAS EXAMPLE 6.6 **Testing Difference in Mean Public Health Awareness Scores Between Males and Females Using SAS**

The following output was generated using SAS Proc Ttest, which conducts a two independent samples test of hypothesis. The same procedure automatically produces a preliminary test of the homogeneity of variances. A brief interpretation appears after the output.

SAS Output for Example 6.6

The TTEST Procedure

Statistics

Variable	Class	N	Lower CL Mean	Mean	Upper CL Mean	Lower CL Std Dev	Std Dev
test	female	12	471.93	554.17	636.41	91.692	129.44
test	male	11	470.57	560	649.43	93.011	133.12
test	Diff (1-2)		-119.7	-5.833	108.06	100.94	131.2

Statistics

Variable	Class	Upper CL Std Dev	Std Err	Minimum	Maximum
test	female	219.77	37.365	260	750
test	male	233.61	40.136	370	770
test	Diff (1-2)	187.5	54.767		

T-Tests

Variable	Method	Variances	DF	t Value	Pr > |t|
test	Pooled	Equal	21	-0.11	0.9162
test	Satterthwaite	Unequal	20.7	-0.11	0.9163

Equality of Variances

Variable	Method	Num DF	Den DF	F Value	Pr > F
test	Folded F	10	11	1.06	0.9215

Interpretation of SAS Output for Example 6.6

In the top section of the output, SAS provides summary statistics on the analytic variable (test score) for each comparison group (females and males) and then for the differences in means (females – males). The summary statistics include the sample sizes, the sample mean (Mean), and 95% confidence intervals (CI) for the population means of each group and for the difference in means (the limits of the CI for the mean are labeled "lower CL mean" and "upper CL mean"), standard deviations, and 95% confidence intervals for the population standard deviations of each group and for the differences (the limits of the CI for the standard deviation are labeled "lower CL Std Dev" and "Upper CL Std Dev"), standard errors $(s/\sqrt{n})$, minimums, and maximums.

In the next section of the output, SAS performs the test of hypothesis for equality of means. SAS actually carries out two different tests, one in which the population variances are assumed to be equal (we called this Case 2) and one in which the population variances are assumed to be unequal (we called this Case 3). SAS uses the formulas we summarized in Tables 6.3 and 6.5 for equal and unequal variances, respectively. The values of the test statistics appear under the column headed "t Value," and just before these SAS displays the degrees of freedom. Again, these are computed using the formulas from Tables 6.3 and 6.5. Finally, SAS provides two-sided p values (assuming that the alternative hypothesis is $H_1: \mu_1 \neq \mu_2$).

The user must decide which analysis (equal or unequal variances) is most appropriate. To aid in this decision, SAS provides a preliminary test of the homogeneity of variances (i.e., $H_0: \sigma_1^2 = \sigma_2^2$ versus $H_1: \sigma_1^2 \neq \sigma_2^2$). The results of the preliminary test appear at the bottom of the SAS Ttest output in the section titled "Equality of Variances." SAS provides an F statistic (computed by taking the ratio of the sample variances). It computes F by dividing the larger sample variance by the smaller, regardless of the group (1 or 2) designation. Therefore, the F statistic produced by SAS is always greater than or equal to 1. In this case, since the sample variance among males is larger: $F = (133.1)^2/(129.4)^2 = 1.06$. The degrees of freedom associated with F are given by $\mathrm{df}_1 = n_1 - 1 = 11 - 1 = 10$ and $\mathrm{df}_2 = n_2 - 1 = 12 - 1 = 11$. For the preliminary test, SAS produces the probability of observing a value of F more extreme than the value of the test statistic (denoted $\mathrm{Pr} > F$); the following rule should be applied to draw a conclusion: Reject H_0 if p value $\leq \alpha_F$, where p value is the p value produced by SAS and α_F is the level of significance selected for the preliminary test. In this case, we do not reject H_0 since the p value, 0.9215, is larger than 0.05. Therefore, the population variances are not significantly different and this example is considered an example of Case 2.

Because the preliminary test suggested that this example is an example of Case 2 (i.e., equal variances), we look at the output for the equal variances case; the test statistic is $t = -0.11$ with 21 degrees of freedom ($n_1 + n_2 - 2$). For the main test, SAS produces a two-sided p value. In this example, the p value is 0.9162. Because we are interested in a one-sided test, the following rule should be applied: Reject H_0 if (p value/2) $\leq \alpha$. In this example, we do not reject H_0 since (p value/2) = (0.9162/2) = 0.4581 > 0.05. We do not have significant evidence to show that male students score higher than female students on the public health awareness test within this school district. (*Note:* The test statistic produced by SAS differs from that computed in Example 6.6 due to the fact that SAS orders the groups alphabetically and calls females group 1.) ■

EXAMPLE 6.7   **Estimating Difference in Mean Number of Emergency Room Visits Between Children 5 and Under and 6–10 Years of Age**

Suppose we wish to estimate the difference in the mean numbers of emergency room (ER) visits in 12 months among children with asthma age 5 and under as compared to children aged 6–10. For the purposes of this investigation, our analyses are restricted to children who are free from any other chronic conditions (i.e., they suffer from asthma alone). The following data are collected on random samples of 65 children age 5 and under and 50 children aged 6–10:

Age	Number of Patients	Mean Number of ER Visits	Variance in Number of ER Visits
≤ 5 years	65	6.4	4.1
6–10 years	50	3.2	2.6

Use the data to construct a 95% confidence interval for the difference in the mean numbers of ER visits in 12 months among children with asthma age 5 and under as compared to children aged 6–10.

To select the appropriate formula for the confidence interval for $(\mu_1 - \mu_2)$, we need to determine if the population variances are equal (Case 2) or not (Case 3) using the preliminary test for equality of variances.

1. Set up hypotheses.

$$H_0: \sigma_1^2 = \sigma_2^2$$
$$H_1: \sigma_1^2 \neq \sigma_2^2, \quad \alpha = 0.05$$

2. Select the appropriate test statistic.

$$F = \frac{s_1^2}{s_2^2}$$

3. Decision rule.

$df_1 = n_1 - 1 = 65 - 1 = 64$ (numerator degrees of freedom)

$df_2 = n_2 - 1 = 50 - 1 = 49$ (denominator degrees of freedom)

$F_{0.975}(64, 49) = 1.76$ and $F_{0.975}(49, 64) = 1.65$

Reject H_0 if $F \leq 1/1.65 = 0.61$ or if $F \geq 1.76$

Do not reject H_0 if $0.61 < F < 1.76$

4. Test statistic.

$$F = \frac{s_1^2}{s_2^2} = \frac{4.1}{2.6} = 1.58$$

5. Conclusion.

Do not reject H_0 since $0.61 < 1.58 < 1.76$. We do not have significant evidence to show that $\sigma_1^2 \neq \sigma_2^2$. Therefore, the appropriate formula for the confidence interval for $(\mu_1 - \mu_2)$ is given under Case 2:

$$(\overline{X}_1 - \overline{X}_2) \pm Z_{1-(\alpha/2)} S_p \sqrt{\frac{1}{n_1} + \frac{1}{n_2}}$$

We first compute S_p:

$$S_p = \sqrt{\frac{(n_1 - 1)s_1^2 + (n_2 - 1)s_2^2}{n_1 + n_2 - 2}}$$

$$= \sqrt{\frac{64(4.1) + 49(2.6)}{65 + 50 - 2}} = \sqrt{3.45} = 1.86$$

The confidence interval is

$$(6.4 - 3.2) \pm 1.96(1.86) \sqrt{\frac{1}{65} + \frac{1}{50}}$$

$$3.2 \pm 1.96(0.348)$$

$$3.2 \pm 0.682$$

$$(2.52, 3.88)$$

We are 95% confident that the difference in the mean numbers of emergency room visits in 12 months is between 2.52 and 3.88 for children age 5 and under with asthma as compared to children aged 6–10. If we conducted a test of H_0: $\mu_1 = \mu_2$ against H_1: $\mu_1 \neq \mu_2$, would we reject H_0 (use the confidence interval)? ■

EXAMPLE 6.8

Testing Difference in Mean Number of Visits to Health Center Between Freshmen and Sophomores

The health department at a major university is interested in whether there is a difference in the mean number of visits to the student health center between college freshmen and sophomores. The following data are collected on random samples of freshmen and sophomores, respectively, over the course of one academic year:

Year in School	Number of Students	Mean Number of Visits to the Student Health Center	Variance in Number of Visits to the Student Health Center
Freshmen	50	3.4	4.5
Sophomores	60	4.5	1.9

Test if there is a significant *difference* in the mean number of visits to the student health center between university freshmen and sophomores using a 5% level of significance. In this example, we consider the number of visits a continuous variable.

1. Set up hypotheses.

$$H_0: \mu_1 = \mu_2$$
$$H_1: \mu_1 \neq \mu_2, \quad \alpha = 0.05$$

where $\mu_1 =$ mean number of visits to the student health center among university freshmen

$\mu_2 =$ mean number of visits to the student health center among university sophomores

2. Select the appropriate test statistic.

In order to determine if this example falls into Case 2 or Case 3, a preliminary test of the equality of population variances must be conducted.

1. Set up hypotheses.

$$H_0: \sigma_1^2 = \sigma_2^2$$
$$H_1: \sigma_1^2 \neq \sigma_2^2, \quad \alpha = 0.05$$

2. Select the appropriate test statistic.

$$F = \frac{s_1^2}{s_2^2}$$

3. Decision rule.

$\text{df}_1 = n_1 - 1 = 50 - 1 = 49$ (numerator degrees of freedom)
$\text{df}_2 = n_2 - 1 = 60 - 1 = 59$ (denominator degrees of freedom)
$F_{0.975}(49, 59) = 1.70$ and $F_{0.975}(59, 49) = 1.76$

Reject H_0 if $F \leq 1/1.76 = 0.568$ or if $F \geq 1.70$

Do not reject H_0 if $0.571 < F < 1.70$

4. Test statistic.

$$F = \frac{s_1^2}{s_2^2} = \frac{4.5}{1.9} = 2.37$$

5. Conclusion.

Reject H_0 since $2.37 > 1.70$. We have significant evidence to show that $\sigma_1^2 \neq \sigma_2^2$. Therefore, for the purposes of this test of means, we apply the test statistic given under Case 3:

$$Z = \frac{\overline{X}_1 - \overline{X}_2}{\sqrt{\dfrac{s_1^2}{n_1} + \dfrac{s_2^2}{n_2}}}$$

3. Decision rule.
   Using Table B.2B, the decision rule is

   Reject H_0 if $Z \geq 1.960$ or if $Z \leq -1.960$
   Do not reject H_0 if $-1.960 < Z < 1.960$

4. Test statistic.

$$Z = \frac{\overline{X}_1 - \overline{X}_2}{\sqrt{\dfrac{s_1^2}{n_1} + \dfrac{s_2^2}{n_2}}} = \frac{3.4 - 4.5}{\sqrt{\dfrac{4.5}{50} + \dfrac{1.9}{60}}} = \frac{-1.10}{0.349} = -3.15$$

5. Conclusion.
   Reject H_0 since $-3.15 \leq -1.960$. We have significant evidence, $\alpha = 0.05$, to show that there is a difference in the mean number of visits to the student health center between university freshmen and sophomores. For this test, $p = 0.001$ (see Table B.2B). ■

SAS Example 6.8 Testing Difference in Mean Number of Visits to Health Center Between Freshmen and Sophomores Using SAS

The following output was generated using SAS Proc Ttest. A brief interpretation appears after the output.

SAS Output for Example 6.8

```
                    The TTEST Procedure
                        Statistics
                    Lower CL              Upper CL Lower CL
Variable  Class       N    Mean     Mean     Mean  Std Dev  Std Dev
visits    freshman    50  2.8046   3.4071  4.0096  1.7709    2.12
visits    sophomore   60  4.1457   4.5059  4.8661  1.1819   1.3944
visits    Diff (1-2)      -1.767  -1.099   -0.43  1.5543    1.761
```

```
                            Statistics
                             Upper CL
    Variable   Class         Std Dev    Std Err   Minimum    Maximum
    visits     freshman       2.6418     0.2998    -1.548     7.7675
    visits     sophomore      1.7007       0.18    1.4388     8.0719
    visits     Diff (1-2)     2.0318     0.3372


                              T-Tests
   Variable   Method        Variances       DF    t Value    Pr > |t|
   visits     Pooled        Equal          108      -3.26      0.0015
   visits     Satterthwaite Unequal       81.9      -3.14      0.0023


                    Equality of Variances
      Variable   Method      Num DF   Den DF   F Value    Pr > F
      visits     Folded F        49       59      2.31    0.0022
```

Interpretation of SAS Output for Example 6.8

In the top section of the output, SAS provides summary statistics on the analytic variable (Number of Visits) for each comparison group (freshmen and sophomores) and then for the differences in means (freshmen − sophomores). The summary statistics include the sample sizes, the sample mean (Mean), and 95% confidence intervals (CI) for the population means of each group and for the difference in means (the limits of the CI for the mean are labeled "lower CL mean" and "upper CL mean"), standard deviations, and 95% confidence intervals for the population standard deviations of each group and for the differences (the limits of the CI for the standard deviation are labeled "lower CL Std Dev" and "Upper CL Std Dev"), standard errors ($s/\sqrt{n}$), minimums, and maximums.

In the next section of the output, SAS performs the test of hypothesis for equality of means. SAS actually carries out two different tests, one in which the population variances are assumed to be equal (we called this Case 2) and one in which the population variances are assumed to be unequal (we called this Case 3). SAS uses the formulas we summarized in Tables 6.3 and 6.5 for equal and unequal variances, respectively. The values of the test statistics appear under the column headed "t Value," and just before these SAS displays the degrees of freedom. Again, these are computed using the formulas from Tables 6.3 and 6.5. Finally, SAS provides two-sided p values (assuming that the alternative hypothesis is H_1: $\mu_1 \neq \mu_2$).

The user must decide which situation (equal or unequal variances) is most appropriate. To aid in this decision, SAS provides a preliminary test of the homogeneity of variances (i.e., $H_0: \sigma_1^2 = \sigma_2^2$ versus $H_1: \sigma_1^2 \neq \sigma_2^2$). The results of the preliminary test appear at the bottom of the SAS Ttest output in the section titled "Equality of Variances." SAS provides an F statistic (computed by taking the ratio of the sample variances). It computes F by dividing the larger sample variance by the smaller, regardless of the group (1 or 2) designation. Therefore, the F statistic produced by SAS is always greater than or equal to 1. In this case, since the sample variance among freshmen is larger: $F = (2.12)^2/(1.39)^2 = 2.31$. The degrees of freedom associated with F are given by: $df_1 = n_1 - 1 = 50 - 1 = 49$ and $df_2 = n_2 - 1 = 60 - 1 = 59$. For the preliminary test, SAS produces the probability of observing a value of F more extreme than the value of the test statistic (denoted $Pr > F$); the following rule should be applied to draw a conclusion: Reject H_0 if p value $\leq \alpha_F$, where p value is the p value produced by SAS and α_F is the level of significance selected for the preliminary test. In this case, we reject H_0 since the p value, 0.0022, is smaller than 0.05. Therefore, the population variances are significantly different and this example is considered an example of Case 3.

Because the preliminary test suggested that this example is an example of Case 3 (i.e., unequal variances), we use the output for the unequal variances case, and the test statistic is $t = -3.14$ with 81.9 degrees of freedom (we computed $df = 83$; the difference is due to rounding). For the main test, SAS produces a two-sided p value. In this example, the p value is 0.0023. We are interested in a two-sided test, so the following rule should be applied: Reject H_0 if p value $\leq \alpha$. We reject H_0 in this example since $0.0023 < 0.05$. We have significant evidence to show that there is a significant difference in the mean number of visits to the student health center between university freshmen and sophomores. (*Note:* Some of the calculations produced by SAS differ somewhat from those we computed in Example 6.8 due to the fact that SAS is using more decimal places in computations.) ■

6.1.4 Case 4: Two Dependent Populations— The Data Are Matched or Paired

The techniques presented in the previous sections were for estimating or testing a difference between two independent population means. In each application, random samples were selected from each population and sample statistics were compared to draw inferences about the difference in population means $(\mu_1 - \mu_2)$. Case 4 is concerned with the comparison of two dependent (matched or paired) population means. Example 6.9 illustrates the distinction between applications involving two independent as compared to two dependent populations.

EXAMPLE 6.9

Two Independent versus Two Dependent Populations

An investigator is concerned that fraudulent claims have been made about new diet pills. The manufacturers of the pills claim that the pills are 100% effective in inducing weight loss within 4 weeks. To evaluate the claim, the investigator randomly selects a sample of individuals who are overweight and interested in weight loss. Subjects are randomly assigned to either the active treatment group or to a control group. Subjects in the active treatment group take the new diet pills as directed. Subjects in the control group take a *placebo* (an inert substance designed to look exactly like the active treatment; in this case, exactly like the new diet pills). The subjects in the control group are instructed to take the placebo according to the same protocol as subjects in the treatment group. After 4 weeks, subjects' weights are measured and compared. The hypotheses of interest are H_0: $\mu_{treatment} = \mu_{control}$ versus H_1: $\mu_{treatment} < \mu_{control}$. The alternative hypothesis reflects the situation in which the diet pills (treatment) are effective for weight loss. It is possible that the mean weight among subjects in the control group ($\mu_{control}$) will be less than the mean weight among subjects in the treatment group ($\mu_{treatment}$) after 4 weeks for a variety of reasons that may be unrelated to the effectiveness or ineffectiveness of the treatment.

A more efficient design for an application of this type is the two dependent samples test in which a single sample of subjects is drawn and evaluated. Each subject is weighed at the outset, instructed to follow the treatment (i.e., take the new diet pills as directed), and then weighed again after 4 weeks of treatment. Similar to the applications presented in Sections 6.1.1 through 6.1.3, two samples of data are analyzed. In this example, two weights are measured on each subject, one *pretreatment* and the second at 4 weeks *posttreatment*. However, the samples are dependent or matched by subject. ■

The dependent samples test, a technique used to remove extraneous variation, is illustrated in Examples 6.10 and 6.11. The formulas for the test statistics and confidence intervals when the samples are matched or paired are given in Table 6.6.

Table 6.6 *Case 4: Two Dependent Populations—The Data Are Matched or Paired*

Attributes	Test Statistic	Confidence Interval
Samples are matched or paired, n(# pairs) ≥ 30	$Z = \dfrac{\overline{X}_d - \mu_d}{\dfrac{s_d}{\sqrt{n}}}$	$\overline{X}_d \pm Z_{1-(\alpha/2)} \dfrac{s_d}{\sqrt{n}}$
Samples are matched or paired, n(# pairs) < 30	$t = \dfrac{\overline{X}_d - \mu_d}{\dfrac{s_d}{\sqrt{n}}}$ df $= n - 1$	$\overline{X}_d \pm t_{1-(\alpha/2)} \dfrac{s_d}{\sqrt{n}}$ df $= n - 1$

where $\overline{X}_d, s_d$ are the mean and standard deviation of the *difference* scores

EXAMPLE 6.10   **Testing Effectiveness of a Weight Loss Program**

A nutrition expert is examining a weight-loss program to evaluate its effectiveness (i.e., if participants lose weight on the program). Ten subjects are randomly selected for the investigation. The subjects' initial weights are recorded, they follow the program for 6 weeks, and they are weighed again. The data are as follows:

Subject	Initial Weight	Final Weight
1	180	165
2	142	138
3	126	128
4	138	136
5	175	170
6	205	197
7	116	115
8	142	128
9	157	144
10	136	130

The test of interest is

$$H_0: \mu_{\text{initial}} = \mu_{\text{final}}$$

$$H_1: \mu_{\text{initial}} > \mu_{\text{final}}$$

where μ_{initial} = mean initial weight
μ_{final} = mean final weight

The samples of initial and final weights are matched by participants. That is, for each participant an initial *and* a final weight is recorded and these two measurements are specific to each individual.

The two samples (initial weights and final weights) cannot be analyzed separately as in the two independent samples case (Sections 6.1.1–6.1.3). When the samples are matched or paired, we instead analyze difference scores. Difference scores can be computed by subtracting the first measurement from the second, or vice versa. In this example, it is more intuitive to compute difference scores by subtracting the final weight from the initial weight. Computing differences in this fashion allows for a more intuitive interpretation of the differences—in this case, differences reflect the weight lost by each participant over 6 weeks.

The following table displays the sample data, along with difference scores (*d*) and differences squared (d^2), which are required to compute the variance in difference scores:

Subject	Initial Weight	Final Weight	Difference (d)	Difference² (d²)
1	180	165	15	225
2	142	138	4	16
3	126	128	−2	4
4	138	136	2	4
5	175	170	5	25
6	205	197	8	64
7	116	115	1	1
8	142	128	14	196
9	157	144	13	169
10	136	130	6	36
			66	740

The summary statistics on the difference scores (d) are

$$\overline{X}_d = \frac{\sum d}{n} = \frac{66}{10} = 6.6$$

$$s_d^2 = \frac{\sum d^2 - (\sum d)^2/n}{n-1} = \frac{740 - (66)^2/10}{9}$$

$$= 33.82$$

$$s_d = \sqrt{33.82} = 5.82$$

The test of hypotheses concerning the mean difference of two dependent populations follows the same five steps illustrated in previous applications.

1. Set up hypotheses.

$$H_0: \mu_d = 0$$

$$H_1: \mu_d > 0, \quad \alpha = 0.05$$

where μ_d = mean difference in weights (or mean weight loss)

2. Select the appropriate test statistic.
   Because $n = 10$, the test statistic is

$$t = \frac{\overline{X}_d - \mu_d}{\dfrac{s_d}{\sqrt{n}}}$$

$$df = n - 1 = 10 - 1 = 9$$

3. Decision rule.

$$\text{Reject } H_0 \text{ if } t \geq 1.833$$

$$\text{Do not reject } H_0 \text{ if } t < 1.833$$

4. Compute the test statistic.

$$t = \frac{\overline{X}_d - \mu_d}{\frac{s_d}{\sqrt{n}}} = \frac{6.6 - 0}{\frac{5.82}{\sqrt{10}}} = 3.59$$

5. Conclusion.

Reject H_0 since $3.59 \geq 1.833$. We have significant evidence, $\alpha = 0.05$, to show that the mean weight loss following 6 weeks of the program is greater than zero. For this test, $p = 0.005$ (see Table B.3). A point estimate for the mean weight loss in the program is 6.6 pounds. ◼

SAS EXAMPLE 6.10 Testing Effectiveness of a Weight Loss Program Using SAS

SAS will conduct a two dependent samples test of hypothesis in its Means procedure using an approach similar to that used in the one-sample test of hypotheses illustrated in Chapter 5. Recall, in the one-sample test SAS assumes that the test of interest is H_0: $\mu = 0$ versus H_1: $\mu \neq 0$. In this application, we are interested in testing if the mean *difference* in weights is significantly greater than zero. In order to use the SAS Means procedure, we first create a new variable for each subject, which is simply the difference in weights computed by subtracting the final weight from the initial weight, that is, the difference (we call this variable diff in the SAS program) = initial − final weight. The difference score can be interpreted as the number of pounds lost over the course of the investigation—6 weeks in this example. The output from the SAS means procedure is shown here. A brief interpretation appears after the output.

SAS Output for Example 6.10

```
                    The MEANS Procedure
                  Analysis Variable : diff
  N        Mean       Std Dev      Std Error     t Value      Pr > |t|
 -----------------------------------------------------------------------
 10     6.6000000    5.8156876    1.8390819       3.59        0.0059
 -----------------------------------------------------------------------
```

Interpretation of SAS Output for Example 6.10

SAS provides the number of observations ($N = 10$). Again, SAS uses uppercase, but this does not denote a population size. The mean of diff (i.e., the mean difference in initial and final weights, or the mean number of pounds lost) is 6.6. The standard deviation in diff is 5.8157 and the standard error (i.e., $s_d/\sqrt{n}$) is 1.8391. The test statistic (computing by taking the ratio of the mean diff to its standard error) is 3.590. The two-sided p value is 0.0059. Because we are interested in a one-sided test, the following rule should be applied: Reject H_0 if p value/2 $< \alpha$. We reject H_0 since 0.0059/2 = 0.003 $< \alpha = 0.05$. Therefore, we have significant evidence, $\alpha = 0.05$, to show that the mean weight loss following 6 weeks of the program is greater than zero. ■

EXAMPLE 6.11

Testing Effect of Antihistamine and Alcohol on Functioning

Several years ago there were concerns about patients' appropriate use of over-the-counter medications, in particular the effects on functional abilities of over-the-counter antihistamines taken (inappropriately) in combination with alcohol. An investigation of these effects was undertaken according to a *crossover design*. In a crossover design, each participant is given each treatment under investigation (as opposed to only a single treatment, as was the case in previous examples).

Each subject had functional ability measured (defined as time in seconds to complete a physical task) in the presence of alcohol and the antihistamine and also had functional ability measured in the presence of alcohol and a placebo. The order of treatments (i.e., alcohol and antihistamine, alcohol and placebo) was randomly assigned to eliminate *carryover effects*. Carryover effects occur when subjects learn from one treatment and therefore improve under the second treatment solely due to the carryover from the first treatment and not due to the second treatment itself.

One hundred subjects were involved in the investigation. Differences in times to complete the task were taken as follows: Time under the influence of alcohol and antihistamine − Time under the influence of alcohol and placebo. Summary statistics on the differences in times to complete the task are as follows:

Number of subjects (n)	100
Mean difference in times *(i.e., increase in time due to antihistamine)*	25 sec
Standard deviation in difference in times	20 sec

The test of interest is as follows.

1. Set up hypotheses.

$$H_0: \mu_d = 0$$

$$H_1: \mu_d \neq 0, \quad \alpha = 0.05$$

2. Select the appropriate test statistic (since the sample size is large, $n = 100$).

$$Z = \frac{\overline{X}_d - \mu_d}{\frac{s_d}{\sqrt{n}}}$$

3. Decision rule.

$$\text{Reject } H_0 \text{ if } Z \geq 1.96 \text{ or if } Z \leq -1.96$$
$$\text{Do not reject } H_0 \text{ if } -1.96 < Z < 1.96$$

4. Compute the test statistic.

$$Z = \frac{\overline{X}_d - \mu_d}{\frac{s_d}{\sqrt{n}}} = \frac{25}{\frac{20}{\sqrt{100}}} = 12.5$$

5. Conclusion.

Reject H_0 since $12.5 \geq 1.96$. We have significant evidence, $\alpha = 0.05$, to show that there is a difference in the time to complete a physical task under the influence of alcohol and antihistamine as compared to alcohol and a placebo. For this example, $p = 0.0001$. ■

6.2 Power and Sample Size Determination

In Chapter 5 we introduced the concepts of precision (in estimation) and statistical power (in hypothesis testing). In the two-sample applications concerning $(\mu_1 - \mu_2)$, these same concepts are of interest.

As in the one-sample applications, power (i.e., Power $= 1 - \beta = P$(Reject $H_0 | H_0$ false)) depends on three components:

1. $n_i =$ sample sizes $(i = 1, 2)$
2. $\alpha =$ level of significance $= P$(Type I error)
3. ES (effect size) $=$ the standardized difference in means specified under H_0 and H_1

The relationship between each component 1–3 and statistical power is the same in the two-sample case as it was in the one-sample case. In the two-sample applications, the hypotheses are (assuming a two-sided test):

$$H_0: \mu_1 = \mu_2$$
$$H_1: \mu_1 \neq \mu_2$$

or equivalently,

$$H_0: \mu_1 - \mu_2 = 0$$
$$H_1: \mu_1 - \mu_2 \neq 0$$

where $\mu_1 =$ mean of population 1 and $\mu_2 =$ mean of population 2

The effect size (*ES*) is defined as the standardized difference in the values of the parameter of interest specified under H_0 and H_1. In the two-sample applications, the parameter of interest is $(\mu_1 - \mu_2)$. The *ES* is defined as follows:

$$ES = \frac{|(\mu_1 - \mu_2)_{H_1} - (\mu_1 - \mu_2)_{H_0}|}{\sigma} \tag{6.4}$$

Under H_0: $\mu_1 - \mu_2 = 0$, therefore the *ES* reduces to

$$ES = \frac{|(\mu_1 - \mu_2)_{H_1}|}{\sigma} = \frac{|(\mu_1 - \mu_2)|}{\sigma} \tag{6.5}$$

where σ = the common standard deviation (i.e., $\sigma_1 = \sigma_2 = \sigma$)

The power of a test is higher (or better) with larger sample sizes (n_1 and n_2) a larger level of significance and relative to a larger effect size. The following formula is used to determine the power of a two independent samples test with a two-sided alternative (i.e., H_0: $\mu_1 = \mu_2$ versus H_1: $\mu_1 \neq \mu_2$):

$$Power = P\left(Z > Z_{1-(\alpha/2)} - \frac{|\mu_1 - \mu_2|}{\sqrt{2\sigma^2/n}}\right) \tag{6.6}$$

where $\mu_1 - \mu_2$ is the difference in means under H_1
σ is the common standard deviation of the characteristic of interest
n is the common sample size (i.e., $n_1 = n_2 = n$)
$Z_{1-(\alpha/2)}$ is the Z value with lower-tail area $1 - \alpha/2$

We restrict our attention to situations where the sample sizes are equal ($n_1 = n_2$); there are formulas that can compute power for samples of unequal size.

EXAMPLE 6.12

Power in Test of Hypothesis for Differences in Means

Suppose we wish to conduct the following test:

$$H_0: \mu_1 = \mu_2$$
$$H_1: \mu_1 \neq \mu_2, \quad \alpha = 0.05$$

Find the power of the test if the difference in means is 2 units. Suppose that the standard deviation of the characteristic of interest is 3, and equal samples of size 20 are available.

$$Power = P\left(Z > Z_{1-(\alpha/2)} - \frac{|\mu_1 - \mu_2|}{\sqrt{2\sigma^2/n}}\right) = P\left(Z > 1.96 - \frac{2}{\sqrt{2(3)^2/20}}\right)$$

$$= P(Z > 1.96 - 2.11) = P(Z > -0.15) = 1 - 0.4404 = 0.5596$$

There is a 56% probability that this test will detect a difference of 2 units in means with samples of size 20 at a 5% level of significance.

Suppose for the same problem we are interested in the probability of detecting a difference of 3 units in means. Find the power of the test with all other conditions the same.

$$\text{Power} = P\left(Z > Z_{1-(\alpha/2)} - \frac{|\mu_1 - \mu_2|}{\sqrt{2\sigma^2/n}}\right) = P\left(Z > 1.96 - \frac{3}{\sqrt{2(3)^2/20}}\right)$$

$$= P(Z > 1.96 - 3.16) = P(Z > -1.20) = 1 - 0.1151 = 0.8849$$

There is an 88% probability that this test will detect a difference of 3 units in means with samples of size 20 at a 5% level of significance. ▪

EXAMPLE 6.13 **Estimating the Probability of a Type II Error**

Suppose we conducted a two-sided test of the equality of mean cholesterol levels under two competing treatments and failed to reject the null hypothesis at $\alpha = 0.05$. The sample data and an outline of the test are shown next. What is the probability that we committed a Type II error (i.e., find β) if the true difference in means is 15 units?

Summary Statistics	*Treatment 1*	*Treatment 2*
Sample size	20	20
Mean cholesterol level	205	190
Standard deviation	42	36

$$H_0: \mu_1 = \mu_2$$
$$H_1: \mu_1 \neq \mu_2, \quad \alpha = 0.05$$

Assuming equal variances (Case 2), reject H_0 if $t \geq 2.042$ or if $t \leq -2.042$.

$$S_p = \sqrt{\frac{42^2 + 36^2}{2}} = 39.1$$

$$t = \frac{205 - 190}{39.1\sqrt{\frac{1}{20} + \frac{1}{20}}} = 1.2$$

Do not reject H_0 because $-2.042 \leq 1.2 \leq 2.042$.

To find β we first compute power:

$$\text{Power} = P\left(Z > Z_{1-(\alpha/2)} - \frac{|\mu_1 - \mu_2|}{\sqrt{2\sigma^2/n}}\right) = P\left(Z > 1.96 - \frac{15}{\sqrt{2(39.1)^2/20}}\right)$$

$$= P(Z > 1.96 - 1.21) = P(Z > 0.75) = 1 - 0.7734 = 0.2266$$

So, $\beta = 1 - 0.2266 = 0.7734$. There is a very high probability that we committed a Type II error (relative to a difference of 15 units in mean cholesterol levels). Are the two treatments the same with respect to their effect on cholesterol levels? ■

In many applications, the number of subjects (i.e., n_1 and n_2) that can be involved depends on financial, logistic, and/or time constraints. In other cases, the sample size required to ensure a certain level of power can be determined relative to alternative hypotheses of importance. The sample size required to ensure a specific level of power in a *two-sided, two independent samples test* is

$$n_i = 2\left(\frac{Z_{1-(\alpha/2)} + Z_{1-\beta}}{ES}\right)^2 \tag{6.7}$$

where n_i is the minimum number of subjects required in sample
 $i(i = 1, 2)$
 $Z_{1-(\alpha/2)}$ is the value from the standard normal distribution with
  lower-tail area equal to $1 - \alpha/2$
 $Z_{1-\beta}$ is the value from the standard normal distribution with
  lower-tail area equal to $1 - \beta$
 ES is the effect size (6.5)

The sample size required to ensure a specific level of power in a *one-sided two independent samples test* is

$$n_i = 2\left(\frac{Z_{1-\alpha} + Z_{1-\beta}}{ES}\right)^2 \tag{6.8}$$

where n_i is the minimum number of subjects required in sample
 $i(i = 1, 2)$
 $Z_{1-\alpha}$ is the value from the standard normal distribution with
  lower-tail area equal to $1 - \alpha$
 $Z_{1-\beta}$ is the value from the standard normal distribution with
  lower-tail area equal to $1 - \beta$
 ES is the effect size (6.5)

The following example illustrates the use of formula (6.7).

EXAMPLE 6.14

Sample Size Determination in Tests for Differences in Means

A certain lung capacity measurement varies from day to day with a standard deviation of 0.3 liters. As individuals age from 30 to 50 years, their lung capacities decrease. Mean lung capacities decrease 0.02 liters/year for nonsmokers and 0.04 liters/year for smokers. Over 20 years, mean lung capacities decrease 0.4 (0.02 × 20) liters for nonsmokers and 0.8 (0.04 × 20) liters for smokers.

How many 30-year-old smokers and nonsmokers should be followed for 20 years for there to be a 90% chance of recognizing the difference at $\alpha = 0.05$? Assume a two-sided test will be conducted.

The formula to determine sample sizes is given by (6.7):

$$n_i = 2 \left(\frac{Z_{1-(\alpha/2)} + Z_{1-\beta}}{ES} \right)^2$$

The *ES* (6.5) is

$$ES = \frac{|\mu_1 - \mu_2|}{\sigma} = \frac{0.4}{0.3} = 1.33$$

At $\alpha = 0.05$, $Z_{1-(\alpha/2)} = Z_{0.975} = 1.96$. Similarly, for power $= 0.90$, $\beta = 0.10$, therefore $Z_{1-\beta} = 1.282$.

Substituting,

$$n_i = 2 \left[\frac{(1.96 + 1.282)}{1.33} \right]^2 = 11.88$$

Thus, $n_1 = n_2 = 12$ subjects (24 total) are needed. ■

EXAMPLE 6.15

Consider the application described in Example 6.13 where two treatments were compared for their effect on cholesterol levels. How many subjects would have been required in the study to ensure an 80% chance of detecting a difference of 20 units in cholesterol levels between treatments at $\alpha = 0.05$? Assume that a two-sided test will be conducted and use S_p to estimate the variation in cholesterol levels.

The *ES* (6.5) is

$$ES = \frac{|\mu_1 - \mu_2|}{\sigma} = \frac{20}{39.1} = 0.5$$

Since $\alpha = 0.05$, $Z_{1-(\alpha/2)} = Z_{0.975} = 1.96$. Similarly, for power $= 0.80$, $\beta = 0.20$; therefore, $Z_{1-\beta} = 0.84$.

Substituting,

$$n_i = 2\left[\frac{(1.96 + 0.84)}{0.5}\right]^2 = 62.72$$

Thus, $n_1 = n_2 = 63$ subjects (126 total) are needed. ■

SAS Example 6.14 Sample Size Determination in Tests for Differences in Means Using SAS

The following output was generated using SAS to determine the sample sizes (per group) required to ensure a specified power (we considered scenarios with 80% and 90% power) for differences in means of 0.4 and 0.3 units with a standard deviation of 0.3 units. A brief interpretation appears after the output.

SAS Output for Example 6.14

OBS	ALPHA	BETA	Z_ALPHA2	Z_BETA	MU1	MU2	SIGMA	POWER	ES	N_2
1	0.05	0.1	1.95996	1.28155	0.8	0.4	0.3	0.9	1.33333	12
2	0.05	0.1	1.95996	1.28155	0.8	0.5	0.3	0.9	1.00000	22
3	0.05	0.2	1.95996	0.84162	0.8	0.4	0.3	0.8	1.33333	9
4	0.05	0.2	1.95996	0.84162	0.8	0.5	0.3	0.8	1.00000	16

Interpretation of SAS Output for Example 6.14

There is no SAS procedure specifically designed to determine the number of subjects required to detect a specific difference in the means of two independent populations. But as we did in Chapter 5 to determine the sample size required to detect a specific effect size in the one-sample test of hypothesis, we use SAS to program appropriate formulas. Once the formulas are implemented, users can evaluate different scenarios easily. In the output shown, four scenarios are considered (denoted OBS 1–4). Scenario 1 corresponds to the situation presented in Example 6.14. Five variables are input into the program, the level of significance (alpha), the power (power), the mean for group 1 (mu1), the mean for group 2 (mu2), and the standard deviation (sigma). Several variables are created in the program and the values of all variables are printed in the output. A description of the variables and an interpretation of results follows.

In scenario 1 (OBS = 1), the level of significance (alpha) is set at 0.05 (5%); the probability of Type II error, β, is computed to be 0.10 (10%); the mean in group 1 (mu1) was specified as 0.8; the mean in group 2 was specified as

0.4 (mu2); the standard deviation (sigma) was specified at 0.3; and the power (power) was specified at 0.90 (90%). The effect size (es) was computed by dividing the absolute value of the difference in means between groups by the standard deviation. In scenario 1 (OBS = 1), the effect size is 1.33. Twelve subjects (n_2) are required per group to ensure that the probability of detecting a 0.4 unit difference in means is 90% (i.e., power = 0.90), with a *two-sided* level of significance of 5%. In scenario 2 (OBS = 2), we change the mean in group 2 to 0.5. The effect size is reduced to 1.00, and a total of 22 subjects are required per group to ensure that the probability of detecting a 0.3 unit difference in means is 90% (i.e., power = 0.90), with a *two-sided* level of significance of 5%.

In scenarios 3 and 4 (OBS = 3 and 4), we consider the same scenarios and reduce the power to 0.80 (80%). The result is that fewer subjects are required. Nine and sixteen subjects are required per group, respectively, to ensure that the probability of detecting a 0.4 and 0.3 unit difference in means is 80%, with a *two-sided* level of significance of 5%. ■

6.3 Key Formulas

Application	Notation/Formula	Description
Confidence interval estimate for $(\mu_1 - \mu_2)$	$(\overline{X}_1 - \overline{X}_2) \pm Z_{1-(\alpha/2)} S_p \sqrt{\dfrac{1}{n_1} + \dfrac{1}{n_2}}$	CI for two independent samples; see Tables 6.2, 6.3, and 6.5 for alternate formulas (find critical Z in Table B.2A)
Test statistic for $H_0: \mu_1 = \mu_2$	$Z = \dfrac{\overline{X}_1 - \overline{X}_2}{S_p \sqrt{\dfrac{1}{n_1} + \dfrac{1}{n_2}}}$	Test statistic for two independent samples; see Tables 6.2, 6.3, and 6.5 for alternate formulas (find critical Z in Table B.2B)
Confidence interval estimate for μ_d	$\overline{X}_d \pm Z_{1-(\alpha/2)} \dfrac{s_d}{\sqrt{n}}$	CI for two dependent samples; see Table 6.6
Test statistic for $H_0: \mu_d = 0$	$Z = \dfrac{\overline{X}_d}{s_d/\sqrt{n}}$	Test statistic for two dependent samples; see Table 6.6
Find power for test of $H_0: \mu_1 = \mu_2$	$\text{Power} = P\left(Z > Z_{1-(\alpha/2)} - \dfrac{|\mu_1 - \mu_2|}{\sqrt{2\sigma^2/n}}\right)$	Power of two-sided test of the equality of means
Find n_i to test $H_0: \mu_1 = \mu_2$	$n_i = 2\left[\dfrac{Z_{1-(\alpha/2)} + Z_{1-\beta}}{ES}\right]^2$, where $ES = \dfrac{|\mu_1 - \mu_2|}{\sigma}$	Sample sizes to detect ES with power $= 1 - \beta$

6.4 Statistical Computing

Following are the SAS programs used to run the two independent samples tests (Cases 2 and 3) and the two dependent samples test (Case 4) and to estimate the sample sizes to detect a specified effect size. The SAS procedures and brief descriptions are noted in the header to each example. Notes are provided to the right of the SAS programs (in blue) for orientation purposes and are not part of the programs. In addition, the blank lines in the programs are solely to accommodate the notes. Blank lines and spaces can be used throughout SAS programs to enhance readability. A summary of the SAS procedures used in the examples is provided at the end of this section.

SAS EXAMPLE 6.6 **Two Independent Samples Test**

Compare Mean Public Health Awareness Scores Between Men and Women (Example 6.6)

Random samples of 11 male high school students and 12 female high school students are selected within a particular school district and their scores on a public health awareness test are recorded. Use the following data to test if the male students score significantly *higher* than the female students on the test. Use SAS to run the appropriate test at a 5% level of significance.

Men	540	520	510	640	720	440	370	600	670	770	380	
Women	420	630	750	260	470	520	63	540	30	620	610	670

Program Code

```
options ps=62 ls=80;          Formats the output page to 62 lines in length and 80 columns in width
data in;                      Beginning of Data Step
  input gender $ test;        Inputs variables gender (a character variable) and test score (numeric).

cards;                        Beginning of Raw Data section.
male 540                      actual observations
male 520
male 510
male 640
male 720
male 440
male 370
male 600
male 670
male 770
male 380
female 420
```

```
female 630
female 750
female 260
female 470
female 520
female 630
female 540
female 530
female 620
female 610
female 670
run;
```

```
proc ttest;            Procedure call Proc Ttest to run two independent samples test of means.
  class gender;        Specification of grouping variable gender.
  var test;            Specification of analytic variable test.
run;                   End of procedure section.                                    ■
```

SAS EXAMPLE 6.8 Two Independent Samples Test

Compare Mean Numbers of Visits to Health Center Between Freshmen and Sophomores (Example 6.8)

The health department at a major university is interested in whether there is a difference in the mean number of visits to the student health center between college freshmen and sophomores. Data are collected from each of 50 freshmen and 60 sophomores reflecting the number of visits each made to the health center during the academic year. Use the data to test if there is a significant *difference* in the mean number of visits to the student health center between university freshmen and sophomores. Use SAS to run the appropriate test at a 5% level of significance. (*Note:* The raw data are not shown here and are abbreviated in the SAS program.)

Program Code

```
options ps=62 ls=80;        Formats the output page to 62 lines in length
                                and 80 columns in width

data in;                    Beginning of Data Step
  input year $ visits;      Inputs variables year (a character variable)
                                and number of visits (numeric).

cards;                      Beginning of Raw Data section.
freshman 4                  actual observations
freshman 0
```

```
freshman 6
.
.
.
sophomore 2
sophomore 1
sophomore 0
.
.
.
run;
```

`proc ttest;`	Procedure call Proc Ttest to run two independent samples test of means.
`  class year;`	Specification of grouping variable **year**.
`  var visits;`	Specification of analytic variable **visits**.
`run;`	End of procedure section. ■

SAS Example 6.10 Two Dependent Samples Test

Test the Effectiveness of a Weight-Loss Program (Example 6.10)

A nutrition expert is examining a weight-loss program to evaluate its effectiveness (i.e., whether participants lose weight on the program). Ten subjects are randomly selected for the investigation. Each subject's initial weight is recorded, they follow the program for 6 weeks, and they are weighed again. The data are

Subject	Initial Weight	Final Weight
1	180	165
2	142	138
3	126	128
4	138	136
5	175	170
6	205	197
7	116	115
8	142	128
9	157	144
10	136	130

Use SAS to test if there is evidence of a significant weight loss. Run the appropriate test at a 5% level of significance.

Program Code

```
options ps=62 ls=80;
```
Formats the output page to 62 lines in length and 80 columns in width

```
data in;
  input initial final;
```
Beginning of Data Step

Inputs variables **initial** and **final** weight for each subject. Because both weights were recorded on the same subject (matched/paired data), they are input from the same record. Distinct records reflect different (independent) observations.

```
diff=final-initial;
```
Creates new variable **diff** by taking the difference between the **final** and **initial** weights.

```
cards;
180 165
142 138
126 128
138 136
175 170
205 197
116 115
142 128
157 144
136 130
run;
```
Beginning of Raw Data section.

actual observations

```
proc means n mean std stderr t prt;
```
Procedure call Proc Means to run one sample test. Here we request specific options for testing (t and prt generate the test statistic and p value, respectively).

```
  var diff;
```
Specification of analytic variable **diff**.

```
run;
```
End of procedure section. ■

SAS EXAMPLE 6.14 Determine the Number of Subjects Required to Detect a Specific Effect Size in a Test of Hypothesis About $(\mu_1 - \mu_2)$

Sample Size Requirements (Example 6.14)

We wish to conduct the following test at the 5% level of significance:

$$H_0: \mu_1 = \mu_2 \quad \text{vs.} \quad H_1: \mu_1 \neq \mu_2$$

If the mean of group 1 is 0.8, how many subjects would be required to ensure that the probability of detecting a 0.4 (or a 0.3) unit difference is 80% (i.e., power = 0.80)? Also consider scenarios with 90% power and assume that $\sigma = 0.3$.

Program Code

```
options ps=62 ls=80;
```
Formats the output page to 62 lines in length and 80 columns in width

```
data in;
```
Beginning of Data Step.

```
  input alpha power mu1 mu2 sigma;
```
Inputs 5 variables **alpha, power, mu1, mu2** and **sigma.**

```
z_alpha2=probit(1-(alpha/2));
```
Determines the value from the standard normal distribution with lower-tail area 1—**alpha/2** (see $Z_{a/2}$ above)

```
beta=1-power;
```
Computes **beta.**

```
z_beta=probit(1-beta);
```
Determines the value from the standard normal distribution with lower-tail area 1-**beta** (see Z_b above)

```
es=abs(mu1-mu2)/sigma;
```
Computes the effect size (**es**)

```
tempn_2=2*((z_alpha2+z_beta)/es)**2;
```
Creates a temporary variable, called **tempn_2,** determined by formula (6.7).

```
n_2=ceil(tempn_2);
```
Computes a variable **n_2** using the ceil function which computes the smallest integer greater than **tempn_2** (i.e., rounds up).

```
/*
```
Beginning of comment section

```
 Input the following information (required)
 alpha: Level of Significance: range 0.0 to 1.0 (e.g., 0.05),
 power: Power: range 0.0 to 1.0 (e.g., 0.80), and
 mu1: Mean in Group 1,
 mu2: Mean in Group 2, and
 sigma: Standard Deviation
*/
```
End of comment section

```
cards;
```
Beginning of Raw Data section

```
0.05 0.90 0.8 0.4 0.3
```
actual observations

```
0.05 0.90 0.8 0.5 0.3
0.05 0.80 0.8 0.4 0.3
0.05 0.80 0.8 0.5 0.3
run;
```

```
proc print;
```
Procedure call. Print to display

```
  var alpha beta mu1 mu2 sigma power es n_2;
```
computed variables.

```
run;
```
■

6.4.1 Summary of SAS Procedures and Functions

The SAS Ttest procedure is used to run a two independent samples test of means. SAS also provides a preliminary test so that the user can determine if the population variances are equal or unequal. The Means procedure is used to run a one-sample test. Here we use the Means procedure to test if the mean difference (in the two dependent samples problem) is significantly different from zero. The procedures and specific options to perform the tests of hypothesis are shown in the table. Users should refer to the examples in this section for complete descriptions of the procedure and specific options.

Procedure	*Sample Procedure Call*	*Description*
proc ttest	proc ttest; class class_var; var analytic_var;;	Conducts a two independent samples test of equality of means. Class_var is the name of the (dichotomous) variable that distinguishes the two groups, and analytic_var is the name of the outcome or analytic variable (continuous).
proc means	proc means *n mean std stderr t prt*; var diff;	Conducts a one-sample test of hypothesis ($H_0: \mu_d = 0$ vs. $H_1: \mu_d \neq 0$)

The following SAS function was also used in the programs in this section to determine sample size requirements.

Function	*Sample Call Statement*	*Description*
probit	Z_alpha = probit(*alpha*);	Determines the value from the standard normal distribution with lower-tail area equal to *alpha*

6.5 Analysis of Framingham Heart Study Data

The Framingham data set includes data collected from the original cohort. Participants contributed up to three examination cycles of data. First we analyze data collected in the first examination cycle (called the period = 1 examination) and use SAS Proc ttest to test for differences in means between two independent groups. We then construct a data set that contains risk factors measured at period 1 and then again at period 2 (approximately 6 years later) and evaluate changes in risk factors over time using two dependent samples procedures.

Framingham Data Analysis—SAS Code

```
data in;
 set in.frmgham;
 if period=1;
run;

proc ttest data=in;
 class bpmeds;
 var age sysbp diabp;
run;
```

Framingham Data Analysis—SAS Output

The TTEST Procedure
Statistics

Variable	BPMEDS	N	Lower CL Mean	Mean	Upper CL Mean	Lower CL Std Dev	Std Dev
AGE	N	4229	49.398	49.658	49.918	8.434	8.6138
AGE	Y	144	54.938	56.215	57.493	6.9502	7.7542
AGE	Diff (1-2)		-7.984	-6.557	-5.131	8.4107	8.587
SYSBP	N	4229	131.11	131.76	132.4	20.923	21.369
SYSBP	Y	144	160.65	165.14	169.62	24.383	27.204
SYSBP	Diff (1-2)		-36.97	-33.38	-29.79	21.142	21.585
DIABP	N	4229	82.266	82.619	82.972	11.465	11.71
DIABP	Y	144	94.431	96.646	98.86	12.05	13.444
DIABP	Diff (1-2)		-15.98	-14.03	-12.07	11.529	11.771

Statistics

Variable	BPMEDS	Upper CL Std Dev	Std Err	Minimum	Maximum
AGE	N	8.8014	0.1325	32	70
AGE	Y	8.7702	0.6462	38	70
AGE	Diff (1-2)	8.7709	0.7277		
SYSBP	N	21.835	0.3286	83.5	295
SYSBP	Y	30.768	2.267	110	248
SYSBP	Diff (1-2)	22.047	1.8291		
DIABP	N	11.965	0.1801	48	142.5
DIABP	Y	15.205	1.1203	55	141
DIABP	Diff (1-2)	12.023	0.9974		

T-Tests

Variable	Method	Variances	DF	t Value	Pr > |t|
AGE	Pooled	Equal	4371	-9.01	<.0001
AGE	Satterthwaite	Unequal	155	-9.94	<.0001
SYSBP	Pooled	Equal	4371	-18.25	<.0001
SYSBP	Satterthwaite	Unequal	149	-14.57	<.0001
DIABP	Pooled	Equal	4371	-14.06	<.0001
DIABP	Satterthwaite	Unequal	150	-12.36	<.0001

```
                    Equality of Variances

Variable    Method      Num DF    Den DF    F Value    Pr > F

AGE         Folded F      4228       143      1.23     0.0980
SYSBP       Folded F       143      4228      1.62     <.0001
DIABP       Folded F       143      4228      1.32     0.0149
```

We used SAS in the output to test for significant differences in mean ages (variable name AGE), systolic blood pressures (SYSBP), and diastolic blood pressures (DIABP), considered separately, between participants taking and not taking antihypertensive medications (BPMEDS). The bpmeds variable is a dichotomous variable that distinguishes our two independent comparison groups. SAS provides descriptive statistics on each outcome variable and then information to test if the means are significantly different between groups. The preliminary tests for equality of variances are provided at the bottom of the output.

In terms of ages, are the variances in ages between groups similar or different? Because $p = 0.0980$, we would not reject the null hypothesis of equality (assuming we are using $\alpha = 0.05$) and would conclude that the variances are equal. In that case, the t statistic for testing if the mean ages are different is $t = -9.01$ with $p < 0.0001$. Thus, there is a highly significant difference in mean ages between persons taking and not taking antihypertensive medications. The mean age of participants taking antihypertensives is 56.2 as compared to 49.7 for participants not taking antihypertensives.

Framingham Data Analysis—SAS Code

```
data fhs1;
  set in.frmgham;
  if period=1;

bmi1=bmi;
sysbp1=sysbp;
diabp1=diabp;
totchol1=totchol;
keep randid bmi1 sysbp1 diabp1 totchol1;
run;

data fhs2;
  set in.frmgham;
  if period=2;

bmi2=bmi;
sysbp2=sysbp;
diabp2=diabp;
```

```
totchol2=totchol;
keep randid bmi2 sysbp2 diabp2 totchol2;
run;

proc sort data=fhs1; by randid; run;
proc sort data=fhs2; by randid; run;
data fhs12;
 merge fhs1(in=a) fhs2(in=b);
 by randid;
 if a and b;

diffbmi=bmi2-bmi1;
diffsysbp=sysbp2-sysbp1;
diffdiabp=diabp2-diabp1;
difftotchol=totchol2-totchol1;
run;

proc means n mean std stderr t prt data=fhs12;
 var bmi1 bmi2 diffbmi;
run;

proc means n mean std stderr t prt data=fhs12;
 var diffsysbp diffdiabp difftotchol;
run;
```

We created two data sets in the preceding code, called fhs1 and fhs2, which contain the risk factors measured at periods 1 and 2, respectively. We then merged them into a single data set so that the period 1 and period 2 values are together (one SAS record per participant). In the merged data set, we created difference variables by subtracting the risk factors measured at period 1 from the risk factors measured at period 2. The differences reflect how much the risk factors changed over time (specifically how they increased from period 1 to period 2). In the following, we used SAS Proc Means to test for differences in risk factors over time.

Framingham Data Analysis—SAS Output

The MEANS Procedure

Variable	N	Mean	Std Dev	Std Error	t Value	Pr > |t|
bmi1	3920	25.8407449	4.0685278	0.0649822	397.66	<.0001
bmi2	3914	25.8980710	4.1225299	0.0658951	393.02	<.0001
diffbmi	3909	0.0678306	1.8015165	0.0288141	2.35	0.0186

Recall that SAS Proc Means assumes that the test of interest is H_0: $\mu = 0$. For illustration, we included the BMI measured at period 1 (BMI1), the BMI measured at period 2 (BMI2), and the difference between them (DIFFBMI). The mean BMI at period 1 was 25.8; the mean BMI at period 2 was 25.9. The difference, or increase in BMI over time, is 0.07 units. The p values in the rightmost column are used to test if the mean of each variable is zero. Both BMI1 and BMI2 have means that are highly significantly different from zero. These tests are not useful. The test of interest is in the last row—the test of whether the mean difference in BMI over time is significantly different from zero. We found that BMI increased by 0.07 units over time and that difference is statistically significant at $p = 0.0186$. Why is such a small difference statistically significant?

One other item of importance is the sample size (denoted N). Notice that the sample size goes from $n = 3920$ in period 1 to $n = 3914$ in period 2 and that for the difference variable $n = 3909$. Some BMI measurements are missing at one or both exams. The difference variable can only be created in persons with both measurements. Analysts must always take notice of the sample size in analysis, especially in longitudinal studies where participants may drop out over time. Missing data can be problematic in that if there is a pattern to the missing data, the results of analyses can be biased. There is very little missing data in this example.

```
                 The MEANS Procedure

Variable       N       Mean       Std Dev   Std Error  t Value  Pr > |t|
-----------------------------------------------------------------------
diffsysbp     3930   5.2012723   16.6766197  0.2660184   19.55   <.0001
diffdiabp     3930   1.3819338   10.0109283  0.1596901    8.65   <.0001
difftotchol   3759  13.1657356   33.3060531  0.5432341   24.24   <.0001
-----------------------------------------------------------------------
```

How much did systolic blood pressure, diastolic blood pressure, and total cholesterol change over time? Are the differences in these risk factors over time statistically significant?

6.6 Problems

1. In recent years, there have been numerous incidents of mass layoffs, due in part to the changing political and economic climate. In 15 facilities within the medical field sampled, the mean number of employees laid off was 138 ($s = 15.4$). In 20 facilities in the government contracting business sampled, the mean number of employees laid off was 175 ($s = 20.9$). Assume that the medical and contracting facilities are of approximately equal size and that the variations in numbers of layoffs

are comparable. Test if there is a difference in the mean number of employee layoffs using a 5% level of significance.

2. A pediatrician is interested in the long-term effects of an experimental medical treatment designed to improve joint flexibility in children affected with arthritis. A measure of joint flexibility is taken on 12 children randomly selected for the study. After using the experimental treatment for 12 months, a second measure of joint flexibility is taken. The mean increase in flexibility is 4.6 units with a standard deviation of 2.1 units. Construct a 90% confidence interval for the true mean increase in joint flexibility. (Assume that the difference scores are approximately normally distributed.)

3. To investigate the numbers of hours that graduate students work in addition to a full-time class load, a random sample of 20 male graduate students is selected who work a mean of 16.4 hours per week with a variance of 4.7 hours. A second random sample of 20 female graduate students is selected who work a mean of 13.8 hours per week with a variance of 6.1 hours. Construct a 95% confidence interval for the difference in the mean numbers of hours worked between male and female graduate students. Assume that the variances are equal.

4. A pharmaceutical company is running a test comparing two forms of advertising for the same product, one of which involves a television campaign and the other a print campaign. Market sectors are randomly assigned to receive a particular form of advertising, and product sales are recorded during the month following the ad campaigns. Using the data provided, test if there is a significant difference in product sales by method of advertisement. Use $\alpha = 0.05$.

	Number of Markets	Mean Sales	SD
Television:	80	125	24.5
Print:	120	150	46.1

5. A randomized trial is conducted to evaluate the effectiveness of a newly developed treatment for joint pain in patients with arthritis. The treatment is compared to an established treatment that has been shown to be effective. Two hundred patients with arthritis agree to participate and are randomly assigned to either the newly developed treatment or to the established treatment. Among the several clinical outcomes in the investigation we focus on a secondary outcome, quality of life (QOL). QOL scores (ranging from 0 to 100, with higher scores indicative of better QOL) are measured on each patient. The

following data are observed:

Treatment	No. of Patients	Mean QOL	SD QOL
Newly developed	100	80.2	5.7
Established	100	75.4	6.1

a. Construct a 95% confidence interval for the difference in mean QOL scores between treatments.

b. Based on (a), is there a significant difference in mean QOL scores between treatments? Justify your answer.

6. Following head and/or neck trauma, patients are recommended for long-term physical therapy. An observational study is conducted to assess whether there is a significant difference in the mean numbers of physical therapy sessions attended by male and female patients suffering head and/or neck trauma.

Gender	Number of Patients	Mean Number of Physical Therapy Sessions Attended	Standard Deviation
Male	20	14.6	6.3
Female	15	18.1	5.9

a. Construct a 95% confidence interval for the difference in mean numbers of physical therapy sessions attended between male and female patients suffering head and/or neck trauma.

b. Based on (a), is there a significant difference in the mean numbers of physical therapy sessions attended between male and female patients? Justify your answer briefly.

7. A randomized trial is run to compare two competing medications for peripheral vascular disease. One of the outcomes is self-reported physical functioning. After taking the assigned medication for 6 weeks, patients provide data on their abilities to perform various physical activities, and a score is computed for each individual. The physical functioning scores range from 0 to 100, with higher scores indicative of better functioning. The data are

Medication	n	$\overline{X}$	s
1	25	70.5	24.3
2	25	76.3	21.6

Generate a 95% confidence interval for the difference in mean physical functioning scores between medications.

8. A clinical trial is conducted to compare an intervention to a control with respect to medication adherence. The intervention treatment consists of personalized scheduling and education regarding medication adherence. The control treatment reflects standard care. For the purposes of the study, medication adherence is measured as the percent of prescribed doses of medication taken over a 3-month period (e.g., a score of 100 indicates that all medications were taken as prescribed, 50 indicates that only half of all medications were taken as prescribed). Use the data given to test whether medication adherence is significantly higher in the intervention group. Run the appropriate test at $\alpha = 0.05$. Assume equal variances between groups.

	Intervention	*Control*
Number of patients	75	75
Mean medication adherence	85	72
Standard deviation	12.6	14.3

9. In a study comparing two competing medications for asthma, 16 subjects are randomized to one of the competing treatments. The data shown reflect asthma symptom scores for patients assigned to each treatment. Higher scores are indicative of worse asthma symptoms. Test if there is a significant difference in the mean asthma symptom scores between medications. Run the appropriate test at a 5% level of significance.

Treatment A:	55	60	80	65	72	78	68	71
Treatment B:	80	82	86	89	76	81	90	76

10. Suppose we are interested in whether there is a difference between the mean numbers of sick days taken by men and women in a local company. The numbers of sick days taken by men and women in the study sample are shown. Use the data to run the appropriate test at a 5% level of significance.

Men:	5	10	2	0	6	4	5	15
Women:	8	9	3	5	0	4	15	

11. A study reports that a newly developed medication reduces systolic blood pressure in patients with hypertension. When the medication ultimately makes it to market, a research group in a particular hospital wishes to test its effectiveness among their hypertensive patients. A sample of 12 patients with hypertension agree to participate, and their systolic blood pressures are measured before initiating treatment (baseline) and after 6 weeks of medication treatment. The mean

reduction in systolic blood pressures is 15 units (baseline − posttreatment) with a standard deviation of 12 units. Test if the medication significantly reduces systolic blood pressure. Run the appropriate test at a 5% level of significance.

12. A research group wishes to conduct a randomized trial to compare a new medication to a medication they consider standard care. One hundred patients with hypertension are enrolled and randomized to one of the two comparison treatments. After taking the assigned medication for 6 weeks, their systolic blood pressure (SBP) is measured. Summary statistics are given here. Use the data to test if there is a significant difference in systolic blood pressures between medication groups. Run the appropriate test at a 5% level of significance.

Treatment	Number of Patients	Mean SBP	Standard Deviation
New	50	130	12
Standard	50	135	10

13. The following table describes a randomized trial comparing an experimental medication to a placebo for treatment of reflux.

Patient Characteristics	Experimental Treatment (n = 100)	Placebo (n = 100)	p
Mean (SD) age (years)	52 (5.6)	54 (5.2)	0.0090
Gender: % Female	63%	58%	0.4363
Mean (SD) educational level (years)	13 (3.2)	10 (2.1)	0.0453
Mean (SD) annual income ($000s)	$42 ($4.5)	$39 ($5.6)	0.0918
Clinical characteristics			
Mean (SD) number of episodes per week	8 (3.2)	6 (2.9)	0.0001
Severity of symptoms: % Minimal	21%	15%	0.0021
% Moderate	45%	21%	
% Severe	24%	64%	

a. Are there any statistically significant differences between the patients assigned to the experimental treatment and placebo? Justify your answer briefly.

b. Consider the comparison of treatments with respect to patient age. Run the appropriate test using the data provided.

c. Consider the comparison of treatments with respect to the number of episodes per week. Run the appropriate test using the data provided.

14. We wish to investigate whether there is a significant increase in CD4 cell count following a course of experimental antiretroviral therapy. A sample of 20 patients agree to participate and as a comparative measure, each patient's CD4 cell count is measured prior to taking the experimental therapy. After taking the therapy, each patient's CD4 count is again measured. Using the following data, test whether there is a significant increase in CD4 cell counts following the experimental therapy. Run the appropriate test at $\alpha = 0.05$.

Statistic	CD4 Count Before Therapy	CD4 Count After Therapy	Increase
n	20	20	20
Mean	474	479	5
SD	8.1	10.2	6.7

15. A study is conducted to assess whether females spend significantly more money out-of-pocket on prescription medications than males. The study is restricted to male and female patients in the same health plan who are 65 years of age or older. Samples of men and women are selected at random and asked the total number of dollars they spent on prescription medications over the last year. Use the following data to run the appropriate test. Use a 5% level of significance.

Gender	Number of Patients	Mean Dollars Spent	Standard Deviation in Dollars Spent
Male	25	390	57
Female	40	425	65

16. The following table describes the results of a research project in which two different pain medications were compared.

Characteristic	Pain Medication 1	Pain Medication 2	p
Age (years)	48	51	0.0854
Educational level (years)	13.1	12.9	0.3425
Annual income	$38,456	$41,254	0.4351
No. visits to MD in past year	4.2	5.6	0.0211

Are there any statistically significant differences between groups with respect to the characteristics shown? Justify your answer (be brief but complete).

17. The following table summarizes data collected in a study comparing patients between hospitals. The variable summarized is body mass

index (BMI), computed as the ratio of weight in kilograms to height in meters squared.

BMI	Enrollment Site	
	Hospital 1	Hospital 2
n	100	100
Mean	21.6	24.8
SD	2.1	1.8

Test if there is a significant difference in the mean BMI scores between hospitals. Run the appropriate test at a 5% level of significance.

18. The following data reflect body mass index scores for men and women who are considered at high risk for coronary heart disease. Construct a 95% confidence interval for the true difference in BMI scores between men and women at high risk for cardiovascular disease.

BMI	High Risk	
	Men	Women
n	20	10
Mean	31.6	28.1
SD	1.7	2.1

19. We wish to design a study to compare two antihypertensive medications. Two outcome variables will be considered: systolic and diastolic blood pressure.

   a. How many subjects would be required to detect with 80% power and $\alpha = 0.05$ a difference of 20 units in mean systolic blood pressures between groups? Assume that the standard deviation in systolic blood pressure is 25.

   b. How many subjects would be required to detect with 80% power and $\alpha = 0.05$ a difference of 10 units in mean diastolic blood pressures between groups? Assume that the standard deviation in diastolic blood pressure is 21.

   c. What is the minimum number of subjects we need to satisfy both (a) and (b)?

   d. Suppose that this study involves enrolling patients and measuring their blood pressures after 6 months—at which time the medications should show an effect. If 10% of the patients drop out of the study within 6 months, how many patients need to be enrolled to satisfy the requirements of (c)?

20. We wish to design a study to compare an experimental therapy to a standard therapy in patients with diabetes. Subjects will be randomly assigned to one of the two therapy groups, and outcomes of treatment will be measured at 6 months and 12 months after initiation of treatment. One of the primary outcomes is blood sugar level, which will be measured at 6 months. The other primary outcome is total cholesterol level, measured at 12 months. Summary statistics on the two outcome measures from a similar study are as follows: mean (SD) of blood sugar level = 6.7 (2.1); mean (SD) of total cholesterol level = 215 (38). We anticipate losing 10% of the enrolled subjects within 6 months and an additional 10% in the period from 6 to 12 months.

    a. How many subjects would we need to enroll in order to detect with 80% power (assume $\alpha = 0.05$) a difference of 1.8 units in mean blood sugar levels between treatment groups?

    b. How many subjects would we need to enroll in order to detect with 80% power (assume $\alpha = 0.05$) a difference of 18 units in mean total cholesterol levels between treatment groups?

    c. What is the minimum number of subjects we would need to enroll to satisfy both (a) and (b)?

21. A nutritionist is investigating the effects of a rigorous walking program on systolic blood pressure (SBP) in patients with mild hypertension. Subjects who agree to participate have their systolic blood pressure measured at the start of the study and then after completing the 6-week walking program. The data are

Subject	Starting SBP	Ending SBP
1	140	128
2	130	125
3	150	140
4	160	162
5	135	137
6	128	130
7	142	135
8	151	140

Based on the data, is there evidence that SBP is significantly reduced by the walking program? Run the appropriate test at the 5% level of significance.

22. A medical treatment is compared to a surgical treatment for gallstones. The comparison of treatments is based on a patient-reported symptom score measured 3 weeks posttreatment. The symptom score ranges from 0 to 40, with higher scores indicative of worse symptoms. Suppose the

20 patients who received the medical treatment reported a mean symptom score of 20 with a standard deviation of 8, and the 20 patients who received the surgical treatment reported a mean symptom score of 14 with a standard deviation of 6. Construct a 95% confidence interval estimate for the difference in mean symptom scores between the medical and surgical treatments.

23. An antismoking campaign is being evaluated prior to its implementation in high schools across the state. A pilot study involving 6 volunteers who smoke is conducted. Each volunteer reports the number of cigarettes smoked the day before enrolling in the study, and each then is subjected to the antismoking campaign, which involves educational material, support groups, formal programs designed to reduce or quit smoking, and so on. After 4 weeks, each volunteer again reports the number of cigarettes smoked the day before. Based on the following pilot data, does it appear that the program is effective?

| At enrollment | 21 | 15 | 8 | 6 | 12 | 20 |
| After campaign | 12 | 10 | 10 | 6 | 10 | 20 |

24. We wish to test the effects of a new diet on weights in animals. Two groups of 20 animals each are to be compared. If the standard deviation in weights is assumed to be 10 pounds in each group, what is the probability that a two independent sample t test will recognize a difference of 1 pound between groups, using a 5% level of significance? What is the probability that a two independent sample t test will recognize a difference of 2 pounds between groups, using a 5% level of significance?

25. The following table displays the background characteristics of subjects who participated in a trial to compare two different medications.

Background Characteristic	Medication 1 ($n = 100$)	Medication 2 ($n = 100$)	p
Mean (SD) age (years)	45 (7.2)	43 (8.1)	0.6453
Gender: % Male	65%	48%	0.0253
Educational level			
% Less than high school	8%	6%	0.0502
% High school graduate	12%	14%	
% Some college	36%	38%	
% College graduate	34%	36%	
% Postgraduate	10%	6%	
Mean (SD) annual income ($)	$41,352 ($8754)	$37,459 ($9687)	0.2736
Race: % Nonwhite	36%	39%	0.5342

a. Are there any statistically significant differences in background characteristics between patients receiving medications 1 and 2? Justify your answer briefly.

b. Write out the hypotheses tested and give the formula for the test statistic (no calculations) for comparing ages and incomes of the participants.

c. What is the power of this test to detect a difference in mean ages on the order observed between medication groups? Use $\alpha = 0.05$.

d. What is the power of this test to detect a 10-year difference in mean ages between medication groups? Use $\alpha = 0.05$.

SAS Problems

Use SAS to solve the following problems.

1. In a study comparing two competing medications for asthma 16 subjects are randomized to one of the competing medication treatments. The data shown reflect asthma symptom scores for patients assigned to each treatment. Higher scores are indicative of worse asthma symptoms. Use SAS Proc ttest to generate summary statistics on the symptom scores for each treatment, and run the appropriate test at a 5% level of significance.

| Treatment A: | 55 | 60 | 80 | 65 | 72 | 78 | 68 | 71 |
| Treatment B: | 80 | 82 | 86 | 89 | 76 | 81 | 90 | 76 |

2. Suppose we are interested in whether there is a difference between the mean numbers of sick days taken by men and women in a local company. The numbers of sick days taken by men and women are shown. Use SAS Proc ttest to generate summary statistics on the numbers of sick days taken by men and women, and run the appropriate test at a 5% level of significance.

| Men: | 5 | 10 | 2 | 0 | 6 | 4 | 5 | 15 |
| Women: | 8 | 9 | 3 | 5 | 0 | 4 | 15 | |

Use SAS to generate a 95% confidence interval for the difference in the mean number of sick days taken by men and women.

3. We wish to evaluate a program designed to improve health status in older patients with Type II diabetes. Health status is measured on a scale of 0–100, with higher scores indicative of better health. Health status measures are taken on each subject at baseline and then again after participating in the program. Based on the following data, is there

evidence that the program significantly improves health status? Use SAS Proc Means to run the appropriate test at a 5% level of significance.

Baseline	80	55	63	76	88	45	65	77
Postprogram	85	50	65	78	82	55	68	90

4. We wish to design a study in which a two independent samples t test will be run, and we wish to consider several scenarios and to estimate the required sample sizes for each. Use SAS to determine the sample sizes required for each scenario to ensure 80% power. Suppose the mean for group 1 is 80. Consider means for group 2 of 90, 95, and 120; levels of significance of 0.05 and 0.01; standard deviations of 7 and 10 (for a total of $3 \times 2 \times 2 = 12$ scenarios).

5. An antismoking campaign is being evaluated prior to its implementation in high schools across the state. A pilot study involving 6 volunteers who smoke is conducted. Each volunteer reports the number of cigarettes smoked the day before enrolling in the study. Each then is subjected to the antismoking campaign, which involves educational material, support groups, formal programs designed to reduce or quit smoking, and so on. After 4 weeks, each volunteer again reports the number of cigarettes smoked the day before. Based on the following pilot data, does it appear that the program is effective?

At enrollment	21	15	8	6	12	20
After campaign	12	10	10	6	10	20

Use SAS Proc Means to run the appropriate test at a 5% level of significance.

Descriptive Statistics
(Ch. 2)

Probability
(Ch. 3)

Sampling Distributions
(Ch. 4)

Statistical Inference
(Chapters 5–13)

OUTCOME VARIABLE	GROUPING VARIABLE(S)/ PREDICTOR(S)	ANALYSIS	CHAPTER(S)
Continuous	—	Estimate μ; Compare μ to Known, Historical Value	5/12
Continuous	Dichotomous (2 groups)	Compare Independent Means (Estimate/Test ($\mu_1 - \mu_2$)) or the Mean Difference (μ_d)	6/12
Continuous	Discrete (>2 groups)	Test the Equality of k Means using Analysis of Variance ($\mu_1 = \mu_2 = \cdots = \mu_k$)	9/12
Continuous	Continuous	Estimate Correlation or Determine Regression Equation	10/12
Continuous	Several Continuous or Dichotomous	Multiple Linear Regression Analysis	10
Dichotomous	—	Estimate p; Compare p to Known, Historical Value	7
Dichotomous	Dichotomous (2 groups)	Compare Independent Proportions (Estimate/Test ($p_1 - p_2$))	7/8
Dichotomous	Discrete (>2 groups)	Test the Equality of k Proportions (Chi-Square Test)	7
Dichotomous	Several Continuous or Dichotomous	Multiple Logistic Regression Analysis	11
Discrete	Discrete	Compare Distributions Among k Populations (Chi-Square Test)	7
Time Event	Several Continuous or Dichotomous	Survival Analysis	13

7

Categorical Data

In Chapter 2 we defined variables as either *continuous* or *discrete*. Continuous (or measurement) variables assume, in theory, any value between the minimum and maximum value on a given measurement scale. Discrete variables take on a limited number of values or categories and can be either *ordinal* or *categorical* (sometimes called *nominal*) variables. Ordinal variables take on a limited number of values or categories and the categories are ordered. For example, socioeconomic status (SES) is an ordinal variable with the following response categories: lower SES, lower-middle, middle, upper-middle, and upper SES. Categorical variables take on a limited number of categories and the categories are unordered. For example, hospital type is a categorical variable with the following response categories: teaching, nonteaching. Statistical inference techniques applied to continuous (and sometimes ordinal) variables are concerned with means (μ) of those variables. Statistical inference techniques applied to discrete variables are concerned, instead, with the proportions of subjects in each response category.

In this chapter, we present statistical inference techniques for categorical variables. We begin the discussion with the case of dichotomous variables from the binomial distribution (i.e., each observation takes on one of two possible values, usually denoted success and failure) and then consider applications involving variables from multinomial distributions (i.e., each observation takes on one of several—more than two—possible values). We discuss one-sample and two-sample techniques for proportions analogous to the techniques presented in Chapters 5 and 6 relative to means.

In Section 7.1, we present statistical inference techniques for the one-sample case in which the analytic variable is dichotomous. In Section 7.2, we discuss cross-tabulation tables and several measures used to compare proportions between two independent populations. In Section 7.3, we discuss the evaluation of diagnostic tests. In Section 7.4, we present statistical inference techniques for the two-sample case in which the analytic variable is dichotomous. In Section 7.5, we consider analytic variables from the multinomial distribution and introduce chi-square tests. In Section 7.6, we discuss power and sample size determination. Key formulas are summarized in Section 7.7 and statistical computing applications are presented in Section 7.8. In Section 7.9, we use data from the Framingham Heart Study to illustrate the applications presented here.

7.1 Statistical Inference Concerning p

Recall the binomial distribution, in which each observation takes on one of two possible values, called success and failure. Suppose for analytic purposes, successes are coded as 1s and failures are coded as 0s. The parameter of interest is the proportion of successes in the population, or the *population proportion,* denoted p. The population proportion is defined as

$$p = \frac{\text{Number of successes}}{\text{Population size}} = \frac{X}{N} \qquad (7.1)$$

The two areas of statistical inference concerning p are estimation and hypothesis testing. The goal in estimation is to make valid inferences about the population proportion based on a single random sample from the population. There are two types of estimates for the population proportion: *the point estimate* and *the confidence interval estimate*. The point estimate is the "best" single-number estimate of the population proportion and is given by the sample proportion, denoted $\hat{p}$ (shown next). The confidence interval estimate is a range of plausible values for the population proportion.

$$\hat{p} = \frac{\text{Number of successes in the sample}}{n} = \frac{X}{n} \qquad (7.2)$$

EXAMPLE 7.1

Point Estimate for Proportion of Patients with Osteoarthritis

Suppose we wish to estimate the proportion of patients in a particular physician's practice with diagnosed osteoarthritis. The medical records of a random sample of 200 patients are reviewed for the diagnosis of osteoarthritis. Suppose that 38 patients are observed with diagnosed osteoarthritis. A point estimate for the proportion of all patients in this physician's practice with osteoarthritis (7.2) is given by: $\hat{p} = 38/200 = 0.19$, or 19%. ▪

As in applications concerning means, it is useful to know the standard error in order to assess the variation in the point estimate—in this case, the sample proportion. The standard error of the sample proportion is given by the following:

$$s.e.(\hat{p}) = \sqrt{\frac{p(1-p)}{n}} \qquad (7.3)$$

where *s.e.* = standard error
p = the population proportion

In most applications, the population proportion, p, is unknown. For large samples, the following can be used to estimate the standard error of the sample proportion:

$$s.e.(\hat{p}) = \sqrt{\frac{\hat{p}(1-\hat{p})}{n}} \qquad (7.4)$$

For applications involving binomial variables, a large sample is one with at least 5 successes and 5 failures. A large sample is defined as one that satisfies the following: $\min(n\hat{p}, n(1-\hat{p})) \geq 5$ (i.e., the smaller of $n\hat{p}$ and $n(1-\hat{p})$ must be greater than or equal to 5).

In Example 7.1, $\min(n\hat{p}, n(1-\hat{p})) = \min(200(0.19), 200(1-0.19)) = \min(38, 162) = 38 \geq 5$; therefore, the sample is sufficiently large. The standard error of the sample proportion is $\sqrt{0.19 * 0.81/200} = \sqrt{0.0007695} = 0.028$.

Table 7.1 *Statistical Inference Concerning p*

Attributes	Test Statistic*	Confidence Interval[†]
Simple random sample from binomial population, large sample	$Z = \dfrac{\hat{p} - p_0}{\sqrt{\dfrac{p_0(1 - p_0)}{n}}}$ where p_0 = value specified under H_0	$\hat{p} \pm Z_{1-(\alpha/2)}\sqrt{\dfrac{\hat{p}(1 - \hat{p})}{n}}$

*$\min(np_0, n(1 - p_0)) \geq 5$;
[†]$\min(n\hat{p}, n(1 - \hat{p})) \geq 5$

Confidence intervals for the population proportion can be generated using the techniques described in previous chapters. For large samples, we can appeal to the Central Limit Theorem for the derivation of the appropriate confidence interval (and test statistic in the test of hypothesis applications). Table 7.1 contains the formulas for the confidence interval for p and the test statistic for tests concerning p.

Table B.2A in the Appendix contains the values from the standard normal distribution for commonly used confidence levels ($Z_{1-(\alpha/2)}$). When the sample size is large (i.e., if and only if $\min(n\hat{p}, n(1 - \hat{p})) \geq 5$), the confidence interval formula given in Table 7.1 is appropriate. If the sample size is not sufficiently large, alternative formulas are available that are based on the binomial distribution and not the normal approximation, which is given here.

EXAMPLE 7.2

Estimating Proportion of Patients with Osteoarthritis

Consider the data from Example 7.1 and compute a 95% confidence interval for the proportion of all patients in the physician's practice with diagnosed osteoarthritis. The appropriate formula is given in Table 7.1:

$$\hat{p} \pm Z_{1-(\alpha/2)} \sqrt{\frac{\hat{p}(1 - \hat{p})}{n}}$$

Substituting the sample data and the appropriate value from Table B.2A for 95% confidence:

$$0.19 \pm 1.96\sqrt{\frac{0.19(1 - 0.19)}{200}}$$

$$0.19 \pm 1.96(0.028)$$

$$0.19 \pm 0.0549$$

$$(0.135, 0.245)$$

Thus, we are 95% confident that the true proportion of patients in this physician's practice with diagnosed osteoarthritis is between 13.5% and 24.5%. ■

SAS EXAMPLE 7.2 **Estimating Proportion of Patients with Osteoarthritis Using SAS**

The following output was generated using SAS Proc Freq, which generates a frequency distribution table for a categorical (or ordinal) variable. In this example, we record whether each subject has been diagnosed with osteoarthritis (or not). The usual convention is to assign scores of 1 to successes (i.e., diagnosis of osteoarthritis) and scores of 0 to failures (i.e., free of osteoarthritis). The input data consists of designations (0 or 1) for each subject. A brief interpretation appears after the output.

SAS Output for Example 7.2

```
                 The FREQ Procedure


                                   Cumulative     Cumulative
   x      Frequency     Percent     Frequency       Percent
   ---------------------------------------------------------
   0         162         81.00         162           81.00
   1          38         19.00         200          100.00


              Binomial Proportion for x = 1
              -------------------------------
              Proportion                 0.1900
              ASE                        0.0277
              95% Lower Conf Limit       0.1356
              95% Upper Conf Limit       0.2444


              Exact Conf Limits
              95% Lower Conf Limit       0.1381
              95% Upper Conf Limit       0.2513


               Test of H0: Proportion = 0.5

              ASE under H0               0.0354
              Z                         -8.7681
              One-sided Pr <  Z          <.0001
              Two-sided Pr > |Z|         <.0001


                   Sample Size = 200
```

Interpretation of SAS Output for Example 7.2

Of interest in the frequency distribution table produced by SAS are the frequencies (i.e., the total numbers) of respondents with 0s and 1s and the percent of respondents with 0s and 1s. There are 38 (out of 200) respondents scored as 1 (success = diagnosis of osteoarthritis). The percent of respondents with the diagnosis is 19%. SAS then provides a 95% confidence interval for the proportion as 13.56% to 24.44%. SAS also provides exact confidence limits, which are used when the sample size is small (i.e., when we do not satisfy $\min(n\hat{p}, n(1 - \hat{p})) \geq 5$). In the last part of the output, SAS provides a test of hypothesis, specifically, the test of H_0: $p = 0.5$, which may or may not be of interest. In the following examples, we illustrate the procedure for a test of hypothesis for a population proportion. ■

EXAMPLE 7.3

Testing Proportion of Cases with Abnormality Correctly Detected Against a Referent

Suppose that a diagnostic test has been shown to be 80% effective in detecting a genetic abnormality in human cells. An investigator modifies the diagnostic testing protocol and wishes to test if the new protocol has a detection rate that is significantly different from 80% in specimens known to possess the abnormality. The new protocol is applied to 300 independent specimens of human cells known to possess the abnormality. The abnormality is detected in 222 specimens. Run the appropriate test at a 5% level of significance.

1. Set up hypotheses.

$$H_0: p = 0.80$$
$$H_1: p \neq 0.80, \quad \alpha = 0.05$$

2. Select the appropriate test statistic.
   First, we check whether or not the sample size is sufficiently large:

$$\min(np_0, n(1 - p_0)) = \min(300(0.8), 300(1 - 0.8))$$
$$= \min(240, 60) = 60 \geq 5 \ ✔$$

The appropriate test statistic is given in Table 7.1:

$$Z = \frac{\hat{p} - p_0}{\sqrt{\dfrac{p_0(1 - p_0)}{n}}}$$

3. Decision rule (see Table B.2B in the Appendix for the appropriate critical value).

$$\text{Reject } H_0 \text{ if } Z \leq -1.960 \text{ or if } Z \geq 1.960$$
$$\text{Do not reject } H_0 \text{ if } -1.960 < Z < 1.960$$

4. Test statistic.

Substituting the sample data and the value of p specified in H_0 (i.e., $p_0 = 0.80$):

$$\hat{p} = \frac{222}{300} = 0.74$$

$$Z = \frac{\hat{p} - p_0}{\sqrt{\dfrac{p_0(1 - p_0)}{n}}}$$

$$Z = \frac{0.74 - 0.80}{\sqrt{\dfrac{0.80(1 - 0.80)}{300}}} = \frac{-0.06}{0.023} = -2.61$$

5. Conclusion.

Reject H_0 since $-2.61 \leq -1.960$. We have significant evidence, $\alpha = 0.05$, to show that the modified protocol has a significantly different detection rate than 80% (the rate for the original diagnostic test). The detection rate is lower with the modified protocol (74%) as compared to the original. For this example, $p < 0.010$ (see Table B.2B). ■

EXAMPLE 7.4

Testing Proportion of Cases with Flu Following Vaccination Against Referent

In the winter of 2002, 15% of all pediatric outpatient visits at a particular clinic were due to a single strain of flu. An investigator hypothesizes that the proportion of visits due to flu will decrease if patients are provided with flu shots. Suppose that flu shots were given to a random sample of 125 pediatric patients in the fall of 2003. These patients were tracked over the following winter to assess whether or not they come to clinic for flu (visits for other illnesses or injuries were not counted). Of these patients, 12% were seen in the winter of 2003–2004 for flu. Based on the data, is there evidence of a significant reduction in the proportion of patients seen in the clinic for flu after receiving the flu shot? Use a 5% level of significance.

1. Set up hypotheses.

$$H_0\colon p = 0.15$$
$$H_1\colon p < 0.15, \quad \alpha = 0.05$$

2. Select the appropriate test statistic.

First, check whether or not the sample size is sufficiently large:

$$\min(np_0, n(1 - p_0)) = \min(125(0.15), 125(1 - 0.15))$$
$$= \min(18.75, 106.25) = 18.75 \geq 5 \; ✔$$

The appropriate test statistic is given in Table 7.1:

$$Z = \frac{\hat{p} - p_0}{\sqrt{\dfrac{p_0(1 - p_0)}{n}}}$$

3. Decision rule (see Table B.2B in the Appendix for the appropriate critical value).

$$\text{Reject } H_0 \text{ if } Z \leq -1.645$$
$$\text{Do not reject } H_0 \text{ if } Z > -1.645$$

4. Test statistic.

$$\hat{p} = 0.12$$

$$Z = \frac{\hat{p} - p_0}{\sqrt{\dfrac{p_0(1 - p_0)}{n}}}$$

Substituting the sample data and the value of p specified in H_0 (i.e., $p_0 = 0.15$):

$$Z = \frac{0.12 - 0.15}{\sqrt{\dfrac{0.15(1 - 0.15)}{125}}} = \frac{-0.03}{0.032} = -0.938$$

5. Conclusion.

   Do not reject H_0 since $-0.938 > -1.645$. We do not have significant evidence, $\alpha = 0.05$, to show a reduction in the proportion of patients seen in the clinic for flu after receiving the vaccine. ■

Example 7.4 brings up an important issue of clinical versus statistical significance. In the formal test of hypothesis, we failed to reach statistical significance. However, we may have committed a Type II error (e.g., a larger sample size may be required to detect an effect). In any statistical application it is extremely important to look at the direction and magnitude of the observed effect. In Example 7.4, there is a reduction in the proportion of flu cases seen following the vaccines. The point estimate is 0.12, or 12%. Our test did not indicate that this was statistically significantly lower than 15%; however, there is a reduction and it should be evaluated carefully. Is this reduction clinically important? On a different note, was the study design we used optimal to address the question of effectiveness of the flu shots in pediatric patients? A concurrent comparison group might have provided a better comparison than historical data (i.e., $p_0 = 0.15$). We will discuss tests with a concurrent comparison group in the following sections.

7.2 Cross-Tabulation Tables

In applications involving discrete variables, *cross-tabulation* tables are often constructed to display the data. Cross-tabulation tables are also called $R \times C$ ("*R* by *C*") tables, where *R* denotes the number of rows in the table and *C* denotes the number of columns. A 2×2 table is illustrated in Example 7.5.

EXAMPLE 7.5 **Cross-Tabulation to Summarize Proportions in Two Populations**

A longitudinal study is conducted to evaluate the long-term complications in diabetic patients treated under two competing treatment regimens. Complications are measured by incidence of foot disease, eye disease, *or* cardiovascular disease within a 10-year observation period. The following 2×2 cross-tabulation table summarizes the data:

	Long-Term Complications		
Treatment	*Yes*	*No*	*Total*
Treatment 1	12	88	100
Treatment 2	8	92	100
Total	20	180	200

The estimate of the population proportion of all patients who develop complications under treatment 1 (p_1) is $\hat{p}_1 = 12/100 = 0.12$, by (7.2). This is equivalent to the estimate of the probability that a single patient develops complications under treatment 1. The estimate of the probability that a single patient develops complications under treatment 2 is $\hat{p}_2 = 8/100 = 0.08$. ▪

The probability of success or outcome (in Example 7.5, the outcome of interest is the development of complications) is often called the *risk* of outcome. There are a number of statistics used to compare risks of outcomes between populations (or between treatments). These statistics are called *effect measures* and are described in detail in Chapter 8.

SAS EXAMPLE 7.5 **Generating Cross-Tabulations Using SAS**

The following output was generated using SAS Proc Freq, which generates a contingency table (or cross-tabulation) when two variables are specified. A brief interpretation appears after the output.

SAS Output for Example 7.5

```
The FREQ Procedure
                       Table of trt by compl
                  Frequency|
                  Percent  |
                  Row Pct  |        compl
                  Col Pct  |yes      |z_no     |  Total
                  ---------+---------+---------+
                  trt_1    |    12 |      88 |    100
                           |  6.00 |   44.00 |  50.00
                           | 12.00 |   88.00 |
                           | 60.00 |   48.89 |
          trt     ---------+---------+---------+
                  trt_2    |     8 |      92 |    100
                           |  4.00 |   46.00 |  50.00
                           |  8.00 |   92.00 |
                           | 40.00 |   51.11 |
                  ---------+---------+---------+
                  Total         20       180       200
                              10.00     90.00    100.00

                     Sample Size = 200
```

Interpretation of SAS Output for Example 7.5

SAS generates a contingency table and in each cell of the table displays the Frequency, the Percent, the Row Percent, and the Column Percent (see legend in top left corner of table). The Frequency is the number of subjects in each cell, and the Percent is the percent of all subjects in each cell. For example, there are 12 patients in the top left cell (i.e., subjects on treatment 1 who also had complications). These patients reflect 6% of the total sample (12/200 = 0.06). The Row Percent is the percent of subjects in the particular row that fall in that cell. For example, there are 100 patients on treatment 1, or 100 patients in the first row of the contingency table. The 12 patients who had complications reflect 12% of all patients on treatment 1 (12/100 = 0.12). The Column Percent is the percent of subjects in the particular column who fall in that cell. For example, there are 20 patients who report complications. These 20 patients appear in the first column of the contingency table. The 12 patients in treatment 1 who had complications reflect 60% of all patients who had complications (12/20 = 0.60). The row total and column total (called the marginal totals) are displayed to the right and at the bottom of the contingency table, respectively. Both row and column frequencies and percents (of total) are displayed. ■

7.3 Diagnostic Tests: Sensitivity and Specificity

A diagnostic test is a tool used to detect outcomes or events that are not directly observable. For example, an individual may have a condition or disease that is not directly observable by a physician. A diagnostic test designed to detect such a condition can be used as a tool to assist the physician in detection. Desirable properties in diagnostic tests include the following:

■ The diagnostic test will indicate an event when the event is present, and

■ The diagnostic test will indicate a nonevent when the event is absent.

EXAMPLE 7.6   **Estimating Sensitivity and Specificity**

A clinical trial is conducted to evaluate a diagnostic screening test designed to detect chromosomal fetal abnormalities. Chromosomal fetal abnormalities are confirmed using amniocentesis. The diagnostic test is performed on a random sample of 200 pregnant women, who later undergo an amniocentesis. The following 2 × 2 cross-tabulation table summarizes the data:

| | Diagnostic Test | | |
Amniocentesis	Positive	Negative	Total
Abnormal (Disease)	14	6	20
Normal (No Disease)	64	116	180
Total	78	122	200

Based on amniocentesis, the estimate of the population proportion of all women carrying fetuses with chromosomal abnormalities (p) is $\hat{p} = 20/200 = 0.10$, by (7.2). ∎

The following statistics are used to describe diagnostic tests: the *sensitivity* of the test, the *specificity* of the test, the *predictive value positive* (PV$^+$) and the *predictive value negative* (PV$^-$). These statistics are defined as follows:

$$Sensitivity = P(\text{Positive test} \mid \text{Disease}) \tag{7.5}$$
$$Specificity = P(\text{Negative test} \mid \text{No disease})$$
$$Predictive\ value\ positive = P(\text{Disease/Positive test})$$
$$Predictive\ value\ negative = P(\text{No disease/Negative test})$$

In Example 7.6, the estimate of the sensitivity of the test is $14/20 = 0.70$. The estimate of the specificity is $116/180 = 0.64$. The estimate of the predictive value positive is $PV^+ = 14/78 = 0.18$, and the estimate of the predictive negative is $PV^- = 116/122 = 0.95$.

In most cases, higher sensitivities and higher specificities are desirable. There are instances, however, where a better test is determined by only one criterion (e.g., higher sensitivity).

SAS EXAMPLE 7.6 **Estimating Sensitivity and Specificity Using SAS**

The following output was generated using SAS Proc Freq, which generates a contingency table (or cross-tabulation) when two variables are specified. SAS does not produce the estimates of sensitivity, specificity, false positive rate, and false negative rate directly, but these can be extracted from the contingency table. The statistics of interest are described after the output.

SAS Output for Example 7.6

```
                     The FREQ Procedure
                  Table of amnio by diagtest
              amnio       diagtest
              Frequency|
              Percent   |
              Row Pct   |
              Col Pct   |positive|negative|  Total
              ---------+--------+--------+
              abnormal |     14 |      6 |     20
                       |   7.00 |   3.00 |  10.00
                       |  70.00 |  30.00 |
                       |  17.95 |   4.92 |
              ---------+--------+--------+
              normal   |     64 |    116 |    180
                       |  32.00 |  58.00 |  90.00
                       |  35.56 |  64.44 |
                       |  82.05 |  95.08 |
              ---------+--------+--------+
              Total          78      122      200
                           39.00    61.00    100.
```

Interpretation of SAS Output for Example 7.6

The sensitivity is the proportion of abnormal cases correctly classified by the test as positive, $14/20 = 0.70$. This is the Row Percent in the top left cell of the table. The specificity is the proportion of normal cases that are correctly classified by the test as negative, $116/180 = 0.644$. This is the Row Percent of the bottom right cell of the table. The predictive positive value is the proportion of normal cases classified as positive that are, in fact diseased, PV^+. This is the Column Percent of the top left cell of the table. The predictive value Negative is the proportion of cases classified as Negative that are, in fact, normal, $PV^- = 116/122 = 0.95$. This is the Column Percent of the bottom right cell of the table. ■

7.4 Statistical Inference Concerning $(p_1 - p_2)$

We often compare two independent populations with respect to the proportion of successes in each. A better study design to evaluate the effectiveness of flu shots in pediatric patients (Example 7.4) would involve two comparison groups. One group would receive the flu shots and the other would receive a placebo shot (to maintain blinding—why is this important?). The analysis would then compare the groups with respect to the proportions of children who developed flu. In the two independent samples situation, one parameter of interest is the difference in proportions, the risk difference: $(p_1 - p_2)$, where $p_1 =$ the proportion of successes in population 1 and $p_2 =$ the proportion of successes in population 2.

The point estimate for the risk difference, or difference in independent proportions, is given by

$$\hat{p}_1 - \hat{p}_2 \tag{7.6}$$

where $\hat{p}_i =$ the sample proportion in population i $(i = 1, 2)$

If samples from both populations are sufficiently large (see criteria in Table 7.2), then the confidence interval formula shown in Table 7.2 can be used to estimate $(p_1 - p_2)$.

Table B.2A contains the values from the standard normal distribution for commonly used confidence levels. When the sample sizes are adequate (i.e., if and only if $\min(n_1 \hat{p}_1, n_1(1 - \hat{p}_1)) \geq 5$ *and* $\min(n_2 \hat{p}_2, n_2(1 - \hat{p}_2)) \geq 5$), the confidence interval formula given in Table 7.2 is appropriate. If either (or both) sample size(s) are not adequate, alternative formulas are available that are based on the binomial distribution and not the normal approximation given here.

Table 7.2 *Confidence Interval for $(p_1 - p_2)$*

Attributes	Confidence Interval
Simple random samples from binomial populations Independent populations Large samples:	$(\hat{p}_1 - \hat{p}_2) \pm Z_{1-(\alpha/2)}\sqrt{\left(\dfrac{\hat{p}_1(1 - \hat{p}_1)}{n_1}\right) + \left(\dfrac{\hat{p}_2(1 - \hat{p}_2)}{n_2}\right)}$
$\min(n_1 \hat{p}_1, n_1(1 - \hat{p}_1)) \geq 5$ *and*	where $\hat{p}_1 = X_1/n_1$ and $\hat{p}_2 = X_2/n_2$
$\min(n_2 \hat{p}_2, n_2(1 - \hat{p}_2)) \geq 5$	

EXAMPLE 7.7

Estimating Difference in Proportions of Children Who Use the Emergency Room Between Treatments

We want to evaluate the effectiveness of a new treatment for asthma. The new treatment is administered in an inhaler and will be compared to a standard treatment administered in the same way. Because asthma is a serious condition, it would be unethical to use a placebo comparator in this trial. Suppose our outcome variable is emergency room (ER) use for complications of asthma during a 6-month follow-up period. A random sample of 375 asthmatic children are selected from a registry, of which 250 are randomized to the new treatment group and 125 are randomized to the comparison group (standard treatment). Both groups are provided instruction on the proper use of their inhalers. This allocation scheme is called 2-to-1, where twice as many participants are randomized to the investigational treatment as the control. Both groups of children are followed for 6 months and monitored for ER use. Of the children on the new treatment, 60 used the ER during the 6 months for complications of asthma, and 19 of the children on the standard treatment used the ER for complications of asthma during the same period. Construct a 95% confidence interval for the difference in the proportions of asthmatic children on the new and standard treatments who used the ER during the 6-month follow-up period.

The data layout is as follows:

The sample proportions are

$$\hat{p}_1 = \frac{60}{250} = 0.24, \qquad \hat{p}_2 = \frac{19}{125} = 0.15$$

A point estimate for the difference in proportions is given by (7.6):

$$\hat{p}_1 - \hat{p}_2 = 0.24 - 0.15 = 0.09$$

Now, check whether or not the sample sizes are sufficiently large:

$$\min(n_1 \hat{p}_1, n_1(1 - \hat{p}_1)) = \min(250(0.24), 250(1 - 0.24))$$
$$= \min(60, 190) = 60 \geq 5 \ ✔$$

and

$$\min(n_2 \hat{p}_2, n_2(1 - \hat{p}_2)) = \min(125(0.15), 125(1 - 0.15))$$
$$= \min(19, 106) = 19 \geq 5 \ ✔$$

The formula from Table 7.2 is appropriate:

$$(\hat{p}_1 - \hat{p}_2) \pm Z_{1-(\alpha/2)} \sqrt{\left(\frac{\hat{p}_1(1 - \hat{p}_1)}{n_1}\right) + \left(\frac{\hat{p}_2(1 - \hat{p}_2)}{n_2}\right)}$$

Substituting the sample data and the appropriate value from Table B.2A for 95% confidence:

$$0.09 \pm 1.960 \sqrt{\frac{0.24(1 - 0.24)}{250} + \frac{0.15(1 - 0.15)}{125}}$$

$$0.09 \pm 1.960(0.042)$$

$$0.09 \pm 0.082$$

$$(0.008, 0.172)$$

Thus, we are 95% confident that the true difference in the population proportions of asthmatic children on the new treatment as compared to children on standard treatment who used the ER during a 6-month period is between 0.8% and 17.2%. Based on the confidence interval estimate, can we say that there is a significant difference in the proportions of asthmatic children on the new treatment as compared to the standard treatment who used the ER during a 6-month period? (*Hint:* Does the confidence interval estimate include 0?) Is the new treatment effective? Notice the direction of the effect. ■

> NOTE: The two-sample confidence interval concerning $(p_1 - p_2)$ estimates the *difference* in proportions, as opposed to the value of either proportion (as was the case in the one-sample applications).

In some applications, it is of interest to compare two populations on the basis of the proportions of successes in each using a formal test of hypothesis. Table 7.3 contains the test statistic for tests concerning $(p_1 - p_2)$.

Table 7.3 *Test Statistic for* $(p_1 - p_2)$

Attributes	Test Statistic
Large samples* Simple random samples from binomial populations Independent populations	$Z = \dfrac{\hat{p}_1 - \hat{p}_2}{\sqrt{\hat{p}(1 - \hat{p})\left(\dfrac{1}{n_1} + \dfrac{1}{n_2}\right)}}$
	where $\hat{p}_1 = X_1/n_1$ and $\hat{p}_2 = X_2/n_2$
	$\hat{p} = \dfrac{(X_1 + X_2)}{(n_1 + n_2)}$

*$\min(n_1\hat{p}_1, n_1(1 - \hat{p}_1)) \geq 5$ *and* $\min(n_2\hat{p}_2, n_2(1 - \hat{p}_2)) \geq 5$

EXAMPLE 7.8

Testing Difference in Proportions of Patients Who Experience Pain Relief Between Treatments

A new drug is being compared to an existing drug for its effectiveness in relieving headache pain. One hundred subjects who suffer from chronic headaches are randomly assigned to either Group 1: Existing Drug, or Group 2: New Drug. Subjects do not know which drug they are taking in this experiment. Subjects are provided with a single dose of the assigned drug and instructed to take the full dose as soon as they experience headache pain and to record whether or not they experience relief from headache pain within 60 minutes. Among the 50 subjects assigned to Group 1: Existing Drug, 28 reported relief from headache pain within 60 minutes. Among the 50 subjects assigned to Group 2: New Drug, 34 reported relief from headache pain within 60 minutes. Based on the data, is the proportion of subjects reporting relief from headache pain within 60 minutes under the New Drug significantly different from the proportion of subjects reporting relief within 60 minutes under the Existing Drug? Use a 5% level of significance.

The data layout is as follows:

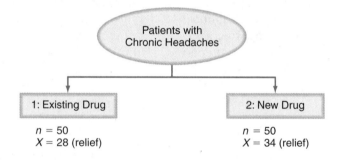

1. Set up hypotheses.

$$H_0: p_1 = p_2$$
$$H_1: p_1 \neq p_2, \quad \alpha = 0.05$$

where p_1 = the proportion of patients who experience relief from headache pain using the Existing Drug
p_2 = the proportion of patients who experience relief from headache pain using the New Drug

2. Select the appropriate test statistic.
The sample proportions are

$$\hat{p}_1 = \frac{28}{50} = 0.56, \qquad \hat{p}_2 = \frac{34}{50} = 0.68$$

First, check whether or not the sample sizes are sufficiently large:

$$\min(n_1 \hat{p}_1, n_1(1 - \hat{p}_1)) = \min(50(0.56), 50(1 - 0.56))$$
$$= \min(28, 22) = 22 \geq 5 \ \checkmark$$

and

$$\min(n_2 \hat{p}_2, n_2(1 - \hat{p}_2)) = \min(50(0.68), 50(1 - 0.68))$$
$$= \min(34, 16) = 16 \geq 5 \ \checkmark$$

The appropriate test statistic is given in Table 7.3:

$$Z = \frac{\hat{p}_1 - \hat{p}_2}{\sqrt{\hat{p}(1 - \hat{p})\left(\dfrac{1}{n_1} + \dfrac{1}{n_2}\right)}}$$

3. Decision rule.

Reject H_0 if $Z \leq -1.960$ or if $Z \geq 1.960$
Do not reject H_0 if $-1.960 < Z < 1.960$

4. Test statistic.

$$Z = \frac{\hat{p}_1 - \hat{p}_2}{\sqrt{\hat{p}(1 - \hat{p})\left(\dfrac{1}{n_1} + \dfrac{1}{n_2}\right)}}$$

We compute the estimate of the common proportion:

$$\hat{p} = \frac{X_1 + X_2}{n_1 + n_2} = \frac{28 + 34}{50 + 50} = 0.62$$

Note that $\hat{p}$ lies between $\hat{p}_1$ and $\hat{p}_2$.
   Now substituting the sample data:

$$Z = \frac{0.56 - 0.68}{\sqrt{0.62(1 - 0.62)\left(\dfrac{1}{50} + \dfrac{1}{50}\right)}} = \frac{-0.12}{0.097} = -1.24$$

5. Conclusion.
   Do not reject H_0 since $-1.960 < -1.24 < 1.960$. We do not have significant evidence, $\alpha = 0.05$, to show a difference in the proportions of subjects experiencing relief from headache pain with the New Drug within 60 minutes. ■

7.5 Chi-Square Tests

In the previous sections, we focused attention on variables from the binomial distribution (i.e., observations took on one of two possible values, called success and failure). We now consider variables from the multinomial distribution in which each observation takes on one of several (more than two) possible values. Some of the techniques described in the previous sections can be applied to sample data from a multinomial distribution. For example, instead of estimating the proportion of successes in the population, we estimate the proportion of subjects in response category 1, the proportion of subjects in response category 2, and so on.

Here we describe two tests, called the goodness-of-fit test and the test of independence, which are used for tests of hypotheses in the presence of multinomial data in one-sample and two-or-more sample applications, respectively. Both tests involve a test statistic that follows a chi-square distribution (χ^2).

7.5.1 Goodness-of-Fit Test

We begin with the case of a categorical variable measured in a single sample. The following example illustrates the goodness-of-fit test.

EXAMPLE 7.9

Goodness of Fit Test for Patient Preferences

Following coronary artery bypass graft (CABG) surgery, patients are encouraged to participate in a cardiac rehabilitation program. The program lasts approximately 14 weeks and includes exercise training, nutritional information, and general lifestyle guidelines. One particular hospital is offering three cardiac rehabilitation programs that are identical in content but are offered at three different times. The administration is interested in whether the three time slots are equally popular or convenient for the patients. A total of 100 patients are involved in the investigation, and each patient is asked to select the one day and time (from three options) that is most convenient. The following data are observed:

Time slot:	Mondays 6:00–7:30 PM	Thursdays 4:00–5:30 PM	Saturdays 8:00–9:30 AM
Number of patients:	47	32	21

The day and time variable follows a multinomial distribution with three response categories, denoted $k = 3$ (i.e., Mon. 6:00–7:30 PM, Thurs. 4:00–5:30 PM, Sat. 8:00–9:30 AM). The proportions of patients in the sample selecting each day and time are given by (7.2): $\hat{p}_1 = 47/100 = 0.47$, $\hat{p}_2 = 32/100 = 0.32$, and $\hat{p}_3 = 21/100 = 0.21$.

We wish to use the data to test the hypothesis that the three day and time options are equally popular. Mathematically, this is represented as follows:

1. Set up hypotheses.

H_0: $p_1 = p_2 = p_3$, $\alpha = 0.05$ (Population proportions are equal)
H_1: H_0 is false (Population proportions are not all equal)

NOTE: We do not write H_1: $p_1 \neq p_2 \neq p_3$, as we will reject H_0 in favor of H_1 if *any* of the three proportions are not equal, not only if *all* three are not equal (e.g., we will reject H_0 if $p_1 \neq p_2$, regardless of whether p_3 is equal to either p_1 or p_2).

The following are equivalent to the preceding hypotheses:

H_0: $p_1 = 0.33$, $p_2 = 0.33$, $p_3 = 0.33$
H_1: H_0 is false

2. Select the appropriate test statistic.

In tests involving multinomial distributions, the test statistic is no longer based on the sample proportion, but on the *observed frequencies*, or numbers of subjects in each response category. In this example, we observed 47 patients in the first response category, 32 in the second, and 21 in the third. If the null hypothesis were true—that is, if the true proportions of patients in each response category were equal (i.e., if $p_1 = p_2 = p_3 = 0.33$)—then we would have *expected* approximately 33 patients to select each time slot (since $n = 100$).

χ^2 *tests are based on the agreement between expected (under H_0) and observed (sample) frequencies.* The χ^2 statistic for testing whether the distribution of a single multinomial variable is as specified under H_0 is given by

$$\chi^2 = \sum \frac{(O - E)^2}{E} \tag{7.7}$$

where $\sum$ indicates summation over the k response categories
O = observed frequencies
E = expected frequencies (i.e., if H_0 is true, or under H_0)

The statistic follows a χ^2 distribution and has df $= k - 1$, where $k =$ the number of response categories. The general form of the χ^2 distribution is shown in Figure 7.1.

Notice that all of the values in the χ^2 distribution are greater than or equal to zero. In order to test the hypotheses of interest, we need to determine an appropriate critical value. In χ^2 tests, we reject the null hypothesis in favor of the alternative hypothesis if the value of the test statistic is large. The test statistic (7.7) is large when the observed and expected frequencies are not similar. In such a case, we reject H_0.

Figure 7.1 *Chi-Square Distribution*

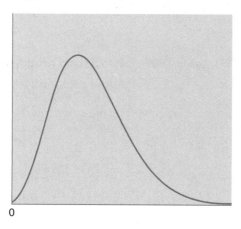

0

3. Decision rule.

In order to select the appropriate critical value, we first determine the degrees of freedom.

$$df = k - 1 = 3 - 1 = 2$$

The appropriate critical value of χ^2 is $\chi^2 = 5.99$ from Table B.5 in the Appendix. The decision rule is

$$\text{Reject } H_0 \text{ if } \chi^2 \geq 5.99$$
$$\text{Do not reject } H_0 \text{ if } \chi^2 < 5.99$$

4. Test statistic.

To organize the computations of the test statistic (7.7), the following table is used:

Time Slot:	Mondays 6:00–7:30 PM	Thursdays 4:00–5:30 PM	Saturdays 8:00–9:30 AM	Total
O = observed frequency:	47	32	21	100
E = expected frequency:	33.3	33.3	33.3	100
(O − E):	13.7	−1.3	−12.3	0
(O − E)²/E:	$(13.7)^2/33.3$ $= 5.64$	$(-1.3)^2/33.3$ $= 0.05$	$(-12.3)^2/33.3$ $= 4.54$	10.23

NOTE: The sum of the expected frequencies is equal to the sum of the observed frequencies ($n = 100$).

The test statistic is $\chi^2 = 10.23$.

5. Conclusion.
   Reject H_0 since $10.23 \geq 5.99$. We have significant evidence, $\alpha = 0.05$, to show that the three time slots are not equally popular or convenient for the patients. In fact, almost half (47%) of the patients selected the Monday 6:00–7:30 PM slot. For this example, $p < 0.01$ (see Table B.5). ▪

SAS EXAMPLE 7.9 **Goodness of Fit Test for Patient Preferences Using SAS**

The following output was generated using SAS Proc Freq with an option to run a goodness-of-fit test. Specifically, the user specifies the distribution of responses under the null hypothesis (see Section 7.9 for the SAS code).

SAS Output for Example 7.9

```
                    The FREQ Procedure
                                 Test     Cumulative    Cumulative
day        Frequency   Percent   Percent   Frequency     Percent
-----------------------------------------------------------------
Mon           47        47.00     33.00        47         47.00
Sat           21        21.00     33.00        68         68.00
Thurs         32        32.00     33.00       100        100.00

                 Chi-Square Test
             for Specified Proportions
             -------------------------
             Chi-Square          10.3333
             DF                        2
             Pr > ChiSq          0.0057

             Sample Size = 100
```

Interpretation of SAS Output for Example 7.9

SAS first generates a frequency distribution table and provides the number and percent of respondents in each response category. SAS then lists the Test Percent in each category (these are supplied by the user and reflect the expected proportions). The last two columns contain the cumulative frequencies and cumulative percents for the sample data. SAS then produces the chi-square statistic for the goodness-of-fit test along with degrees of freedom ($df = k - 1$) and a p value. Here $\chi^2 = 10.33$ and $p = 0.0057$. We therefore reject H_0 because $p = 0.0057 < 0.05$ and conclude that the three time slots are not equally popular or convenient for the patients. ▪

EXAMPLE 7.10   **Goodness of Fit Test for Teen Issues**

Volunteers at a teen hotline have been assigned based on the assumption that 40% of all calls are drug related, 25% are sex related (e.g., date rape), 25% are stress related, and 10% concern educational issues. For this investigation, each call is classified into one category based on the primary issue raised by the caller. To test the hypothesis, the following data are collected from 120 randomly selected calls placed to the teen hotline. Based on the data, is the assumption regarding the distribution of topic issues appropriate?

Topic Issue:	Drugs	Sex	Stress	Education
Number of calls:	52	38	21	9

1. Set up hypotheses.

H_0: $p_1 = 0.40$, $p_2 = 0.25$, $p_3 = 0.25$, $p_4 = 0.10$
H_1: H_0 is false

or

H_0: Distribution across categories is 0.40, 0.25, 0.25, 0.10
H_1: H_0 is false, $\alpha = 0.05$

2. Select the appropriate test statistic.

$$\chi^2 = \sum \frac{(O - E)^2}{E}$$

where $\sum$ indicates summation over the k response categories
O = observed frequencies
E = expected frequencies (i.e., if H_0 is true, or under H_0)

3. Decision rule.
   In order to select the appropriate critical value, we first determine the degrees of freedom.

$$df = k - 1 = 4 - 1 = 3$$

The appropriate critical value of χ^2 is $\chi^2 = 7.815$ from Table B.5. The decision rule is

Reject H_0 if $\chi^2 \geq 7.815$
Do not reject H_0 if $\chi^2 < 7.815$

4. Test statistic.

To organize the computations of the test statistic (7.7), the following table is used:

Topic Issue:	Drugs	Sex	Stress	Education	TOTAL
O = observed frequency:	52	38	21	9	120
E = expected frequency:	48	30	30	12	120
(O − E):	4	8	−9	−3	0
(O − E)²/E:	$(4)^2/48 =$ 0.33	$(8)^2/30 =$ 2.13	$(-9)^2/30 =$ 2.70	$(-3)^2/12 =$ 0.75	5.913

NOTE: The expected frequencies are computed assuming that H_0 is true, or under H_0: $120(0.40) = 48$, $120(0.25) = 30$, $120(0.25) = 30$, and $120(0.10) = 12$.

The test statistic is $\chi^2 = 5.913$.

5. Conclusion.

Do not reject H_0 since $5.913 < 7.815$. We do not have significant evidence, $\alpha = 0.05$, to show that the distribution of topic issues in the calls placed to the teen hotline is not as assumed (i.e., 40% drug related, 25% sex related, 25% stress related, and 10% education related). ■

EXAMPLE 7.11

Goodness of Fit Test for Genetic Abnormalities

Genetic counselors work with pregnant women (usually women at high risk of fetal abnormalities or those who might not be at high risk but screen positive for abnormalities based on standard screening tests) and hypothesize that about one-half of all abnormalities are trisomy 21, one-third are trisomy 18, and the remainder are trisomy 13. (Trisomy indicates three copies of a particular chromosome, e.g., 21, and reflects a particular abnormality associated with that chromosome.) To test the hypothesis, a sample of 200 pregnant women who deliver babies with abnormalities are studied. The specific abnormalities are recorded and are summarized here. Based on the data, is the assumption regarding the distribution of abnormalities appropriate?

	Trisomy 21	Trisomy 18	Trisomy 13	Total
Number of women:	107	70	23	200

1. Set up hypotheses.

$$H_0: p_1 = 0.50, p_2 = 0.33, p_3 = 0.17$$
$$H_1: H_0 \text{ is false}, \quad \alpha = 0.05$$

2. Select the appropriate test statistic.

$$\chi^2 = \sum \frac{(O - E)^2}{E}$$

where $\sum$ indicates summation over the k response categories
O = observed frequencies
E = expected frequencies (i.e., if H_0 is true, or under H_0)

3. Decision rule.
   In order to select the appropriate critical value, we first determine the degrees of freedom:

$$df = k - 1 = 3 - 1 = 2$$

The appropriate critical value of χ^2 is $\chi^2 = 5.99$ from Table B.5. The decision rule is

Reject H_0 if $\chi^2 \geq 5.99$
Do not reject H_0 if $\chi^2 < 5.99$

4. Test statistic.
   To organize the computations of the test statistic (7.7), the following table is used:

	Trisomy 21	Trisomy 18	Trisomy 13	TOTAL
O = observed frequency:	107	70	23	200
E = expected frequency:	100	66	34	200
(O − E):	7	4	−11	0
(O − E)²/E:	$(7)^2/100 = 0.49$	$(4)^2/66 = 0.24$	$(-11)^2/34 = 3.55$	4.29

NOTE: The expected frequencies are computed assuming that H_0 is true.

The test statistic is $\chi^2 = 4.29$.

5. Conclusion.
   Do not reject H_0 since $4.29 < 5.99$. We do not have significant evidence, $\alpha = 0.05$, to show that the distribution of abnormalities is not as assumed (i.e., 50% trisomy 21, 33% trisomy 18, and 17% trisomy 13). ■

7.5.2 Tests of Independence

We now consider applications involving two or more samples or two categorical variables, where interest lies in evaluating whether these two categorical variables are related (dependent) or unrelated (independent). The following example illustrates the use of the χ^2 test of independence.

EXAMPLE 7.12

Testing Independence Between Site and Treatment Regimen

The following data were collected in a multisite observational study of medical effectiveness in Type II diabetes. Three sites were involved: a health maintenance organization (HMO), a university teaching hospital (UTH), and an independent practice association (IPA). Type II diabetic patients were enrolled in the study from each site and monitored over a 3-year observation period. The data shown display the treatment regimens of patients measured at baseline by site.

Site	Diet & Exercise	Treatment Regimen Oral Hypoglycemics	Insulin	TOTAL
HMO:	294	827	579	1700
UTH:	132	288	352	772
IPA:	189	516	404	1109
TOTAL:	615	1631	1335	3581

The table is a 3×3 *cross-tabulation table* or a *contingency table*. Both site and treatment regimen are categorical variables. Site is called the *row variable* and treatment regimen is called the *column variable*. The number of rows in the table is denoted R and the number of columns is denoted C. In this example, $R = 3$ and $C = 3$. The *row totals* are shown on the right side of the table, and the *column totals* are shown at the bottom of the table. The row and column totals are called the *marginal totals*. The 9 combinations of site and treatment regimen are called the *cells* of the table (e.g., patients in the HMO treated by diet and exercise denote one cell of the table, patients in the HMO treated by oral hypoglycemics denote another, and so on).

We wish to use the data to test the hypothesis that the two variables (site and treatment regimen) are independent (i.e., no difference in treatment regimens across sites). The hypotheses are written as follows:

1. Set up hypotheses.

   H_0: Site and Treatment Regimen are independent

   (No relationship between site and treatment regimen)

   H_1: H_0 is false, $\alpha = 0.05$

   (Site and treatment regimen are related)

NOTE: In the test of independence, the hypotheses are generally expressed in words as opposed to mathematical symbols.

2. Select the appropriate test statistic.

   The test statistic in the test of independence is similar to the test statistic used in the goodness-of-fit test illustrated in the previous section. It is based on the *observed frequencies,* or numbers of subjects in each cell of the contingency table. In this example, we involved 1700 patients from the HMO, 772 from the UTH, and 1109 from the IPA. If the null hypothesis, H_0, is true (i.e., if there is no relationship between site and treatment regimen), we would expect the distribution of patients by treatment regimen to be similar across sites (i.e., the proportions of patients in each treatment regimen would be approximately equal in each of the sites).

   Again, χ^2 tests are based on the agreement between expected (under H_0) and observed (sample) frequencies. Recall from probability theory, that when two events are independent, the probability of their intersection is given by

   $$P(A \text{ and } B) = P(A) \cdot P(B) \tag{7.8}$$

For example, if site and treatment regimen are independent, then the probability that a patient is in the HMO *and* treated by diet and exercise is given by

$$P(\text{HMO and Diet/Exercise}) = P(\text{HMO}) \cdot P(\text{Diet/Exercise})$$
$$= (1700/3581)(615/3581) = (0.4747)(0.1717) = 0.0815$$

Similarly, if site and treatment regimen are independent, then the probability that a patient is in the UTH and treated by insulin is given by

$$P(\text{UTH and Insulin}) = P(\text{UTH}) \cdot P(\text{Insulin})$$
$$= (772/3581)(1335/3581) = (0.216)(0.373) = 0.0806$$

Therefore, if site and treatment regimen are independent, the probabilities (or proportions) of patients in each cell of the table can be computed using (7.8). To compute the test statistic, we must compute the *expected frequencies* (i.e., the *expected numbers* of patients in each cell of the table if site and treatment regimen are independent). Formula (7.8) yields the proportions of patients in each cell. To convert these proportions to frequencies, we use the following:

$$\text{Expected cell frequency} = n \cdot P(\text{cell}) \qquad (7.9)$$

This is equivalent to

$$\text{Expected cell frequency} = (\text{Row total} \cdot \text{Column total})/n \qquad (7.10)$$

For example, if site and treatment regimen are independent, then the expected number of patients in the HMO and treated by diet and exercise is given by (7.9) as

$$\text{Expected frequency (HMO and Diet/Exercise)} = 3581(0.0815) = 291.9$$

Equivalent to this by (7.10) is

$$\begin{aligned} &\text{Expected frequency (HMO and Diet/Exercise)} \\ &= (1700)(615)/3581 = 291.9 \end{aligned}$$

The χ^2 statistic for tests of independence is given by

$$\chi^2 = \sum \frac{(O - E)^2}{E} \qquad (7.11)$$

where $\sum$ indicates summation over all cells of the contingency table
O = observed frequencies
E = expected frequencies (i.e., if H_0 is true, or under H_0)

The statistic follows a χ^2 distribution and has df $= (R - 1)(C - 1)$, where R = the number of rows in the contingency table and C = the number of columns in the contingency table.

  Similar to the goodness-of-fit tests, we reject the null hypothesis in favor of the alternative hypothesis if the value of the test statistic is large. The test statistic (7.11) is large when the observed and expected frequencies are not similar.

3. Decision rule.
  In order to select the appropriate critical value, we first determine the degrees of freedom:

$$\text{df} = (R - 1)(C - 1) = (3 - 1)(3 - 1) = 2(2) = 4$$

The appropriate critical value of χ^2 is $\chi^2 = 9.49$ from Table B.5. The decision rule is

$$\text{Reject } H_0 \text{ if } \chi^2 \geq 9.49$$
$$\text{Do not reject } H_0 \text{ if } \chi^2 < 9.49$$

4. Test statistic.

   To organize the computations of the test statistic (7.11), we use the contingency table given earlier. The observed frequencies in each cell are shown. We compute the expected frequencies for each cell using (7.10) and display expected frequencies in parentheses to distinguish them from the observed frequencies. The computations are shown in detail for a few sample cells.

| Site | Treatment Regimen | | | |
	Diet & Exercise	*Oral Hypoglycemics*	*Insulin*	*TOTAL*
HMO:	294 ((1700*615)/ 3581 = 291.9)	827 ((1700*1631)/ 3581 = 774.3)	579 ((1700*1335)/ 3581 = 633.8)	1700
UTH:	132 ((772*615/ 3581 = 132.6)	288 (351.6)	352 (287.8)	772
IPA:	189 (190.5)	516 (505.1)	404 (413.4)	1109
TOTAL:	615	1631	1335	3581

NOTE: The marginal totals of the expected frequencies = the marginal totals of the observed frequencies. For example, $291.9 + 774.3 + 633.8 = 1700$. Similarly, $291.9 + 132.6 + 190.5 = 615$.

Using the observed and expected frequencies, we compute the test statistic (7.11):

$$\chi^2 = \sum \frac{(O - E)^2}{E} = \frac{(294 - 291.9)^2}{291.9} + \frac{(827 - 774.3)^2}{774.3} + \frac{(579 - 633.8)^2}{633.8}$$

$$+ \frac{(132 - 132.6)^2}{132.6} + \frac{(288 - 351.6)^2}{351.6} + \frac{(352 - 287.8)^2}{287.8}$$

$$+ \frac{(189 - 190.5)^2}{190.5} + \frac{(516 - 505.1)^2}{505.1} + \frac{(404 - 413.3)^2}{413.4}$$

$$\chi^2 = 0.014 + 3.359 + 4.732 + 0.003 + 11.509 + 14.320$$
$$+ 0.011 + 0.235 + 0.215 = 34.629$$

5. Conclusion.

   Reject H_0 since $34.629 \geq 9.49$. We have significant evidence, $\alpha = 0.05$, to show that site and treatment regimen are not independent (i.e., they are related). For this example, $p < 0.005$ (see Table B.5). Notice in the table that there are discrepancies between the observed and expected frequencies, particularly among the university teaching hospital patients. ■

SAS EXAMPLE 7.12 **Testing Independence Between Site and Treatment Regimen Using SAS**

The following output was generated using SAS Proc Freq. We requested a chi-square test of independence. In this example, we also requested some additional statistics, which are described following the output.

SAS Output for Example 7.12

```
                    The FREQ Procedure
                     Table of site by trt
        site            trt
        Frequency        |
        Expected         |
        Cell Chi-Square|
        Percent          |
        Row Pct          |
        Col Pct          |diet     |insulin |oral     |  Total
        ---------------+--------+--------+--------+
        hmo            |    294 |    579 |    827 |   1700
                       | 291.96 | 633.76 | 774.28 |
                       | 0.0143 | 4.7318 | 3.5895 |
                       |   8.21 |  16.17 |  23.09 |   47.47
                       |  17.29 |  34.06 |  48.65 |
                       |  47.80 |  43.37 |  50.71 |
        ---------------+--------+--------+--------+
        ipa            |    189 |    404 |    516 |   1109
                       | 190.46 | 413.44 | 505.1  |
                       | 0.0112 | 0.2154 | 0.235  |
                       |   5.28 |  11.28 |  14.41 |   30.97
                       |  17.04 |  36.43 |  46.53 |
                       |  30.73 |  30.26 |  31.64 |
        ---------------+--------+--------+--------+
        uth            |    132 |    352 |    288 |    772
                       | 132.58 | 287.8  | 351.61 |
                       | 0.0026 | 14.32  | 11.509 |
                       |   3.69 |   9.83 |   8.04 |   21.56
                       |  17.10 |  45.60 |  37.31 |
                       |  21.46 |  26.37 |  17.66 |
        ---------------+--------+--------+--------+
        Total               615     1335     1631     3581
                          17.17    37.28    45.55   100.00
```

```
          Statistics for Table of site by trt
Statistic                        DF       Value        Prob
------------------------------------------------------------
Chi-Square                        4      34.6291      <.0001
Likelihood Ratio Chi-Square       4      34.4975      <.0001
Mantel-Haenszel Chi-Square        1      10.5953      0.0011
Phi Coefficient                           0.0983
Contingency Coefficient                   0.0979
Cramer's V                                0.0695
               Sample Size = 3581
```

Interpretation of SAS Output for Example 7.12

SAS generates a contingency table and in each cell of the table displays a number of statistics. We requested some additional statistics here (compare this output with the outputs of SAS Examples 7.5 and 7.6). In each cell, SAS produces the Frequency, the Expected frequency (computed by formula (7.10)), the Cell Chi-Square (computed by formula (7.11) in each cell), the Percent, the Row Percent, and the Column Percent. (See legend in top left corner of the table.)

SAS generates a number of statistics that can be used to assess relationships between variables in a contingency table. We are concerned with the test of H_0: Site and Treatment Regimen are independent versus H_1: H_0 is false. SAS produces the χ^2 statistic $= 34.629$ and the associated p value, $p < 0.0001$. Assuming $\alpha = 0.05$, we reject H_0 since $p = 0.0001 < \alpha = 0.05$. We have significant evidence, $\alpha = 0.05$, to show that site and treatment regimen are not independent (i.e., they are related).

To understand the nature of the relationship, we evaluate the Row Percents. For example, among HMO patients, 17% are on Diet, 34% on Insulin, and 49% on Oral Hypoglycemics. Among the IPA patients, 17% are on Diet, 36% on Insulin, and 47% on Oral Hypoglycemics. There is little difference between the treatment regimens of patients at these two sites. The difference appears to be with the UTH patients. Among the UTH patients, 17% are on Diet, 46% on Insulin, and 37% on Oral Hypoglycemics. A higher proportion of IPA patients are on Insulin as compared to the other sites. This resulted in a statistically significant difference. Is it clinically meaningful? ■

EXAMPLE 7.13

Testing Independence Between Gender and Treatment Regimen

Consider the study described in Example 7.12. Suppose an investigator is interested in evaluating whether or not treatment regimens differ by gender. Restricting our analyses to the HMO patients, the following table displays the numbers of male and female patients according to their treatment regimens. Based on the data, is there evidence of a significant relationship between gender and treatment regimen among the HMO patients?

| | Treatment Regimen | | | |
Gender	Diet & Exercise	Oral Hypoglycemics	Insulin	TOTAL
Female:	147	435	256	838
Male:	147	392	323	862
TOTAL:	294	827	579	1700

In this contingency table, gender is the row variable ($R = 2$) and treatment regimen is the column variable ($C = 3$).

1. Set up hypotheses.

   H_0: Gender and Treatment Regimen are independent   (No relationship between gender and treatment regimen)

   H_1: H_0 is false, $\alpha = 0.05$ (Gender and treatment regimen are related)

2. Select the appropriate test statistic.

$$\chi^2 = \sum \frac{(O - E)^2}{E}$$

   where $\sum$ indicates summation over all cells of the contingency table
   O = observed frequencies
   E = expected frequencies (i.e., if H_0 is true, or under H_0)

3. Decision rule.

   In order to select the appropriate critical value, we first determine the degrees of freedom:

$$df = (R - 1)(C - 1) = (2 - 1)(3 - 1) = 1(2) = 2$$

   The appropriate critical value of χ^2 is $\chi^2 = 5.99$ from Table B.5. The decision rule is

   Reject H_0 if $\chi^2 \geq 5.99$

   Do not reject H_0 if $\chi^2 < 5.99$

4. Test statistic.

To organize the computations of the test statistic (7.11), we use the contingency table given previously. The observed frequencies in each cell are shown along with the expected frequencies computed by (7.10) and displayed in parentheses.

	Treatment Regimen			
Gender	*Diet & Exercise*	*Oral Hypoglycemics*	*Insulin*	*TOTAL*
Female:	147 (144.9)	435 (407.7)	256 (285.4)	838
Male:	147 (149.1)	392 (419.3)	323 (293.6)	862
TOTAL:	294	827	579	1700

Using the observed and expected frequencies, we compute the test statistic (7.11):

$$\chi^2 = \sum \frac{(O - E)^2}{E} = \frac{(147 - 144.9)^2}{144.9} + \frac{(435 - 407.7)^2}{407.7}$$

$$+ \frac{(256 - 285.4)^2}{285.4} + \frac{(147 - 149.1)^2}{149.1}$$

$$+ \frac{(392 - 419.3)^2}{419.3} + \frac{(323 - 293.6)^2}{293.6}$$

$$\chi^2 = 0.030 + 1.833 + 3.031 + 0.029 + 1.782 + 2.947 = 9.652$$

5. Conclusion.

Reject H_0 since $9.652 \geq 5.99$. We have significant evidence, $\alpha = 0.05$, to show that gender and treatment regimen are not independent (i.e., they are related) in the sample of HMO patients. For this example, $p < 0.01$ (see Table B.5). ▪

SAS EXAMPLE 7.13 Testing Independence Between Gender and Treatment Regimen Using SAS

The following output was generated using SAS Proc Freq. Again, we requested a chi-square test of independence but not as many statistics as were requested for SAS Example 7.12. A brief interpretation appears after the output.

SAS Output for Example 7.13

```
                    The FREQ Procedure
                   Table of gender by trt
        gender       trt
        Frequency|
        Percent  |
        Row Pct  |
        Col Pct  |diet     |insulin |oral    |   Total
        ---------+--------+--------+--------+
        female   |    147 |    256 |    435 |     838
                 |   8.65 |  15.06 |  25.59 |   49.29
                 |  17.54 |  30.55 |  51.91 |
                 |  50.00 |  44.21 |  52.60 |
        ---------+--------+--------+--------+
        male     |    147 |    323 |    392 |     862
                 |   8.65 |  19.00 |  23.06 |   50.71
                 |  17.05 |  37.47 |  45.48 |
                 |  50.00 |  55.79 |  47.40 |
        ---------+--------+--------+--------+
        Total         294      579      827     1700
                    17.29    34.06    48.65   100.00
```

```
          Statistics for Table of gender by trt
Statistic                           DF       Value       Prob
---------------------------------------------------------------
Chi-Square                           2       9.6519      0.0080
Likelihood Ratio Chi-Square          2       9.6684      0.0080
Mantel-Haenszel Chi-Square           1       2.6751      0.1019
Phi Coefficient                              0.0753
Contingency Coefficient                      0.0751
Cramer's V                                   0.0753
```

```
              Sample Size = 1700
```

Interpretation of SAS Output for Example 7.13

SAS generates a contingency table and in each cell of the table displays the Frequency, the Percent, the Row Percent, and the Column Percent. (See legend in top left corner of table.) The Frequency is the number of subjects in each cell; the Percent is the percent of all subjects in each cell. For example, there are 147 patients in the top left cell (i.e., females on diet and exercise treatment). These patients reflect 8.65% of the total sample (i.e., $147/1700 = 0.0865$). The Row Percent is the percent of subjects in the particular row that fall in that cell. For example, there are 838 female patients or 838 patients in the first row of the contingency table. The 147 female patients on diet and exercise treatment reflect 17.54% of all female patients (i.e., 147/838 = 0.1754). The Column Percent is the percent of subjects in the particular column that fall in that cell. For example, there are 294 patients treated by diet and exercise. These 294 patients appear in the first column of the contingency table. The 147 female patients on diet and exercise treatment reflect 50% of all patients treated by diet and exercise (i.e., 147/294 = 0.50). The row total and column total (called the marginal totals) are displayed to the right and at the bottom of the contingency table, respectively. Both row and column frequencies and percents (of total) are displayed.

SAS generates a number of statistics that can be used to assess relationships between variables in a contingency table. We are concerned with the test of H_0: Gender and Treatment Regimen are independent versus H_1: H_0 is false. SAS produces the χ^2 statistic = 9.652 and the associated p value, $p = 0.008$. Assuming $\alpha = 0.05$, we reject H_0 since $p = 0.008 < \alpha = 0.05$. We have significant evidence, $\alpha = 0.05$, to show that gender and treatment regimen are not independent (i.e., they are related) in the sample of HMO patients. What is the nature of the relationship between gender and treatment? ■

NOTE: Chi-square tests are valid when the expected frequencies in each cell (or response category) are greater than or equal to 5. If an expected frequency falls below 5, then alternative techniques should be used.

EXAMPLE 7.14

Testing Independence Between Center and Genetic Abnormalities

Consider the study described in Example 7.11 involving the distribution of abnormalities among women who deliver babies with abnormalities. Suppose we wish to test if there is a difference in the distribution of abnormalities among clinical centers. Based on the following data, is there evidence of a significant relationship between the types of abnormalities and clinical center?

	Type of Abnormality			
	Trisomy 21	*Trisomy 18*	*Trisomy 13*	*Total*
Center A:	107	70	23	200
Center B:	65	62	23	150
Center C:	60	32	8	100
Total:	232	164	54	450

1. Set up hypotheses.

H_0: Center and Type of Abnormality are independent
H_1: H_0 is false, $\alpha = 0.05$

2. Select the appropriate test statistic.

$$\chi^2 = \sum \frac{(O - E)^2}{E}$$

3. Decision rule.

In order to select the appropriate critical value, we first determine the degrees of freedom:

$$df = (R - 1)(C - 1) = (3 - 1)(3 - 1) = 2(2) = 4$$

The appropriate critical value of χ^2 is $\chi^2 = 9.49$ from Table B.5. The decision rule is

$$\text{Reject } H_0 \text{ if } \chi^2 \geq 9.49$$
$$\text{Do not reject } H_0 \text{ if } \chi^2 < 9.49$$

4. Test statistic.

To organize the computations of the test statistic (7.11), we use the contingency table given earlier. The observed frequencies in each cell are shown. We compute the expected frequencies for each cell using (7.10) and display expected frequencies in parentheses to distinguish them from the observed frequencies.

Center:	Trisomy 21	Trisomy 18	Trisomy 13	TOTAL
Center A:	107 (103.1)	70 (72.9)	23 (24)	200
Center B:	65 (77.3)	62 (54.7)	23 (18)	150
Center C:	60 (51.6)	32 (36.4)	8 (12)	100
TOTAL:	232	164	54	450

Treatment Regimen

Using the observed and expected frequencies, we compute the test statistic (7.11). Only the results are shown here.

$$\chi^2 = 0.15 + 0.12 + 0.04 + 1.96 + 0.97 + 1.39 + 1.37 + 0.53 + 1.33$$
$$= 7.86$$

5. Conclusion.

Do not reject H_0 since $7.86 < 9.49$. We do not have significant evidence, $\alpha = 0.05$, to show that there is an association between clinical center and type of trisomy. ▪

7.6 Precision, Power, and Sample Size Determination

In Section 7.2, we illustrated estimation techniques for the population proportion p. The following formula is used to determine the sample size requirements to produce an estimate for p with a certain level of precision:

$$n = p(1 - p) \left(\frac{Z_{1-(\alpha/2)}}{E} \right)^2 \tag{7.12}$$

where $Z_{1-(\alpha/2)}$ reflects the desired level of confidence (e.g., 95%)
$p =$ the population proportion
$E =$ margin of error

Equation (7.12) produces the *minimum* number of subjects required to ensure a margin of error equal to E in the confidence interval for p with the specified level of confidence. Recall that in estimating the sample size to make an inference about μ, an estimate of σ was required. Several alternatives were suggested to estimate σ. In the binomial case, our goal is to make an inference about p. However, the formula for estimating the sample size (7.12) involves p. An estimate from a previous study or an estimate based on pilot data may be used in (7.12). If such an estimate is not available, it can be shown that $p(1 - p)$ is maximized at $p = 0.50$ (see Figure 7.2). Therefore, the most

Figure 7.2 *Relationship Between p and p(1 − p):*
p(1 − p) Maximized at p = 0.5

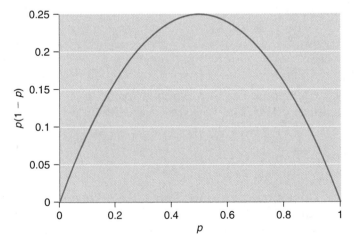

conservative estimate of n is produced by substituting $p = 0.5$ into the formula (7.12):

$$n = 0.5(1 - 0.5)\left(\frac{Z_{1-(\alpha/2)}}{E}\right)^2 \qquad (7.13)$$

which is equivalent to

$$n = 0.25\left(\frac{Z_{1-(\alpha/2)}}{E}\right)^2 \qquad (7.14)$$

where $Z_{1-(\alpha/2)}$ reflects the desired level of confidence (e.g., 95%)
E = margin of error

EXAMPLE 7.15

Sample Size Determination to Estimate Proportion of Patients Who Favor New Policy

An investigator wishes to estimate the proportion of patients in a particular health plan in favor of a new policy (e.g., reimbursement for services) and wants the estimate to be within 3% of the true proportion of patients in favor of the new policy. How many subjects would be required to produce such an estimate with 95% confidence?

Suppose that there is no available data on the issue, that is, no estimate for p that can be used in (7.12). Therefore (7.14) is used:

$$n = 0.25\left(\frac{Z_{1-(\alpha/2)}}{E}\right)^2$$

Substituting $E = 0.03$ (the margin of error in the estimate of the proportion) and the appropriate value for 95% confidence:

$$n = 0.25\left(\frac{1.96}{0.03}\right)^2 = 1067.1$$

Since (7.14) produces the minimum number of subjects to satisfy our criteria, 1068 patients must be sampled in order to produce a 95% confidence interval estimate for the proportion of patients in favor of the new policy with a margin of error of 3%. ▪

SAS EXAMPLE 7.15 **Sample Size Determination to Estimate Proportion of Patients Who Favor New Policy Using SAS**

SAS does not have a procedure to generate sample size requirements. However, we can program formulas (7.12) and (7.14) as we did in previous chapters. In the following, we generated sample size requirements for various scenarios, including the one described in Example 7.15. A brief interpretation appears after the output.

SAS Output for Example 7.15

Obs	c_level	z	p	e	n
1	0.95	1.95996	0.5	0.03	1068
2	0.95	1.95996	0.1	0.03	385
3	0.95	1.95996	0.2	0.03	683
4	0.95	1.95996	0.3	0.03	897
5	0.95	1.95996	0.4	0.03	1025
6	0.95	1.95996	0.5	0.03	1068
7	0.95	1.95996	0.6	0.03	1025
8	0.95	1.95996	0.7	0.03	897
9	0.95	1.95996	0.8	0.03	683
10	0.95	1.95996	0.9	0.03	385

Interpretation of SAS Output for Example 7.15

We considered several scenarios. The first scenario reflects the situation in Example 7.15 where p is not known and we use the most conservative $p = 0.5$. SAS estimates that $n = 1068$ subjects are required. To illustrate the effect of the estimate of p on the computations, we considered various values of p in scenarios 2–10. Specifically, we considered $p = 0.10, 0.20, \ldots, 0.90$. Notice that the sample size estimate is largest when $p = 0.5$. ■

EXAMPLE 7.16

Sample Size Determination to Estimate Proportion of High School Seniors Who Smoke

A study is run to estimate the proportion of high school seniors who smoke. The study involved a random sample of 100 high school seniors, and 28 reported that they were smokers. A 95% confidence interval for the true proportion of all high school seniors who smoke was computed as 0.28 ± 0.088. Suppose that we wanted to estimate the true proportion with a margin of error not exceeding 0.05 (the analysis based on the sample of size 100 had a margin of error of 0.088). How many subjects would be required to ensure a margin of error of 0.05 with 95% confidence?

Because we have data from the initial study, we use (7.12):

$$ n = p(1 - p) \left(\frac{Z_{1-(\alpha/2)}}{E} \right)^2 = 0.28(1 - 0.28) \left(\frac{1.96}{0.05} \right)^2 = 309.8 $$

In order to estimate the true proportion of high school seniors who smoke with a margin of error of 0.05, we need a sample of at least 310 high school seniors. ■

In Chapter 5, we introduced the concepts of precision (in estimation) and statistical power (in hypothesis testing). We now present formulas for determining sample size to achieve a specified power in the two-sample test for proportions. Recall, power (Power $= 1 - \beta = P(\text{Reject } H_0 | H_0 \text{ false})$)) depends on three components:

1. $n_i =$ sample sizes ($i = 1, 2$)
2. $\alpha =$ level of significance $= P(\text{Type I error})$
3. $ES =$ the effect size $=$ the standardized difference in proportions specified under H_0 and H_1

In the two-sample applications the hypotheses are

$$H_0: p_1 = p_2$$
$$H_1: p_1 \neq p_2$$

or equivalently,

$$H_0: p_1 - p_2 = 0$$
$$H_1: p_1 - p_2 \neq 0$$

where $p_1 =$ proportion of successes in population 1
$p_2 =$ proportion of successes in population 2

The effect size (ES) is defined as the difference in the proportions under H_0 and H_1. In the two-sample applications, the parameter of interest is ($p_1 - p_2$). The ES is defined as follows:

$$ES = |p_2 - p_1| \tag{7.15}$$

In many applications, the number of subjects (i.e., n_1 and n_2) that can be involved depends on financial, logistic, and/or time constraints. In other cases, the sample size required to ensure a certain level of power can be determined relative to alternative hypotheses of importance. The sample size required to ensure a specific level of power in a *two-sided, two independent samples test for proportions* is

$$n_i = \left(\frac{\sqrt{\bar{p}\bar{q}\,2}\ Z_{1-(\alpha/2)} + \sqrt{p_1 q_1 + p_2 q_2}\ Z_{1-\beta}}{ES} \right)^2 \tag{7.16}$$

where n_i is the minimum number of subjects required in sample
i ($i = 1, 2$)
$Z_{1-(\alpha/2)}$ is the value from the standard normal distribution with lower-tail area equal to $1 - \alpha/2$
$Z_{1-\beta}$ is the value from the standard normal distribution with lower-tail area equal to $1 - \beta$
ES is the effect size (7.15)

In addition, $p_1 =$ proportion of successes in population 1, $q_1 = 1 - p_1$
$p_2 =$ proportion of successes in population 2, $q_2 = 1 - p_2$

$$\bar{p} = \frac{p_1 + p_2}{2}, \qquad \bar{q} = 1 - \bar{p}$$

The following example illustrates the use of formula (7.16).

EXAMPLE 7.17   **Sample Size Determination in Tests for Differences in Proportions**

Suppose we wish to design a study to compare two treatments with respect to the proportion of successes in each. A two-sided test is planned at a 5% level of significance. Based on a review of the literature, $p_1 = 0.20$. How many subjects would be required per group to detect $p_2 = 0.10$ with 90% power?

The formula to determine sample sizes is given by (7.16):

$$n_i = \left(\frac{\sqrt{\bar{p}\bar{q}\, 2}\, Z_{1-(\alpha/2)} + \sqrt{p_1 q_1 + p_2 q_2}\, Z_{1-\beta}}{ES} \right)^2$$

The *ES* is (7.15)

$$ES = |p_2 - p_1| = |0.10 - 0.20| = 0.10$$

Since $\alpha = 0.05$, $Z_{1-(\alpha/2)} = Z_{0.975} = 1.96$. Similarly, for power $= 0.90$, $\beta = 0.10$; therefore, $Z_{1-\beta} = 1.282$.

$$\bar{p} = \frac{0.10 + 0.20}{2} = 0.15, \qquad \bar{q} = 1 - 0.15 = 0.85$$

Substituting,

$$n_i = \left(\frac{\sqrt{0.15(0.85)2}(1.96) + \sqrt{0.20(0.80) + (0.10)(0.90)}(1.282)}{0.10} \right)^2 = 265.9$$

Thus, $n_1 = n_2 = 266$ subjects (532 total) are needed.   ■

SAS EXAMPLE 7.17

SAS does not have a procedure to generate sample size requirements. We can program formulas (7.15) and (7.16) as we did in previous chapters. In the following, we generated sample size requirements for various scenarios, including the one described in Example 7.17. A brief interpretation appears after the output.

SAS Output for Example 7.17

Obs	alpha	beta	z_alpha2	z_beta	p1	p2	power	es	n_2
1	0.05	0.1	1.95996	1.28155	0.2	0.1	0.9	0.1	266
2	0.05	0.1	1.95996	1.28155	0.2	0.3	0.9	0.1	392
3	0.05	0.2	1.95996	0.84162	0.2	0.1	0.8	0.1	199
4	0.05	0.2	1.95996	0.84162	0.2	0.3	0.8	0.1	294

Interpretation of SAS Output for Example 7.17

We considered several scenarios. The first scenario reflects the situation in Example 7.17. SAS estimates that 266 subjects are required per group to detect the specified difference with 90% power. In scenario 2, we specify $p_2 = 0.3$. Notice that the effect size is the same as the effect size in scenario 1. However, SAS estimates that 392 subjects are required per group. The values of p_1 and p_2 affect the sample size computation (and not just the difference between them). In the last two scenarios, we specify power $= 80\%$. Notice that fewer cases are required to detect the same differences with lower power. ▪

7.7 Key Formulas

Application	Notation/Formula	Description
Confidence interval estimate for p	$\hat{p} \pm Z_{1-(\alpha/2)}\sqrt{\dfrac{\hat{p}(1-\hat{p})}{n}}$	See Table 7.1 for necessary conditions (find Z in Table B.2A)
Test H_0: $p = p_0$	$Z = \dfrac{\hat{p} - p_0}{\sqrt{\dfrac{p_0(1-p_0)}{n}}}$	See Table 7.1 for necessary conditions
Confidence interval estimate for $(p_1 - p_2)$	$(\hat{p}_1 - \hat{p}_2) \pm Z_{1-(\alpha/2)}\sqrt{\dfrac{\hat{p}_1(1-\hat{p}_1)}{n_1} + \dfrac{\hat{p}_2(1-\hat{p}_2)}{n_2}}$	See Table 7.2 for necessary conditions (find Z in Table B.2A)
Test H_0: $p_1 = p_2$	$Z = \dfrac{\hat{p}_1 - \hat{p}_2}{\sqrt{\hat{p}(1-\hat{p})\left(\dfrac{1}{n_1} + \dfrac{1}{n_2}\right)}}$	See Table 7.3 for necessary conditions and definitions of components of Z
Test H_0: distribution of responses follows specified pattern	$\chi^2 = \sum \dfrac{(O-E)^2}{E}$, df $= k - 1$	Chi-square goodness-of-fit test
Test H_0: two variables are independent	$\chi^2 = \sum \dfrac{(O-E)^2}{E}$, df $= (r-1)*(c-1)$	Chi-square test of independence
Find n to estimate p	$n = p(1-p)\left[\dfrac{Z_{1-(\alpha/2)}}{E}\right]^2$	Sample size to estimate p with margin of error E
Find n_1, n_2 to test H_0: $p_1 = p_2$	$n_i = \left(\dfrac{\sqrt{\bar{p}\bar{q}}\, 2\, Z_{1-(\alpha/2)} + \sqrt{p_1 q_1 + p_2 q_2}\, Z_{1-\beta}}{ES}\right)^2$	Sample sizes to detect effect size ES with power $1 - \beta$

7.8 Statistical Computing

Following are the SAS programs that were used to generate the frequency distribution tables, confidence interval estimates for the proportion of successes in a population, the contingency tables, and the chi-square goodness-of-fit and tests of independence and to determine the required sample size to estimate p and to compare to proportions using a test of hypothesis. The SAS procedures used and brief descriptions are noted in the header to each example. Notes are provided to the right of the SAS programs (in blue) for orientation purposes and are not part of the programs. In addition, there are blank lines in the programs that are solely to accommodate the notes. Blank lines and spaces can be used throughout SAS programs to enhance readability. A summary of the SAS procedures used in the examples is provided at the end of this section.

SAS Example 7.2 **Frequency Distribution Table for Single Categorical Variable, CI for p**
Estimate Proportion of Patients with Osteoarthritis (Example 7.2)

Suppose we wish to estimate the proportion of patients in a particular physician's practice with diagnosed osteoarthritis. A random sample of 200 patients is selected and each patient's medical record is reviewed for the diagnosis of osteoarthritis. Suppose that 38 patients are observed with diagnosed osteoarthritis. Generate a frequency distribution table and determine a point estimate and 95% confidence interval using SAS for the proportion of all patients with diagnosed osteoarthritis.

Program Code

`options ps=62 ls=80;`	Formats the output page to 62 lines in length and 80 columns in width
`data in;`	Beginning of Data Step
`  input x count;`	Inputs two variables x and **count**, here $x = 0$ if the patient does not have osteoarthritis and $x = 1$ if the patient has osteoarthritis. Count reflects the number of patients in each category.
`cards;`	Beginning of Raw Data section.
`0 162`	actual observations (value of x and
`1 38`	**count** on each line)
`run;`	
`proc freq;`	Procedure call. Proc Freq generates a frequency distribution table for a categorical (or ordinal) variable.
`  tables x/binomial` `    (level='1');`	Specification of analytic variable x. The binomial option requests a confidence interval for the proportion and the level='1' indicates that a success is coded as 1.
`  weight count;`	Specification of variable containing the number of subjects in each category, **count.**
`run;`	End of procedure section.　■

SAS EXAMPLE 7.5   Cross-Tabulation Table

Cross-Tabulation of Treatment by Long-Term Complications (Example 7.5)

A longitudinal study is conducted to evaluate the long-term complications in diabetic patients treated under two competing treatment regimens. Complications are measured by incidence of foot disease, eye disease, *or* cardiovascular disease within a 10-year observation period. The following 2×2 cross-tabulation table summarizes the data. Estimate the relative risk and odds ratio using SAS.

	Long-Term Complications		
Treatment	YES	NO	*Total*
Treatment 1	12	88	100
Treatment 2	8	92	100
Total	20	180	200

Program Code

`options ps=62 ls=80;`	Formats the output page to 62 lines in length and 80 columns in width
`data in;`	Beginning of Data Step
`  input trt $ compl $ count;`	Inputs three variables *trt* (treatment 1 or 2), *compl* (complications yes or no) and *count,* where count reflects the number of patients in each cell of the table. Here we input both trt and compl using character labels, and we indicate this to SAS using the "$" symbol.
`cards;`	Beginning of Raw Data section.
`trt_1 z_no 88`	actual observations (value of *trt, compl* and *count* on
`trt_1 yes 12`	each line)
`trt_2 z_no 92`	Because SAS alphabatizes labels, we use z_no to ensure
`trt_2 yes 8`	that the yes response is in column 1.
`run;`	
`proc freq;`	Procedure call. Proc Freq generates a contingency table for two categorical (or ordinal) variables.
`  tables trt*compl;`	Specification of analytic variables *trt* and *compl.*
`  weight count;`	Specification of variable containing the number of subjects in each cell of the table, *count.*
`run;`	End of procedure section. ■

SAS Example 7.9 **Frequency Distribution Table and Chi-Square Goodness-of-Fit Test**
Chi-Square Goodness-of-Fit Test (Example 7.9)

Following coronary artery bypass graft (CABG) surgery, patients are encouraged to participate in a cardiac rehabilitation program. The program lasts approximately 14 weeks and includes exercise training, nutritional information, and general lifestyle guidelines. One particular hospital is offering three cardiac rehabilitation programs that are identical in content but are offered at three different times. The administration is interested in whether the three time slots are equally popular among or convenient for the patients. A total of 100 patients are involved in the investigation, and each patient is asked to select the one day and time (from three options) that is most convenient. The following data are observed:

Time slot:	Mondays 6:00–7:30 PM	Thursdays 4:00–5:30 PM	Saturdays 8:00–9:30 AM
Number of patients:	47	32	21

Program Code

`options ps=62 ls=80;`	Formats the output page to 62 lines in length and 80 columns in width
`data in;`	Beginning of Data Step
`  input day $ count;`	Inputs two variables *day* (which contains the 3 options) and *count,* which contains the number of patients in each response category. Here we input *day* using character data, and we indicate this to SAS using the "$" symbol.
`cards;`	Beginning of Raw Data section.
`Mon 47`	actual observations (value of *day*
`Thurs 32`	and *count* on each line)
`Sat 21`	
`run;`	
`proc freq;`	Procedure call. Proc Freq generates a frequency distribution table for a categorical (or ordinal) variable.
`  tables day/chisq`	Specification of the analytic variable *day* and the chisq
`    testp = (0.33 0.33 0.33);`	option along with the expected proportions (under H_o) in parentheses following the testp option.
`   weight count;`	Specification of variable containing the number of subjects in each cell of the table, *count.*
`run;`	End of procedure section. ■

SAS EXAMPLE 7.12 Contingency Table and Chi-Square Test of Independence

Test If There Is a Relationship Between Site and Treatment Regimen (Example 7.12)

The following data were collected in a multisite observational study of medical effectiveness in Type II diabetes. Three sites were involved in the study: a health maintenance organization (HMO), a university teaching hospital (UTH), and an independent practice association (IPA). Type II diabetic patients were enrolled in the study from each site and monitored over a 3-year observation period. The data shown display the treatment regimens of patients measured at baseline by site. Test if there is a relationship between site and treatment regimen using the chi-square test of independence in SAS.

Site	Treatment Regimen			
	Diet & Exercise	*Oral Hypoglycemics*	*Insulin*	TOTAL
HMO:	294	827	579	1700
UTH:	132	288	352	772
IPA:	189	516	404	1109
TOTAL:	615	1631	1335	3581

Program Code

```
options ps=62 ls=80;
```
Formats the output page to 62 lines in length and 80 columns in width

```
data in;
  input site $ trt $ count;
```
Beginning of Data Step

Inputs three variables *site* (HMO, UTH, or IPA), *trt* (diet, oral hypoglycemics, or insulin) and *count,* where count reflects the number of patients in each cell of the table. Here we input both site and trt using character labels, and we indicate this to SAS using the "$" symbol.

```
cards;
hmo diet 294
hmo oral 827
hmo insulin 579
uth diet 132
uth oral 288
uth insulin 352
ipa diet 189
ipa oral 516
ipa insulin 404
run;
```
Beginning of Raw Data section.

actual observations (value of *site,* *trt* and *count* on each line)

```
proc freq;
```
Procedure call. Proc Freq generates a contingency table for two categorical (or ordinal) variables.

```
  tables site*trt/expected cellchi2 chisq;
```
Specification of analytic variables *site* and *trt*. We also request that SAS produce expected frequencies in each cell, the chi-square statistic for each cell and run a chi-square test of independence with the expected, cellchi2, and chisq options, respectively.

```
  weight count;
```
Specification of variable containing the number of subjects in each cell of the table, *count*.

```
run;
```
End of procedure section.   ■

SAS EXAMPLE 7.13 **Contingency Table and Chi-Square Test of Independence**

Test If There Is a Relationship Between Gender and Treatment Regimen (Example 7.13)

Suppose an investigator is interested in evaluating whether or not treatment regimens differ by gender. Restricting our analyses to the HMO patients, the following table displays the numbers of male and female patients according to their treatment regimens. Based on the data, is there evidence of a significant relationship between gender and treatment regimen among the HMO patients? Test if there is a relationship between site and treatment regimen using the chi-square test of independence in SAS.

	Treatment Regimen			
Gender	Diet & Exercise	Oral Hypoglycemics	Insulin	TOTAL
Female:	147	435	256	838
Male:	147	392	323	862
TOTAL:	294	827	579	1700

Program Code

```
options ps=62 ls=80;
```
Formats the output page to 62 lines in length and 80 columns in width

```
data in;
```
Beginning of Data Step

```
  input gender $ trt $ count;
```
Inputs three variables gender (male or female), *trt* (diet, oral hypoglycemics or insulin), and *count,* where count reflects the number of patients in each cell of the table. Here we input both gender and trt using character labels, and we indicate this to SAS using the "$" symbol.

```cards;```	Beginning of Raw Data section.
```male diet 147```	actual observations (value of *gender,*
```male oral 392```	*trt,* and *count* on each line)
```male insulin 323```	
```female diet 147```	
```female oral 435```	
```female insulin 256```	
```run;```	
```proc freq;```	Procedure call. Proc Freq generates a contingency table for two categorical (or ordinal) variables.
``` tables gender*trt/chisq;```	Specification of analytic variables *gender* and *trt.* We request that SAS runs a chi-square test of independence with the chisq option.
``` weight count;```	Specification of variable containing the number of subjects in each cell of the table, *count.*
```run;```	End of procedure section. ■

SAS Example 7.15 Determine the Number of Subjects Required to Generate a Confidence Interval for *p*

Patients in Favor of a New Policy (Example 7.15)

An investigator wishes to estimate the proportion of patients in a particular health plan in favor of a new policy (e.g., reimbursement for services) and wants the estimate to be within 3% of the true proportion of patients in favor of the new policy. How many subjects would be required to produce such an estimate with 95% confidence?

Program Code

```options ps=62 ls=80;```	Formats the output page to 55 lines in length and 80 columns in width
```data in;```	Beginning of Data Step
``` input c_level e p;```	Inputs 3 variables *c_level, e,* and *p.*
```z=probit((1-c_level)/2);```	Determines the value from the standard normal distribution (*z*) with lower-tail area $(1-c\_level)/2$
```if p=. then p=0.5;```	If *p* is not available (input as missing), then SAS recodes *p* to 0.5 (see (7.13))
```tempn=(p*(1-p))*(z/e)**2;```	Creates a temporary variable, called *tempn,* determined by formula (7.12).
```n=ceil(tempn);```	Computes a variable *n* using the ceil function, which computes the smallest integer greater than *tempn* (i.e., rounds up)

```
/*    Input the following information (required)
   c_level: Confidence Level: range 0.0 to 1.0 (e.g., 0.95),
   e: Margin of Error, and
   p  Proportion of Successes: range 0.0 to 1.0
   NOTE: p may not be available, in which case enter . to indicate
         p is missing
*/
cards;
0.95   0.03   .
0.95   0.03 0.10
0.95   0.03 0.20
0.95   0.03 0.30
0.95   0.03 0.40
0.95   0.03 0.50
0.95   0.03 0.60
0.95   0.03 0.70
0.95   0.03 0.80
0.95   0.03 0.90
run;


proc print;
  var c_level z p e n;
run;
```

Beginning of comment section

End of comment
Beginning of Raw Data section
actual observations
values of *c_level*, *e*, and **p**
on each line

Procedure call. Print to display
input and computed variables.
End of procedure section.   ■

SAS Example 7.17 Determine the Number of Subjects Required to Detect a Specific Effect Size in a Test of Hypothesis About $(p_1 - p_2)$

Sample Size Requirements (Example 7.17)

Suppose we wish to design a study to compare two treatments with respect to the proportion of successes in each. A two sided test is planned at a 5% level of significance. Based on a review of the literature, $p_1 = 0.20$. How many subjects would be required per group to detect $p_2 = 0.10$ with 90% power?

Program Code

```
options ps=62 ls=80;
data in;
  input alpha power p1 p2;
z_alpha2=probit(1-alpha/2);


beta=1-power;
z_beta=probit(1-beta);
```

Formats the output page to 62 lines in length and 80 columns in width
Beginning of Data Step.
Inputs 4 variables *alpha, power, p1,* and *p2*.
Determines the value from the standard normal distribution with lower-tail area *1-alpha/2*.

Computes *beta*.
Determines the value from the standard normal distribution with lower-tail area *1-beta*

```
q1=1-p1;                                          Compute components for (7.16).
q2=1-p2;
pbar=(p1+p2)/2;
qbar=1-pbar;
es=abs(p2-p1);                                    Computes the effect size (es)
                                                  Compute (7.16).

tempn_2=(sqrt(pbar*qbar*2)*z_alpha2+sqrt(p1*q1+p2*q2)*z_beta)**2/
        (es)**2;


n_2=ceil(tempn_2);                                Computes a variable n_2 using the ceil function,
                                                  which computes the smallest integer greater
                                                  than tempn_2 (i.e., rounds up).


/*                                                Beginning of comment section

   Input the following information (required)
   alpha: Level of Significance: range 0.0 to 1.0 (e.g., 0.05),
   power: Power: range 0.0 to 1.0 (e.g., 0.80),
   p1: Proportion of Successes in Group 1, and
   p2: Proportion of Successes in Group 2.

*/                                                End of comment section
cards;                                            Beginning of Raw Data section
0.05 0.90 0.20 0.10                               actual observations
0.05 0.90 0.20 0.30
0.05 0.80 0.20 0.10
0.05 0.80 0.20 0.30
run;

proc print;                                       Procedure call. Print to display
   var alpha beta z_alpha2 z_beta p1 p2           computed variables.
      power es n_2;
run;                                              End of procedure section.              ■
```

7.8.1 Summary of SAS Procedures

The SAS Freq Procedure is used to generate a frequency distribution table and to generate contingency tables. Specific options can be requested to produce estimates of relative risks and odds ratios, and to run a chi-square test of independence. The specific options are shown in italics below. Users should refer to the examples in this section for complete descriptions of the procedure and specific options. A general description of the procedure and options is provided in the table on the next page.

Procedure	Sample Procedure Call	Description
proc freq	proc freq; tables x/*binomial (level = '1');* *weight count;*	Generates a frequency distribution table for a categorical (or ordinal) variable. When x is dichotomous, the binomial option can be specified to generate a CI for the proportion. The level = '1' statement indicates that a success is coded as 1. The weight option indicates that summary data are input and the variable count contains the numbers of subjects in each response category.
	proc freq; tables x/*chisq testp=(p0 p1 p2 ... pk);*	Generates a frequency distribution table for a categorical (or ordinal) variable. The chisq option is used to request a goodness-of-fit test and the values following the testp option indicate the expected proportions for the k response categories
	proc freq; tables a*b;	Generates a contingency table (a by b).
	proc freq; tables a*b/expected cellchi2 chisq;	Generates a contingency table and produces the expected frequency in each cell, the value of the chi-square statistic in each cell, and runs the chi-square test of independence.

7.9 Analysis of Framingham Heart Study Data

The Framingham data set includes data collected from the original cohort. Participants contributed up to three examination cycles of data. Here we analyze data collected in the first examination cycle (called the period = 1 examination) and use SAS Proc Freq to generate frequency distribution tables and 95% confidence intervals for several dichotomous variables. We then run a goodness-of-fit test and a test of independence. For the last two illustrations, we created an ordinal variable with four levels from the continuous BMI variable as follows: BMI < 18.5 (underweight), 18.5–24.9 (normal weight), 25.0–29.9 (overweight), and $\geq$ 30.0 (obese). The SAS code to create this variable and to attach a format for better interpretability is given next.

Framingham Data Analysis—SAS Code

```
proc format;
 value bmifmt 1='<18.5     '2='18.5-24.9' 3='25.0-29.9' 4='30.0+';
run;
```

```
data fhs;
 set in.frmgham;
 if period=1;

if 0 le mbi lt 18.5 then bmi_grp=1;
else if 18.5 le bmi lt 25.0 then bmi_grp=2;
else if 25.0 le bmi lt 30.0 then bmi_grp=3;
else if bmi ge 30.0 then bmi_grp=4;
format bmi_grp bmifmt.;
run;

proc freq data=fhs;
 tables bpmeds cursmoke diabetes/binomial (level='1');
run;

proc freq data=fhs;
 tables bmi_grp/chisq testp=(0.01 0.39 0.40 0.20);
run;

proc freq data=fhs;
 tables sex*bmi_grp/chisq;
run;
```

Framingham Data Analysis—SAS Output

The FREQ Procedure

Anti-hypertensive meds Y/N

BPMEDS	Frequency	Percent	Cumulative Frequency	Cumulative Percent
0	4229	96.71	4229	96.71
1	144	3.29	4373	100.00

Frequency Missing = 61

Binomial Proportion
for BPMEDS = 1

Proportion	0.0329
ASE	0.0027
95% Lower Conf Limit	0.0276
95% Upper Conf Limit	0.0382

```
          Exact Conf Limits
          95% Lower Conf Limit        0.0278
          95% Upper Conf Limit        0.0387

          Test of H0: Proportion = 0.5

          ASE under H0                0.0076
          Z                         -61.7735
          One-sided Pr <  Z          <.0001
          Two-sided Pr > |Z|         <.0001

          Effective Sample Size = 4373
             Frequency Missing = 61

          Current Cig Smoker Y/N

                                              Cumulative   Cumulative
CURSMOKE    Frequency      Percent             Frequency      Percent
-----------------------------------------------------------------------
       0         2253        50.81                  2253        50.81
       1         2181        49.19                  4434       100.00

                   The SAS System                              9

                   The FREQ Procedure

                   Binomial Proportion
                     for CURSMOKE = 1
          ---------------------------------
          Proportion                  0.4919
          ASE                         0.0075
          95% Lower Conf Limit        0.4772
          95% Upper Conf Limit        0.5066

          Exact Conf Limits
          95% Lower Conf Limit        0.4771
          95% Upper Conf Limit        0.5067

          Test of H0: Proportion = 0.5

          ASE under H0                0.0075
          Z                          -1.0813
          One-sided Pr <  Z           0.1398
          Two-sided Pr > |Z|          0.2796

                  Sample Size = 4434
                   Diabetic Y/N
```

DIABETES	Frequency	Percent	Cumulative Frequency	Cumulative Percent
0	4313	97.27	4313	97.27
1	121	2.73	4434	100.00

The SAS System   10

The FREQ Procedure

Binomial Proportion
for DIABETES = 1

Proportion	0.0273
ASE	0.0024
95% Lower Conf Limit	0.0225
95% Upper Conf Limit	0.0321

Exact Conf Limits	
95% Lower Conf Limit	0.0227
95% Upper Conf Limit	0.0325

Test of H0: Proportion = 0.5

ASE under H0	0.0075
Z	-62.9540
One-sided Pr < Z	<.0001
Two-sided Pr > |Z|	<.0001

Sample Size = 4434

For presentation, the preceding data could be summarized as follows:

	Percent	*95% CI*
Antihypertensive medication:	3.3%	(2.8%–3.9%)
Current smoking:	49.1%	(47.7%–50.7%)
Diabetes:	2.7%	(2.3%–3.3%)

Suppose that national data are reported to suggest that 1% of Americans are underweight, 39% are in the normal BMI range, 40% are overweight and 20% are obese. We are interested in whether the Framingham data follow that distribution. Using SAS, we perform a goodness-of-fit with that referent distribution and produce the following:

```
                    The FREQ Procedure
                                Test      Cumulative    Cumulative
bmi_grp     Frequency    Percent    Percent    Frequency     Percent
-------------------------------------------------------------------
<18.5              76       1.71       1.00           76        1.71
18.5-24.9        1936      43.66      39.00         2012       45.38
25.0-29.9        1845      41.61      40.00         3857       86.99
30.0+             577      13.01      20.00         4434      100.00

                     Chi-Square Test
                  for Specified Proportions
                  -------------------------
                  Chi-Square        158.4245
                  DF                       3
                  Pr > ChiSq          <.0001

                    Sample Size = 4434
```

Do the Framingham data follow the reported distribution? How do the Framingham data line up with the national figures?

Using the same BMI categories, we now use SAS to test for a difference in the distribution of weight categories by sex.

```
                        The FREQ Procedure
                      Table of SEX by bmi_grp

SEX(SEX)        bmi_grp

Frequency|
Percent  |
Row Pct  |
Col Pct  |<18.5    |18.5-24.|25.0-29.|30.0+    |   Total
         |         |9        |9        |         |
---------+--------+--------+--------+--------+
M        |      17 |     703 |     992 |     232 |    1944
         |    0.38 |   15.85 |   22.37 |    5.23 |   43.84
         |    0.87 |   36.16 |   51.03 |   11.93 |
         |   22.37 |   36.31 |   53.77 |   40.21 |
---------+--------+--------+--------+--------+
F        |      59 |    1233 |     853 |     345 |    2490
         |    1.33 |   27.81 |   19.24 |    7.78 |   56.16
         |    2.37 |   49.52 |   34.26 |   13.86 |
         |   77.63 |   63.69 |   46.23 |   59.79 |
---------+--------+--------+--------+--------+
Total          76      1936      1845      577      4434
             1.71     43.66     41.61     13.01    100.00
```

```
               Statistics for Table of SEX by bmi_grp

      Statistic                      DF        Value          Prob
      --------------------------------------------------------------
      Chi-Square                      3       135.7296        <.0001
      Likelihood Ratio Chi-Square     3       136.8781        <.0001
      Mantel-Haenszel Chi-Square      1        43.7383        <.0001
      Phi Coefficient                            0.1750
      Contingency Coefficient                    0.1723
      Cramer's V                                 0.1750

                      Sample Size = 4434
```

Is there a significant difference in the distribution of weight categories by sex? If so, what is the nature of the difference?

7.10 Problems

1. Recent attention has focused on the health care system, particularly on managed care plans. A study was undertaken within one managed care plan to assess whether patients' reports of satisfaction with the plan were related to their leaving the plan within 1 year. In a random sample of 120 patients who reported that they were satisfied with the plan, 30 left that plan within 1 year. In a second random sample of 150 patients who reported that they were not satisfied with the plan, 62 left within 1 year. Compute a 95% confidence interval for the difference in the proportions of patients who left the plan within 1 year relative to their satisfaction reports.

2. An investigator wishes to estimate the proportion of Type II diabetic patients who take insulin to manage their diabetes. A large, national database of Type II diabetic patients is used to generate the estimate. The database involves a random sample of 1200 Type II diabetic patients, 423 of whom report taking insulin to manage their diabetes.
   a. Compute a point estimate for the proportion of all Type II diabetics who take insulin to manage their diabetes.
   b. Compute the standard error of the point estimate.
   c. Compute a 95% confidence interval for the proportion of all Type II diabetics who take insulin to manage their diabetes.

3. Recent studies have investigated the relationship between gender and career advancement. The following data represent a random sample of

mathematicians classified by gender and academic rank. Using the following data, test for a relationship between gender and academic rank.

		Academic Rank		
	Instructor	Assistant Professor	Associate Professor	Full Professor
Female:	12	15	18	7
Male:	21	25	30	22

4. An investigator wishes to test if patients tend to use the middle response in 5-point ordinal scales more frequently than other response options. Scales such as these are used in health status and satisfaction measures. One hundred patients were asked to rate their own health on the following 5-point ordinal scale. Based on the data, is there evidence that the distribution of responses is 10%, 20%, 40%, 20%, 10%, respectively? Use $\alpha = 0.05$.

Health status:	Excellent	Very Good	Good	Fair	Poor
# Patients:	12	18	50	10	10

5. A manufacturer of medical devices has two plants and wants to compare them on the proportion of defective items produced. In a random sample of 250 items from plant I, 28 were defective. In a random sample of 220 items from plant II, 38 were defective. Is there any significant evidence to support the claim that plant II produces more defective items? Use a 5% level of significance.

6. As the first step in a study to evaluate highway fatalities, the traffic department wants to evaluate whether the left-hand lane of a three-lane highway services twice the traffic volume as the other two lanes. The following data were collected, reflecting the numbers of vehicles traveling each of the three lanes during the rush hour. Use the data to test the claim that twice as many vehicles travel the left lane as compared to the other two lanes ($\alpha = 0.05$).

	Right Lane	Center Lane	Left Lane
# Vehicles:	205	220	575

7. A corporation offers six different health plans to its employees. Each year the corporation offers them an opportunity to switch plans. In 2002, 15% of all employees switched from one plan to another. With all of the emphasis on health plans this year, the corporation thinks that a higher proportion of employees will switch. A random sample of 125 employees is selected, and 25 indicate that they will switch plans in 2003. Based on the data, is the proportion of employees who switch plans significantly higher in 2003? Run the appropriate test at a 5% level of significance.

8. A study is conducted over 10 years to investigate long-term complication rates in patients treated with two different therapies. The data are shown here:

	Complications	
Therapy	*No*	*Yes*
1:	911	89
2:	873	127

   a. Generate a point estimate for the difference in proportions.
   b. Compute a 95% confidence interval for the difference in proportions.
   c. Based on (a) and (b), is there a significant difference in the therapies?

9. A study is conducted comparing two experimental treatments to a control treatment with respect to their effectiveness in reducing joint pain in patients with arthritis. One hundred and fifty patients are randomly assigned to one of the three treatments, and the following data are collected, representing the numbers of patients reporting improvement in joint pain after the assigned treatment is administered:

	Control	*Experimental 1*	*Experimental 2*
Number of subjects:	50	50	50
Number reporting improvement:	21	28	34

   Is there a significant difference in the proportions of patients reporting improvement in joint pain among the treatments? Run the appropriate test at the 5% level of significance.

10. Using the data in Problem 9,

   a. Construct a 95% confidence interval for the difference in the proportions of patients reporting improvement in joint pain between the Control and Experimental 1 treatments.

   b. Based on (a), is there a significant difference in the proportions of patients reporting improvement in joint pain between the Control and Experimental 1 treatments? Justify your answer (be brief but complete).

11. An investigator wished to design a study to estimate the proportion of patients in a particular hospital whose primary insurance coverage is Medicaid.

   a. How many subjects would be required to estimate the true proportion within 4% with 95% confidence?

   b. Suppose a similar study was conducted in 2003 and produced the following 95% confidence interval for the proportion of patients in the same hospital whose primary insurance coverage was Medicaid: 27% ± 6%. If the point estimate from the 2003 study is used, how does that affect the answer given in (a)?

   c. Which answer is more appropriate, (a) or (b)? Be brief.

12. Prior to being randomized to one of two competing therapies, the severity of participants' migraines is clinically assessed. The following table displays the severity classifications for patients assigned to the medical and nontraditional therapies. Is there a significant association between severity and assigned therapy? Run the appropriate test at $\alpha = 0.05$.

	Severity Classification			
Therapy	*Minimal*	*Moderate*	*Severe*	*Total*
Medical	90	60	50	200
Nontraditional	50	60	90	200
Total	140	120	140	400

13. A survey is conducted among current MPH students to assess their knowledge of safe (alcohol) drinking limits. Two hundred students are randomly selected for the investigation, and 60% correctly identified safe drinking limits.

   a. Construct a 95% confidence interval for the proportion of all MPH students who could correctly identify safe drinking limits.

b. It has been reported that 70% of practicing clinicians can correctly identify safe drinking limits when asked in a survey format. One hundred of the MPH students involved in the survey are former or current practicing clinicians, and 72% of these individuals correctly identified safe drinking limits in the survey. Is the proportion of current MPH students who are former or current practicing clinicians significantly different from the proportion reported in the literature? Run the appropriate test at $\alpha = 0.05$.

14. Suppose we wish to design a study to estimate the proportion of patients who received pain therapy following a particular surgical procedure in a local hospital. How many subjects would be required in the study to ensure that the estimate was within 4 percentage points of the true proportion with 95% confidence?

15. The following table was derived from a study of HIV patients, and the data reflect the numbers of subjects classified by their primary HIV risk factor and gender. Test if there is a relationship between HIV risk factor and gender using a 5% level of significance.

HIV Risk Factor	Gender	
	Male	Female
IV drug user:	24	40
Homosexual:	32	18
Other:	15	25

16. Under standard care, 10% of all patients suffering their first MI are readmitted to the hospital within 6 weeks. A new protocol for care following the first MI is proposed and evaluated. Two hundred patients receive the new protocol following an MI and 12 are readmitted to the hospital within 6 weeks. Is there a significant reduction in the proportion of patients readmitted to the hospital within 6 weeks under the new protocol? Run the appropriate test at the 5% level of significance.

17. Suppose that the data in Problem 16 are criticized based on the use of a 10% readmission rate under standard care. A new study is mounted to directly compare standard care to the new protocol using a randomized trial. Based on the following data, is there a significant reduction in the readmission rate under the new protocol? Run the appropriate test at the 5% level of significance.

Treatment	Number of Patients	Number Readmitted Within 6 Weeks
Standard care	125	16
New protocol	125	11

18. Some investigators suggest that medication adherence exceeding 85% is sufficient to classify a patient as "adherent" and medication adherence below 85% suggests that the patient is "not adherent." Suppose in a clinical trial that each patient is classified as either adherent or not based on this definition.

   a. Construct a 95% confidence interval for the difference in proportions of adherent patients between the intervention and control groups using the data shown here.

	Intervention	Control
Number of patients:	75	75
Number with medication adherence ≥ 85%:	47	36

   b. Based on (a), is there a significant difference in proportions of adherent patients between the intervention and control groups? Justify your answer briefly.

19. Patients were enrolled into the study described in Problem 18 from three clinical centers. Based on the following data, is there a significant difference in the proportions of adherent patients among the clinical centers? Run the appropriate test at $\alpha = 0.05$.

	Enrollment Sites		
	Center 1	*Center 2*	*Center 3*
Adherent (>85%):	28	30	25
Not adherent (<85%):	21	29	17

20. Suppose an observational study is conducted to investigate the smoking behaviors of male and female patients with a history of coronary heart disease. Among 220 men surveyed, 80 were smokers. Among 190 women surveyed, 95 were smokers.

   a. Construct a 95% confidence interval for the difference in the proportions of males and females who smoke.

b. Based on (a), would you reject $H_0: p_1 = p_2$ in favor of $H_1: p_1 \neq p_2$? Justify your answer briefly.

21. Suppose we wish to estimate what proportion of patients in an HMO spend more than $500 on prescription medications over 12 months. In a random sample of 150 patients, 34% spent more than $500.

    a. Construct a 95% confidence interval for the true proportion of patients who spend more than $500 on prescription medications per year.

    b. How many subjects would be required to estimate the true proportion who spend more than $500 on prescription medications per year with a margin of error no more than 2%, with 95% confidence?

22. Suppose a cross-sectional study is conducted to investigate cardio-vascular risk factors among a sample of patients seeking medical care at one of two local hospitals. A total of 200 patients are enrolled. Construct a 95% confidence interval for the difference in proportions of patients with a family history of cardiovascular disease (CVD) between hospitals.

	Enrollment Site	
Family History of CVD	Hospital 1	Hospital 2
Yes:	24	14
No:	76	86
Total:	100	100

23. The following table was constructed based on a comparison of various sociodemographic characteristics between men and women enrolled in a study of cardiovascular risk factors:

Characteristic	Men (n = 160)	Women (n = 140)	p
Mean age (SD)	45 (7.8)	46 (8.6)	0.7256
% High school graduate	78	64	0.0245
Mean annual income (SD)	47,345 (8,456)	31.987 (9,645)	0.0001
% with no insurance	8	9	0.9876

    a. Which, if any, of the characteristics shown are significantly different between men and women? Assume $\alpha = 0.05$. Justify your answer.

b. Write the hypotheses tested and show the formula of the test statistic used to compare educational levels between men and women.

c. Suppose we wish to test whether the study population (men and women combined) had a significantly higher-than-average proportion of patients with no insurance. The reported proportion is 7%. Write the hypotheses we would test, and show the formula of the test statistic we would use to conduct such a test.

24. Investigators want to use a self-reported measure of pain to evaluate the effectiveness of a new medication designed to reduce postoperative pain. Before using the measure, they examine its distributional properties in a sample of patients recently undergoing knee surgery. Postprocedure, each patient is asked to rate pain on the following scale: no pain, minimal pain, some pain, moderate pain, or severe pain. The investigators hypothesize that the distribution of responses will be approximately normal, on the order of 1 to 2 to 4 to 2 to 1 across the response categories. Use the following data to test the claim at a 5% level of significance:

	Pain				
	None	Minimal	Some	Moderate	Severe
Number of patients:	15	25	46	36	18

25. Suppose the measure described in Problem 24 is used to compare a newly developed pain medication against a standard medication with respect to self-reported pain. Using the following data, is there a difference in the pain levels of patients on the different medications? Run the appropriate test at a 5% level of significance.

	Pain				
	None	Minimal	Some	Moderate	Severe
New:	20	35	41	15	6
Standard:	15	25	46	36	18

26. In the application described in Problem 25, suppose instead of using the five-level pain measure, the investigators collapse the responses as follows: {None, Minimal, and Some = Low Pain} and {Moderate and

Severe = High Pain}. Is there a significant difference in the proportions of patients with High Pain between medication groups? Run the appropriate test at a 5% level of significance.

27. A study is conducted comparing three different prenatal care programs among women at high risk for preterm delivery. The programs differ in intensity of medical intervention. Women who meet the criteria for high risk of preterm delivery are asked to participate in the study and are randomly assigned to one of the three prenatal care programs. At the time of delivery, they are classified as preterm or term delivery. Based on the following data, is there a relationship between the prenatal care programs and preterm delivery? Run the appropriate test at $\alpha = 0.05$.

	Program 1: Minimal Intensity	Program 2: Moderate Intensity	Program 3: High Intensity
Preterm:	34	18	12
Term:	36	52	58

28. A national study reports that 28% of high school students smoke. We wish to test if the smoking rate is higher than the national rate among freshmen at a local university. To run the test, we take a random sample of 200 freshmen, and 32% report that they smoke. Run the appropriate test at a 5% level of significance.

SAS Problems

Use SAS to solve the following problems.

1. Recent studies have investigated the relationship between gender and career advancement. The following data represent a random sample of mathematicians classified by gender and academic rank:

	Academic Rank			
	Instructor	Assistant Professor	Associate Professor	Full Professor
Female:	12	15	18	7
Male:	21	25	30	22

Use SAS Proc Freq to generate a contingency table and run a chi-square test of independence. Use a 5% level of significance.

2. A study is conducted over 10 years to investigate long-term complication rates in patients treated with two different therapies. The data are

Therapy	Complications	
	No	Yes
1	911	89
2	873	127

Use SAS Proc Freq to generate a contingency table and test if there is a significant difference in the proportions of patients with complications between therapies.

3. Prior to being randomized to one of two competing therapies, the severity of participants' migraines is clinically assessed. The following table displays the severity classifications for patients assigned to the medical and the nontraditional therapies:

Therapy	Severity Classification			Total
	Minimal	Moderate	Severe	
Medical:	90	60	50	200
Nontraditional:	50	60	90	200
Total:	140	120	140	400

Use SAS Proc Freq to generate a contingency table and run a chi-square test of independence. Use a 5% level of significance.

4. Use SAS to estimate the sample sizes required in each group in Problem 3 to test the hypothesis $H_0: p_1 = p_2$. Consider a two-sided test. Suppose that $p_1 = 0.4$, and consider the following values for p_2: 0.20, 0.25, 0.30. How many subjects would be required to detect the specified differences with a power of 80% at a 5% level of significance?

5. Suppose we wish to estimate what proportion of patients in an HMO spend more than $500 on prescription medications over 12 months. In a random sample of 150 patients, 34% spent more than $500.

   a. Use SAS to determine the number of subjects that would be required to estimate the true proportion who spend more than $500 on prescription medications per year with a margin of error no more than 2% with 95% confidence.

b. Consider scenarios where the margin of error is 1%, 3%, and 5%. What is the effect of changing the margin of error on the number of subjects required?

c. Suppose that the sample described was not available. What would the estimate be in (a) without that information?

Descriptive Statistics
(Ch. 2)

Probability
(Ch. 3)

Sampling Distributions
(Ch. 4)

Statistical Inference
(Chapters 5–13)

OUTCOME VARIABLE	GROUPING VARIABLE(S)/ PREDICTOR(S)	ANALYSIS	CHAPTER(S)
Continuous	—	Estimate μ; Compare μ to Known, Historical Value	5/12
Continuous	Dichotomous (2 groups)	Compare Independent Means (Estimate/Test $(\mu_1 - \mu_2)$) or the Mean Difference (μ_d)	6/12
Continuous	Discrete (>2 groups)	Test the Equality of k Means Using Analysis of Variance ($\mu_1 = \mu_2 = \cdots = \mu_k$)	9/12
Continuous	Continuous	Estimate Correlation or Determine Regression Equation	10/12
Continuous	Several Continuous or Dichotomous	Multiple Linear Regression Analysis	10
Dichotomous	—	Estimate p; Compare p to Known, Historical Value	7
Dichotomous	Dichotomous (2 groups)	Compare Independent Proportions (Estimate/Test $(p_1 - p_2)$)	7/8
Dichotomous	Discrete (> 2 groups)	Test the Equality of k Proportions (Chi-Square Test)	7
Dichotomous	Several Continuous or Dichotomous	Multiple Logistic Regression Analysis	11
Discrete	Discrete	Compare Distributions Among k Populations (Chi-Square Test)	7
Time of Event	Several Continuous or Dichotomous	Survival Analysis	13

Comparing Risks in Two Populations

In Chapter 7 we introduced estimation and statistical inference procedures for categorical variables. We presented statistical inference techniques for both one- and two-sample cases where the analytic variable was dichotomous. We also considered multinomial distributions and introduced chi-square tests.

In this chapter we extend our discussion of the two-sample case with a dichotomous analytic variable and consider the specific application of comparing risks in two populations. This situation is very common in biostatistical research. For example, consider a clinical trial of a proposed stroke-prevention medication. Participants are randomly assigned to one of two treatments and are followed over the course of a 5-year study period. The analytic outcome variable is dichotomous: The participant either had a stroke during the follow-up period or did not. Another example is a longitudinal study of two populations in which the outcome variable of interest is the development of cardiovascular disease. In both of these situations, the interest is in comparing the risks of the outcome, or the proportion who develop disease in the two populations.

In Section 8.1, we introduce several effect measures used to compare risks in two populations, and in Section 8.2, we present confidence intervals for these effect measures. In Section 8.3, we review the use of the chi-square test in comparing risks in two populations and present a computational formula for the test statistic. We describe in Section 8.4 a test to be used when the assumptions of the chi-square test are not met. In Section 8.5, we present a method for calculating an adjusted relative risk and performing an adjusted chi-square test. In Section 8.6, we review the power and sample size calculations presented in Chapter 7. We summarize key formulas and present statistical computing applications in Sections 8.7 and 8.8, respectively. In Section 8.9, we use data from the Framingham Heart Study to illustrate the applications presented here.

8.1 Effect Measures

Dichotomous outcome variables in biostatistical applications usually represent events such as the development of a disease, a change in disease severity, or mortality. The parameters of interest are the population proportions, or the conditional probabilities of the event in the two populations. For example, consider the example in which one group receives treatment 1 and the other receives treatment 2 and the event of interest is stroke. The two parameters are the probability of having a stroke during a 5-year follow-up period conditional on receiving treatment 1 (p_1) and the probability of having a stroke during the 5-year follow-up period conditional on receiving treatment 2 (p_2). The probability of having the outcome of interest is often called the *risk* of that outcome. As described in Chapter 7, we estimate these probabilities, or risks, using the sample proportions in each treatment group, $\hat{p}_1$ and $\hat{p}_2$.

EXAMPLE 8.1

Risk of Coronary Abnormalities on Gamma Globulin and Aspirin

A clinical trial of gamma globulin in the treatment of children with Kawasaki syndrome randomized approximately half of the patients to receive gamma globulin. The standard treatment for Kawasaki syndrome was a regimen of aspirin; however, about one-quarter of these patients developed coronary abnormalities even under the standard treatment. The outcome of interest was the development of coronary abnormalities (CA) over a 7-week follow-up period. The following 2×2 cross-tabulation table summarizes the results of the trial.

Treatment Group	Coronary Abnormalities, CA = 1	No Coronary Abnormalities, CA = 0	Total
Gamma Globulin, GG = 1:	5	78	83
Aspirin, GG = 0:	21	63	84
Total:	26	141	167

Overall, 26 of the 167 patients developed coronary abnormalities. The estimated conditional probability or risk of developing coronary abnormalities given the standard treatment (aspirin) is $\hat{p}_0 = 21/84 = 0.25$. The risk of CA given treatment with gamma globulin is $\hat{p}_1 = 5/83 = 0.06$. ■

A number of statistics are used to compare conditional probabilities of outcomes between populations (or treatments). Measures of difference in risk are known as *effect* measures. Three of these measures are the risk difference, the relative risk, and the odds ratio.

The simple difference between risks is called the *risk difference* and is estimated as follows:

$$\text{Estimate of risk difference} = \hat{R}D = \hat{p}_1 - \hat{p}_0 \qquad (8.1)$$

A risk difference of zero indicates no difference in risks, a positive *RD* indicates higher risk in group 1, and a negative *RD* indicates lower risk in group 1. In Example 8.1, the risk difference is estimated as $\hat{R}D = 0.06 - 0.25 = -0.19$. The risk of CA in patients treated with gamma globulin is 0.19 lower than the risk in patients on the standard treatment. Note that you can reverse the roles of the groups and calculate the risk difference as $RD = \hat{p}_0 - \hat{p}_1$. In Example 8.1, this would yield an estimate of 0.19 (instead of -0.19), indicating a higher risk in group 0. The convention is to estimate the risk difference using (8.1), where $\hat{p}_0$ is the estimate of risk in the control or standard treatment group.

The relative risk is the ratio of the risks and is estimated as follows:

$$\text{Estimate of relative risk} = \hat{R}R = \hat{p}_1/\hat{p}_0 \qquad (8.2)$$

A relative risk of 1 indicates no difference in risks, a RR greater than 1 indicates a higher risk in group 1, and a RR less than 1 indicates a lower risk in group 1. In Example 8.1, the relative risk is estimated as $\hat{RR} = (0.06/0.25) = 0.24$. The risk of CA in patients treated with gamma globulin is 24% (less than a quarter) of the risk of CA in those on standard treatment. Again, we can reverse the groups and estimate the relative risk as $\hat{RR} = \hat{p}_0/\hat{p}_1$. However, in many cases, such as in clinical trials, it is easier to interpret a relative risk in which group 1 is compared to group 0. For example, in a clinical trial, the relative risk as defined in (8.2) compares the treated group to the control group. In Example 8.1, the estimated $\hat{RR} = 0.24$, which is the risk in the gamma globulin treatment group relative to the risk in the standard treatment group. Reversing the comparison groups would yield $\hat{RR} = (0.25/0.06) = 4.17$, which is interpreted (awkwardly) as lack of treatment with gamma globulin more than quadruples the risk of coronary abnormalities.

The third measure commonly used to compare risks in two populations is the odds ratio, based on a comparison of the odds of the outcome in the two groups. Suppose that $x =$ the number of outcome events in a trial of size n. We estimate the risk of the event as $\hat{p} = x/n$. The odds of the event are defined as $\hat{o} = x/(n - x)$, the ratio of events to nonevents.

In Example 8.1, the odds of CA in the standard treatment group (GG $= 0$) are estimated at $\hat{o}_0 = 21/63 = 0.333$. The odds of CA in the gamma globulin group are estimated at $\hat{o}_1 = 5/78 = 0.064$. The odds can also be estimated using the sample proportion, as follows:

$$\text{Estimate of odds} = \hat{o} = \frac{x}{(n - x)} = \frac{x/n}{(n - x)/n} = \frac{x/n}{(1 - x/n)} = \frac{\hat{p}}{(1 - \hat{p})}$$

The odds ratio is estimated by the ratio of the odds in group 1 to the odds in group 2:

$$\text{Estimate of odds ratio} = \hat{OR} = \frac{\hat{o}_1}{\hat{o}_0} = \frac{\hat{p}_1/(1 - \hat{p}_1)}{\hat{p}_0/(1 - \hat{p}_0)} \tag{8.3}$$

In Example 8.1, the odds ratio is estimated as $\hat{OR} = (5/78)/(21/63) = 0.064/0.333 = 0.192$. The odds of CA in the gamma globulin group are less than a fifth the odds of CA in the standard treatment group.

The interpretation of the odds ratio is not as intuitive as is the relative risk. However, there are some research designs in which the relative risk cannot be estimated, and in those situations, the odds ratio can be used to estimate a relative risk. If the event is rare (i.e., p is small), the odds ratio is a very good estimate of relative risk, because when p is small, $(1 - p)$ is very close to 1, so the odds $o = p/(1 - p)$ are very close to p. For example, if $p = 0.1$, then $(1 - p) = 0.9$ and $o = (0.1/0.9) = 0.11$, which is close to $p = 0.1$. For more common events, the odds are not a good estimate of the probability of that event. For example, if $p = 0.5$, then $(1 - p) = 0.5$ and $o = (0.5/0.5) = 1.0$, which is not close to p.

When the prevalence is less than 0.10, the odds of an event generally provide a good estimate of the probability of the event. In Example 8.1, the risk of CA is low in the GG group (0.06) and the estimated odds $\hat{o}_1 = 0.064$, is very close to $\hat{p}_1 = 0.060$. In the standard treatment group, the risk is higher (0.250) and $\hat{o}_0 = 0.333$ is larger than $\hat{p}_0 = 0.250$. Overall, the risk of CA is 0.156. The estimate of the odds ratio, $\hat{OR} = 0.192$, is smaller than the estimate of the relative risk, $\hat{RR} = 0.240$. ■

8.2 Confidence Intervals for Effect Measures

In Chapter 7, we presented techniques for statistical inference concerning the difference in proportions and introduced the chi-square goodness-of-fit tests and tests of independence. In this section, we provide formulas for calculating confidence intervals around the three effect measures used to compare risks in two populations.

The general case involving two populations with a dichotomous outcome variable can be represented by the general 2×2 table in Table 8.1. This framework facilitates many of the calculations needed for statistical inference. In this framework, the letter a represents x_1, the number of successes in group 1, and $(a + b)$ represents n_1, the sample size in group 1. Likewise, the letter c represents x_0, the number of successes in group 0, and $(c + d)$ represents n_0, the sample size in group 0.

The convention is to label events as 1 if present and 0 if absent. Similarly, the active or experimental treatment is usually labeled 1 and the control (e.g., placebo) is usually labeled 0. The estimated risks and effect measures can be expressed as follows using the notation of Table 8.1:

$$\hat{p}_1 = \frac{a}{(a + b)}, \qquad \hat{o}_1 = a/b$$

$$\hat{p}_0 = \frac{c}{(c + d)}, \qquad \hat{o}_0 = c/d$$

Table 8.1 *General Form of a 2 × 2 Table*

Treatment or Comparison Group	Outcome		
	1	*0*	
1	a	b	$a + b$
0	c	d	$c + d$
	$a + c$	$b + d$	N

Table 8.2 *Confidence Intervals for RD, RR, and OR*

Risk Difference:	$\hat{R}D \pm Z_{1-(\alpha/2)}\sqrt{\left(\dfrac{\hat{p}_0(1-\hat{p}_0)}{n_0}\right) + \left(\dfrac{\hat{p}_1(1-\hat{p}_1)}{n_1}\right)}$
Relative Risk:	$\exp\left(\ln(\hat{R}R) \pm Z_{1-(\alpha/2)}\sqrt{\dfrac{(d/c)}{n_0} + \dfrac{(b/a)}{n_1}}\right)$
Odds Ratio:	$\exp\left(\ln(\hat{O}R) \pm Z_{1-(\alpha/2)}\sqrt{\left(\dfrac{1}{a} + \dfrac{1}{b} + \dfrac{1}{c} + \dfrac{1}{d}\right)}\right)$

$$\text{Estimate of risk difference} = \hat{R}D = (\hat{p}_1 - \hat{p}_0) = \frac{a}{a+b} - \frac{c}{c+d} \qquad (8.4)$$

$$\text{Estimate of relative risk} = \hat{R}R = \hat{p}_1/\hat{p}_0 = \frac{a/(a+b)}{c/(c+d)} \qquad (8.5)$$

$$\text{Estimate of odds ratio} = \hat{O}R = \frac{\hat{o}_1}{\hat{o}_0} = \frac{a/b}{c/d} \qquad (8.6)$$

The risk difference defined in (8.1) and (8.4) is simply the point estimate for the difference in proportions $(p_1 - p_0)$ introduced in Chapter 7. The confidence interval estimate of the risk difference is given in Table 7.3 and again in Table 8.2. Confidence interval formulas for the other two effect measures are also given in Table 8.2. Notice that the formulas for the confidence intervals for *RR* and *OR* are given in terms of the notation in Table 8.1.

As in Table 7.3, the three confidence intervals in Table 8.2 all assume that $\min\left(n_0\hat{p}_0, n_0(1-\hat{p}_0)\right) \geq 5$ *and* $\min\left(n_1\hat{p}_1, n_1(1-\hat{p}_1)\right) \geq 5$. This assumption is satisfied if the cell frequencies a, b, c, and d are each at least 5.

Recall Example 8.1 in which we compared gamma globulin to standard treatment with respect to the development of coronary abnormalities. The results are summarized next using the notation of Table 8.1.

EXAMPLE 8.1

Estimating Difference in Risk of Coronary Abnormalities on Gamma Globulin and Aspirin (continued)

Treatment Group	Coronary Abnormalities, CA = 1	No Coronary Abnormalities, CA = 0	Total
Gamma Globulin, GG = 1:	$a = 5$	$b = 78$	83
Aspirin, GG = 0:	$c = 21$	$d = 63$	84
Total:	26	141	167

We now calculate 95% confidence intervals for the *RD, RR,* and *OR* using the formulas in Table 8.2. First we check the sample size assumption, and note that each of the cell frequencies is at least 5. Next, we substitute the following into the formulas in Table 8.2:

$$\hat{RD} = -0.19, \qquad \hat{RR} = 0.24, \qquad \hat{OR} = 0.19$$

$$\hat{p}_0 = 0.25, \qquad \hat{p}_1 = 0.06, \qquad n_0 = 84, \qquad n_1 = 83$$

$$a = 5, \qquad b = 78, \qquad c = 21, \qquad d = 63$$

A 95% confidence interval for the *risk difference* is given by

$$\hat{RD} \pm Z_{1-(\alpha/2)} \sqrt{\left(\frac{\hat{p}_0(1 - \hat{p}_0)}{n_0}\right) + \left(\frac{\hat{p}_1(1 - \hat{p}_1)}{n_1}\right)}$$

$$-0.19 \pm 1.96 \sqrt{\left(\frac{0.25(0.75)}{84}\right) + \left(\frac{0.06(0.94)}{83}\right)}$$

$$-0.19 \pm 1.96(.054)$$

$$-0.19 \pm 0.106$$

$$(-0.296, -0.084)$$

The risk of coronary abnormalities in patients treated with gamma globulin is *lower* than that in patients on the standard treatment by 19%. We are 95% confident that the true difference in risk of coronary abnormalities in patients treated with gamma globulin as compared to patients on the standard treatment is between −29.6% and −8.4% (or is from 8.4% to 29.6% lower).

A 95% confidence interval for the *relative risk* is given by

$$\exp\left(\ln(\hat{RR}) \pm Z_{1-(\alpha/2)} \sqrt{\frac{(d/c)}{n_0} + \frac{(b/a)}{n_1}}\right)$$

$$\exp\left(\ln(0.24) \pm 1.96 \sqrt{\frac{(63/21)}{84} + \frac{(78/5)}{83}}\right)$$

$$\exp(-1.423 \pm 1.96(0.473))$$

$$\exp(-1.423 \pm 0.927)$$

$$(0.095, 0.609)$$

The risk of coronary abnormalities in patients treated with gamma globulin is less than a quarter the risk in patients on the standard treatment ($RR = 0.24$). We are 95% confident that the true relative risk of coronary abnormalities in patients treated with gamma globulin as compared to patients on the standard treatment is between 0.095 and 0.609.

A 95% confidence interval for the *odds ratio* is given by

$$\exp\left(\ln(\hat{OR}) \pm Z_{1-(\alpha/2)}\sqrt{\left(\frac{1}{a} + \frac{1}{b} + \frac{1}{c} + \frac{1}{d}\right)}\right)$$

$$\exp\left(\ln(0.19) \pm 1.96\sqrt{\left(\frac{1}{5} + \frac{1}{78} + \frac{1}{21} + \frac{1}{63}\right)}\right)$$

$$\exp(-1.649 \pm 1.96(0.526))$$

$$\exp(-1.649 \pm 1.030)$$

$$(0.069, 0.539)$$

The odds of coronary abnormalities in patients treated with gamma globulin are less than one-fifth the odds in patients on the standard treatment ($OR = 0.19$). We are 95% confident that the true relative risk of coronary abnormalities in patients treated with gamma globulin as compared to patients on the standard treatment is between 0.069 and 0.539. ■

SAS EXAMPLE 8.1 Estimating Difference in Risk of Coronary Abnormalities on Gamma Globulin and Aspirin Using SAS

The following output was generated for the data in Example 8.1 using SAS Proc Freq, which produces a contingency table (or cross-tabulation) when two variables are specified. Notice that the row percentages in the "CA" column are 6.02% and 25% for those in the gamma globulin and standard treatment groups, respectively. These are the estimated risks we calculated as 0.06 and 0.25.

We requested that SAS provide the odds ratio and relative risk and 95% confidence intervals. It is important that the event of interest is reported in the first column, as SAS calculates the relative risk and odds ratios assuming that is the case. The relative risks and odds ratios compare the row 1 group to the row 2 group. For that reason, we make sure that the GG = 1 group is in the first row so that the effect measures will compare the gamma globulin group to the standard treatment group. A brief interpretation appears after the output.

SAS Output for Example 8.1

```
                      The FREQ Procedure
                     Table of group by event

group         event

Frequency|
Percent   |
Row Pct   |
Col Pct  |CA       |noCA     |  Total
---------+--------+--------+
GG       |       5 |      78 |      83
         |    2.99 |   46.71 |   49.70
         |    6.02 |   93.98 |
         |   19.23 |   55.32 |
---------+--------+--------+
noGG     |      21 |      63 |      84
         |   12.57 |   37.72 |   50.30
         |   25.00 |   75.00 |
         |   80.77 |   44.68 |
---------+--------+--------+
Total            26      141      167
              15.57    84.43   100.00
```

Statistics for Table of group by event

Estimates of the Relative Risk (Row1/Row2)

Type of Study	Value	95% Confidence Limits	
Case-Control (Odds Ratio)	0.1923	0.0686	0.5388
Cohort (Col1 Risk)	0.2410	0.0954	0.6089
Cohort (Col2 Risk)	1.2530	1.0948	1.4340

Sample Size = 167

Interpretation of SAS Output for Example 8.1

We requested that SAS provide the odds ratio and the relative risk. The OR is given in the first row and the RR is given in the next row, labeled Col1 Risk, as it is the relative risk of the event in the first column (CA) of the 2×2 table. Note that the values SAS provides for the OR and RR agree with those we calculated earlier. The 95% confidence limits are also given for the odds ratio and relative risk. ■

8.3 The Chi-Square Test of Homogeneity

The chi-square test was introduced in Chapter 7 as a technique to be used in tests for goodness of fit and independence. The chi-square test of homogeneity can be used to test hypotheses concerning risks in two populations. The null hypothesis is that the risks are the same in the two populations, or that the two populations are *homogeneous* with respect to risk.

$$H_0: \ p_0 = p_1$$
$$H_1: \ p_0 \neq p_1$$

Now consider the relationship between the effect measures and the risks. If the two risks are the same, then the risk difference must be equal to zero and both the relative risk and odds ratios must be equal to 1. Thus, the null hypothesis may equivalently be written in terms of the risk difference,

$$H_0: \ RD = 0$$
$$H_1: \ RD \neq 0$$

the relative risk,

$$H_0: \ RR = 1$$
$$H_1: \ RR \neq 1$$

or the odds ratio,

$$H_0: \ OR = 1$$
$$H_1: \ OR \neq 1$$

As stated in Chapter 7, χ^2 tests are based on the agreement between observed (sample) and expected (under H_0) frequencies. We therefore calculate expected frequencies using the fact that the risks in the two groups are equal under the null hypothesis.

EXAMPLE 8.2

Testing Difference in Risk of Stroke Between Treatments

Consider again the example of a clinical trial of a proposed stroke-prevention medication. Suppose that 250 participants are randomly assigned to receive either the new treatment ($n_1 = 120$) or a placebo ($n_0 = 130$) and are followed over the course of a 5-year follow-up period. During the 5-year follow-up period, 12 of the 120 subjects in the new treatment group (T = 1) and 28 of the 130 subjects in the placebo group (T = 0) had strokes. Thus, $\hat{p}_1 = 0.100$ and $\hat{p}_0 = 0.215$. These results are given in the following table:

Treatment Group	Event Stroke	No Stroke	Total
T = 1	12	108	120
T = 0	28	102	130
	40	210	250

Suppose we want to test the hypothesis that the risk of stroke is different between treatment groups.

1. Set up hypotheses.

$$H_0: \ p_0 = p_1$$
$$H_1: \ p_0 \neq p_1$$

Alternatively, the hypotheses can be phrased in terms of the relative risk of stroke in the new treatment group (T = 1) as compared to the placebo group (T = 0):

$$H_0: \ RR = 1$$
$$H_1: \ RR \neq 1$$

2a. Select the appropriate test statistic.

We test the null hypothesis of homogeneity using the χ^2 test of homogeneity:

$$\chi^2 = \sum \frac{(O - E)^2}{E} \tag{8.7}$$

where $\sum$ indicates summation over all four cells of the table
  O = observed frequency
  E = expected frequency

The test statistic χ^2 follows a χ^2 distribution with df = 1 as long as the expected frequencies in each cell are at least 5. The critical value (see Table B.5 in the Appendix) for a test with $\alpha = 0.05$ is $\chi^2 = 3.84$.

2b. Check that the test is valid.

We must calculate the expected frequencies and check that each is at least 5 so that we can be sure the test statistic follows the χ^2 distribution. Combining the two treatment groups, we see that 40 of the 250 subjects had strokes in the 5-year study period. The estimated 5-year risk of stroke is thus $\hat{p} = 40/250 = 0.16$. Assuming that the null hypothesis is true, the risk of stroke is the same in both treatment groups, and we would expect that 16% of the subjects in each group would have a stroke during the study period. Thus, among the

120 subjects in the new treatment group (T = 1), we expect that 16%, or (0.16)(120) = 19.2 subjects, would have a stroke. Similarly, we expect that 16% of the 130 subjects in the placebo group (T = 0), or (0.16)(130) = 20.8 subjects, would have a stroke. The following table contains the expected cell frequencies under the null hypothesis. The expected frequencies in the "No Stroke" column are obtained by subtraction (e.g., 120 − 19.2 = 100.8). Notice that we do not round the expected frequencies to integers.

	Event		
Treatment Group	Stroke	No Stroke	Total
T = 1	19.2	100.8	120
T = 0	20.8	109.2	130
	40	210	250

Each of the expected frequencies (19.2, 100.8, 20.8, and 109.2) is at least 5, so we can compare the test statistic to the critical value $\chi^2 = 3.84$.

3. Decision rule.

$$\text{Reject } H_0 \text{ if } \chi^2 \geq 3.84$$

$$\text{Do not reject } H_0 \text{ if } \chi^2 < 3.84$$

4. Test statistic.

We now calculate the test statistic using equation (8.7) and the expected frequencies.

$$\chi^2 = \frac{(12 - 19.2)^2}{19.2} + \frac{(108 - 100.8)^2}{100.8} + \frac{(28 - 20.8)^2}{20.8}$$
$$+ \frac{(102 - 109.2)^2}{109.2} = 6.181$$

5. Conclusion.

Reject H_0 since $6.181 \geq 3.84$. There is significant evidence of a difference in risk, or that the relative risk is not equal to 1. ■

A computational formula for the χ^2 statistic in the special situation of a 2 × 2 table is

$$\chi^2 = \frac{(ad - bc)^2(N)}{(a + b)(c + d)(a + c)(b + d)} \tag{8.8}$$

This formula does not require the calculation of expected frequencies in a separate step. However, it is still necessary to check that the cell sizes are sufficient (that each expected frequency is at least 5). We can use this formula to calculate the test statistic in Example 8.2.

$$\chi^2 = \frac{(ad - bc)^2(N)}{(a + b)(c + d)(a + c)(b + d)}$$

$$= \frac{\{(12)(102) - (108)(28)\}^2(250)}{(120)(130)(40)(210)}$$

$$= \frac{810,000,000}{131,040,000}$$

$$= 6.181$$

SAS EXAMPLE 8.2 **Testing Difference in Risk of Stroke Between Treatments Using SAS**

The following output was generated for the data in Example 8.2 using SAS Proc Freq. We requested that SAS provide the expected cell frequencies and the row percents and the odds ratio, relative risk, and 95% confidence intervals. We also requested the chi-square test of homogeneity. Again, we make sure that the event of interest (stroke) is reported in the first column and the treatment group is in the first row. A brief interpretation appears after the output.

SAS Output for Example 8.2

```
                   The FREQ Procedure
               Table of group by stroke

group        stroke

Frequency |
Expected  |
Row Pct   |yes      |no       |  Total
----------+--------+--------+
Treatment |     12 |    108 |    120
          |   19.2 |  100.8 |
          |  10.00 |  90.00 |
----------+--------+--------+
Placebo   |     28 |    102 |    130
          |   20.8 |  109.2 |
          |  21.54 |  78.46 |
----------+--------+--------+
Total            40      210      250
```

```
           Statistics for Table of group by stroke
     Statistic                      DF         Value        Prob
     --------------------------------------------------------------
     Chi-Square                      1         6.1813       0.0129
     Likelihood Ratio Chi-Square     1         6.3540       0.0117
     Continuity Adj. Chi-Square      1         5.3526       0.0207
     Mantel-Haenszel Chi-Square      1         6.1566       0.0131
     Phi Coefficient                           -0.1572
     Contingency Coefficient                    0.1553
     Cramer's V                                -0.1572


                      Fisher's Exact Test
           --------------------------------------
           Cell (1,1) Frequency (F)          12
           Left-sided Pr <= F            0.0097
           Right-sided Pr >= F           0.9965


           Table Probability (P)         0.0062
           Two-sided Pr <= P             0.0154


          Estimates of the Relative Risk (Row1/Row2)
     Type of Study                   Value      95% Confidence Limits
     --------------------------------------------------------------
     Case-Control (Odds Ratio)      0.4048      0.1954        0.8386
     Cohort (Col1 Risk)             0.4643      0.2475        0.8710
     Cohort (Col2 Risk)             1.1471      1.0296        1.2779


                      Sample Size = 250
```

Interpretation of SAS Output for Example 8.2

The first number in each cell is the observed frequency, the second is the expected frequency, and the final number is the row percent. For example, in the "Treatment"–"yes" cell (the "a" cell), the observed frequency is 12, the expected frequency is 10.2, and the row percent is 10%. The estimated relative risk is $\hat{R}R = 0.464$ with 95% confidence limits (0.248, 0.871). The risk of stroke in subjects on the new treatment is estimated to be less than half the risk of stroke in those in the placebo group. Notice that the confidence interval does not include the null value 1.

The requested chi-square procedure allows us to test the null hypothesis that the risk of stroke is the same in the two groups. Here we will phrase the hypotheses in terms of the relative risk, $H_0: RR = 1$ and $H_1: RR \neq 1$. The value of the chi-square statistic is 6.181, as we calculated earlier, and the associated p value is $p = 0.0129$. Assuming $\alpha = 0.05$, we reject H_0 since $p = 0.0129 < \alpha = 0.05$. There is significant evidence ($p = 0.0129$) that the relative risk of stroke in treated subjects as compared to those on placebo is not equal to 1. ■

EXAMPLE 8.3

Testing Difference in Risk of Institutionalization Between Male and Female Stroke Survivors

A study of stroke patients who survived 6 months after the stroke found that 6/45 men and 22/63 women lived in an institution (such as an assisted living or nursing facility). Is there evidence of a difference in risk of living in an institution after stroke for men versus women?

The null hypothesis is that men and women stroke survivors are equally likely to live in an institution, or that the relative risk for men of living in an institution compared to women is equal to 1. First, we set up the 2×2 table.

Comparison Group: Gender	Institution Yes	No	Total
Men:	6	39	45
Women:	22	41	63
	28	80	108

We estimate the probabilities of living in an institution for men and women separately: $\hat{p}_M = 0.133$ and $\hat{p}_W = 0.349$. The estimated relative risk comparing men to women is $\hat{R}R = 0.381$; among stroke survivors, men are only 38% as likely as women to live in an institution. Suppose that we want to test the hypothesis that the risk of living in an institution is different for men and women.

1. Set up hypotheses.

$$H_0: p_M = p_W$$
$$H_1: p_M \neq p_W$$

Again, the hypotheses can be phrased in terms of the relative risk of living in an institution for men as compared to women:

$$H_0: \ RR = 1$$
$$H_1: \ RR \neq 1$$

2a. Select the appropriate test statistic.

We test the null hypothesis of homogeneity using the χ^2 test of homogeneity.

$$\chi^2 = \sum \frac{(O - E)^2}{E}$$

where $\sum$ indicates summation over all four cells of the table
O = observed frequency
E = expected frequency

The test statistic χ^2 follows a χ^2 distribution with df $= 1$ as long as the expected frequencies in each cell are at least 5.

2b. Check that the test is valid.

Before we can use the chi-square test, we must check that each of the expected frequencies is at least 5. Under the null hypothesis we assume that the probability of living in an institution is the same for men and women ($p_M = p_W = p$). The best estimate of this probability is obtained by combining men and women and calculating the overall sample proportion. Based on this combined sample, the estimated probability of living in an institution is $\hat{p} = (28/108) = 0.259$. Thus, we expect about 25.9% of men and 25.9% of women to be living in an institution. Under the null hypothesis, in this sample of 45 men, we expect $(0.259)(45)$ or 11.67 men to be living in an institution, while the remaining 33.33 live at home. Similarly, among the women, we expect $(.259)(63) = 16.33$ to be living in an institution and 46.67 to be living at home. The expected frequencies are 11.67, 33.33, 16.33, and 46.67, and each is of sufficient size (i.e., each is at least 5), so we can use the chi-square test to test H_0.

3. Decision rule.

$$\text{Reject } H_0 \text{ if } \chi^2 \geq 3.84$$
$$\text{Do not reject } H_0 \text{ if } \chi^2 < 3.84$$

4. Test statistic.

We now calculate the test statistic using the computational formula (8.7).

$$\chi^2 = \frac{(ad - bc)^2(N)}{(a + b)(c + d)(a + c)(b + d)}$$

$$= \frac{\{(6)(41) - (39)(22)\}^2(108)}{(45)(63)(28)(80)}$$

$$= \frac{40,450,752}{6,350,400}$$

$$= 6.3698$$

5. Conclusion.

Reject H_0 since $\chi^2 = 6.37 > 3.84$. There is significant evidence that the probability of living in an institution is different for men and women stroke survivors. Another interpretation is that there is significant evidence that the relative risk of living in an institution comparing men and women is different from 1. ■

8.4 Fisher's Exact Test

In Chapter 7 and in Section 8.3, we stressed that the chi-square tests are valid if each of the expected cell frequencies is at least 5. If one or more of the four expected cell frequencies in a 2×2 table is less than 5, the chi-square test is not valid and we must use another method. One such method is Fisher's Exact test. A discussion of the mechanics of this test is beyond the scope of this text, so we simply present examples using SAS.

SAS EXAMPLE 8.4 **Testing Differences in Functional Ability Between Stroke Survivors and Controls**

A substudy of the stroke study we have described involved a comparison of the functional ability of stroke patients to patients of the same age and gender who did not suffer stroke. As many as possible of the stroke survivors were matched to controls who were the same age and sex. Cases (stroke patients) and controls (persons who did not suffer stroke) were questioned with respect to various measures of disability. Among the 24 stroke cases, only 16 were able to walk unassisted, whereas 23 of the 24 controls were able to walk unassisted. Using the data shown in the SAS output, we test whether the stroke cases were more likely to need assistance walking as compared to age- and sex-matched controls. The analysis is conducted using SAS Proc Freq; a brief interpretation appears after the output.

SAS Output for Example 8.4

```
                      The FREQ Procedure
                   Table of group by walk

group            walk

Frequency     |
Expected      |
Row Pct       |Needs As|Able to |  Total
              |ssistanc|Walk    |
              |e       |        |
      --------------+--------+--------+
Stroke Cases  |      8 |     16 |    24
              |    4.5 |   19.5 |
              |  33.33 |  66.67 |
      --------------+--------+--------+
Controls      |      1 |     23 |    24
              |    4.5 |   19.5 |
              |   4.17 |  95.83 |
      --------------+--------+--------+
Total                9       39       48
```

```
           Statistics for Table of group by walk

Statistic                      DF       Value      Prob
-----------------------------------------------------------
Chi-Square                      1       6.7009    0.0096
Likelihood Ratio Chi-Square     1       7.4609    0.0063
Continuity Adj. Chi-Square      1       4.9231    0.0265
Mantel-Haenszel Chi-Square      1       6.5613    0.0104
Phi Coefficient                         0.3736
Contingency Coefficient                 0.3500
Cramer's V                              0.3736
```

```
WARNING: 50% of the cells have expected counts less
         than 5. Chi-Square may not be a valid test.
```

```
                   Fisher's Exact Test
         -----------------------------------
         Cell (1,1) Frequency (F)          8
         Left-sided Pr <= F           0.9992
         Right-sided Pr >= F          0.0113

         Table Probability (P)        0.0105
         Two-sided Pr <= P            0.0226
```

```
                        Estimates of the Relative Risk (Row1/Row2)

Type of Study                        Value        95% Confidence Limits
--------------------------------------------------------------------------
Case-Control (Odds Ratio)          11.5000          1.3071        101.1816
Cohort (Col1 Risk)                  8.0000          1.0823         59.1349
Cohort (Col2 Risk)                  0.6957          0.5180          0.9343

                           Sample Size = 48
```

Interpretation of SAS Output for Example 8.4

The expected cell frequencies in the four cells are 4.5, 4.5, 19.5, and 19.5. Two of these are less than the required 5. Notice that SAS provides a "Warning" statement, indicating that 50% of the cells have expected counts less than 5 and that the chi-square test may not be valid. Below this, SAS provides results of the two-sided Fisher's Exact test, which is recommended when the expected counts are small (i.e., < 5).

One-third (33.3%) of the stroke cases need assistance to walk whereas only 4.2% of their age and sex-matched controls need assistance. The estimated RR is 8.0 (0.333/0.0417). The null hypothesis is H_0: $RR = 1$ and H_1: $RR \neq 1$. We use the two-sided Fisher's Exact test to test the hypothesis. Assuming $\alpha = 0.05$, we reject H_0 since $p = 0.0226 < \alpha = 0.05$. There is significant evidence ($p = 0.0226$) that stroke cases are more likely to need assistance walking as compared to age- and sex-matched controls. ■

8.5 Cochran–Mantel–Haenszel Method

The Cochran–Mantel–Haenszel method is an extension of the chi-square method and is applied when interest lies in comparing two groups in terms of a dichotomous outcome over several levels of a third variable. For example, suppose there are a series of 2×2 tables, one for each of several strata (maybe the strata reflect different levels of severity of the index condition or different clinical centers), and interest lies in combining the information to take into account the differences in the strata. In this section we will describe a method used to adjust the estimate of relative risk and the test statistic to account for a third variable in the analysis.

EXAMPLE 8.5   **Testing Differences in Risk of Hypertension Between Drugs in Older and Younger Patients**

Suppose we want to compare two medications, call them Drug A and Drug B, with respect to the development of or risk of hypertension (HTN). A total of 200 subjects are involved in the analysis, and half are randomized to receive Drug A and half are randomized to receive Drug B. The following table summarizes the results of the study:

Medication	HTN Yes	HTN No	
Drug A	56	44	100
Drug B	32	68	100
Total	88	112	200

Overall, 88/200 subjects, or 44%, developed hypertension (HTN). The risk of HTN among subjects taking Drug A is $\hat{p}_A = 56/100 = 0.56$ and the risk of HTN among subjects taking Drug B is $\hat{p}_B = 32/100 = 0.32$, and the estimated relative risk comparing subjects on Drug A to those on Drug B is $\hat{R}R = (0.56/0.32) = 1.75$. A test of the hypothesis $H_0: RR = 1$ yields $\chi^2 = 11.688$ with $p < 0.001$. We reject H_0 and conclude that there is significant evidence that the relative risk of HTN comparing subjects taking Drug A to those taking Drug B is different from 1.

Half of the subjects were randomized to receive Drug A and the other half received Drug B. However, further inspection of the data revealed that this balance was not maintained within age groups. In fact, it turns out that 60% of the 100 older subjects (age 65 and older) received Drug A, and only 40% of the 100 younger subjects (age less than 65) received Drug A. We can repeat the analysis within the two age groups (i.e., two separate replications).

	Age 65 + HTN Yes	No			Age < 65 HTN Yes	No	
Drug A:	32	8	40	*Drug A:*	24	36	60
Drug B:	24	36	60	*Drug B:*	8	32	40
Total:	56	44	100	*Total:*	32	68	100

$$\hat{R}R = 0.8/0.4 = 2.0 \qquad\qquad \hat{R}R = 0.4/0.2 = 2.0$$

The overall risk of hypertension is 56% among those age 65 and older, and it is only 32% for those less than 65 years of age. However, the relative risk of HTN comparing those taking Drug A to those taking Drug B is the same in the two age groups. The baseline level of HTN is different, but the relative impact of the two drugs is the same.

This would appear to indicate that combining the two age groups and reporting results based on the combined data are fine. In fact, since the estimated relative risk is 2.0 in each age group, the relative risk based on the combined data should surely also be 2.0. However, recall that we actually obtained an estimate of $\hat{R}R = 1.75$. This is caused by the combination of different baseline risks in the age groups and the imbalance in the treatment group allocation in the age groups, which is called *confounding* by age group. ■

In this section, we present the Cochran–Mantel–Haenszel method used to adjust the relative risk and the chi-square statistic to remove the effect of confounding. In Chapter 11, we introduce another method to adjust for confounding.

With a series of 2 × 2 tables, each table reflecting the relationship between group and event for a different strata (e.g., different level of a confounder), the Cochran–Mantel–Haenszel method can be used to calculate an *adjusted chi-square test statistic and an adjusted relative risk*. The procedure is as follows.

First, create a 2 × 2 table *within* each stratum of the stratification variable. As before, the basic notation we use for the 2 × 2 table is as follows:

	Stratum 1 Event		
Group	1	0	
1:	a	b	$a + b$
0:	c	d	$c + d$
	$a + c$	$b + d$	N

Next, calculate:

$$\chi^2_{CMH} = \frac{\left(\sum \frac{(ad - bc)}{N}\right)^2}{\sum \frac{(a + b)(c + d)(a + c)(b + d)}{(N - 1)N^2}} \tag{8.9}$$

where the summations are over the strata. This test statistic has a χ^2 distribution with 1 df, so we compare the test statistic to the critical value of 3.84, assuming a 5% level of significance.

EXAMPLE 8.5

Testing Differences in Risk of Hypertension Between Drugs in Older and Younger Patients (continued)

Returning to Example 8.5, the adjusted test statistic is computed as follows:

$$\chi^2_{CMH} = \frac{\left(\sum \frac{(ad - bc)}{N}\right)^2}{\sum \frac{(a + b)(c + d)(a + c)(b + d)}{(N - 1)N^2}}$$

$$= \frac{\left(\frac{(32)(36) - (8)(24)}{100} + \frac{(24)(32) - (36)(86)}{100}\right)^2}{\frac{(40)(60)(56)(44)}{(99)100^2} + \frac{(60)(40)(32)(68)}{(99)100^2}}$$

$$= \frac{207.36}{11.248} = 18.435$$

The null hypothesis is that the relative risk of hypertension comparing subjects on Drug A to subjects on Drug B, adjusted for age, is equal to 1. We reject H_0 since $\chi^2_{\text{MH}} = 18.435 > 3.84$ and conclude that, after adjusting for age, the risk of HTN in subjects taking Drugs A and B are not equal.

We can also calculate the adjusted estimate of the relative risk. The crude relative risk, $\hat{R}R_C$, is the relative risk obtained by combining all the data and estimating RR. In Example 8.5, $\hat{R}R_C = (0.56/0.32) = 1.75$. The Cochran–Mantel–Haenszel adjusted RR is estimated by the following:

$$\hat{R}R_{\text{CMH}} = \frac{\sum \dfrac{a(c+d)}{N}}{\sum \dfrac{c(a+b)}{N}} \qquad (8.10)$$

where the summations are over the strata. In Example 8.5,

$$\hat{R}R_{\text{CMH}} = \frac{\dfrac{(32)(60)}{100} + \dfrac{(24)(40)}{100}}{\dfrac{(24)(40)}{100} + \dfrac{(8)(60)}{100}}$$

$$= \frac{28.8}{14.4}$$

$$= 2.0$$

The relative risk adjusted for age is $\hat{R}R_{\text{CMH}} = 2.0$, which is the same relative risk we obtained in each of the age groups. ■

SAS Example 8.5 Testing Differences in Risk of Hypertension Between Drugs in Older and Younger Patients Using SAS

We can request that SAS perform the Cochran–Mantel–Haenszel adjusted chi-square test and produce a Cochran–Mantel–Haenszel adjusted relative risk. The analysis is conducted using SAS Proc Freq; a brief interpretation appears after the output.

SAS Output for Example 8.5

The FREQ Procedure

Table 1 of group by htn
Controlling for age=65+

group   htn

Frequency|
Row Pct |HTN |noHTN | Total
---------+--------+--------+
Drug A | 32 | 8 | 40
         | 80.00 | 20.00 |
---------+--------+--------+
Drug B | 24 | 36 | 60
         | 40.00 | 60.00 |
---------+--------+--------+
Total   56   44   100

Table 2 of group by htn
Controlling for age=<65

group   htn

Frequency|
Row Pct |HTN |noHTN | Total
---------+--------+--------+
Drug A | 24 | 36 | 60
         | 40.00 | 60.00 |
---------+--------+--------+
Drug B | 8 | 32 | 40
         | 20.00 | 80.00 |
---------+--------+--------+
Total   32   68   100

Summary Statistics for group by htn
Controlling for age

Cochran-Mantel-Haenszel Statistics (Based on Table Scores)

Statistic	Alternative Hypothesis	DF	Value	Prob
1	Nonzero Correlation	1	18.4345	<.0001
2	Row Mean Scores Differ	1	18.4345	<.0001
3	General Association	1	18.4345	<.0001

```
                  Estimates of the Common Relative Risk (Row1/Row2)

Type of Study     Method                 Value     95% Confidence Limits
------------------------------------------------------------------------
Case-Control      Mantel-Haenszel        4.0000     2.0791         7.6955
  (Odds Ratio)    Logit                  4.0000     2.0707         7.7268

Cohort            Mantel-Haenszel        2.0000     1.4427         2.7727
  (Col1 Risk)     Logit                  2.0000     1.4670         2.7266

Cohort            Mantel-Haenszel        0.5714     0.4351         0.7505
  (Col2 Risk)     Logit                  0.6722     0.5286         0.8546

                        Breslow-Day Test for
                     Homogeneity of the Odds Ratios
                     -----------------------------
                     Chi-Square               1.4664
                     DF                            1
                     Pr > ChiSq              0.2259

                     Total Sample Size = 200
```

Interpretation of SAS Output for Example 8.5

SAS produces individual 2×2 tables for each of the strata (age groups). The Cochran–Mantel–Haenszel adjusted chi-square statistic and its associated p value is given under "Cochran–Mantel–Haenszel Statistics (Based on Table Scores)"; in this situation, all three given statistics are identical. The Cochran–Mantel–Haenszel adjusted relative risk and the 95% confidence interval are given under "Estimates of the Common Relative Risk (Row1/Row2)" in the row labeled "Type of Study = Cohort (Col 1 Risk)" and "Method = Mantel–Haenszel."

This method is only valid if the relative risks are homogeneous across strata. SAS provides the p value associated with the Breslow–Day test for the hypothesis of homogeneity. If this p value is > 0.05, we assume that the relative risks are homogeneous and that the Cochran–Mantel–Haenszel statistics are valid.

In this example, the p value for the Breslow–Day test is $p = 0.226 > 0.05$, so the Cochran–Mantel–Haenszel method is valid. From the output, we see that $\hat{R}R_{MH} = 2.0$, with 95% confidence interval (1.44, 2.77). The $\chi^2_{CMH} = 18.435$, with $p = < 0.001$. ■

8.6 Precision, Power, and Sample Size Determination

The formulas given in Section 7.7 may be used to calculate the minimum number of subjects needed to ensure a specified level of power to detect a difference in proportions (or risks) between two groups. Equation (7.16) is given here using the terminology of this chapter.

$$n_i = \left(\frac{\sqrt{2\bar{p}\bar{q}}\,Z_{1-(\alpha/2)} + \sqrt{p_1 q_1 + p_0 q_0}\,Z_{1-\beta}}{\hat{R}D} \right)^2 \tag{8.11}$$

where $\bar{p} = (\hat{p}_1 + \hat{p}_0)/2$ and $\bar{q} = 1 - \bar{p}$

8.7 Key Formulas (using 2 × 2 table notation in Table 8.1)

Application	Notation/Formula	Description
Risk difference	$\hat{R}D = \hat{p}_1 - \hat{p}_0 = \dfrac{a}{(a+b)} - \dfrac{c}{(c+d)}$	
Relative risk	$\hat{R}R = \hat{p}_1/\hat{p}_0 = \dfrac{a/(a+b)}{c/(c+d)}$	
Odds ratio	$\hat{O}R = \dfrac{\hat{o}_1}{\hat{o}_0} = \dfrac{\hat{p}_1/(1-\hat{p}_1)}{\hat{p}_0/(1-\hat{p}_0)} = \dfrac{a/b}{c/d}$	
Confidence interval for risk difference	$\hat{R}D \pm Z_{1-(\alpha/2)}\sqrt{\left(\dfrac{\hat{p}_0(1-\hat{p}_0)}{n_0}\right) + \left(\dfrac{\hat{p}_1(1-\hat{p}_1)}{n_1}\right)}$	
Confidence interval for relative risk	$\exp\left(\ln(\hat{R}R) \pm Z_{1-(\alpha/2)}\sqrt{\dfrac{(d/c)}{n_0} + \dfrac{(b/a)}{n_1}} \right)$	
Confidence interval for odds ratio	$\exp\left(\ln(\hat{O}R) \pm Z_{1-(\alpha/2)}\sqrt{\left(\dfrac{1}{a} + \dfrac{1}{b} + \dfrac{1}{c} + \dfrac{1}{d}\right)} \right)$	
$H_0: RD = 0$ or $H_0: RR = 1$ or $H_0: OR = 1$	$\chi^2 = \sum \dfrac{(O-E)^2}{E}, \quad df = 1$	Chi-square test of homogeneity. Assumes all expected cell counts are ≥ 5
$H_0: RD = 0$ or $H_0: RR = 1$ or $H_0: OR = 1$	$\chi^2 = \dfrac{(ad-bc)^2(N)}{(a+b)(c+d)(a+c)(b+d)}$	Chi-square test of homogeneity: computational formula. Assumes all expected cell counts are ≥ 5
$H_0: RD = 0$ or $H_0: RR = 1$ or $H_0: OR = 1$ (each adjusted for confounder)	$\chi^2_{MH} = \dfrac{\left(\sum\dfrac{(ad-bc)}{N}\right)^2}{\sum\dfrac{(a+b)(c+d)(a+c)(b+d)}{(N-1)N^2}}$	Cochran–Mantel–Haenszel chi-square test of homogeneity, adjusted for confounder
Estimate of relative risk, adjusted for confounder	$RR_{MH} = \left(\sum\dfrac{a(c+d)}{N}\right) \Big/ \left(\sum\dfrac{c(a+b)}{N}\right)$	Cochran–Mantel–Haenszel estimate of relative risk, adjusted for confounder
Sample size required per group to test $H_0: RD = 0$ with power $1 - \beta$	$n_i = \left(\dfrac{\sqrt{2\bar{p}\bar{q}}\,Z_{1-(\alpha/2)} + \sqrt{p_1 q_1 + p_0 q_0}\,Z_{1-\beta}}{\hat{R}D} \right)^2$	

8.8 Statistical Computing

Following are the SAS programs used to generate the analyses presented in this chapter. The SAS procedures and brief descriptions of their use are noted in the header to each example. Notes are provided to the right of the SAS programs (in blue) for orientation purposes and are not part of the programs. In addition, there are blank lines in the programs that are solely to accommodate the notes. Blank lines and spaces can be used throughout SAS programs to enhance readability. A summary of the SAS procedures used in the examples is provided at the end of this section.

SAS EXAMPLE 8.1 **Estimate Relative Risk and Odds Ratio**

Compare Risk of Coronary Abnormalities on Gamma Globulin as Compared to Aspirin (Example 8.1)

A trial of gamma globulin in the treatment of children with Kawasaki syndrome randomized approximately half of the patients to receive gamma globulin plus aspirin; the other half received standard treatment of aspirin. The outcome of interest was the development of coronary abnormalities (CA) within seven weeks of treatment. The following 2×2 cross-tabulation table summarizes the results of the trial. Use SAS to estimate the relative risk (and odds ratio) of CA in patients treated with gamma globulin as compared to those on standard treatment.

Treatment Group	Coronary Abnormalities CA = 1	No Coronary Abnormalities CA = 0	Total
Gamma Globulin, GG = 1	5	78	83
Aspirin, GG = 0	21	63	84
Total	26	141	167

Program Code

```
options ps=62 ls=80;

data in;
  input group $ event $ f;
```

Formats the output page to 62 lines in length and 80 columns in width

Beginning of Data Step

Inputs three variables, **group** (noGG or GG), **event** (CA or noCA), and *f*, the cell frequency. **Group** and **event** are defined as character variables using the $.

```
cards;
noGG noCA 63
noGG CA 21
GG noCA 78
GG CA 5
run;
proc freq;


  tables group*event/relrisk;




  weight f;

run;
```

Beginning of Raw Data section.
Actual observations (values of
  group, event, and **f** on each line)

Procedure call. Proc freq generates
  a 2 × 2 table for two categorical
  variables.
Specifies the analytic variables to
  form the rows (**group**) and
  columns (**event**) of the table. The
  relrisk option requests that SAS
  generate estimates of the relative
  risk and odds ratio.
Specifies the variable that contains
  the cell counts, **f.**
End of procedure section.   ■

SAS Example 8.2 Estimate Relative Risk and Test H_0: $RR = 1$ Versus H_1: $RR \neq 1$

Estimate the Relative Risk of Stroke Comparing a Proposed Stroke Medication to Placebo, and Test the Hypothesis That the Relative Risk Is Equal to 1 (Example 8.2)

In a clinical trial 250 participants are randomly assigned to receive either a proposed stroke prevention medication (T_1) or a placebo (T_0) and are followed over the course of a 5-year study period. These results are given in the following table. Using SAS, estimate the relative risk of stroke comparing those on the medication to those on placebo, and test the hypothesis that the relative risk is not equal to 1.

	Event		
Treatment Group	*Stroke*	*No Stroke*	*Total*
T = 1	12	108	120
T = 0	28	102	130
	40	210	250

Program Code

```options ps=62 ls=80;```	Formats the output page to 62 lines in length and 80 columns in width
```proc format;```	Proc format defines formats to label the values of the variables.
``` value eventf 1='yes' 2='no';``` ``` value grpf 1='Treatment' 2='Placebo';```	Variables with values of 1 and 2 will be labeled as 'yes' and 'no,' respectively, if formatted with the **eventf** format. Variables with values of 1 and 2 will be labeled as 'Treatment' and 'Placebo,' respectively, if formatted with the **grpf** format.
```run;```	End of format procedure.
```data in;``` ``` input group stroke f;```	Beginning of Data Step Inputs three variables **group** (1 or 2), **stroke** (1 or 2), and *f* (the cell frequency).
``` format stroke eventf. group grpf.;```	The format statement assigns the format **eventf** to the variable **stroke,** and the format **grpf** to the variable **group.** Note that the format names are followed by a period to distinguish them from variable names.
```cards;```	Beginning of Raw Data section.

```1 1 12```
```1 2 108```
```2 1 28```
```2 2 102```
```run;```
	Actual observations (values of **group, stroke,** and *f* on each line).

```proc freq;```	Procedure call. Proc freq generates a 2×2 table for two categorical variables.
``` tables group*stroke/nocol nopercent``` ``` expected relrisk chisq;```	Specifies the analytic variables to form the rows (**group**) and columns (**stroke**) of the table. The nocol and nopercent options suppress the printing of the column and overall percents, respectively. The relrisk option requests that SAS generate estimates of the relative risk and odds ratio. The chisq option requests the Chi-Square test.
``` weight f;```	Specifies the variable that contains the cell counts, *f*.
```run;```	End of procedure section. ■

SAS EXAMPLE 8.4 **Estimate Relative Risk and Test H_0: $RR = 1$ Versus H_1: $RR \neq 1$**

Are Stroke Cases More Likely to Need Assistance Walking as Compared to Age- and Sex-Matched Controls? (Example 8.4)

As part of the stroke study described earlier, as many as possible of the stroke survivors were matched to controls who were the same age and sex. Cases and controls were questioned with respect to various measures of disability. Among the 24 stroke cases, only 16 were able to walk unassisted, whereas 23 of the 24 controls were able to walk unassisted. Use SAS to test whether stroke cases are more likely to need assistance walking as compared to age- and sex-matched controls.

Program Code

```
options ps=62 ls=80;

proc format;

  value eventf 1='Needs Assistance'
          2='Able to Walk';


  value grpf 1='Stroke Cases'
          2='Controls';

run;

data in;
  input group walk f;

  format walk eventf. group grpf.;




cards;
1 1 8
1 2 16
2 1 1
2 2 23
run;

proc freq;
```

Formats the output page to 62 lines in length and 80 columns in width

Proc format defines formats to label the values of the variables.

Variables with values of 1 and 2 will be labeled as 'Needs Assistance' and 'Able to Walk,' respectively, if formatted with the **eventf** format.

Variables with values of 1 and 2 will be labeled as 'Stroke Cases' and 'Controls,' respectively, if formatted with the **grpf** format.

End of format procedure.

Beginning of Data Step

Inputs three variables **group** (1 or 2), **walk** (1 or 2), and *f* (the cell frequency).

The format statement assigns the format **eventf** to the variable **walk**, and the format **grpf** to the variable **group**. Note that the format names are followed by a period to distinguish them from variable names.

Beginning of Raw Data section.

Actual observations (values of **group**, **walk**, and *f* on each line).

Procedure call. Proc freq generates a 2×2 table for two categorical variables.

```
tables group*walk/nocol nopercent
   expected relrisk chisq;
```

Specifies the analytic variables to form the rows (**group**) and columns (**walk**) of the table. The nocol and nopercent options suppress the printing of the column and overall percents, respectively. The relrisk option requests that SAS generate estimates of the relative risk and odds ratio. The chisq option requests the Chi-Square test.

```
  weight f;
```

Specifies the variable that contains the cell counts, *f*.

```
run;
```

End of procedure section. ■

SAS EXAMPLE 8.5 **Estimate Cochran–Mantel–Haenszel Adjusted Relative Risk and** *Perform Cochran–Mantel–Haenszel Chi-Square Test of Homogeneity (Example 8.5)*

We want to compare the risk of hypertension (HTN) in two groups of subjects, those treated with Drug A and those treated with Drug B, adjusted for age (< 65 versus $65+$). The assignments to drug treatment and the outcome status (Hypertension yes/no) for each age group are summarized in the tables below. Use SAS to estimate a relative risk adjusted for age and perform the Cochran–Mantel–Haenszel test of homogeneity adjusting for age.

	Age 65 + HTN					Age < 65 HTN		
	Yes	No				Yes	No	
Drug A:	32	8	40		*Drug A:*	24	36	60
Drug B:	24	36	60		*Drug B:*	8	32	40
Total:	56	44	100		*Total:*	32	68	100

Program Code

```
options ps=62 ls=80;
```

Formats the output page to 62 lines in length and 80 columns in width

```
proc format;
```

Proc format defines formats to label the values of the variables.

```
  value eventf 1='HTN' 2='noHTN';
```

Variables with values of 1 and 2 will be labeled as 'HTN' and 'noHTN,' respectively, if formatted with the **eventf** format.

`value grpf 1='Drug A' 2='Drug B';`	Variables with values of 1 and 2 will be labeled as 'Drug A' and 'Drug B,' respectively, if formatted with the **grpf** format.
`value agef 1='65+' 2='<65';`	Variables with values of 1 and 2 will be labeled as '65+' and '<65,' respectively, if formatted with the **agef** format.
`run;`	End of format procedure.
`data in;`	Beginning of Data Step.
`  input group htn age f;`	Inputs four variables **group** (1 or 2), **htn** (1 or 2), age (1 or 2), and *f* (the cell frequency).
`  format htn eventf. group grpf. age agef.;`	The format statement assigns the format **eventf** to the variable **htn,** the format **grpf** to the variable **group,** and the format **agef** to the variable **age.** Note that the format names are followed by a period to distinguish them from variable names.
`cards;`	Beginning of Raw Data section.
`1 1 1 32`	
`1 2 1 8`	
`2 1 1 24`	
`2 2 1 36`	
`1 1 2 24`	
`1 2 2 36`	
`2 1 2 8`	
`2 2 2 32`	
`run;`	
`proc freq;`	Procedure call. Proc freq generates a 2×2 table for two categorical variables.
`  tables group*htn/nocol nopercent relrisk` `    chisq;`	Specifies the analytic variables to form the rows (**group**) and columns (**htn**) of the table. The nocol and nopercent options suppress the printing of the column and overall percents, respectively. The relrisk option requests that SAS generate estimates of the crude relative risk and odds ratio. The chisq option requests the Chi-Square test.
`  weight f;`	Specifies the variable that contains the cell counts, *f*.
`run;`	End of procedure section.
`proc freq;`	Procedure call. Proc freq generates a 2×2 table for two categorical variables.

```
tables age*group*htn/cmh nocol nopercent;
```
Specifies the analytic variables to form the rows (**group**) and columns (**htn**) of the table, stratified by the third variable (**age**). The cmh option requests an estimate of the Cochran-Mantel-Haenszel relative risk and requests the Mantel-Haenszel test of homogeneity. Again, the nocol and nopercent options suppress the printing of the column and overall percents, respectively.

```
weight f;
```
Specifies the variable that contains the cell counts, *f*.

```
run;
```
End of procedure section.   ■

8.8.1 Summary of SAS Procedures

The SAS Freq Procedure is used to generate $R \times C$ tables for categorical variables. Specific options can be requested to produce crude and adjusted estimates of relative risks and odds ratios and to run unadjusted and adjusted chi-square tests of homogeneity. The specific options are shown in italics in the following table. Users should refer to the examples in this section for complete descriptions of the procedure and specific options. A general description of the procedure and options is provided in the table.

Procedure	Sample Procedure Call	Description

proc freq	proc freq;
	    tables a*b;
	    weight f;
		Generates a 2 × 2 contingency table with row variable *a* (a dichotomous variable) and column variable *b* (a dichotomous variable). The weight option indicates that summary data are input, and the variable *f* contains the cell frequencies.

	proc freq; tables a*b/*relrisk chisq;*	Generates a 2 × 2 contingency table (*a* by *b*) and requests estimates of the crude or unadjusted relative risk and odds ratio and runs the unadjusted chi-square test of homogeneity.
	proc freq; tables strata*a*b/*cmh;*	Generates 2 × 2 contingency tables (*a* by *b*) stratified by the third variable strata. The cmh option requests estimates of the adjusted relative risk and odds ratio and runs the Cochran–Mantel–Haenszel adjusted chi-square test of homogeneity.

8.9 Analysis of Framingham Heart Study Data

The Framingham data set includes data collected from the original cohort. Participants contributed up to three examination cycles of data. Here we use data collected in the first examination cycle (called the period = 1 examination) to ascertain data on the comparison groups. In this analysis, we will compare participants with diabetes to those without diabetes at baseline in terms of their risk for development of cardiovascular disease over the 24-year follow-up period. We use SAS Proc Freq to generate 2 × 2 tables to assess the effect of diabetes status on the development of CVD for the total sample and then for men and women separately. We then conduct the Cochran–Mantel–Haenszel test to assess whether sex is a confounder. The SAS code to run these analyses is given next.

Framingham Data Analysis—SAS Code

```
data fhs;
 set in.frmgham;
 if period=1;
run;


proc freq data=fhs;
 tables diabetes*cvd/riskdiff relrisk chisq;
run;


proc sort data=fhs;
 by sex;
run;


proc freq data=fhs;
 tables diabetes*cvd/riskdiff relrisk chisq;
 by sex;
run;


proc freq data=fhs;
 tables sex*diabetes*cvd/cmh;
run;
```

Framingham Data Analysis—SAS Output

```
                     The FREQ Procedure
                  Table of DIABETES by CVD

DIABETES(Diabetic Y/N)
            CVD(Incident Hosp MI or Stroke, Fatal or Non)
Frequency|
Percent  |
Row Pct  |
Col Pct  |Yes     |No      |  Total
---------+--------+--------+
Yes      |     74 |     47 |    121
         |   1.67 |   1.06 |   2.73
         |  61.16 |  38.84 |
         |   6.40 |   1.43 |
---------+--------+--------+
No       |   1083 |   3230 |   4313
         |  24.42 |  72.85 |  97.27
         |  25.11 |  74.89 |
         |  93.60 |  98.57 |
---------+--------+--------+
Total        1157     3277     4434
             26.09    73.91   100.00
```

```
           Statistics for Table of DIABETES by CVD
```

Statistic	DF	Value	Prob
Chi-Square	1	79.3024	<.0001
Likelihood Ratio Chi-Square	1	67.7732	<.0001
Continuity Adj. Chi-Square	1	77.4443	<.0001
Mantel-Haenszel Chi-Square	1	79.2845	<.0001
Phi Coefficient		0.1337	
Contingency Coefficient		0.1326	
Cramer's V		0.1337	

```
                    Fisher's Exact Test
           ------------------------------------
            Cell (1,1) Frequency (F)         74
            Left-sided Pr <= F          1.0000
            Right-sided Pr >= F       1.848E-16

            Table Probability (P)     2.220E-16
            Two-sided Pr <= P         2.562E-16
```

The FREQ Procedure
Statistics for Table of DIABETES by CVD

Column 1 Risk Estimates

	Risk	ASE	(Asymptotic) 95% Confidence Limits		(Exact) 95% Confidence Limits	
Row 1	0.6116	0.0443	0.5247	0.6984	0.5187	0.6988
Row 2	0.2511	0.0066	0.2382	0.2640	0.2382	0.2643
Total	0.2609	0.0066	0.2480	0.2739	0.2481	0.2741
Difference	0.3605	0.0448	0.2727	0.4483		

Difference is (Row 1 - Row 2)

Column 2 Risk Estimates

	Risk	ASE	(Asymptotic) 95% Confidence Limits		(Exact) 95% Confidence Limits	
Row 1	0.3884	0.0443	0.3016	0.4753	0.3012	0.4813
Row 2	0.7489	0.0066	0.7360	0.7618	0.7357	0.7618
Total	0.7391	0.0066	0.7261	0.7520	0.7259	0.7519
Difference	-0.3605	0.0448	-0.4483	-0.2727		

Difference is (Row 1 - Row 2)

Estimates of the Relative Risk (Row1/Row2)

Type of Study	Value	95% Confidence Limits	
Case-Control (Odds Ratio)	4.6958	3.2371	6.8118
Cohort (Col1 Risk)	2.4356	2.0941	2.8327
Cohort (Col2 Risk)	0.5187	0.4145	0.6490

Sample Size = 4434

In the output, SAS first generates a 2 × 2 table showing the relationship between diabetes status and incident CVD. Of interest are the row percents; the risk of CVD for diabetics is 61.2% as compared to 25.1% for nondiabetics. The difference is highly significant at $p < 0.0001$ (chi-square test). The Fisher's Exact test gives a similar result, but because the cell sizes are sufficiently large, the chi-square test is appropriate here. We requested that SAS estimate the risk difference, which it does considering column1 as the outcome of interest and again considering column2 as the outcome of interest. Here, column 1 (CVD = yes) is the outcome of interest, so the estimate of the risk difference is 36.1%, with a 95% confidence interval of (27.3%, 44.8%). SAS estimates the relative risk at 2.44 (diabetics are 2.44 times more likely to develop CVD as compared to nondiabetics), with a 95% confidence interval of (2.09, 2.83).

In the following, we perform similar analyses but stratify by sex. Is the effect of diabetes similar for men and women?

```
----------------------- SEX=M -----------------------

                     The FREQ Procedure
                   Table of DIABETES by CVD


DIABETES(Diabetic Y/N)
          CVD(Incident Hosp MI or Stroke, Fatal or Non)
Frequency|
Percent  |
Row Pct  |
Col Pct  |Yes      |No       |  Total
---------+--------+--------+
Yes      |     38 |     21 |     59
         |   1.95 |   1.08 |   3.03
         |  64.41 |  35.59 |
         |   5.54 |   1.67 |
---------+--------+--------+
No       |    648 |   1237 |   1885
         |  33.33 |  63.63 |  96.97
         |  34.38 |  65.62 |
         |  94.46 |  98.33 |
---------+--------+--------+
Total          686     1258     1944
              35.29    64.71   100.00
```

Statistics for Table of DIABETES by CVD

Statistic	DF	Value	Prob
Chi-Square	1	22.5927	<.0001
Likelihood Ratio Chi-Square	1	21.3085	<.0001
Continuity Adj. Chi-Square	1	21.2968	<.0001
Mantel-Haenszel Chi-Square	1	22.5811	<.0001
Phi Coefficient		0.1078	
Contingency Coefficient		0.1072	
Cramer's V		0.1078	

Fisher's Exact Test

Cell (1,1) Frequency (F)	38
Left-sided Pr <= F	1.0000
Right-sided Pr >= F	3.588E-06
Table Probability (P)	2.603E-06
Two-sided Pr <= P	5.109E-06

The FREQ Procedure
Statistics for Table of DIABETES by CVD

Column 1 Risk Estimates

	Risk	ASE	(Asymptotic) 95% Confidence Limits		(Exact) 95% Confidence Limits	
Row 1	0.6441	0.0623	0.5219	0.7662	0.5087	0.7645
Row 2	0.3438	0.0109	0.3223	0.3652	0.3223	0.3657
Total	0.3529	0.0108	0.3316	0.3741	0.3316	0.3746
Difference	0.3003	0.0633	0.1763	0.4243		

Difference is (Row 1 - Row 2)

Column 2 Risk Estimates

	Risk	ASE	(Asymptotic) 95% Confidence Limits		(Exact) 95% Confidence Limits	
Row 1	0.3559	0.0623	0.2338	0.4781	0.2355	0.4913
Row 2	0.6562	0.0109	0.6348	0.6777	0.6343	0.6777
Total	0.6471	0.0108	0.6259	0.6684	0.6254	0.6684
Difference	-0.3003	0.0633	-0.4243	-0.1763		

Difference is (Row 1 - Row 2)

Estimates of the Relative Risk (Row1/Row2)

Type of Study	Value	95% Confidence Limits	
Case-Control (Odds Ratio)	3.4543	2.0103	5.9355
Cohort (Col1 Risk)	1.8736	1.5344	2.2876
Cohort (Col2 Risk)	0.5424	0.3842	0.7657

Sample Size = 1944

------------------------ SEX=F ------------------------

The FREQ Procedure
Table of DIABETES by CVD

DIABETES(Diabetic Y/N)
              CVD(Incident Hosp MI or Stroke, Fatal or Non)

Frequency Percent Row Pct Col Pct	Yes	No	Total
Yes	36 1.45 58.06 7.64	26 1.04 41.94 1.29	62 2.49
No	435 17.47 17.92 92.36	1993 80.04 82.08 98.71	2428 97.51
Total	471 18.92	2019 81.08	2490 100.00

Statistics for Table of DIABETES by CVD

Statistic	DF	Value	Prob
Chi-Square	1	63.5363	<.0001
Likelihood Ratio Chi-Square	1	48.0701	<.0001
Continuity Adj. Chi-Square	1	60.9456	<.0001
Mantel-Haenszel Chi-Square	1	63.5108	<.0001
Phi Coefficient		0.1597	
Contingency Coefficient		0.1577	
Cramer's V		0.1597	

Fisher's Exact Test

Cell (1,1) Frequency (F)	36
Left-sided Pr <= F	1.0000
Right-sided Pr >= F	4.543E-12
Table Probability (P)	3.854E-12
Two-sided Pr <= P	4.543E-12

The FREQ Procedure
Statistics for Table of DIABETES by CVD

Column 1 Risk Estimates

	Risk	ASE	(Asymptotic) 95% Confidence Limits		(Exact) 95% Confidence Limits	
Row 1	0.5806	0.0627	0.4578	0.7035	0.4485	0.7049
Row 2	0.1792	0.0078	0.1639	0.1944	0.1641	0.1950
Total	0.1892	0.0078	0.1738	0.2045	0.1739	0.2051
Difference	0.4015	0.0632	0.2777	0.5253		

Difference is (Row 1 - Row 2)

```
                  Column 2 Risk Estimates

                            (Asymptotic) 95%       (Exact) 95%
              Risk      ASE   Confidence Limits   Confidence Limits
          ----------------------------------------------------------------
Row 1     0.4194   0.0627   0.2965   0.5422   0.2951   0.5515
Row 2     0.8208   0.0078   0.8056   0.8361   0.8050   0.8359
Total     0.8108   0.0078   0.7955   0.8262   0.7949   0.8261

Difference  -0.4015   0.0632   -0.5253   -0.2777

              Difference is (Row 1 - Row 2)

        Estimates of the Relative Risk (Row1/Row2)

  Type of Study                   Value      95% Confidence Limits
  --------------------------------------------------------------------
  Case-Control (Odds Ratio)       6.3438     3.7904         10.6171
  Cohort (Col1 Risk)              3.2409     2.5801          4.0710
  Cohort (Col2 Risk)              0.5109     0.3809          0.6851

                Sample Size = 2490
```

What is the relative risk of CVD for diabetic versus nondiabetic men? What is the relative risk of CVD for diabetic versus nondiabetic women? Are the relative risks similar? Is sex a confounding variable? In the next analysis, we generate an adjusted relative risk.

```
                The FREQ Procedure
              Table 1 of DIABETES by CVD
                 Controlling for SEX=M

DIABETES(Diabetic Y/N)
          CVD(Incident Hosp MI or Stroke, Fatal or Non)
Frequency|
Percent  |
Row Pct  |
Col Pct  |Yes      |No       |  Total
---------+--------+--------+
Yes      |      38 |      21 |     59
         |    1.95 |    1.08 |    3.03
         |   64.41 |   35.59 |
         |    5.54 |    1.67 |
---------+--------+--------+
No       |     648 |    1237 |   1885
         |   33.33 |   63.63 |  96.97
         |   34.38 |   65.62 |
         |   94.46 |   98.33 |
---------+--------+--------+
Total          686     1258     1944
              35.29    64.71   100.00
```

Table 2 of DIABETES by CVD
Controlling for SEX=F

```
DIABETES(Diabetic Y/N)
            CVD(Incident Hosp MI or Stroke, Fatal or Non)
Frequency|
Percent  |
Row Pct  |
Col Pct  |Yes     |No      |  Total
---------+--------+--------+
Yes      |     36 |     26 |     62
         |   1.45 |   1.04 |   2.49
         |  58.06 |  41.94 |
         |   7.64 |   1.29 |
---------+--------+--------+
No       |    435 |   1993 |   2428
         |  17.47 |  80.04 |  97.51
         |  17.92 |  82.08 |
         |  92.36 |  98.71 |
---------+--------+--------+
Total          471     2019    2490
             18.92    81.08  100.00
```

The FREQ Procedure

Summary Statistics for DIABETES by CVD
Controlling for SEX

Cochran-Mantel-Haenszel Statistics (Based on Table Scores)

Statistic	Alternative Hypothesis	DF	Value	Prob
1	Nonzero Correlation	1	76.8912	<.0001
2	Row Mean Scores Differ	1	76.8912	<.0001
3	General Association	1	76.8912	<.0001

```
          Estimates of the Common Relative Risk (Row1/Row2)

Type of Study     Method                 Value    95% Confidence Limits
-----------------------------------------------------------------------
Case-Control      Mantel-Haenszel        4.5914    3.1470      6.6987
  (Odds Ratio)    Logit                  4.7526    3.2726      6.9021

Cohort            Mantel-Haenszel        2.3592    2.0328      2.7380
  (Col1 Risk)     Logit                  2.3766    2.0451      2.7619

Cohort            Mantel-Haenszel        0.5245    0.4194      0.6559
  (Col2 Risk)     Logit                  0.5239    0.4190      0.6551


                     Breslow-Day Test for
                 Homogeneity of the Odds Ratios
                 -----------------------------
                 Chi-Square               2.5936
                 DF                            1
                 Pr > ChiSq               0.1073


                 Total Sample Size = 4434
```

Is the method valid (i.e., are relative risks homogeneous across sexes)? What is the adjusted relative risk? How does it compare to the relative risk for the pooled or combined sample and to the relative risks in men and in women?

8.10 Problems

1. A study is conducted over 10 years to investigate long-term complication rates in patients treated with two different therapies. The data are

Therapy	Complications	
	Yes	*No*
1:	89	911
2:	127	873

a. Estimate the risk difference.

b. Estimate the relative risk.

c. Estimate the odds ratio.

d. Based on (a)–(c), how do the therapies compare?

2. A study is conducted comparing an experimental treatment to a control treatment with respect to effectiveness in reducing joint pain in patients with arthritis. One hundred patients are randomly assigned to one of the competing treatments and the following data are collected, representing the numbers of patients reporting improvement in joint pain after the assigned treatment is administered:

	Control	Experimental
Number Reporting Improvement:	21	28
Total Number of Subjects:	50	50

Is there a significant difference in the proportions of patients reporting improvement in joint pain among the treatments? Run the appropriate test at the 5% level of significance.

3. Using the data in Problem 2,

a. Construct a 95% confidence interval for the risk difference of patients reporting improvement in joint pain between the Control and Experimental treatments.

b. Based on (a), is there a significant difference in the proportions of patients reporting improvement in joint pain between the Control and Experimental treatments? Justify your answer (be brief but complete).

4. Patients were enrolled in a study from two clinical centers. The objective of the study was to assess medication adherence (i.e., taking medications as prescribed). Some investigators suggest that adherence exceeding 85% is sufficient to classify a patient as adherent, whereas medication adherence below 85% suggests that the patient is not adherent to the prescribed schedule. Based on the following data, is there a significant difference in the proportions of adherent patients among the clinical centers? Run the appropriate test at $\alpha = 0.05$.

	Enrollment Site	
	Center 1	Center 2
Adherent (> 85%):	28	30
Not Adherent (< 85%):	21	29

5. Suppose an observational study is conducted to investigate the smoking behaviors of male and female patients with a history of coronary heart disease. Among 220 men surveyed, 80 were smokers. Among 190 women surveyed, 95 were smokers.

   a. Estimate the odds ratio of smoking for male versus female patients with a history of coronary heart disease.

   b. Construct a 95% confidence interval for the odds ratio.

   c. Based on (a) and (b), would you reject H_0: $OR = 1$ in favor of H_1: $OR \neq 1$? Justify your answer briefly.

6. Suppose a cross-sectional study is conducted to investigate cardiovascular risk factors among a sample of patients seeking medical care at one of two local hospitals. A total of 200 patients are enrolled. Construct a 95% confidence interval for the difference in proportions of patients with a family history of cardiovascular disease (CVD) between hospitals.

	Enrollment Site	
Family History of CVD	*Hospital 1*	*Hospital 2*
Yes:	24	14
No:	76	86
Total:	100	100

7. A randomized trial is conducted to compare a newly developed pain medication against a standard medication with respect to self-reported pain. Pain is reported on a 5-point scale. Suppose we are interested in the proportions of patients reporting moderate or severe pain between medications. Organize the data given into a dichotomous outcome {moderate or severe pain} versus {none, minimal or some pain}. Estimate the relative risk of moderate or severe pain between treatment groups. Test the hypotheses H_0: $RR = 1$ versus H_1: $RR \neq 1$ at a 5% level of significance.

	Pain				
	None	*Minimal*	*Some*	*Moderate*	*Severe*
New:	20	35	41	15	6
Standard:	15	25	46	36	18

8. A randomized trial is conducted comparing two different prenatal care programs among women at high risk for preterm delivery. The programs differ in intensity of medical intervention. Women who meet the criteria for high risk of preterm delivery are asked to participate in the study and are randomly assigned to one of the prenatal care programs. At the time

of delivery, they are classified as preterm or term delivery. The results are summarized here.

	Preterm Delivery	Term Delivery
Intensive Prenatal Care:	12	43
Standard Prenatal Care:	18	47

a. Estimate the relative risk of preterm delivery for women in the intensive prenatal care program as compared to the standard.

b. Construct a 95% confidence interval for the relative risk.

c. Based on (a), would you reject H_0: $RR = 1$ in favor of H_1: $RR \neq 1$? Justify your answer briefly.

9. A clinical trial of a new treatment for migraine headaches was conducted using a sample of 300 young adults with documented history of migraine headache. Among the 120 subjects on the new treatment (coded as 1), 62 experienced at least one migraine headache episode (MH = 1) during the study period, and 86 of the 180 subjects on the standard treatment (coded as 0) experienced at least one migraine headache episode during the study period.

a. Complete the following table of results.

b. Estimate the RR and test H_0: $RR = 1$.

	Migraine Headache (MH)	
Treatment	Yes (1)	No (0)
New (1):	62	
Standard (0):	86	
Total:		300

One of the investigators suggested adjustment for age since the risk of MH was higher in older subjects. The following results are given by age (older versus younger):

Age = Old Migraine				Age = Young Migraine			
Trt	1	0		Trt	1	0	
1:	30	30	60	1:	32	28	60
0:	35	5	40	0:	51	89	140
Total:	65	35	100	Total:	83	117	200

c. Estimate the RR adjusted for age, and test H_0: $RR = 1$.

SAS Problems

Use SAS to solve the following problems.

1. A study is conducted over 10 years to investigate long-term complication rates in patients treated with two different therapies. The data are

	Complications	
Therapy	Yes	No
1:	89	911
2:	127	873

Use SAS Proc Freq to generate a 2 × 2 contingency table, and estimate the relative risk and odds ratio of complications. Test whether the risk of complications is different between therapies, using the chi-square test of homogeneity.

2. A randomized trial is conducted comparing two different prenatal care programs among women at high risk for preterm delivery. The programs differ in intensity of medical intervention. Women who meet the criteria for high risk of preterm delivery are asked to participate in the study and are randomly assigned to one of the prenatal care programs. At the time of delivery, they are classified as preterm or term delivery. The results are summarized here.

	Preterm Delivery	Term Delivery
Intensive Prenatal Care:	12	43
Standard Prenatal Care:	18	47

Use SAS Proc Freq to generate a 2 × 2 contingency table, and estimate the relative risk and odds ratio of complications. Generate 95% confidence intervals for the relative risk and odds ratio. Test whether the risk of complications is different between therapies, using the chi-square test of homogeneity.

3. A clinical trial of a new treatment for migraine headaches was conducted using a sample of 300 young adults with documented history of migraine headache. Among the 150 subjects on the new treatment (coded as 1), 62 experienced at least one migraine headache episode (MH = 1) during the study period, and 86 of the 100 subjects on the standard treatment (coded as 0) experienced at least one migraine headache episode during the study period. One of the investigators suggested adjustment for age

since the risk of MH was higher in older subjects. The results are given here by age (older versus younger).

	Age = Old Migraine				Age = Young Migraine		
Trt	1	0		Trt	1	0	
1:	30	30	60	1:	32	28	60
0:	35	5	40	0:	51	89	140
Total:	65	35	100	Total:	83	117	200

Use SAS to estimate the RR of treatment on migraine and test H_0: $RR = 1$. Use SAS to estimate the RR adjusted for age and test H_0: $RR = 1$.

Descriptive Statistics
(Ch. 2)

Probability
(Ch. 3)

Sampling Distributions
(Ch. 4)

Statistical Inference
(Chapters 5–13)

OUTCOME VARIABLE	GROUPING VARIABLE(S)/ PREDICTOR(S)	ANALYSIS	CHAPTER(S)
Continuous	—	Estimate μ; Compare μ to Known, Historical Value	5/12
Continuous	Dichotomous (2 groups)	Compare Independent Means (Estimate/Test $(\mu_1 - \mu_2)$) or the Mean Difference (μ_d)	6/12
Continuous	Discrete (>2 groups)	Test the Equality of k Means Using Analysis of Variance $(\mu_1 = \mu_2 = \cdots = \mu_k)$	9/12
Continuous	Continuous	Estimate Correlation or Determine Regression Equation	10/12
Continuous	Several Continuous or Dichotomous	Multiple Linear Regression Analysis	10
Dichotomous	—	Estimate p; Compare p to Known, Historical Value	7
Dichotomous	Dichotomous (2 groups)	Compare Independent Proportions (Estimate/Test $(p_1 - p_2)$)	7/8
Dichotomous	Discrete (> 2 groups)	Test the Equality of k Proportions (Chi-Square Test)	7
Dichotomous	Several Continuous or Dichotomous	Multiple Logistic Regression Analysis	11
Discrete	Discrete	Compare Distributions Among k Populations (Chi-Square Test)	7
Time to Event	Several Continuous or Dichotomous	Survival Analysis	13

9

Analysis of Variance

Analysis of variance (ANOVA) is one of the most widely used statistical techniques for testing the equality of population means. ANOVA is used to test the equality of more than two treatment means (to test the equality of two treatment means, we use the techniques described in Chapter 6). Specifically, the hypotheses are

$$H_0: \mu_1 = \mu_2 = \mu_3 = \cdots = \mu_k$$

$$H_1: \text{means not all equal}$$

where k = the number of populations under consideration ($k > 2$)

NOTE: The alternative hypothesis is not written $H_1: \mu_1 \neq \mu_2 \neq \mu_3 \neq \cdots \neq \mu_k$, as we want to reject the null hypothesis (H_0) if *any* of the population means are not equal (i.e., if at least one pair of means is not equal).

The assumptions necessary for valid applications of ANOVA are

1. k independent populations
2. Random samples from each of k ($k > 2$) populations under consideration
3. Large samples ($n_i \geq 30$, where $i = 1, 2, \ldots, k$) or normal populations
4. Equal population variances (i.e., $\sigma_1^2 = \sigma_2^2 = \cdots = \sigma_k^2 = \sigma^2$).

The logic of the analysis of variance technique is presented in Section 9.1. Notation and computations are illustrated through examples in Section 9.2. In Section 9.3, we define fixed and random effects models. In Section 9.4, we present a statistic to assess the magnitude of the treatment effect. We define multiple comparison procedures and illustrate the use of two specific procedures in Section 9.5. In Section 9.6, we outline simple analysis of variance procedures for dependent samples, called repeated measures analysis of variance. Key formulas are summarized in Section 9.7, statistical computing applications are presented in Section 9.8, and in Section 9.9 we use data from the Framingham Heart Study to illustrate the applications presented here.

9.1 Background Logic

Consistent with other tests of hypotheses, we take random samples from each population of interest and evaluate sample statistics as a means of assessing the likelihood that H_0 is true. Consider the following examples, which illustrate the logic of the analysis of variance procedure.

EXAMPLE 9.1

Variation in Time to Relief of Symptoms Between and Within Treatments

An experiment is conducted in which three treatments are compared with respect to their effectiveness. For the purposes of this example, effectiveness is evaluated in terms of time to relief of symptoms, reported in minutes. We assume that the distribution of times to relief are approximately normal, and the test of interest is as follows:

$$H_0: \mu_1 = \mu_2 = \mu_3$$
$$H_1: \text{means not all equal}$$

Fifteen subjects are randomly selected to participate in the investigation. Five subjects are randomly assigned to each treatment and each subject reports the time to relief of symptoms, in minutes, following their assigned treatment. The sample data, and summary statistics, are as follows:

Treatment 1	Treatment 2	Treatment 3
29.0	25.1	20.1
29.2	25.0	20.0
29.1	25.0	19.9
28.9	24.9	19.8
28.8	25.0	20.2

Summary Statistics by Treatment

$\overline{X}_1 = 29.0$	$\overline{X}_2 = 25.0$	$\overline{X}_3 = 20.0$
$s_1^2 = 0.025$	$s_2^2 = 0.005$	$s_3^2 = 0.025$
$s_1 = 0.158$	$s_2 = 0.071$	$s_3 = 0.158$

From the summary statistics we see that the sample means are numerically different (i.e., 29.0 versus 25.0 versus 20.0 minutes). In addition, the variability *within* each sample is small (i.e., the standard deviations are 0.158, 0.071, 0.158, respectively), which implies that observations within each sample are tightly clustered about their respective sample means.

Suppose that the population means are equal (i.e., $H_0: \mu_1 = \mu_2 = \mu_3$ is true). We can assume, then, that the three samples are drawn from the same population and can pool all of the observations together (i.e., $n = 15$). The overall sample mean is $\overline{X} = 24.67$ minutes, with a sample variance $s^2 = 14.54$ and standard deviation $s = 3.81$. The variability in the pooled sample is large. In ANOVA, we compare the variation *within* samples to the variation *between* samples to assess the equality of the population means. If the observations within a sample are similar in value (i.e., small within-sample variation) and the means are different across samples (large between-sample variation), then a real difference is said to exist in the population means. Figure 9.1 displays the sample means and ranges of observations within each sample in Example 9.1.

Figure 9.1 *Variation Between and Within Treatments in Example 9.1*

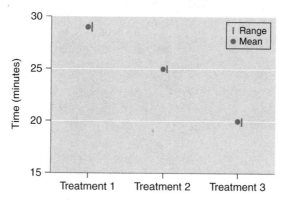

EXAMPLE 9.2

Variation in Time to Relief of Symptoms Between and Within Treatments

Consider the investigation described in Example 9.1. Suppose that the following sample data are observed in a similar investigation. Summary statistics for each sample are also shown.

Treatment 1	Treatment 2	Treatment 3
29.0	33.1	15.2
14.2	7.4	39.3
45.1	17.6	14.8
48.9	44.2	25.5
7.8	22.7	5.2

Summary Statistics by Treatment

$\overline{X}_1 = 29.0$	$\overline{X}_2 = 25.0$	$\overline{X}_3 = 20.0$
$s_1^2 = 330.87$	$s_2^2 = 201.07$	$s_3^2 = 167.96$
$s_1 = 18.19$	$s_2 = 14.18$	$s_3 = 12.96$

From the summary statistics we see that again the sample means are numerically different (i.e., 29.0 versus 25.0 versus 20.0 minutes). However, the variability within each sample is large (i.e., the standard deviations are 18.19, 14.18, 12.96, respectively), which implies that the observations within each sample are not tightly clustered about their respective sample means, but widely spread.

Suppose again we pool all of the observations together ($n = 15$) and compute summary statistics. The overall sample mean is $\overline{X} = 24.67$, the sample variance is $s^2 = 214.49$, and the standard deviation is $s = 14.65$. In

Figure 9.2 *Variation Between and Within Treatments in Example 9.2*

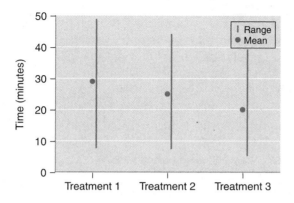

Example 9.2, the variation between sample means is the same as in Example 9.1; however, there is large variation within samples, which suggests that there is no real difference in the population means (see Figure 9.2). ■

In ANOVA, we wish to test the following: H_0: $\mu_1 = \mu_2 = \mu_3 = \cdots = \mu_k$ versus H_1: Means are not all equal (or at least one pair is not equal), where $k =$ the number of populations under consideration. To test H_0, we compute two estimates of the common population variance (σ^2). The first estimate is independent of H_0 (i.e., we do not assume that the population means are equal and we treat each sample separately). This estimate is called the estimate of the *within treatment variation*. The second estimate is based on the assumption that H_0 is true (i.e., population means are equal) and we pool all data together. This second estimate is called the estimate of the *between treatment variation*. The formulas for the estimates of the between and within variation follow for the case of equal sample sizes (i.e., $n_1 = n_2 = \cdots = n_k$). Sample sizes do not have to be equal in practice. The formulas presented in the next section illustrate the computations for both equal and unequal sample size applications.

Our first estimate is the estimate of the *within treatment variation*. To compute this estimate, we assume nothing about the population means. We do, however, assume that $\sigma_1^2 = \sigma_2^2 = \cdots = \sigma_k^2 = \sigma^2$. (This assumption is required for appropriate use of ANOVA.) It follows that each sample variance s_j^2 (where $j = 1, 2, \ldots, k$) is an estimate of σ^2 (the common population variance). Again, to simplify things, we assume that $n_1 = n_2 = \cdots = n_k$. An estimate of σ^2 is obtained by taking the mean of the sample variances (s_j^2) over the k treatments:

$$s_w^2 = \sum_{j=1}^{k} \frac{s_1^2}{k} = \frac{s_1^2 + s_2^2 + \cdots + s_k^2}{k} \tag{9.1}$$

where $s_w^2 =$ denotes the within treatment variance or within variation.

This estimate of σ^2 depends only on the assumptions for ANOVA, specifically on the assumption of equal population variances ($\sigma_1^2 = \sigma_2^2 = \cdots = \sigma_k^2 = \sigma^2$).

The second estimate of σ^2 is called the *between variation*. This second estimate depends on the assumptions for ANOVA (e.g., $\sigma_1^2 = \sigma_2^2 = \cdots = \sigma_k^2 = \sigma^2$) and also on the assumption that H_0 is true (i.e., $\mu_1 = \mu_2 = \mu_3 = \cdots = \mu_k = \mu$). If the population means are all equal, then each sample mean $\overline{X}_j$ ($j = 1, 2, \ldots, k$) is an estimate of the common population mean, μ. The $\overline{X}_j$'s can be viewed as a simple random sample of size k from a population with mean $\mu_{\overline{X}_j} = \mu$ and variance $\sigma^2_{\overline{X}_j} = \sigma^2/n$. Our goal is to generate an estimate of σ^2. We can estimate $\sigma^2_{\overline{X}_j}$, which is equal to σ^2/n, and then use algebra to solve for σ^2.

The variance of the k sample means is

$$s_{\overline{X}_j}^2 = \sum_{j=1}^{k} \frac{(\overline{X}_j - \overline{X})^2}{k - 1} \tag{9.2}$$

where $\overline{X}$ is the overall mean (i.e., based on all observations pooled together)

$s_{\overline{X}_j}^2$ is an estimate of $\sigma^2_{\overline{X}_j} = \sigma^2/n$. Therefore, $ns_{\overline{X}_j}^2$ is an estimate of σ^2:

$$s_b^2 = ns_{\overline{X}_j}^2 \tag{9.3}$$

where s_b^2 denotes the between treatment variance or between variation.

The test statistic in ANOVA is based on the ratio of these two estimates:

$$F = \frac{s_b^2}{s_w^2} \tag{9.4}$$

The test statistic follows an F distribution (see Figure 9.3). The F distribution is an asymmetric distribution that takes on values greater than or equal to zero.

If the two estimates of σ^2 are close in value (i.e., F is approximately equal to 1), then we have no reason to reject H_0. However, if the variation between samples (s_b^2) is large and the variation within samples (s_w^2) is small (i.e., F is large), then we reject H_0. To draw a conclusion regarding H_0, we need a critical

Figure 9.3 *The F Distribution*

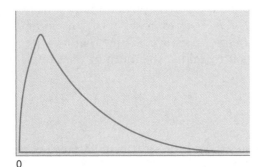

0

value from the F distribution. In order to select the appropriate value, we need 2 degrees of freedom: the numerator degrees of freedom, denoted $df_1 = k - 1$, and the denominator degrees of freedom, denoted $df_2 = k(n - 1)$, where $k =$ the number of populations or treatments under consideration and $n =$ the sample size per treatment. (*Note:* $k(n - 1) = kn - k$, where n is the common sample size or sample size in each group. We sometimes let $N = kn$, where $N =$ the total sample size (all groups combined).) Once the appropriate critical value is selected, we can construct our decision rule, which is of the form: Reject H_0 if $F \geq F_{k-1,N-k}$. Critical values of the F distribution are contained in Table B.4A in the Appendix for $\alpha = 0.05$ (and in Table B.4B for $\alpha = 0.025$).

The ANOVA technique is illustrated in the next section through examples. The computations are based on the logic presented here.

9.2 Notation and Examples

We now illustrate the computations using examples. In each example, the test of interest is a test of hypotheses involving more than two populations. Computations are illustrated for both equal and unequal sample size situations.

EXAMPLE 9.3

Testing Difference in Mean Time to Pain Relief Among 3 Treatments

An investigator wishes to compare the average time to relief of headache pain under three distinct medications, call them Drugs A, B, and C. Fifteen patients who suffer from chronic headaches are randomly selected for the investigation, and five subjects are randomly assigned to each treatment. The following data reflect times to relief (in minutes) after taking the assigned drug:

Drug A	Drug B	Drug C
30	25	15
35	20	20
40	30	25
25	25	20
35	30	20

Notation

X_{ij} denotes the ith observation in the jth treatment (e.g., $X_{11} = 30$, $X_{42} = 25$, $X_{53} = 20$)

A "." in place of a subscript (either i or j) denotes summation over that index: (e.g., $X_{.1} =$ sum over observations in treatment 1

$X_{.1} = 30 + 35 + 40 + 25 + 35 = 165$

$X_{1.} =$ sum first observations over treatments

$X_{1.} = 30 + 25 + 15 = 70$

$\overline{X}_{.1} =$ sample mean in treatment 1

$\overline{X}_{..} =$ overall sample mean (taken over all observations and treatments)

Using the notation, the summary statistics are:

Drug A	Drug B	Drug C
$\overline{X}_{.1} = 33.0$	$\overline{X}_{.2} = 26.0$	$\overline{X}_{.3} = 20.0$
$s_{.1} = 5.7$	$s_{.2} = 4.2$	$s_{.3} = 3.5$

To test whether the true mean times to relief under the three different drugs are equal, we use the same five-step procedure used in other tests of hypotheses.

1. Set up hypotheses.

$$H_0: \mu_1 = \mu_2 = \mu_3$$

$$H_1: \text{means not all equal}, \quad \alpha = 0.05$$

The assumptions for ANOVA are

1. Random samples from each of the three populations under consideration.
2. The times to relief are approximately normally distributed.
3. The three populations are independent.
4. The population variances are equal (i.e., $\sigma_1^2 = \sigma_2^2 = \sigma_3^2 = \sigma^2$).

2. Select the appropriate test statistic.

$$F = \frac{s_b^2}{s_w^2}$$

where s_b^2 = denotes the between variation, which is also denoted MS_b (mean square between)
s_w^2 = denotes the within variation, which is also denoted MS_w (mean square within) (see Table 9.1.)

Table 9.1 *Analysis of Variance Table*

Source of Variation	Sums of Squares (SS)	Degrees of Freedom (df)	Mean Squares (MS)	F
Between	$SS_b = \sum n_j(\overline{X}_{.j} - \overline{X}_{..})^2$	$k - 1$	$s_b^2 = MS_b = \dfrac{SS_b}{k-1}$	$F = \dfrac{MS_b}{MS_w}$
Within	$SS_w = \sum\sum(X_{ij} - \overline{X}_{.j})^2$	$N - k$	$s_w^2 = MS_w = \dfrac{SS_w}{N-k}$	
Total	$SS_{total} = \sum\sum(X_{ij} - \overline{X}_{..})^2$	$N - 1$		

where $X_{ij} = i$th observation in the jth treatment
$\overline{X}_{.j} = $ sample mean of jth treatment
$\overline{X}_{..} = $ overall sample mean
$k = \#$ treatments
$N = \sum n_j = $ total sample size

3. **Decision rule.**

   To select the appropriate critical value from the F distribution (Table B.4A), we first compute the numerator degrees of freedom (df_1) and the denominator degrees of freedom (df_2):

$$\text{df}_1 = k - 1 = 3 - 1 = 2$$
$$\text{df}_2 = N - k = 15 - 3 = 12$$

In ANOVA, we reject H_0 if the test statistic is larger than the critical value. The critical value of F with 2 and 12 degrees of freedom, relative to a 5% level of significance, is denoted $F_{2,12}$ and can be found in Table B.4A: $F_{2,12} = 3.89$. The decision rule is

Reject H_0 if $F \geq 3.89$

Do not reject H_0 if $F < 3.89$

4. **Compute the test statistic.**

   Generally, to organize the computations in ANOVA applications, an ANOVA table is constructed. The table contains all of the computations, the degrees of freedom used to select the appropriate critical value, and the test statistic. The general form of the ANOVA table is shown in Table 9.1.

   Now, using the data in Example 9.3, we construct an ANOVA table. We first compute the sums of squares. The between treatment sums of squares (also called the sums of square due to treatments) is computed by summing the squared differences between each treatment mean and the overall mean. We first compute the overall mean. In Example 9.3, the sample sizes are equal; therefore, the overall mean is the mean of the three treatment means:

$$\overline{X}_{..} = \frac{(33.0 + 26.0 + 20.0)}{3} = 26.3$$

The between treatment sums of squares is

$$SS_b = \sum n_j (\overline{X}_{.j} - \overline{X}_{..})^2 = 5(33 - 26.3)^2 + 5(26 - 26.3)^2$$
$$+ 5(20 - 26.3)^2 = 423.3$$

The within treatment sums of squares (also called the sums of square due to error) is computed by summing the squared differences between each observation and its treatment mean.

To compute the within sums of squares, $SS_w = \sum\sum(X_{ij} - \overline{X}_{.j})^2$, we construct the following table to organize our computations:

Drug A		Drug B		Drug C	
$(X_{i1} - \overline{X}_{.1})$	$(X_{i1} - \overline{X}_{.1})^2$	$(X_{i2} - \overline{X}_{.2})$	$(X_{i2} - \overline{X}_{.2})^2$	$(X_{i3} - \overline{X}_{.3})$	$(X_{i3} - \overline{X}_{.3})^2$
-3	9	-1	1	-5	25
2	4	-6	36	0	0
7	49	4	16	5	25
-8	64	-1	1	0	0
2	4	4	16	0	0
$\sum(X_{i1} - \overline{X}_{.1})$ $= 0$	$\sum(X_{i1} - \overline{X}_{.1})^2$ $= 130$	$\sum(X_{i2} - \overline{X}_{.2})$ $= 0$	$\sum(X_{i2} - \overline{X}_{.2})^2$ $= 70$	$\sum(X_{i3} - \overline{X}_{.3})$ $= 0$	$\sum(X_{i3} - \overline{X}_{.3})^2$ $= 50$

The within treatment sums of squares is

$$SS_w = \sum\sum(X_{ij} - \overline{X}_{.j})^2 = 130 + 70 + 50 = 250$$

The total sums of squares is computed by summing the squared differences between each observation and the overall mean:

$$SS_{total} = \sum\sum(X_{ij} - \overline{X}_{..})^2 = (30 - 26.3)^2 + \cdots + (20 - 26.3)^2 = 673.3$$

However, since $SS_{total} = SS_b + SS_w$, it is not necessary to compute SS_{total} directly: $SS_{total} = 423.3 + 250 = 673.3$.

We now construct the ANOVA table for Example 9.3.

Source of Variation	Sums of Squares (SS)	Degrees of Freedom (df)	Mean Squares (MS)	F
Between	423.3	2	211.67	10.16
Within	250.0	12	20.82	
Total	673.3	14		

5. Conclusion.

Reject H_0, since $10.16 > 3.89$. We have significant evidence, $\alpha = 0.05$, to show that the mean times to relief from headache pain under the three drugs A, B, and C are not all equal. Because we do not have more extensive tables for the F distribution here, we cannot compute a p value. However, when an ANOVA is run using SAS, an exact p value is generated. ■

SAS Example 9.3 Testing Difference in Mean Time to Pain Relief Among 3 Treatments Using SAS

The following output was generated using SAS Proc ANOVA, which runs an analysis of variance test for equality of means. SAS produces the following. A brief interpretation appears after the output.

SAS Output for Example 9.3

The ANOVA Procedure

Dependent Variable: time

Source	DF	Sum of Squares	Mean Square	F Value	Pr >
Model	2	423.3333333	211.6666667	10.16	0.0026
Error	12	250.0000000	20.8333333		
Corrected Total	14	673.3333333			

R-Square	Coeff Var	Root MSE	time Mean
0.628713	17.33299	4.564355	26.33333

Source	DF	ANOVA SS	Mean Square	F Value	Pr > F
trt	2	423.3333333	211.6666667	10.16	0.0026

Interpretation of SAS Output for Example 9.3

SAS has two procedures for analysis of variance applications. The first is the ANOVA procedure, which is used when the sample sizes are equal, and the second is the GLM (general linear models) procedure, which can be used when the sample sizes are unequal or equal. Since the sample sizes are equal in Example 9.3, we used the ANOVA procedure.

SAS produces an ANOVA table, similar to Table 9.1. SAS uses the term "Model" to refer to the between treatment variation and "Error" to refer to the within treatment variation. SAS also presents the degrees of freedom before the sums of squares; otherwise, the table is identical. In Example 9.3, the test statistic is $F = 10.16$ with $p = 0.0026$, which would lead to rejection of H_0 since $p = 0.0026 < \alpha = 0.05$. ■

Example 9.4 Testing Difference in Mean Age at Completion of 8th Grade Among 3 School Districts

The following data, collected from randomly selected students at rural, suburban, and urban schools, reflect ages of students at completion of eighth grade. Test if there is a significant difference in the mean age at completion of eighth grade for rural, suburban, and urban students using a 5% level of significance.

Sample Data:

Rural:	14	14	14	14	13	13	13	12	
Suburban:	14	14	14	13	13	13	13	13	12	12
Urban:	16	16	15	15	15	14	14	14	13	12

1. Set up hypotheses.

$$H_0: \mu_1 = \mu_2 = \mu_3$$
$$H_1: \text{means not all equal}, \quad \alpha = 0.05$$

2. Select the appropriate test statistic.

$$F = \frac{s_b^2}{s_w^2}$$

where s_b^2 = denotes the between variation, which is also denoted MS_b (mean square between)

s_w^2 = denotes the within variation, which is also denoted MS_w (mean square within) (see Table 9.1).

3. Decision rule.

To select the appropriate critical value from the F distribution (Table B.4), we first compute the numerator degrees of freedom (df_1) and the denominator degrees of freedom (df_2):

$$df_1 = k - 1 = 3 - 1 = 2$$
$$df_2 = N - k = 28 - 3 = 25$$

The critical value of F with 2 and 25 degrees of freedom, relative to a 5% level of significance, is found in Table B.4A: $F_{2,25} = 3.39$. The decision rule is

Reject H_0 if $F \geq 3.39$

Do not reject H_0 if $F < 3.39$

4. Compute the test statistic.

Again, we will construct an ANOVA table to organize our computations. In this example, some of the formulas have been modified to accommodate the unequal sample sizes.

The following table contains summary statistics on the age data:

	Rural	Suburban	Urban
n_j	8	10	10
$\sum X_{ij}$	107	131	144
$\overline{X}_{\cdot j}$	13.4	13.1	14.4

The between sums of squares (also called the sums of squares due to treatments) is computed by summing the squared differences between each treatment mean and the overall mean. In Example 9.4, the sample sizes are unequal; therefore, the overall mean is computed by summing all of the observations and dividing by N:

$$\overline{X}_{..} = \frac{(107 + 131 + 144)}{28} = 13.7$$

NOTE: When the sample sizes are equal, the overall mean can be computed by taking the mean of the sample means.

The between sums of squares is

$$SS_b = \sum n_j (\overline{X}_{.j} - \overline{X}_{..})^2 = 8(13.4 - 13.7)^2$$
$$+ 10(13.1 - 13.7)^2 + 10(14.4 - 13.7)^2 = 9.22$$

The within sums of squares (also called the sums of squares due to error) is computed by summing the squared differences between each observation and its treatment mean. To compute the within sums of squares, $SS_w = \sum\sum(X_{ij} - \overline{X}_{.j})^2$, we construct the following table:

Rural		Suburban		Urban	
$(X_{i1} - \overline{X}_{.1})$	$(X_{i1} - \overline{X}_{.1})^2$	$(X_{i2} - \overline{X}_{.2})$	$(X_{i2} - \overline{X}_{.2})^2$	$(X_{i3} - \overline{X}_{.3})$	$(X_{i3} - \overline{X}_{.3})^2$
0.6	0.36	0.9	0.81	1.6	2.56
0.6	0.36	0.9	0.81	1.6	2.56
0.6	0.36	0.9	0.81	0.6	0.36
0.6	0.36	−0.1	0.01	0.6	0.36
−0.4	0.16	−0.1	0.01	0.6	0.36
−0.4	0.16	−0.1	0.01	−0.4	0.16
−0.4	0.16	−0.1	0.01	−0.4	0.16
−1.4	1.96	−0.1	0.01	−0.4	0.16
		−1.1	1.21	−1.4	1.96
		−1.1	1.21	−2.4	5.76
$\sum(X_{i1} - \overline{X}_{.1})$	$\sum(X_{i1} - \overline{X}_{.1})^2$	$\sum(X_{i2} - \overline{X}_{.2})$	$\sum(X_{i2} - \overline{X}_{.2})^2$	$\sum(X_{i3} - \overline{X}_{.3})$	$\sum(X_{i3} - \overline{X}_{.3})^2$
$= 0$	$= 3.88$	$= 0$	$= 4.90$	$= 0$	$= 14.40$

The within sums of squares is

$$SS_w = \sum\sum(X_{ij} - \overline{X}_{.j})^2 = 3.88 + 4.90 + 14.40 = 23.18$$

The total sums of squares is computed by summing the squared differences between each observation and the overall mean:

$$SS_{\text{total}} = \sum\sum(X_{ij} - \overline{X}_{..})^2 = (14 - 13.7)^2 + \cdots + (12 - 13.7)^2 = 32.4$$

Again, since $SS_{total} = SS_b + SS_w$, it is not necessary to compute SS_{total} directly: $SS_{total} = 9.22 + 23.18 = 32.40$.

We now construct the ANOVA table for Example 9.4.

Source of Variation	Sums of Squares (SS)	Degrees of Freedom (df)	Mean Squares (MS)	F
Between	9.22	2	4.63	4.99
Within	23.18	25	0.93	
Total	32.40	27		

5. Conclusion.

Reject H_0, since $4.99 > 3.39$. We have significant evidence, $\alpha = 0.05$, to show that the mean ages at completion of eighth grade are not equal for rural, suburban, and urban students. ■

SAS EXAMPLE 9.4 Testing Difference in Mean Age at Completion of 8th Grade Among 3 School Districts Using SAS

The following output was generated using SAS Proc GLM, which runs an analysis of variance test for equality of means. GLM is used when the sample sizes are unequal. SAS produces the following. A brief interpretation appears after the output.

SAS Output for Example 9.4

```
                        The GLM Procedure
Dependent Variable: age
                             Sum of
Source                DF     Squares     Mean Square   F Value   Pr > F

Model                  2    9.25357143    4.62678571     4.99    0.0150
Error                 25   23.17500000    0.92700000
Corrected Total       27   32.42857143

          R-Square     Coeff Var     Root MSE      age Mean
          0.285352     7.057234      0.962808      13.64286

Source                DF    Type I SS    Mean Square   F Value   Pr > F
school                 2    9.25357143    4.62678571     4.99    0.0150

Source                DF   Type III SS   Mean Square   F Value   Pr > F
school                 2    9.25357143    4.62678571     4.99    0.0150
```

Interpretation of SAS Output for Example 9.4

Here we used the GLM procedure to run the ANOVA because the sample sizes are unequal. SAS produces an ANOVA table similar to Table 9.1. Again, SAS uses the term "Model" to refer to the between treatment variation and "Error" to refer to the within treatment variation. SAS also presents the degrees of freedom before the sums of squares; otherwise, the table is identical. In Example 9.4, the test statistic is $F = 4.99$, with $p = 0.0150$, which would lead to rejection of H_0 since $p = 0.0150 < \alpha = 0.05$. ▪

EXAMPLE 9.5

Testing Difference in Mean Weight Gain Among 4 Different Diets

A study is developed to examine the effects of vitamin and milk supplements on infant weight gain. Four diet plans are considered: Diet A involves a regular diet plus the vitamin supplement, Diet B involves a regular diet plus the special milk formula, Diet C is our control diet (i.e., no restrictions or special considerations), and Diet D involves a regular diet plus the vitamin and the special milk formula. Twenty infants are selected for the investigation and each is randomized to one of the four competing diet programs. The following table displays weight gains, measured in pounds, after 1 month on the assigned diet.

Diet A	Diet B	Diet C	Diet D
2.0	1.6	1.5	2.1
1.5	1.9	2.0	2.4
2.4	2.1	1.8	1.9
1.9	1.1	1.3	1.8
2.6	1.7	1.2	2.2

1. Set up hypotheses.

$$H_0: \mu_1 = \mu_2 = \mu_3 = \mu_4$$

$$H_1: \text{means not all equal,} \quad \alpha = 0.05$$

2. Select the appropriate test statistic.

$$F = \frac{s_b^2}{s_w^2}$$

where s_b^2 = denotes the between variation, which is also denoted MS_b (mean square between)

s_w^2 = denotes the within variation, which is also denoted MS_w (mean square within) (see Table 9.1)

3. Decision rule.

To select the appropriate critical value from the F distribution (Table B.4), we first compute the numerator degrees of freedom (df_1) and the denominator degrees of freedom (df_2):

$$df_1 = k - 1 = 4 - 1 = 3$$
$$df_2 = N - k = 20 - 4 = 16$$

The critical value of F with 3 and 16 degrees of freedom, relative to a 5% level of significance, is found in Table B.4a: $F_{3,16} = 3.24$. The decision rule is

$$\text{Reject } H_0 \text{ if } F \geq 3.24$$

$$\text{Do not reject } H_0 \text{ if } F < 3.24$$

4. Compute the test statistic.

Again, we will construct an ANOVA table to organize our computations. The following table displays summary statistics on the infant weights:

	Diet A	Diet B	Diet C	Diet D
n_j	5	5	5	5
$\sum X_{ij}$	10.4	8.4	7.8	10.4
$\overline{X}_{.j}$	2.1	1.7	1.6	2.1

The between sums of squares (also called the sums of squares due to treatments) is computed by summing the squared differences between each treatment mean and the overall mean. Since the sample sizes are equal, the overall mean can be found by computing the mean of the four treatment means:

$$\overline{X}.. = \frac{(2.1 + 1.7 + 1.6 + 2.1)}{4} = 1.88$$

The between sums of squares is

$$SS_b = \sum n_j (\overline{X}_{.j} - \overline{X}_{..})^2 = 5((2.1 - 1.88)^2 + (1.7 - 1.88)^2$$
$$+ (1.6 - 1.88)^2 + (2.1 - 1.88)^2) = 1.04$$

The within sums of squares (also called the sums of squares due to error) is computed by summing the squared differences between each observation and its treatment mean. To compute the within sums of squares, $SS_w = \sum\sum(X_{ij} - \overline{X}_{.j})^2$, we construct a table similar to those presented in Examples 9.3 and 9.4. Here, we present only the results. The within sums of squares is

$$SS_w = \sum\sum(X_{ij} - \overline{X}_{.j})^2 = 0.75 + 0.57 + 0.46 + 0.23 = 2.01$$

The total sums of squares is computed by adding the between and within sums of squares:

$$SS_{total} = SS_b + SS_w = 1.04 + 2.01 = 3.05$$

We now construct the ANOVA table for Example 9.5.

Source of Variation	Sums of Squares (SS)	Degrees of Freedom (df)	Mean Squares (MS)	F
Between	1.04	3	0.35	2.69
Within	2.01	16	0.13	
Total	3.05	19		

5. Conclusion.

Do not reject H_0, since $2.69 < 3.24$. We do not have significant evidence, $\alpha = 0.05$, to show that the mean weight gains under the four different diets are not equal. ▪

SAS EXAMPLE 9.5 **Testing Difference in Mean Weight Gain Among 4 Different Diets Using SAS**

The following output was generated using SAS Proc GLM, which runs an analysis of variance test for equality of means. GLM can accommodate applications where the sample sizes are equal or unequal. SAS produces the following. A brief interpretation appears after the output.

SAS Output for Example 9.5

```
                         The GLM Procedure
Dependent Variable: gain

                                  Sum of
Source                   DF       Squares      Mean Square    F Value    Pr > F

Model                     3    1.09400000      0.36466667       2.92    0.0659
Error                    16    1.99600000      0.12475000
Corrected Total          19    3.09000000


          R-Square     Coeff Var       Root MSE       gain Mean
          0.354045      19.09187       0.353200        1.850000


Source                   DF     Type I SS      Mean Square    F Value    Pr > F
diet                      3    1.09400000      0.36466667       2.92    0.0659


Source                   DF     Type III SS    Mean Square    F Value    Pr > F
diet                      3    1.09400000      0.36466667       2.92    0.0659
```

Interpretation of SAS Output for Example 9.5

Here we used the GLM procedure to run the ANOVA. In Example 9.5, the test statistic is $F = 2.92$ with $p = 0.0659$. We do not reject H_0 since $p = 0.0659 > \alpha = 0.05$. In this example, the SAS calculations are slightly different from our hand calculations due to rounding. SAS carries many more decimal places in computations and produces a more exact solution. The conclusions of the tests (performed by hand and by SAS) are the same. It should be noted that the test is marginally significant. Although we did not reach the specified significance level of 0.05, these results should be evaluated carefully because $p = 0.0659$, suggesting that there are some differences among the diets. If we evaluate the mean weight gains, Diets A and D produce the largest gains; Diet B is not much different than the control diet, Diet C. ■

9.3 Fixed Versus Random Effects Models

There are two types of analysis of variance applications: fixed effects models and random effects models. In *fixed effects models,* the treatment groups under study—for example, the three headache medications (Drugs A, B, and C), the four infant diets (A, B, C, and D)—represent all treatments of interest. In our concluding statement we say there is (or is not) significant evidence of a difference in means among the treatments studied (e.g., there is a difference in the mean times to relief of headache pain under Drugs A, B, and C). In *random effects models,* we randomly select k treatments for the investigation from a larger pool of available treatments. For example, suppose there are 10 competing treatments for headache pain or 12 well-known infant diets and we randomly select 3 or 4 to study (e.g., Drugs A, B, and C or Diets A, B, C, and D). In our concluding statement we say there is (or is not) significant evidence of a difference in means among ALL treatments (e.g., there is a difference in weight gain among all 12 well-known infant diets), though we studied only a subset.

Basically, in random effects models we can generalize our results to the pool of all treatments since we randomly selected a subset for the investigation. In fixed effects models, our conclusions apply only to the treatments studied. The logic and the computations presented in the preceding sections are appropriate only for fixed effects models. Modifications beyond the scope of this book are necessary for random effects models (see Cobb, G.W. (1998). *Introduction to Design and Analysis of Experiments,* Springer-Verlag, Inc., New York).

9.4 Evaluating Treatment Effects

If an ANOVA is performed and it has been established that a difference in means exists (i.e., we reject H_0), we want to assess the magnitude of the effect *due* to the treatments. That is, we want to address the question, How much variation in the data is due to the treatments?

The following statistic, called "eta-squared," is the ratio of variation due to the treatments (SS_b) to the total variation:

$$\eta^2 = \frac{SS_b}{SS_{total}} \qquad (9.5)$$

where $0 \leq \eta^2 \leq 1$

Values of η^2 that are closer to 1 imply that more variation in the data is attributable to the treatments.

In Example 9.3, comparing the times to relief among the three headache medications,

$$\eta^2 = \frac{423.3}{673.3} = 0.629$$

Thus, 62.9% of the variation in the times to relief is due to the medications (i.e., Drug A, B, or C). In Example 9.4, comparing students' ages,

$$\eta^2 = \frac{9.25}{32.43} = 0.285$$

Thus, only 28.5% of the variation in the students' ages is due to the location of the school (i.e., rural, suburban, urban).

9.5 Multiple Comparisons Procedures

Once we reject $H_0: \mu_1 = \mu_2 = \cdots = \mu_k$ in an ANOVA application, we say there is a significant difference among all of the treatment means (or at least one pair of means are not equal). It is often of interest to then test specific hypotheses comparing certain treatments. For example, in Example 9.3 we concluded that the mean times to relief for the three headache medications were significantly different. Suppose we are particularly interested in comparing only the first two medications (i.e., $H_0: \mu_1 = \mu_2$) or the first and the third (i.e., $H_0: \mu_1 = \mu_3$). Tests of this type are called *pairwise comparisons*, since they involve pairs of treatment means. It is also possible to construct more complicated comparisons. For example, it may be of interest to compare the mean time to relief for patients assigned to either Drug A or B to the mean time to relief for patients assigned to Drug C. The hypotheses would be denoted as follows: $H_0: (\mu_1 + \mu_2)/2 = \mu_3$. Both pairwise (two-at-a-time) and more complicated comparisons are generally called *contrasts*.

There are a number of statistical procedures for handling these applications, which are called *multiple comparison procedures*, or MCPs. Different procedures are recommended for different applications. The procedures differ according to the types of comparisons of interest (e.g., pairwise comparisons), the number of comparisons, and the number of treatments in the ANOVA application. The procedures also differ with respect to their treatment of Type I errors (i.e., $P(\text{Reject } H_0 | H_0 \text{ true})$). Techniques such as the two independent samples t test can be used to test pairs of treatment means, but the Type I error rate over all comparisons of interest is not controlled. In applications involving k treatments, there are as many as $k(k-1)/2$ possible pairwise comparisons. In the worst possible case, the Type I error can be as large as $\alpha(k(k-1)/2)$. For example, if three pairwise comparisons are performed, the Type I error rate could be 15%, or as large as 50% if 10 pairwise comparisons are performed. In most every application, these levels would be unacceptable.

Consider the following definitions:

$$\text{Error rate per comparison (ER\_PC)} = P(\text{Type I error}) \text{ on any one test or comparison} \qquad (9.6)$$

In general, the error rate per comparison is 0.05.

$$\text{Error rate per experiment (ER\_PE)} = \text{the number of Type I errors we expect to make in any experiment under } H_0 \qquad (9.7)$$

For example, suppose we collect data on 100 different variables (e.g., systolic blood pressure, diastolic blood pressure, total cholesterol, age) from a random sample of females and from a random sample of males and wish to test for mean differences between males and females on each variable. The experiment involves 100 two independent samples t tests, each performed at the $\alpha = 0.05$ level. The error rate over the entire experiment (ER_PE) is $100(0.05) = 5$. That is, we expect 5 tests to be significant (i.e., we reject H_0 in 5 tests) solely by chance (i.e., we expect to make five Type I errors in the experiment). Notice that the error rate per experiment is a frequency and not a probability.

$$\text{Familywise error rate (FW\_ER)} = P(\text{at least 1 Type I error}) \text{ in experiment} \qquad (9.8)$$

The FW_ER is computed by: $1 - (1 - \alpha_i)^c$, where α_i is the error rate per comparison (ER_PC) and c represents the number of contrasts, or comparisons, in the experiment. The formula is illustrated in Example 9.6.

EXAMPLE 9.6

Error Rates with Multiple Comparisons

Suppose we test the equality of five treatment means using ANOVA and the null hypothesis is rejected at $\alpha = 0.05$. Suppose that it is of interest to perform all pairwise comparisons. There are $k(k-1)/2 = 5(5-1)/2 = 10$ distinct pairwise comparisons (e.g., $H_0: \mu_1 = \mu_2$ versus $H_1: \mu_1 \neq \mu_2$, $H_0: \mu_1 = \mu_3$ versus $H_1: \mu_1 \neq \mu_3$, $H_0: \mu_1 = \mu_4$ versus $H_1: \mu_1 \neq \mu_4$, ..., $H_0: \mu_4 = \mu_5$ versus $H_1: \mu_4 \neq \mu_5$). Suppose we wish to conduct each comparison at a 5% level of significance. It should be noted that one should perform only tests that are of substantive interest and not just all possible tests.

The error rate per comparison, ER_PC = 0.05.
The error rate per experiment, ER_PE = 10(0.05) = 0.5.
The familywise error rate, FW_ER = $1 - (1 - 0.05)^{10} = 0.401$.

In this example, we expect to make 0.5 Type I errors solely by chance, and the probability of at least one Type I error is large (40.1% chance). ■

Again, there are a number of multiple comparison procedures that differ according to their treatment of the error rates per experiment and familywise error rates. Other MCPs (not discussed here) include the Duncan procedure (also called the multiple range test), Fisher's Least Significant Difference, the Newman–Keuls test, and Dunnett's test (used to compare a control to several active treatments). For further details, see D'Agostino, R. B., Massaro, J., Kwan, H., Cabral, H. (1993). Strategies for dealing with multiple treatment comparisons in confirmatory clinical trials. *Drug Information Journal,* 27, 625–641. We now illustrate the use of two popular multiple comparison procedures, the Scheffe and Tukey procedures.

9.5.1 The Scheffe Procedure

The Scheffe procedure is a MCP that controls the familywise error rate. That is, the P(Type I error) is controlled (and equal to α) over the family of all comparisons. For example, if 6 or 8 or 20 comparisons are performed within a particular experiment, the familywise error rate is 5%. The Scheffe procedure is most desirable in applications involving more than a few contrasts; however, it is a conservative procedure (i.e., has lower statistical power) compared to competing procedures.

The Scheffe procedure is outlined next, for the case of pairwise comparisons in Example 9.7. More complicated contrasts are considered in Example 9.8.

1. Set up hypotheses.

$$H_0: \mu_i = \mu_j$$
$$H_1: \mu_i \neq \mu_j$$

where μ_i and μ_j are two of k treatment means that were found to be significantly different based on ANOVA

2. Select the appropriate test statistic.
    The test statistic for pairwise comparisons takes the following form:

$$F = \frac{(\overline{X}_{.i} - \overline{X}_{.j})^2}{s_w^2 \left(\frac{1}{n_i} + \frac{1}{n_j} \right)} = \frac{(\overline{X}_{.i} - \overline{X}_{.j})^2}{MS_{error} \left(\frac{1}{n_i} + \frac{1}{n_j} \right)}$$

where s_w^2 is the estimate of the within variation and equal to the mean square within or mean square error (from ANOVA table)

3. Decision rule.

The test statistic follows an F distribution; therefore a critical value is selected from the F distribution table (Table B.4). The critical value for the Scheffe procedure (pairwise as well as other contrasts) is the product of $(k-1)$ and the critical value from the ANOVA:

$$\text{Reject } H_0 \text{ if } F \geq (k-1)F_{k-1,N-k}$$

$$\text{Do not reject } H_0 \text{ if } F < (k-1)F_{k-1,N-k}$$

EXAMPLE 9.7

Scheffe Pairwise Comparisons for Differences in Means Between Drugs

Consider Example 9.3 in which we compared the mean times to relief from headache pain under three competing medications. An ANOVA was performed at the 5% level of significance and we rejected $H_0: \mu_1 = \mu_2 = \mu_3$. Suppose we now wish to compare the medications taken two-at-a-time (i.e., pairwise comparisons). The summary statistics are

Drug A	Drug B	Drug C
$\overline{X}_{.1} = 33.0$	$\overline{X}_{.2} = 26.0$	$\overline{X}_{.3} = 20.0$
$s_{.1} = 5.7$	$s_{.2} = 4.2$	$s_{.3} = 3.5$

Drug A Versus Drug B

1. Set up hypotheses.

$$H_0: \mu_1 = \mu_2$$

$$H_1: \mu_1 \neq \mu_2, \quad \alpha = 0.05$$

2. Select the appropriate test statistic.

$$F = \frac{(\overline{X}_{.1} - \overline{X}_{.2})^2}{s_w^2 \left(\dfrac{1}{n_1} + \dfrac{1}{n_2}\right)} = \frac{(\overline{X}_{.1} - \overline{X}_{.2})^2}{MS_{\text{error}} \left(\dfrac{1}{n_1} + \dfrac{1}{n_2}\right)}$$

3. Decision rule.

Reject H_0 if $F \geq (k-1)F_{2,12} = 2(3.89) = 7.78$ (where $F_{2,12} = 3.89$ was used in Example 9.3 in the ANOVA)

4. Compute the test statistic.

$$F = \frac{(33.0 - 26.0)^2}{20.82 \left(\dfrac{1}{5} + \dfrac{1}{5}\right)} = 5.88$$

5. Conclusion.

Do not reject H_0 since $5.88 < 7.78$. We do not have significant evidence, $\alpha = 0.05$, to show that $\mu_1 \neq \mu_2$.

Drug A Versus Drug C

1. Set up hypotheses.

$$H_0: \mu_1 = \mu_3$$
$$H_1: \mu_1 \neq \mu_3, \quad \alpha = 0.05$$

2. Select the appropriate test statistic.

$$F = \frac{(\overline{X}_{.1} - \overline{X}_{.3})^2}{s_w^2 \left(\dfrac{1}{n_1} + \dfrac{1}{n_3} \right)} = \frac{(\overline{X}_{.1} - \overline{X}_{.3})^2}{MS_{error} \left(\dfrac{1}{n_1} + \dfrac{1}{n_3} \right)}$$

3. Decision rule.

Reject H_0 if $F \geq (k-1) F_{2, 12} = 2(3.89) = 7.78$

4. Compute the test statistic.

$$F = \frac{(33.0 - 20.0)^2}{20.82 \left(\dfrac{1}{5} + \dfrac{1}{5} \right)} = 20.28$$

5. Conclusion.

Reject H_0 since $20.28 \geq 7.78$. We have significant evidence, $\alpha = 0.05$, to show that $\mu_1 \neq \mu_3$.

Drug B Versus Drug C

1. Set up hypotheses.

$$H_0: \mu_2 = \mu_3$$
$$H_1: \mu_2 \neq \mu_3, \quad \alpha = 0.05$$

2. Select the appropriate test statistic.

$$F = \frac{(\overline{X}_{.2} - \overline{X}_{.3})^2}{s_w^2 \left(\dfrac{1}{n_2} + \dfrac{1}{n_3} \right)} = \frac{(\overline{X}_{.2} - \overline{X}_{.3})^2}{MS_{error} \left(\dfrac{1}{n_2} + \dfrac{1}{n_3} \right)}$$

3. Decision rule.

Reject H_0 if $F \geq (k-1) F_{2, 12} = 2(3.89) = 7.78$

4. Compute the test statistic.

$$F = \frac{(26.0 - 20.0)^2}{20.82 \left(\dfrac{1}{5} + \dfrac{1}{5} \right)} = 4.32$$

5. Conclusion.

Do not reject H_0 since $4.32 < 7.78$. We do not have significant evidence, $\alpha = 0.05$, to show that $\mu_2 \neq \mu_3$.

▪

SAS EXAMPLE 9.7 **Scheffe Pairwise Comparisons for Differences in Means Between Drugs Using SAS**

The following output was generated using SAS Proc ANOVA. We specified an option to conduct all pairwise comparisons using the Scheffe MCP. SAS produces the following. A brief interpretation appears after the output.

SAS Output for Example 9.7

```
                        The ANOVA Procedure
Dependent Variable: time
                          Sum of
Source              DF       Squares    Mean Square    F Value    Pr > F
Model                2    423.3333333    211.6666667     10.16    0.0026
Error               12    250.0000000     20.8333333
Corrected Total     14    673.3333333


             R-Square      Coeff Var      Root MSE      time Mean
             0.628713       17.33299      4.564355      26.33333


Source              DF      ANOVA SS     Mean Square    F Value    Pr > F
trt                  2    423.3333333    211.6666667     10.16    0.0026


                  Scheffe's Test for time
   NOTE: This test controls the Type I experimentwise error rate.
             Alpha                             0.05
             Error Degrees of Freedom            12
             Error Mean Square             20.83333
             Critical Value of F            3.88529
             Minimum Significant Difference   8.047


   Means with the same letter are not significantly different.
      Scheffe Grouping           Mean       N    trt
                      A          33.000      5    A
                B     A          26.000      5    B
                B                20.000      5    C
```

Interpretation of SAS Output for Example 9.7

Again, SAS produces an ANOVA table, similar to Table 9.1. In Example 9.3, we had a test statistic of $F = 10.16$ with $p = 0.0026$, which would lead to rejection of H_0 since $p = 0.0026 < \alpha = 0.05$.

The next section of the output displays the results of the Scheffe procedure for pairwise comparisons. Along with the results of the Scheffe pairwise comparisons, SAS prints a note that the Scheffe procedure is one that controls the

Type I error rate. This procedure also has a higher Type II error rate than alternative procedures. SAS also copies the mean square error (MS_{error}) from the preceding ANOVA table, which is used in the test statistics. The pairwise comparisons are performed internally and SAS indicates which means are significantly different from others by assigning letters ("Scheffe Grouping") to each treatment. If the same letters are assigned to different treatments, the treatment means are not significantly different. However, if different letters are assigned to the treatments, the treatment means are significantly different. In the output, the treatments (or drugs) A and B are assigned the letter "A," indicating that the means of Drugs A and B are not significantly different. Drug C is assigned the letter "B." Therefore, the means of Drugs A and C are significantly different. Drug B is also assigned the letter "B," indicating that the means of Drugs B and C are not significantly different. ■

EXAMPLE 9.8

Comparison Between Mean Age at Time to Completion of 8th Grade Between Urban versus Rural and Suburban Districts

Consider Example 9.4, comparing the ages of students at the completion of eighth grade from rural, suburban, and urban schools. A one-way ANOVA was performed in which H_0: $\mu_1 = \mu_2 = \mu_3$ was rejected at $\alpha = 0.05$. Suppose we wish to compare the urban students to the rural and suburban students combined. The summary statistics on the students' ages are

	Rural	Suburban	Urban
n_j	8	10	10
$\sum X_{ij}$	107	131	144
$\overline{X}_{.j}$	13.4	13.1	14.4

1. Set up hypotheses.

$$H_0: \tfrac{1}{2}(\mu_1 + \mu_2) = \mu_3$$

$$H_1: \tfrac{1}{2}(\mu_1 + \mu_2) \neq \mu_3, \quad \alpha = 0.05$$

NOTE: In the null hypothesis we state that the mean age for rural and suburban students combined is equal to the mean age for urban students. The null hypothesis is equivalent to $H_0: 1/2(\mu_1 + \mu_2) - \mu_3 = 0$; therefore, the weights for the sample means are 1/2, 1/2, and −1 (see step 2).

2. Select the appropriate test statistic.

$$F = \frac{\left(\dfrac{\overline{X}_{.1} + \overline{X}_{.2}}{2} - \overline{X}_{.3}\right)^2}{MS_{error}\left[\dfrac{1}{2^2}\left(\dfrac{1}{n_1}\right) + \dfrac{1}{2^2}\left(\dfrac{1}{n_2}\right) + \dfrac{1}{n_3}\right]}$$

In this example, we consider a more complicated contrast. Notice that the test statistic is modified accordingly. The numerator of the test statistic compares point estimates, and the denominator involves the reciprocal of each sample size, weighted by the coefficient associated with that population mean squared. For example, in H_0 we weight both μ_1 and μ_2 by $1/2$, which is squared in the denominator of the statistic.

3. Decision rule.
   Reject H_0 if $F \geq (k-1)F_{2,25} = 2(3.39) = 6.78$ (where $F_{2,25} = 3.39$ was used in Example 9.4 in the ANOVA)

4. Compute the test statistic.

$$F = \frac{\left(\dfrac{(13.4 + 13.1)}{2} - 14.4\right)^2}{0.927 \left[\dfrac{1}{4}\left(\dfrac{1}{8}\right) + \dfrac{1}{4}\left(\dfrac{1}{10}\right) + \dfrac{1}{10}\right]} = 9.13$$

5. Conclusion.
   Reject H_0 since $9.13 > 6.78$. We have significant evidence, $\alpha = 0.05$, to show that the mean age for rural and suburban students combined is not equal to the mean age of urban students. ■

9.5.2 The Tukey Procedure

The Tukey procedure, also called the Studentized Range test, is a popular, widely applied MCP that also controls the familywise error rate. The Tukey procedure is appropriate for pairwise comparisons. It does not handle general contrasts. It is a less conservative procedure (i.e., has better statistical power) than the Scheffe procedure when there are a large number of pairwise comparisons.

The Tukey procedure is outlined here and illustrated in Example 9.9. The procedure involves several steps. In the first step, treatments are ordered according to the magnitude of their respective sample means. Let $\overline{X}_{1'}$ denote the largest sample mean, $\overline{X}_{2'}$ denote the second largest sample mean, and so on, to $\overline{X}_{k'}$, which denotes the smallest sample mean. In the Tukey test, pairwise comparisons are made in a specific order. The first comparison involves a comparison of the treatments with the largest and smallest sample means. If this test is significant, then a test comparing the treatment with the largest sample mean to the treatment with the next-to-smallest sample mean is performed. If this test is significant, then one proceeds to compare the treatment with the largest sample mean to the treatment with the third-to-smallest sample mean, and so on. The specifics of each test are outlined here. The example refers to the first test in the sequence of tests that involves the treatments with the largest and smallest sample means.

1. Set up hypotheses.

$$H_0: \mu_{1'} = \mu_{k'}$$
$$H_1: \mu_{1'} \neq \mu_{k'}$$

where $\mu_{1'}$ and $\mu_{k'}$ are the means of the treatments with the largest and smallest sample means, respectively

2. Select the appropriate test statistic.

$$q_k = \frac{(\overline{X}_{1'} - \overline{X}_{k'})}{\sqrt{\dfrac{s_w^2}{n}}} = \frac{(\overline{X}_{1'} - \overline{X}_{k'})}{\sqrt{\dfrac{MS_{error}}{n}}}$$

where s_w^2 is the estimate of the within variation, and equal to the mean square within or mean square error (from the ANOVA table)

3. Decision rule.

The critical value for the Tukey test can be found in Table B.6: Critical Values of the Studentized Range Distribution. The critical value depends upon the level of significance, α, the number of treatments involved in the analysis, k, and the error degrees of freedom from the ANOVA table. Table B.6 contains critical values for $\alpha = 0.05$. The same critical value is used for all pairwise comparisons, and the decision rule is of the form:

$$\text{Reject } H_0 \text{ if } q_k \geq q_\alpha(k, \text{df}_{error})$$
$$\text{Do not reject } H_0 \text{ if } q_k < q_\alpha(k, \text{df}_{error})$$

EXAMPLE 9.9

Tukey Pairwise Comparisons for Differences in Means Between Drugs

Consider Example 9.3 in which we compared the mean times to relief from headache pain under three competing medications. An ANOVA was performed at the 5% level of significance and we rejected $H_0: \mu_1 = \mu_2 = \mu_3$. Suppose we now wish to compare the medications taken two-at-a-time (i.e., pairwise comparisons) using the Tukey procedure. The summary statistics are

Drug A	Drug B	Drug C
$\overline{X}_{.1} = 33.0$	$\overline{X}_{.2} = 26.0$	$\overline{X}_{.3} = 20.0$
$s_{.1} = 5.7$	$s_{.2} = 4.2$	$s_{.3} = 3.5$

The first step is to order the treatments according to the magnitude of their respective sample means: $\overline{X}_{1'} = 33.0$, $\overline{X}_{2'} = 26.0$, and $\overline{X}_{3'} = 20.0$. (*Note:* Only coincidentally do the sample means array from largest to smallest as presented.) The first test in the Tukey procedure involves a comparison of Drugs A and C (largest versus smallest sample means).

Drug A Versus Drug C

1. Set up hypotheses.

$$H_0: \mu_1 = \mu_3$$

$$H_1: \mu_1 \neq \mu_3, \quad \alpha = 0.05$$

2. Select the appropriate test statistic.

$$q_k = \frac{(\overline{X}_{.1} - \overline{X}_{.3})}{\sqrt{\dfrac{MS_{error}}{n}}}$$

where MS_{error} is the mean square error (or mean square within from the ANOVA table)

3. Decision rule.
   The appropriate critical value from Table B.6 for $k = 3$ and $df_{error} = 12$ is 3.77.

$$\text{Reject } H_0 \text{ if } q_k \geq 3.77$$

$$\text{Do not reject } H_0 \text{ if } q_k < 3.77$$

4. Compute the test statistic.

$$q_3 = \frac{(33.0 - 20.0)}{\sqrt{\dfrac{20.82}{5}}} = 6.37$$

5. Conclusion.
   Reject H_0 since $6.37 > 3.77$. We have significant evidence, $\alpha = 0.05$, to show that $\mu_1 \neq \mu_3$.
   Because the first test was significant, we proceed to test the equality of treatments whose sample means reflect the next-largest difference.

Drug A Versus Drug B

1. Set up hypotheses.

$$H_0: \mu_1 = \mu_2$$

$$H_1: \mu_1 \neq \mu_2, \quad \alpha = 0.05$$

2. Select the appropriate test statistic.

$$q_k = \frac{(\overline{X}_{.1} - \overline{X}_{.2})}{\sqrt{\dfrac{MS_{error}}{n}}}$$

where MS_{error} is the mean square error (or mean square within from the ANOVA table)

3. Decision rule.

$$\text{Reject } H_0 \text{ if } q_k \geq 3.77$$

$$\text{Do not reject } H_0 \text{ if } q_k < 3.77$$

4. Compute the test statistic.

$$q_3 = \frac{(33.0 - 26.0)}{\sqrt{\dfrac{20.82}{5}}} = 3.43$$

5. Conclusion.

Do not reject H_0 since $3.43 < 3.77$. We do not have significant evidence, $\alpha = 0.05$, to show that $\mu_1 \neq \mu_2$. Because this test is not significant, we do not go on to test H_0: $\mu_2 = \mu_3$ versus H_1: $\mu_2 \neq \mu_3$. ▪

SAS EXAMPLE 9.9 **Tukey Pairwise Comparisons for Differences in Means Between Drugs Using SAS**

The following output was generated using SAS Proc ANOVA. We specified an option to conduct all pairwise comparisons using the Tukey MCP. A brief interpretation appears after the output.

SAS Output for Example 9.9

```
Analysis of Variance Procedure
Dependent Variable: TIME
                           Sum of          Mean
Source              DF     Squares         Square      F Value   Pr > F
Model                2     423.33333333    211.66666667   10.16   0.0026
Error               12     250.00000000    20.83333333
Corrected Total     14     673.33333333

            R-Square            C.V.        Root MSE            TIME Mean
            0.628713         17.33299      4.5643546            26.333333

Source              DF       ANOVA SS     Mean Square   F Value   Pr > F
TRT                  2     423.33333333   211.66666667    10.16   0.0026

          Tukey's Studentized Range (HSD) Test for time
NOTE: This test controls the Type I experimentwise error rate, but it
generally has a higher Type II error rate than REGWQ.
```

```
Alpha                                         0.05
Error Degrees of Freedom                        12
Error Mean Square                         20.83333
Critical Value of Studentized Range        3.77278
Minimum Significant Difference              7.7012
```

```
Means with the same letter are not significantly different.
     Tukey Grouping         Mean       N    trt
                  A         33.000      5    A
            B     A         26.000      5    B
            B               20.000      5    C
```

Interpretation of SAS Output for Example 9.9

In the second section of the output, along with the results of the Tukey pair-wise comparisons, SAS prints a note that the Tukey procedure is one that controls the Type I error rate but has a higher Type II error rate than alternative procedures. SAS also prints the level of significance, $\alpha = 0.05$, the error degrees of freedom, df $= 12$, and the mean square error (MS_{error}) from the ANOVA table ($MS_{error} = 20.83333$) used in the test statistics. The treatments are ordered according to the magnitude of their respective sample means (largest to smallest). The pairwise comparisons are performed internally and SAS indicates which means are significantly different from others by assigning letters ("Tukey Grouping") to each treatment. If the same letters are assigned to different treatments, the treatment means are not significantly different. However, if different letters are assigned to the treatments, the treatment means are significantly different. Recall, in the Tukey procedure, one first compares the treatments with the largest and smallest sample means. In this application, Drug A has the largest mean and Drug C has the smallest. Drug A is assigned the letter "A" and Drug C is assigned the letter "B." Since the Tukey groupings (assigned letters) are different, the means of Drugs A and C are significantly different. The next test is the test comparing the means of Drugs A and B. In the output, Drugs A and B are assigned the letter "A," indicating that the means of Drugs A and B are not significantly different. Although SAS allows for the test comparing Drugs B and C, the Tukey procedure should be terminated following the nonsignificant result in the second test. ■

9.6 Repeated Measures Analysis of Variance

In some applications, it is of interest to assess changes in a particular measure over time. For example, suppose an intervention is designed to improve patients' medication adherence, which is a particularly important issue in the management of many chronic diseases. The intervention is administered at a point in time and measurements are taken at predetermined intervals to assess adherence. A hypothesis might be that the intervention has a decreasing effect over time. The data layout is as follows:

Subject	Time 1	Time 2	. . .	Time k
1	X_{11}	X_{12}		X_{1k}
2	X_{21}	X_{22}		X_{2k}
.				
.				
.				
n	X_{n1}	X_{n2}		X_{nk}

where X_{sj} represents the measurement (X) on the sth subject
$(s = 1, 2, \ldots, n)$ on the jth occasion $(j = 1, 2, \ldots, k)$

In these applications we have one sample of n subjects and take multiple, or repeated, measurements on each subject (k measurements in total). Without the subject column, the data layout is identical to the layout for analysis of variance procedures described in previous sections (assuming equal sample sizes). It is important to note that in these applications, there is one sample of subjects and repeated measures are taken on these subjects, introducing a dependency among the measurements. Because of this dependency, the test statistic needs modification.

For reference, recall the two independent and two dependent sample applications described in Chapter 6. In the two dependent samples applications, we focused on difference scores. Because there are now more than two measurements ($k > 2$), the appropriate analysis is analysis of variance. The test is outlined next.

1. Set up hypotheses.

$$H_0: \mu_1 = \mu_2 = \cdots = \mu_k$$
$$H_1: \text{means not all equal}$$

2. Select the appropriate test statistic.

$$F = \frac{s_b^2}{s_w^2}$$

where s_b^2 = denotes the between variation, which is also denoted MS_b (mean square between)
s_w^2 = denotes the within variation, which is also denoted MS_w (mean square within)

This test statistic is identical in form to the test statistic used earlier. The analysis of variance table is different, however, and includes an additional

Table 9.2 *Repeated Measures Analysis of Variance Table*

Source of Variation	Sums of Squares (SS)	Degrees of Freedom (df)	Mean Squares (MS)	F
Between Subjects	$SS_{subj} = \sum k(\overline{X}_{s.} - \overline{X}_{..})^2$	$n - 1$		
Between Treatments	$SS_b = \sum n(\overline{X}_{.j} - \overline{X}_{..})^2$	$k - 1$	$S_b^2 = MS_b = \dfrac{SS_b}{k-1}$	$F = \dfrac{MS_b}{MS_w}$
Within	$SS_w = SS_{total} - SS_{subj} - SS_b$	$(n-1)(k-1)$	$S_w^2 = MS_w = \dfrac{SS_w}{(n-1)(k-1)}$	
Total	$SS_{total} = \sum\sum(X_{sj} - \overline{X}_{..})^2$	$nk - 1$		

where X_{sj} = measurement on the *s*th subject in the *j*th treatment (or at the *j*th time point)

$\overline{X}_{s.}$ = sample mean of *s*th subject

$\overline{X}_{.j}$ = sample mean of *j*th treatment (or *j*th time point)

$\overline{X}_{..}$ = overall sample mean

k = # measurements per subject

n = number of subjects

source of variation—variation due to the subjects. Because there are now multiple measurements taken on each subject, we can measure variation within given subjects (in the applications described in previous sections we had only a single measurement on each subject and could not measure variation within a given subject). The analysis of variable table for repeated measures analysis of variance procedures is given in Table 9.2.

3. Decision rule.

To select the appropriate critical value from the *F* distribution (Table B.4), we use the appropriate numerator and denominator degrees of freedom, df$_1$ and df$_2$, respectively (df$_1 = k - 1$ and df$_2 = (n-1)(k-1)$). Again, in ANOVA, we reject H_0 if the test statistic is larger than the critical value. We now illustrate the use of these formulas with an example.

EXAMPLE 9.10

Repeated Measures ANOVA to Test Difference in Mean Completion Times Among 3 Training Courses

An investigator is interested in comparing the cardiovascular fitness of elite runners on three different training courses, each of which covers 10 miles. The courses differ in terms of terrain, Course 1 is flat, Course 2 has graded inclines, and Course 3 includes steep inclines. Each runner's heart rate is monitored at mile 5 of the run on each course. Ten runners are involved, and their heart rates measured on each course are shown next.

Runner Number	Course 1	Course 2	Course 3
1	132	135	138
2	143	148	148
3	135	138	141
4	128	131	139
5	141	141	150
6	150	156	161
7	131	134	138
8	150	156	162
9	142	145	151
10	139	165	160

Is there a significant difference in the mean heart rates of runners on the three courses? Run the appropriate test at a 5% level of significance.

Because we have a single sample of 10 subjects, and three measures taken on each, the appropriate analysis is a repeated measures analysis of variance. The test is carried out next.

1. Set up hypotheses.

$$H_0: \mu_1 = \mu_2 = \mu_3$$

$$H_1: \text{means not all equal}, \quad \alpha = 0.05$$

2. Select the appropriate test statistic.

$$F = \frac{s_b^2}{s_w^2}$$

3. Decision rule.

To select the appropriate critical value from the F distribution (Table B.4), we first compute the numerator degrees of freedom (df_1) and the denominator degrees of freedom (df_2):

$$df_1 = k - 1 = 3 - 1 = 2 \qquad df_2 = (n - 1)(k - 1) = 9(2) = 18$$

The critical value of F with 2 and 18 degrees of freedom, relative to a 5% level of significance, is found in Table B.4A: $F_{2,18} = 3.55$. The decision rule is

$$\text{Reject } H_0 \text{ if } F \geq 3.55$$

$$\text{Do not reject } H_0 \text{ if } F < 3.55$$

4. Compute the test statistic.

Again, we will construct an ANOVA table to organize our computations. We will first compute the *between subjects* sums of squares (necessary to account for the repeated measurements taken on each subject). The between subjects sums of squares is computed by summing the squared differences between each subject's mean heart rate and the overall mean heart rate. The overall mean is computed by summing all of the measurements and dividing by the total number of measurements ($N = nk = 10(3) = 30$):

$$\overline{X}.. = \frac{4328}{30} = 144.3$$

Runner Number	Course 1	Course 2	Course 3	Subject Mean $\overline{X}_s.$	$(\overline{X}_s. - \overline{X}..)^2$
1	132	135	138	135.0	86.5
2	143	148	148	146.3	4.0
3	135	138	141	138.0	39.7
4	128	131	139	132.7	134.6
5	141	141	150	144.0	0.1
6	150	156	161	155.7	130.0
7	131	134	138	134.3	100.0
8	150	156	162	156.0	136.9
9	142	145	151	146.0	2.9
10	139	165	160	154.7	108.2
Course Means $\overline{X}._j$	139.1	144.9	148.8	$\overline{X}.. = 144.3$	742.9

The between subjects sums of squares is

$$SS_{subj} = \sum k(\overline{X}_s. - \overline{X}..)^2 = 3(742.9) = 2228.7$$

The *between treatments* sums of squares is computed by summing the squared differences between each treatment mean and the overall mean. The treatment (or course) means are shown along the bottom of the preceding table. The between treatments sums of squares is

$$SS_b = \sum n(\overline{X}._j - \overline{X}..)^2 = 10[(139.1 - 144.3)^2 + (144.9 - 144.3)^2$$
$$+ (148.8 - 144.3)^2] = 477$$

The total sums of squares is computed by summing the squared differences between each observation and the overall mean. This is equivalent to the numerator of the sample variance. Recall from Chapter 2 the shortcut formula for the sample variance and the following alternative formula for the total sums of squares:

$$SS_{total} = \sum\sum(\overline{X}_{sj} - \overline{X}..)^2 = \sum X_{sj}^2 - \frac{(\sum X_{sj})^2}{N}$$

It is tedious to compute the sum of each observation squared, but for this example, $\sum X_{sj}^2 = 627{,}362$. The sum of the observations is $\sum X_{sj} = 4328$, and the total sums of squares is

$$SS_{\text{total}} = \sum X_{sj}^2 - \frac{(\sum X_{sj})^2}{N} = 627{,}362 - \frac{(4328)^2}{30} = 2975.9$$

The within sums of squares is computed by subtraction: $SS_{\text{within}} = SS_{\text{total}} - SS_{\text{subj}} - SS_{b}$.

$$SS_{\text{within}} = 2975.9 - 2228.7 - 477 = 270.2$$

We now construct the ANOVA table for Example 9.10.

Source of Variation	Sums of Squares (SS)	Degrees of Freedom (df)	Mean Squares (MS)	F
Between Subjects	2228.7	9		
Between Treatments	477	2	238.5	15.9
Within	270.2	18	15.0	
Total	2975.9	29		

5. Conclusion.

Reject H_0, since $15.9 > 3.55$. We have significant evidence, $\alpha = 0.05$, to show that there is a difference in mean heart rates of runners on the three courses. ■

SAS EXAMPLE 9.10 **Repeated Measures ANOVA to Test Difference in Mean Completion Times Among 3 Training Courses Using SAS**

The following output was generated using SAS Proc GLM, which runs a repeated measures analysis of variance test for equality of means. The repeated option (see SAS program code in Section 9.9 for the details) is used to indicate that the measurements are dependent. A brief interpretation appears after the output.

SAS Output for Example 9.10

```
                    The GLM Procedure
            Repeated Measures Analysis of Variance
              Repeated Measures Level Information
       Dependent Variable      course1  course2  course3
            Level of course        1        2        3
```

```
                          The GLM Procedure
                 Repeated Measures Analysis of Variance
           Univariate Tests of Hypotheses for Within Subject Effects

Source            DF      Type III SS      Mean Square    F Value     Pr > F
course             2      476.4666667      238.2333333      15.60     0.0001
Error(course)     18      274.8666667       15.2703704
```

Interpretation of SAS Output for Example 9.10

The GLM procedure generates more output than is shown here for a repeated measures analysis of variance. The section shown is the most relevant for performing the test of interest. SAS generates an abbreviated ANOVA table, similar to Table 9.2. Again, SAS presents the degrees of freedom before the sums of squares. For the repeated measures applications, only the between treatment variation (labeled "course" in the output) and the within treatment variation (labeled "error(course)" in the output) rows of the ANOVA table are shown. These are the most relevant for computing the test statistic, which is $F = 15.60$ with $p = 0.0001$. We computed $F = 15.9$ by hand; the difference is due to rounding. With $p = 0.0001$, we would reject H_0 and conclude that the mean heart rates are different on the three courses. ■

9.7 Key Formulas

Application	Notation/Formula	Description
Test H_0: $\mu_1 = \mu_2 = \cdots = \mu_k$	$F = \dfrac{s_b^2}{s_w^2}$	see Table 9.1 for ANOVA computations (Table 9.2 for repeated measures ANOVA)
Variation explained by treatments	$\eta^2 = \dfrac{SS_b}{SS_{\text{total}}}$	see (9.5)
Test H_0: $\mu_i = \mu_j$ using Scheffe MCP*	$F = \dfrac{(\overline{X}_{.i} - \overline{X}_{.j})^2}{MS_{\text{error}}\left(\frac{1}{n_i} + \frac{1}{n_j}\right)}$	MCP that controls familywise error rate
Test H_0: $\mu_i = \mu_j$ using Tukey MCP*	$q_k = \dfrac{(\overline{X}_{1'} - \overline{X}_{k'})}{\sqrt{\frac{MS_{\text{error}}}{n}}}$	MCP that controls familywise error rate

* These tests should only be performed if H_0: $\mu_1 = \mu_2 = \mu_3 = \cdots = \mu_k$ is rejected.

9.8 Statistical Computing

Following are the SAS programs that were used to conduct tests of equality of k means ($k > 2$) using ANOVA. Also shown are programs to run Scheffe and Tukey multiple comparisons (pairwise comparisons). The SAS procedures and brief descriptions of their use are noted in the header to each example. Notes are provided to the right of the SAS programs (in blue) for orientation purposes and are not part of the programs. In addition, there are blank lines in the programs that are solely to accommodate the notes. Blank lines and spaces can be used throughout SAS programs to enhance readability. A summary of the SAS procedures used in the examples is provided at the end of this section.

SAS EXAMPLE 9.3 Analysis of Variance (ANOVA): Equal Sample Sizes

Compare Mean Times to Relief Among Three Treatments (Example 9.3)

An investigator wishes to compare the average time to relief of headache pain under three distinct medications—call them Drugs A, B, and C. Fifteen patients who suffer from chronic headaches are randomly selected for the investigation, and five subjects are randomly assigned to each treatment. The following data reflect times to relief (in minutes) after taking the assigned drug. Run an ANOVA using SAS.

Drug A	Drug B	Drug C
30	25	15
35	20	20
40	30	25
25	25	20
35	30	20

Program Code

```
options ps=62 ls=80;

data in;
   input trt $ time;


cards;
A 30
A 35
A 40
A 25
A 35
B 25
B 20
B 30
```

Formats the output page to 62 lines in length and 80 columns in width
Beginning of Data Step
Inputs two variables **trt** and **time**, where **trt** is a character variable (A, B, or C).
Beginning of Raw Data section.
actual observations (value of **trt** and **time** on each line)

```
B 25
B 30
C 15
C 20
C 25
C 20
C 20
run;

proc anova;


 class trt;



 model time=trt;




run;
```

Procedure call. Proc ANOVA tests the equality of *k* treatment means when sample sizes are equal.

Specification of grouping variable (i.e., variable that defines *k* comparison groups).

Specification of outcome variable (**time**). SAS requires a model statement relating the outcome to the grouping variable (**trt**).

End of procedure section. ■

SAS EXAMPLE 9.4 ## Analysis of Variance (ANOVA): Unequal Sample Sizes
Compare Mean Ages Among Three Groups of Students (Example 9.4)

The following data reflect ages of students at completion of eighth grade. Test if there is a significant difference in the mean age at completion of eighth grade for rural, suburban, and urban students using SAS and a 5% level of significance. The following data were collected from randomly selected students at rural, suburban, and urban schools.

Rural:	14	14	14	14	13	13	13	12	
Suburban:	14	14	14	13	13	13	13	13	12	12
Urban:	16	16	15	15	15	14	14	14	13	12

Program Code

```
options ps=62 ls=80;


data in;
 input school $ age;
```

Formats the output page to 62 lines in length and 80 columns in width

Beginning of Data Step

Inputs two variables **school** and **age**, where **school** = Rural, Suburban, or Urban.

```
cards;
rural 14
rural 14
rural 14
rural 14
rural 13
rural 13
rural 13
rural 12
suburban 14
suburban 14
suburban 14
suburban 13
suburban 13
suburban 13
suburban 13
suburban 13
suburban 12
suburban 12
urban 16
urban 16
urban 15
urban 15
urban 15
urban 14
urban 14
urban 14
urban 13
urban 12
run;


proc glm;



  class school;



  model age=school;




run;
```

Beginning of Raw Data section.
actual observations (value of **school** and **age** on each line)

Procedure call. Proc Glm tests the equality of *k* treatment means when sample sizes are equal or unequal.

Specification of grouping variable (i.e., variable that defines k comparison groups).

Specification of outcome variable (**age**). SAS requires a model statement relating the outcome to the grouping variable (**school**).

End of procedure section. ■

SAS EXAMPLE 9.5 **Analysis of Variance (ANOVA)**

Compare Mean Weight Gains Among Four Diets (Example 9.5)

A study is developed to examine the effects of vitamin and milk supplements on infant weight gain. Four diet plans are considered: Diet A involves a regular diet plus the vitamin supplement, Diet B involves a regular diet plus the special milk formula, Diet C is our control diet (i.e., no restrictions or special considerations), and Diet D involves a regular diet plus the vitamin and the special milk formula. Twenty infants are selected for the investigation and each is randomized to one of the four competing diet programs. The table displays weight gains, measured in pounds, after 1 month on the respective diet. Run an ANOVA using SAS.

Diet A	Diet B	Diet C	Diet D
2.0	1.6	1.5	2.1
1.5	1.9	2.0	2.4
2.4	2.1	1.8	1.9
1.9	1.1	1.3	1.8
2.6	1.7	1.2	2.2

Program Code

`options ps=62 ls=80;`	Formats the output page to 62 lines in length and 80 columns in width
`data in;`	Beginning of Data Step
`  input diet $ gain;`	Inputs two variables **diet** and **gain,** where **diet** = A, B, C, or D.
`cards;`	Beginning of Raw Data section.
`A 2.0`	actual observations (value of **diet** and
`A 1.5`	**gain** on each line)
`A 2.4`	
`A 1.9`	
`A 2.6`	
`B 1.6`	
`B 1.9`	
`B 2.1`	
`B 1.1`	
`B 1.7`	
`C 1.5`	
`C 2.0`	
`C 1.8`	

```
C 1.3
C 1.2
D 2.1
D 2.4
D 1.9
D 1.8
D 2.2
run;

proc glm;



  class diet;



  model gain=diet;




run;
```

	Procedure call. Proc Glm tests the equality of *k* treatment means when sample sizes are equal or unequal.
	Specification of grouping variable (i.e., variable that defines *k* comparison groups).
	Specification of outcome variable (**gain**). SAS requires a model statement relating the outcome to the grouping variable (**diet**).
	End of procedure section. ■

SAS EXAMPLE 9.7 Analysis of Variance (ANOVA): Pairwise Comparisons Using Scheffe and Tukey

Pairwise Multiple Comparisons (Examples 9.7 and 9.9)

In SAS EXAMPLE 9.3 we ran an ANOVA and rejected H_0. Run pairwise comparisons using the Scheffe and Tukey procedures to determine which means are different.

Program Code

```
options ps=62 ls=80;


data in;
  input trt $ time;



cards;
A 30
A 35
A 40
A 25
A 35
```

Formats the output page to 62 lines in length and 80 columns in width
Beginning of Data Step
Inputs two variables **trt** and **time**, where **trt** = A, B, or C.
Beginning of Raw Data section.
actual observations (value of **trt** and **time** on each line)

```
          B 25
          B 20
          B 30
          B 25
          B 30
          C 15
          C 20
          C 25
          C 20
          C 20
          run;

          proc anova;                          Procedure call. Proc ANOVA tests
                                                   the equality of k treatment means
                                                   when sample sizes are equal.

            class trt;                         Specification of grouping variable
                                                   (i.e., variable that defines k
                                                   comparison groups).

            model time=trt;                    Specification of outcome variable
                                                   (time). SAS requires a model
                                                   statement relating the outcome to
                                                   the grouping variable (trt).

            means trt/scheffe;                 Means option requires a comparison
                                                   of means among the comparison
                                                   groups defined by trt, using the
                                                   Scheffe MCP. (SAS Example 9.7)

            means trt/tukey;                   Means option requires a comparison
                                                   of means among the comparison
                                                   groups defined by trt, using the
                                                   Tukey MCP. (SAS Example 9.9)

          run;                                 End of procedure section.          ■
```

SAS EXAMPLE 9.10 Repeated Measures Analysis of Variance (ANOVA)

Compare Mean Heart Rates on Three Training Courses (Example 9.10)

An investigator is interested in comparing the cardiovascular fitness of elite runners on three different training courses, each of which covers 10 miles. The courses differ in terms of terrain: One includes steep inclines, the other more graded inclines, and the third is flat. Each runner's heart rate is monitored at mile 5 of the run on each course. Ten runners are involved and their heart rates as measured on each course are shown here. Run a repeated measures ANOVA using SAS.

Runner Number	Course 1	Course 2	Course 3
1	132	135	138
2	143	148	148
3	135	138	141
4	128	131	139
5	141	141	150
6	150	156	161
7	131	134	138
8	150	156	162
9	142	145	151
10	139	165	160

Program Code

```
options ps=62 ls=80;

data in;
  input runnerid course1 course2 course3;



  subjmean=mean(course1,course2,course3);


cards;
1     132    135    138
2     143    148    148
3     135    138    141
4     128    131    139
5     141    141    150
6     150    156    161
7     131    134    138
8     150    156    162
9     142    145    151
10    139    165    160
run;
proc print;


 var course1 course2 course3 subjmean;
run;
```

Formats the output page to 62 lines in length and 80 columns in width

Beginning of Data Step

Inputs four variables **runnerid** (subject identifier) and 3 measurements on each subject— **course1**, **course2**, and **course3**.

Computes a new variable, **subjmean**, which is the mean heart rate for each subject.

Beginning of Raw Data section.

actual observations (value of **runnerid**, **course1**, **course2** and **course3** on each line)

Procedure call. Proc Print lists the values of the specified variables.

Specification of variables for listing.

`proc means;`	Procedure call. Proc Means generates summary statistics on specified variables.
`  var course1 course2 course3 subjmean;`	Specification of variables (we are most interested in the means for each course. The mean of **subjmean** is the overall mean).
`run;`	
`proc glm;`	Procedure call. Proc Glm is used for analysis of variance applications.
`  model course1 course2 course3=/nouni;`	Specification of variables for analysis. In repeated measures analysis, the variables listed after the model statement are those measured on each subject. The nouni option indicates that we do not want univariate tests (which test if the mean for each variable is zero or not).
`  repeated course;`	The repeated option indicates a repeated measures ANOVA. After the repeated statement, the user specifies a label for the repeated measurements, which appears on the output (**course**). ■
`run;`	

9.8.1 Summary of SAS Procedures

The SAS Anova procedure is used to run an ANOVA when the sample sizes are equal in all comparison groups. The SAS Glm procedure is used to run an ANOVA when the sample sizes are either equal or unequal. Specific options can be requested in these procedures to run pairwise multiple comparisons (e.g., Scheffe or Tukey). The options are shown in italics. Users should refer to the examples in this section for complete descriptions of the procedure and specific options. A general description of the procedure and options is provided in the table.

Procedure	Sample Procedure Call	Description
proc anova	proc anova; class group; model outcome=group;	Runs an ANOVA comparing the mean outcome scores among groups, when sample sizes are equal in all comparison groups.
proc glm	proc glm; class group; model outcome=group;	Runs an ANOVA comparing the mean outcome scores among groups.

proc anova; class group; model outcome=group; means group/*scheffe*;	Runs an ANOVA comparing the mean outcome scores among groups and generates pairwise comparisons using the Scheffe procedure. Here we assume equal sample sizes.
proc glm; class group; model outcome=group; means group/*tukey*;	Runs an ANOVA comparing the mean outcome scores among groups and generates pairwise comparisons using the Tukey procedure.
proc glm; model outcome1 outcome2 ... outcomek=/nouni; repeated label;	Runs a repeated measures ANOVA comparing the mean outcome scores over time. The nouni option suppresses univariate tests, and the user specifies a label describing the nature of the repeated assessments.

9.9 Analysis of Framingham Heart Study Data

The Framingham data set includes data collected from the original cohort. Participants contributed up to three examination cycles of data. Here we analyze data collected in the first examination cycle (called the period = 1 examination) and use SAS Proc GLM to compare mean systolic blood pressure levels for participants with different body mass indices. We organized BMI into categories as follows: BMI < 18.5 (underweight), 18.5–24.9 (normal weight), 25.0–29.9 (overweight), and ≥ 30.0 (obese). We then performed Scheffe multiple comparisons to assess differences in systolic blood pressure between BMI groups taken two-at-a-time. The SAS code to create this variable and to attach a format for better interpretability is given here along with the procedure calls.

Framingham Data Analysis—SAS Code

```
proc format;
 value bmifmt 1='<18.5    ' 2='18.5-24.9' 3='25.0-29.9' 4='30.0+';
run;

data fhs;
 set in.frmgham;
 if period=1;

if ole bmi lt 18.5 then bmi_grp=1;
else if 18.5 le bmi lt 25.0 then bmi_grp=2;
```

```
else if 25.0 le bmi lt 30.0 then bmi_grp=3;
else if bmi ge 30.0 then bmi_grp=4;
format bmi_grp bmifmt.;
run;

proc glm data=fhs;
  class bmi_grp;
  model sysbp=bmi_grp;
  means bmi_grp;
run;
proc glm data=fhs;
  class bmi_grp;
  model sysbp=bmi_grp;
  means bmi_grp/scheffe cldiff;
run;
```

Framingham Data Analysis—SAS Output

The GLM Procedure

Class Level Information

Class	Levels	Values
bmi_grp	4	18.5-24.9 25.0-29.9 30.0+ <18.5

Number of observations   4434

The GLM Procedure

Dependent Variable: SYSBP   Systolic BP mmHg

Source	DF	Sum of Squares	Mean Square	F Value	Pr > F
Model	3	181059.901	60353.300	130.58	<.0001
Error	4430	2047533.372	462.197		
Corrected Total	4433	2228593.273			

R-Square	Coeff Var	Root MSE	SYSBP Mean
0.081244	16.17571	21.49877	132.9078

Source	DF	Type I SS	Mean Square	F Value	Pr > F
bmi_grp	3	181059.9012	60353.3004	130.58	<.0001

Source	DF	Type III SS	Mean Square	F Value	Pr > F
bmi_grp	3	181059.9012	60353.3004	130.58	<.0001

The SAS System   3

The GLM Procedure

Level of bmi_grp	N	-----------SYSBP----------- Mean	Std Dev
<18.5	76	123.927632	23.9394659
18.5-24.9	1936	126.790289	20.3968744
25.0-29.9	1845	135.873171	21.4538212
30.0+	577	145.134315	24.6783770

There is a highly significant difference in mean systolic blood pressure across the BMI groups. The means option produces the mean and standard deviation in systolic blood pressure for each BMI level. These statistics are very useful in understanding the nature of the difference (or trend, in this case) across comparison groups.

In the following output, we generate Scheffe comparisons to test for differences in mean systolic blood pressures between BMI groups taken two-at-a-time. Here we asked SAS to produce confidence intervals for each pair of comparison groups using the cldiff option. SAS indicates which BMI groups are significantly different in terms of systolic blood pressure.

The GLM Procedure
Class Level Information

Class	Levels	Values
bmi_grp	4	18.5-24.9 25.0-29.9 30.0+ <18.5

Number of observations   4434

The GLM Procedure
Dependent Variable: SYSBP   Systolic BP mmHg

Source	DF	Sum of Squares	Mean Square	F Value	Pr > F
Model	3	181059.901	60353.300	130.58	<.0001
Error	4430	2047533.372	462.197		
Corrected Total	4433	2228593.273			

R-Square	Coeff Var	Root MSE	SYSBP Mean
0.081244	16.17571	21.49877	132.9078

Source	DF	Type I SS	Mean Square	F Value	Pr > F
bmi_grp	3	181059.9012	60353.3004	130.58	<.0001

Source	DF	Type III SS	Mean Square	F Value	Pr > F
bmi_grp	3	181059.9012	60353.3004	130.58	<.0001

Scheffe's Test for SYSBP
NOTE: This test controls the Type I experimentwise error rate, but it generally
has a higher Type II error rate than Tukey's for all pairwise comparisons.

Alpha		0.05
Error Degrees of Freedom		4430
Error Mean Square		462.1971
Critical Value of F		2.60691

Comparisons significant at the 0.05 level are indicated by ***.

bmi_grp Comparison	Difference Between Means	Simultaneous 95% Confidence Limits	
30.0+ - 25.0-29.9	9.2611	6.3934 12.1289	***
30.0+ - 18.5-24.9	18.3440	15.4924 21.1957	***
30.0+ - <18.5	21.2067	13.8700 28.5434	***
25.0-29.9 - 30.0+	-9.2611	-12.1289 -6.3934	***
25.0-29.9 - 18.5-24.9	9.0829	7.1268 11.0390	***
25.0-29.9 - <18.5	11.9455	4.9084 18.9827	***
18.5-24.9 - 30.0+	-18.3440	-21.1957 -15.4924	***
18.5-24.9 - 25.0-29.9	-9.0829	-11.0390 -7.1268	***
18.5-24.9 - <18.5	2.8627	-4.1679 9.8933	
<18.5 - 30.0+	-21.2067	-28.5434 -13.8700	***
<18.5 - 25.0-29.9	-11.9455	-18.9827 -4.9084	***
<18.5 - 18.5-24.9	-2.8627	-9.8933 4.1679	

9.10 Problems

1. A pharmaceutical company is interested in the effectiveness of a new preparation designed to relieve arthritis pain. Three variations of the compound have been prepared for investigation, which differ according to the proportion of the active ingredients: T15 contains 15% active ingredients, T40 contains 40% active ingredients, and T50 contains 50% active ingredients. A sample of 20 patients is selected to participate in a study comparing the three variations of the compound. A control compound, which is currently available over the counter, is also included in the investigation. Patients are randomly assigned to one of the four treatments (control, T15, T40, T50) and the time (in minutes) until pain relief is recorded on each subject. The data are

Control:	12	15	18	16	20
T15:	20	21	22	19	20
T40:	17	16	19	15	19
T50:	14	13	12	14	11

a. Test if there is a difference in the mean time to relief among the four treatments. Use a 5% level of significance.

b. Using Scheffe multiple comparisons, test if there is a significant difference in the mean time to relief between the control and each of the experimental treatments (i.e., T15, T40, and T50), considered separately.

2. A study is performed to compare mean numbers of primary-care visits over 3 years among four different health maintenance organizations (HMOs). Fifty patients are randomly sampled from each HMO.

a. Write the hypotheses to be tested.

b. Complete the following ANOVA table:

Source of Variation	Sums of Squares	Degrees of Freedom	Mean Squares	F
Between	574.3			
Within				
Total	2759.8			

c. Is there a significant difference in mean numbers of primary-care visits among the four HMOs? Use a 5% level of significance.

3. Six different doses of a particular drug are compared in an effectiveness study. The study involves 30 subjects, and equal numbers of subjects are randomly assigned to each dose group.

a. What are the null and alternative hypotheses in the comparison?

b. Complete the following ANOVA table:

Source of Variation	Sums of Squares	Degrees of Freedom	Mean Squares	F
Between	189.85			
Within				
Total	352.57			

c. Is there a significant difference in the effectiveness among the doses? Use $\alpha = 0.05$.

d. Suppose we wish to compare dose groups 1 and 2 and the following are available: $\overline{X}_1 = 21.4$, $\overline{X}_2 = 27.6$. Run the appropriate test to assess whether there is a significant difference in effectiveness between dose groups 1 and 2 at the 5% level of significance.

4. The following data were collected as part of a study comparing a control treatment to an active treatment. Three doses of the active

treatment were considered in the study. The following table displays summary statistics on the ages of participants enrolled in the study classified by treatment group:

Treatment	# Participants	Mean Age	SD
Control	8	29.5	3.74
Low dose	8	34.5	2.88
Moderate dose	8	15.9	3.72
High dose	8	44.0	6.65

a. Test if there is a significant difference in the mean ages of participants across treatment groups. Complete the following table and show all parts of the test. Use $\alpha = 0.05$.

Source	Sums of Squares	Degrees of Freedom	Mean Squares	F
Between				
Within				
Total	3860.97			

b. Is there a significant difference in the mean ages of participants in the control and low dose groups? Run the appropriate test at $\alpha = 0.05$.

5. An investigation is performed to evaluate two new experimental treatments for allergies. Eighteen subjects who suffer from allergies are enrolled in the study and randomly assigned to one of three treatments: control treatment, experimental treatment 1, or experimental treatment 2. Each subject is instructed to take the assigned treatment, and symptoms of allergies are recorded on a scale of 1 to 20, with higher scores indicating worse symptoms. The data are

Control Treatment	Experimental Treatment 1	Experimental Treatment 2
19	12	5
18	14	6
16	13	3
15	10	2
12	9	3
17	10	4
$\overline{X}_1 = 16.2$	$\overline{X}_2 = 11.3$	$\overline{X}_3 = 3.8$

a. Test if there is a significant difference in mean symptom scores among the three treatments under investigation. Use a 5% level of significance. (*Hint:* $SS_{total} = 524.4$.)

b. Which treatment is the "best" of those investigated? (Justify your answer briefly.)

6. A randomized trial is conducted to compare four treatment regimens for HIV. Each regimen is based on a combination of medications. Twenty patients are involved in the investigation and are randomly assigned to one of four treatments: A, B, C, or D (5 subjects per treatment group). An overall symptom score, ranging from 0 to 20, is computed for each individual by aggregating their reports of the frequency and severity of an array of symptoms. Higher symptom scores indicate more frequent and/or severe symptoms. The summary statistics are

Treatment Regimen	Symptom Score
A	7.9
B	5.4
C	12.4
D	5.9

Test whether the mean symptom scores are equal among treatment regimens. Run the appropriate test at the 5% level of significance. (*Hint:* $SS_{within} = SS_{error} = 135.4$.)

7. A study is conducted among college students who smoke to assess whether there is a relationship between the number of cigarettes they smoke per day and the smoking status of their parents. For the purposes of the study, smoking status is defined as follows: never smoked, former smokers, or current smokers. The number of cigarettes smoked per day by each student is recorded and the data are

Parent's Smoking Status	Number of Students	Mean Number of Cigarettes Smoked/Day	SD
Never smoked	30	12.6	3.1
Former smokers	34	14.1	3.7
Current smokers	53	18.3	4.3
ALL	117	15.6	6.5

Based on the data, is there a significant difference in the mean numbers of cigarettes smoked per day according to the parent's smoking status? Run the appropriate test at $\alpha = 0.05$. (*Hint:* $SS_{total} = 3428.4$.)

8. A study is conducted comparing birthweights (in pounds) of infants born to mothers of various ages. The data are

	Mother's Age	
<20 Years	*20–29 Years*	*30+ Years*
8.4	7.5	6.9
7.3	6.3	7.1
9.1	6.9	5.7
7.8	5.4	6.5
8.4	7.1	6.6

Is there a significant difference in the mean birthweights for mothers of different ages? Run the appropriate test at $\alpha = 0.05$.

9. Consider a continuous measure of medication adherence (scores from 0 to 100), and use the following data to test if there is a significant difference in the mean medication adherence scores in patients of different age groups. Use the following data to run the appropriate test at $\alpha = 0.05$. (*Hint:* $\sum\sum(X_{ij} - \overline{X})^2 = 29,159.26.$)

	Age Group			
	20–29	*30–39*	*40–49*	*50–59*
Number of patients:	48	53	37	12
Mean medication adherence:	69.2	83.3	82.1	83.4
Standard deviation:	11.9	13.2	12.8	11.3

10. A clinical trial is performed comparing a test (newly developed) drug to an active control (a drug already proven effective) and to a placebo. Persons with diagnosed hypertension who are at least 18 years of age are eligible for the trial. Persons agreeing to participate are randomly assigned to one of the competing drugs. The outcome measure is systolic blood pressure (SBP), which is measured 4 weeks postrandomization. Summary statistics are

Drug	*n*	*Mean (SD) SBP*
Test	15	125 (25)
Control	15	135 (21)
Placebo	15	160 (22)

a. Is there a significant difference in mean SBPs 4 weeks postrandomization? Run the appropriate test at a 5% level of significance. (*Hint:* $SS_{total} = 31,450$.)

b. Is there a significant difference in mean SBPs 4 weeks postrandomization between the test drug and placebo? Run the appropriate test at a 5% level of significance.

c. Is there a significant difference in mean SBPs 4 weeks postrandomization between the test and control drugs? Run the appropriate test at a 5% level of significance.

11. The following output was generated from SAS:

```
            Analysis of Variance Procedure
               Class Level Information
          Class       Levels      Values
          GROUP          3        a b c
      Number of observations in data set = 24
```

Dependent Variable: SCORE

Source	DF	Sum of Squares	Mean Square	F Value	Pr > F
Model	2	839.58333333	419.79166667	43.16	0.0001
Error	21	204.25000000	9.72619048		
Corrected Total	23	1043.83333333			

a. Test if there is a significant difference in mean scores between treatments. Write the hypotheses and justify your conclusion (use the SAS output).

b. How much of the variation in scores is explained by the groups?

12. The following table summarizes data collected in a multicenter study. The variable summarized is body mass index (BMI) computed as the ratio of weight in kilograms to height in meters squared.

		Enrollment Site		
BMI	Overall	Hospital 1	Hospital 2	Hospital 3
n:	300	100	100	100
Mean:	24.8	21.6	24.8	27.9
SD:	2.5	2.1	1.8	1.3

Test if there is a significant difference in the mean BMI scores among hospitals. Show all parts of the test and use a 5% level of significance. (*Hint:* $MS_{within} = MS_{error} = 3.1$.)

13. A research group wishes to conduct a randomized trial to compare a newly approved medication, an experimental medication (one not yet available on market), and a medication they consider standard care. One hundred and fifty patients with hypertension are enrolled and randomized to one of the three comparison treatments. After taking the assigned medication for 6 weeks, each patient's systolic blood pressure (SBP) is measured. Summary statistics are given here. Use the data to test if there is a significant difference in systolic blood pressures among medication groups. Run the appropriate test at a 5% level of significance.

Treatment	Number of Patients	Mean SBP	SD
New	50	130	12
Standard	50	135	10
Experimental	50	115	17

Is there a difference in systolic blood pressures among the treatments? Run the appropriate test using a 5% level of significance. (*Hint:* $SS_{within} = SS_{error} = 26,117$, and $SS_{total} = 36,951$.)

14. The following data represent birthweights of siblings born to six different mothers:

Mother	Birthweights Child 1	Child 2	Child 3	Child 4
1	6.4	6.9	6.7	7.1
2	8.5	7.8	7.8	8.3
3	7.6	8.7	9.9	8.2
4	5.3	6.7	7.5	6.4
5	6.2	5.6	6.4	5.5
6	7.0	7.8	8.6	6.6

Is there a significant difference in birthweights of siblings? Run the appropriate test at $\alpha = 0.05$.

15. Suppose in a clinical trial that systolic blood pressures are measured at 4, 8, and 12 weeks postrandomization in a subgroup of patients receiving a test drug. We wish to test if there are significant differences in systolic blood pressures over time.

Systolic Blood Pressure

Subject	4 weeks	8 weeks	12 weeks
1	120	125	130
2	110	115	118
3	105	110	100
4	140	130	140
5	150	145	140

Test if there is a significant difference in mean systolic blood pressures over time using a 5% level of significance.

16. Suppose the data in the previous problem were collected from three independent random samples of subjects (i.e., $n_1 = 5, n_2 = 5, n_3 = 5$, total sample size $= 15$). Test if there is a significant difference in mean systolic blood pressures over time using a 5% level of significance. (*Hint:* You do not need to recalculate the sums of squares—compare formulas first!)

SAS Problems

Use SAS to solve the following problems.

1. A pharmaceutical company is interested in the effectiveness of a new preparation designed to relieve arthritis pain. Three variations of the compound have been prepared for investigation, which differ according to the proportion of the active ingredients: T15 contains 15% active ingredients, T40 contains 40% active ingredients, and T50 contains 50% active ingredients. A sample of 20 patients is selected to participate in a study comparing the three variations of the compound. A control compound, which is currently available over the counter, is also included in the investigation. Patients are randomly assigned to one of the four treatments (control, T15, T40, T50) and the time (in minutes) until pain relief is recorded on each subject. The data are

Control:	12	15	18	16	20
T15:	20	21	22	19	20
T40:	17	16	19	15	19
T50:	14	13	12	14	11

Use SAS Proc Anova to test the equality of means. In addition, run all pairwise comparisons using the Scheffe procedure. Use a 5% level of significance.

2. An investigation is performed to evaluate two new experimental treatments for allergies. Eighteen subjects who suffer from allergies are enrolled in the study and randomly assigned to one of three treatments: control treatment, experimental treatment 1, or experimental treatment 2. Each subject is instructed to take the assigned treatment, and symptoms of allergies are recorded on a scale of 1 to 20, with higher scores indicating worse symptoms. The data are

Control Treatment	Experimental Treatment 1	Experimental Treatment 2
19	12	5
18	14	6
16	13	3
15	10	2
12	9	3
17	10	4

Use SAS Proc Anova to test the equality of means. In addition, run all pairwise comparisons using the Tukey procedure. Use a 5% level of significance.

3. A study is conducted comparing birthweights (in pounds) of infants born to mothers of various ages. The data are

Mother's Age		
<20 Years	20–29 Years	30+ Years
8.4	7.5	6.9
7.3	6.3	7.1
9.1	6.9	5.7
7.8	5.4	6.5
8.4	7.1	6.6

Use SAS Proc Glm to test if there is a significant difference in the mean birthweights for mothers of different ages. Run the appropriate test at $\alpha = 0.05$.

4. The following data represent birthweights of siblings born to six different mothers:

	Birthweights			
Mother	Child 1	Child 2	Child 3	Child 4
1	6.4	6.9	6.7	7.1
2	8.5	7.8	7.8	8.3
3	7.6	8.7	9.9	8.2
4	5.3	6.7	7.5	6.4
5	6.2	5.6	6.4	5.5
6	7.0	7.8	8.6	6.6

Use SAS Proc Glm (with a repeated option) to test if there is a significant difference in birthweights of siblings. Run the appropriate test at $\alpha = 0.05$.

5. Suppose in a clinical trial that systolic blood pressures are measured at 4, 8, and 12 weeks postrandomization in a subgroup of patients receiving a test drug. We wish to test if there are significant differences in systolic blood pressures over time. The following data were collected at 4 weeks, 8 weeks, and 12 weeks postrandomization in a sample of five patients receiving the test drug.

	Systolic Blood Pressure		
Subject	4 weeks	8 weeks	12 weeks
1	120	125	130
2	110	115	118
3	105	110	100
4	140	130	140
5	150	145	140

Use SAS Proc Glm (with a repeated option) to test if there is a significant difference in birthweights of siblings. Run the appropriate test at $\alpha = 0.05$.

Probability
(Ch. 3)

Sampling Distributions
(Ch. 4)

Statistical Inference
(Chapters 5–12)

OUTCOME VARIABLE	GROUPING VARIABLE(S)/ PREDICTOR(S)	ANALYSIS	CHAPTER(S)
Continuous	—	Estimate μ; Compare μ to Known, Historical Value	5/12
Continuous	Dichotomous (2 groups)	Compare Independent Means (Estimate/Test $(\mu_1 - \mu_2)$) or the Mean Difference (μ_d)	6/12
Continuous	Discrete (>2 groups)	Test the Equality of k Means using Analysis of Variance $(\mu_1 = \mu_2 \cdots = \mu_k)$	9/12
Continuous	Continuous	Estimate Correlation or Determine Regression Equation	10/12
Continuous	Several Continuous or Dichotomous	Multiple Linear Regression Analysis	10
Dichotomous	—	Estimate p; Compare p to Known, Historical Value	7
Dichotomous	Dichotomous (2 groups)	Compare Independent Proportions (Estimate/Test $(p_1 - p_2)$)	7/8
Dichotomous	Discrete (>2 groups)	Test the Equality of k Proportions (Chi-Square Test)	7
Dichotomous	Several Continuous or Dichotomous	Multiple Logistic Regression Analysis	11
Discrete	Discrete	Compare Distributions Among k Populations (Chi-Square Test)	7
Time to Event	Several Continuous or Dichotomous	Survival Analysis	13

Correlation
and Regression

In correlation and regression applications, we consider the relationship between two continuous (or measurement) variables. For example, we might be interested in the relationship between age and systolic blood pressure, or the relationship between the number of hours of aerobic exercise per week and percent body fat. In correlation and regression applications, we measure both variables on each of n randomly selected subjects. We denote the variables X and Y. (In prior applications, we used the variable name X to denote the single characteristic of interest.) The variable name X is used to represent the *independent, or predictor, variable*, and the variable name Y is used to represent the *dependent, outcome or response, variable*. In many applications, it will be clear which is the independent or predictor variable and which is the dependent or outcome variable. In some applications, the investigator must specify explicitly which variable should be considered the independent and which should be considered the dependent variable.

In Section 10.1 we discuss correlation analysis. We present and illustrate the computation of the sample correlation coefficient and then present techniques for statistical inference concerning the correlation coefficient. In Section 10.2 we present and illustrate simple linear regression analysis. In Section 10.3 we provide an introduction to multiple regression analysis, and in Section 10.4 we introduce logistic regression analysis. In Section 10.5 we summarize key formulas, and in Section 10.6 we display the statistical computing programs used to generate correlation and regression analyses presented in this chapter. In Section 10.7 we use data from the Framingham Heart Study to illustrate the applications presented here.

10.1 Correlation Analysis

The goal of correlation analysis is to understand the nature and strength of the association between the two measurement variables, denoted X and Y. A first step in understanding the relationship between two variables is through a *scatter diagram*. A scatter diagram is a plot of the (X,Y) pairs recorded on each of the n subjects. The X (independent or predictor) variable is plotted on the horizontal axis and the Y (dependent or outcome) variable is plotted on the vertical axis. The scatter diagram shown in Figure 10.1 indicates a positive or direct linear relationship between X and Y.

Figure 10.2 displays an negative or inverse linear association between two variables X and Y.

The population correlation coefficient, ρ (rho), quantifies the nature and strength of the linear relationship between X and Y. The population correlation coefficient takes on values in the range of -1 to 1 ($-1 \leq \rho \leq +1$). The sign of the correlation coefficient indicates the nature of the relationship between X and Y (i.e., positive or direct, negative or inverse), and the magnitude of the correlation coefficient indicates the strength of the linear association

Figure 10.1 *Direct Linear Relationship Between X and Y*

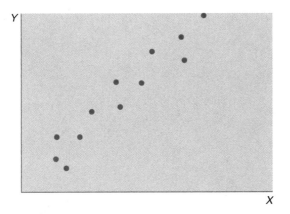

Figure 10.2 *Inverse Linear Relationship Between X and Y*

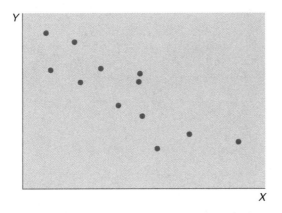

between the two variables. Figures 10.3a–d display scatter diagrams corresponding to the following values of the correlation coefficient: $\rho = -1$, $\rho = -0.5$, $\rho = 0.5$, and $\rho = 1$. These values represent, respectively, a perfect inverse relationship between X and Y ($\rho = -1.0$), a moderate inverse relationship between X and Y ($\rho = -0.5$), a moderate direct relationship between X and Y ($\rho = 0.5$), and a perfect direct relationship between X and Y ($\rho = 1.0$). A correlation of zero indicates that there is no linear association between X and Y.

The correlation coefficient is a measure of the *linear* association between X and Y. One must be cautious in interpreting the value of the correlation coefficient. It is always useful to generate a scatter diagram to assist in the interpretation. For example, the data displayed in both Figures 10.4a and 10.4b

Figure 10.3 *The Correlation Coefficient*

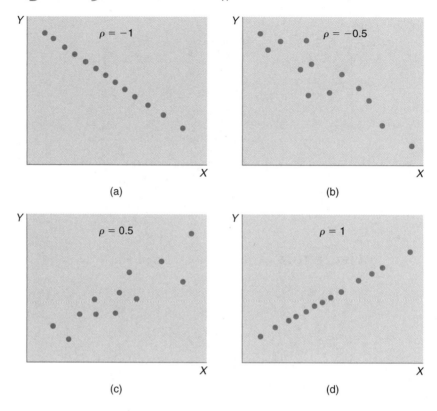

(a)   (b)

(c)   (d)

Figure 10.4 *Correlation Coefficient = 0*

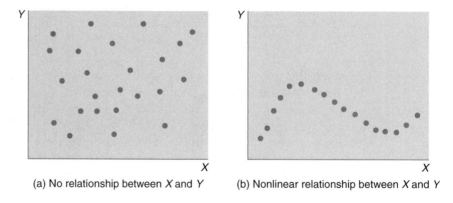

(a) No relationship between *X* and *Y*   (b) Nonlinear relationship between *X* and *Y*

produce correlation coefficients of zero. In Figure 10.4a, there is no association between *X* and *Y*. However, in Figure 10.4b there is an association
between *X* and *Y*, though not a linear one.

The correlation coefficient can be affected by truncation. For example, suppose we analyze the relationship between SAT scores measured during the senior year of high school and grade point averages (GPAs) measured at the completion of the freshman year in college. Because individuals who do poorly on the SAT are less likely to attend college, the correlations might be distorted due to the fact that these individuals are not included.

In other situations, a correlation between two variables, X and Y, may turn out to be zero due to a confounding variable (i.e., a variable that affects either X or Y or both). For example, suppose we investigate the relationship between the size of a home and its selling price. Selling prices might be different each year depending on the market and the economy, but if we pool data from several years, there may appear to be no linear association. In other applications, a correlation may be large due to a confounding variable. For example, say we investigate the association between age of first job and starting salary. Observed salaries may be higher for individuals who start working later. This might be due to the fact that these individuals completed a college education, thus delaying their entry into the workforce.

10.1.1 The Sample Correlation Coefficient r

As with other applications, we will have data measured on a sample of subjects. The sample correlation coefficient, denoted r, is computed as follows:

$$r = \frac{\text{Cov}(X, Y)}{\sqrt{\text{Var}(X)\,\text{Var}(Y)}} \tag{10.1}$$

where $\text{Cov}(X, Y)$ is the covariance of X and Y (defined next), and $\text{Var}(X)$ and $\text{Var}(Y)$ are the sample variances of X and Y, respectively. Recall $\text{Var}(X) = \sum(X - \overline{X})^2/(n - 1)$, and $\text{Var}(Y) = \sum(Y - \overline{Y})^2/(n - 1)$.

$$\text{Cov}(X, Y) = \frac{\sum(X - \overline{X})(Y - \overline{Y})}{n - 1} \tag{10.2}$$

We now use examples to illustrate the computation of the sample correlation coefficient.

EXAMPLE 10.1

Correlation Between Body Mass Index and Systolic Blood Pressure

Suppose we are interested in the relationship between body mass index (computed as the ratio of weight in kilograms to height in meters squared) and systolic blood pressure in males 50 years of age. A random sample of 10 males 50 years of age is selected and their weights, heights, and systolic blood pressures are measured. Their weights and heights are transformed into body mass index scores and are given in the following table. In this analysis, the independent (or predictor) variable is body mass index and the dependent (or outcome) variable is systolic blood pressure.

X = Body Mass Index	Y = Systolic Blood Pressure
18.4	120
20.1	110
22.4	120
25.9	135
26.5	140
28.9	115
30.1	150
32.9	165
33.0	160
34.7	180

The first step in assessing the relationship between body mass index and systolic blood pressure is through a scatter diagram (Figure 10.5). Notice that the independent variable (body mass index) is displayed on the horizontal axis and the dependent variable (systolic blood pressure) is displayed on the vertical axis. From the scatter diagram, it appears that there is a positive (direct) association between body mass index and systolic blood pressure. We now compute the sample correlation coefficient, r, to quantify the degree of linear association between body mass index and systolic blood pressure. In order to compute r, we must first compute the variances of X and Y as well as the covariance between X and Y.

Figure 10.5 *Scatter Diagram Relating Body Mass Index and Systolic Blood Pressure*

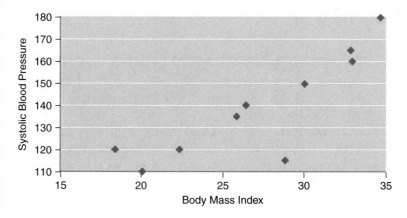

We first compute the variance of X: $\text{Var}(X) = \sum(X - \overline{X})^2/(n - 1)$. The following table summarizes the computations:

X = Body Mass Index	$(X - \overline{X})$	$(X - \overline{X})^2$
18.4	−8.89	79.032
20.1	−7.19	51.696
22.4	−4.89	23.912
25.9	−1.39	1.932
26.5	−0.79	0.624
28.9	1.61	2.592
30.1	2.81	7.896
32.9	5.61	31.472
33.0	5.71	32.604
34.7	7.41	54.908
$\sum X = 272.9, \overline{X} = 27.29$	0	286.669

So, $\text{Var}(X) = \sum(X - \overline{X})^2/(n - 1) = 286.669/9 = 31.852$.

We now compute the variance of Y: $\text{Var}(Y) = \sum(Y - \overline{Y})^2/(n - 1)$. The following table summarizes the computations:

Y = Systolic Blood Pressure	$(Y - \overline{Y})$	$(Y - \overline{Y})^2$
120	−19.5	380.25
110	−29.5	870.25
120	−19.5	380.25
135	−4.5	20.25
140	0.5	0.25
115	−24.5	600.25
150	10.5	110.25
165	25.5	650.25
160	20.5	420.25
180	40.5	1640.25
$\sum Y = 1395.0, \overline{Y} = 139.5$	0	5072.50

So, $\text{Var}(Y) = \sum(Y - \overline{Y})^2/(n - 1) = 5072.50/9 = 563.611$.

Finally, we compute the covariance of X and Y: $\text{Cov}(X, Y) = \sum(X - \bar{X})(Y - \bar{Y})/(n-1)$. The following table summarizes the computations:

$(X - \bar{X})$	$(Y - \bar{Y})$	$(X - \bar{X})(Y - \bar{Y})$
−8.89	−19.5	173.355
−7.19	−29.5	212.105
−4.89	−19.5	95.355
−1.39	−4.5	6.255
−0.79	0.5	−0.395
1.61	−24.5	−39.455
2.81	10.5	29.505
5.61	25.5	143.055
5.71	20.5	117.055
7.41	40.5	300.105
		1036.95

So, $\text{Cov}(X, Y) = \sum(X - \bar{X})(Y - \bar{Y})/(n-1) = 1036.95/9 = 115.22$. Substituting into (10.1), the sample correlation coefficient is

$$r = \frac{115.22}{\sqrt{(31.852)(563.611)}} = 0.859$$

Based on the sign and the magnitude of r, there is a strong, positive association between body mass index and systolic blood pressure. In fact, this correlation is artificially large (the data are hypothetical). In practice, correlations on the order of 0.3 or larger in absolute value are usually indicative of meaningful or important relationships. ■

10.1.2 Statistical Inference Concerning ρ

It is often of interest to draw inferences about the correlation between two variables in the population through a formal test of hypothesis. The sample correlation coefficient, r, is a point estimate for the population correlation coefficient, ρ. In general, tests of hypothesis concerning ρ address whether there is a linear association in the population ($\rho \neq 0$) or not ($\rho = 0$). The hypotheses are of the form:

$$H_0: \rho = 0 \quad \text{(no linear association)}$$
$$H_1: \rho \neq 0 \quad \text{(linear association)}$$

The test statistic follows a t distribution with $n - 2$ degrees of freedom:

$$t = r\sqrt{\frac{n-2}{1-r^2}}, \quad df = n - 2 \tag{10.3}$$

NOTE: The t statistic can be used regardless of the sample size (even when n is large), as the critical value of t (Table B.3) reflects the exact sample size. For example, when n is large, the two-sided critical value of t for $\alpha = 0.05$ is 1.96.

We now illustrate the test using the data from Example 10.1. The issue at hand is whether there is a significant correlation between systolic blood pressure and body mass index among all males 50 years of age. In a random sample of 10 males 50 years of age we observed a sample correlation coefficient of 0.859.

1. Set up hypotheses.

$$H_0: \rho = 0$$

$$H_1: \rho \neq 0, \quad \alpha = 0.05$$

2. Select the appropriate test statistic.

$$t = r\sqrt{\frac{n-2}{1-r^2}}$$

3. Decision rule (two-sided test, $\alpha = 0.05$).
   To determine the appropriate value from the t distribution table, we first compute the degrees of freedom. For Example 10.1, df $= n - 2 = 10 - 2 = 8$. The critical value is $t = 2.306$. The decision rule is

$$\text{Reject } H_0 \text{ if } t \geq 2.306 \text{ or if } t \leq -2.306$$

$$\text{Do not reject } H_0 \text{ if } -2.306 < t < 2.306$$

4. Test statistic.

$$t = r\sqrt{\frac{n-2}{1-r^2}} = 0.859\sqrt{\frac{10-2}{1-0.859^2}} = 4.75$$

5. Conclusion.
   Reject H_0 since $4.75 \geq 2.306$. We have significant evidence, $\alpha = 0.05$, to show that $\rho \neq 0$. For this example, $p < 0.01$ (see Table B.3). Therefore, there is evidence of a significant linear association between systolic blood pressure and body mass index among all males 50 years of age.

EXAMPLE 10.2

Correlation Between Number of Hours of Exercise and Systolic Blood Pressure

Consider the application described in Example 10.1. Suppose for the same sample, we also recorded the number of hours of vigorous exercise in a typical week. We wish to investigate the relationship between the number of hours of exercise per week and systolic blood pressure in males 50 years of age. In this analysis, the independent (or predictor) variable is number of hours of exercise per week and the dependent (or outcome) variable is systolic blood pressure. A scatter diagram is shown in Figure 10.6.

X = Number of Hours of Exercise per Week	Y = Systolic Blood Pressure
4	120
10	110
2	120
3	135
3	140
5	115
1	150
2	165
2	160
0	180

In order to compute r, we must first compute the variances of X and Y as well as the covariance between X and Y. We first compute the variance

Figure 10.6 *Scatter Diagram Relating Hours of Exercise and Systolic Blood Pressure*

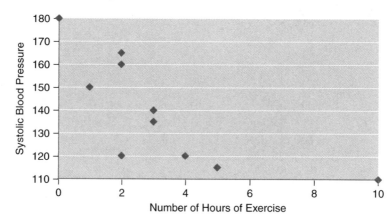

of X: $\text{Var}(X) = \sum (X - \overline{X})^2/(n-1)$. The following table summarizes the computations:

$X = $ Number of Hours of Exercise per Week	$(X - \overline{X})$	$(X - \overline{X})^2$
4	0.8	0.64
10	6.8	46.24
2	−1.2	1.44
3	−0.2	0.04
3	−0.2	0.04
5	1.8	3.24
1	−2.2	4.84
2	−1.2	1.44
2	−1.2	1.44
0	−3.2	10.24
$\sum X = 32, \overline{X} = 3.2$	0	69.60

So, $\text{Var}(X) = \sum (X - \overline{X})^2/(n-1) = 69.60/9 = 7.73$.

We computed the variance of Y in Example 10.1, $\text{Var}(Y) = \sum (Y - \overline{Y})^2/(n-1) = 5072.50/9 = 563.611$. We now need to compute the covariance of X and Y: $\text{Cov}(X, Y) = \sum (X - \overline{X})(Y - \overline{Y})/(n-1)$. The following table summarizes the computations:

$(X - \overline{X})$	$(Y - \overline{Y})$	$(X - \overline{X})(Y - \overline{Y})$
0.8	−19.5	−15.6
6.8	−29.5	−200.6
−1.2	−19.5	23.4
−0.2	−4.5	0.9
−0.2	0.5	−0.1
1.8	−24.5	−44.1
−2.2	10.5	−23.1
−1.2	25.5	−30.6
−1.2	20.5	−24.6
−3.2	40.5	−129.6
		−444.0

So, $\text{Cov}(X, Y) = \sum (X - \overline{X})(Y - \overline{Y})/(n-1) = -444.0/9 = -49.33$.

Substituting into (10.1), the sample correlation coefficient is

$$r = \frac{-49.33}{\sqrt{(7.73)(563.611)}} = -0.75$$

Based on the magnitude and sign of r, there is a strong, negative association between the number of hours of exercise per week and systolic blood pressure (i.e., more hours of exercise per week are associated with lower systolic blood pressures).

Using the same data (Example 10.2), we now test whether there is a significant correlation between systolic blood pressure and the number of hours of exercise per week among all males 50 years of age. In a random sample of 10 males 50 years of age we observed a sample correlation coefficient of -0.75.

1. Set up hypotheses.

$$H_0: \rho = 0$$

$$H_1: \rho \neq 0, \quad \alpha = 0.05$$

2. Select the appropriate test statistic.

$$t = r\sqrt{\frac{n-2}{1-r^2}}$$

3. Decision rule (two-sided test, $\alpha = 0.05$).

To determine the appropriate value from the t distribution table, we first compute the degrees of freedom. For Example 10.1, $df = n - 2 = 10 - 2 = 8$. The critical value is $t = 2.306$. The decision rule is

Reject H_0 if $t \geq 2.306$ or if $t \leq -2.306$

Do not reject H_0 if $-2.306 < t < 2.306$

4. Test statistic.

$$t = r\sqrt{\frac{n-2}{1-r^2}} = -0.75\sqrt{\frac{10-2}{1-(-0.75)^2}} = -3.18$$

5. Conclusion.

Reject H_0 since $-3.18 \leq -2.306$. We have significant evidence, $\alpha = 0.05$, that $\rho \neq 0$. For this example, $p < 0.02$ (see Table B.3). Therefore, there is evidence of a significant linear association between systolic blood pressure and the number of hours of exercise per week among all males 50 years of age. ■

10.2 Simple Linear Regression

In regression analysis, we develop the mathematical equation that best describes the relationship between two variables, X and Y. In correlation analysis, it is actually not necessary to specify which of the two variables is the independent and which is the dependent variable. In contrast, with regression analysis the independent and dependent variables must be specified. In regression analysis we address the following issues:

1. What mathematical equation best describes the relationship between X and Y (e.g., a line or a curve of some form)?

2. How do we estimate the equation that describes the relationship between X and Y?

3. Is the form specified in (1) appropriate?

We begin with the simplest situation, the one in which the relationship between X and Y is linear. The equation of the line relating Y to X is called the *simple linear regression equation* and is given here:

$$Y = \beta_0 + \beta_1 X + \varepsilon \tag{10.4}$$

where Y is the dependent variable
X is the independent variable
β_0 is the Y-intercept (i.e., the value of Y when $X = 0$)
β_1 is the slope (i.e., the expected change in Y relative to one unit change in X)
ε is the random error

The estimates of β_0 and β_1, denoted $\hat{\beta}_0$ and $\hat{\beta}_1$, respectively, are determined using the following equations:

$$\hat{\beta}_1 = r\sqrt{\frac{\text{Var}(Y)}{\text{Var}(X)}} \tag{10.5}$$

$$\hat{\beta}_0 = \overline{Y} - \hat{\beta}_1 \overline{X}$$

These estimates (10.5) are called the *least squares estimates* of the slope and intercept. The formulas shown are those that minimize the squared errors (i.e., minimize $\sum \varepsilon^2$); the derivations of the formulas will not be shown.

The estimate of the simple linear regression equation is given by substituting the least squares estimates (10.5) into equation (10.4):

$$\hat{Y} = \hat{\beta}_0 + \hat{\beta}_1 X \tag{10.6}$$

where $\hat{Y}$ is the expected value of Y for a given value of X

EXAMPLE 10.1   **Regression Analysis of Body Mass Index on Systolic Blood Pressure (continued)**

We now estimate the simple linear regression equation for the data given in Example 10.1. The dependent variable (Y) is systolic blood pressure, the independent variable, X, is body mass index, and the correlation was estimated at 0.859. We first estimate the slope by substituting the appropriate statistics into (10.5):

$$\hat{\beta}_1 = r\sqrt{\frac{\text{Var}(Y)}{\text{Var}(X)}} = 0.859\sqrt{\frac{563.611}{31.852}} = 3.61$$

Now, substituting again, we compute the Y-intercept:

$$\hat{\beta}_0 = \overline{Y} - \hat{\beta}_1\overline{X} = 139.5 - (3.61)(27.29) = 40.98$$

The simple linear regression equation for Example 10.1 is

$$\hat{Y} = 40.98 + 3.61X$$

The Y-intercept is 40.98, which is the expected systolic blood pressure (Y) when body mass index (X) is zero. In this example, the Y-intercept is not meaningful since it is not possible to observe a body mass index equal to zero. However, the estimated slope, $\hat{\beta}_1 = 3.61$, indicates that a 1-unit increase in body mass index is associated with an expected increase of 3.61 units in systolic blood pressure. If we compare two males 50 years of age, the first with a body mass index 10 units higher than the second, we would expect the first subject's systolic blood pressure to be $10(3.61) = 36.1$ units higher than the second subject's.

We can also can use the equation to generate an estimate of systolic blood pressure (Y) for a person with a specific body mass index (X). For example, suppose we wish to estimate the systolic blood pressure of a male aged 50 whose body mass index is 20. Using the simple linear regression equation:

$$\hat{Y} = 40.98 + 3.61X = 40.98 + 3.61(20) = 113.81$$

This can be interpreted in two ways. It is the expected systolic blood pressure of *a* male age 50 whose body mass index is 20. It is also the expected systolic blood pressure for *all* males age 50 with body mass index of 20.   ■

SAS EXAMPLE 10.1   **Correlation and Regression Analysis of Body Mass Index on Systolic Blood Pressure Using SAS**

The following output was generated using SAS Proc Plot, SAS Proc Corr, and SAS Proc Reg, which generate, respectively, a scatter diagram and perform correlation and regression analyses. A brief interpretation appears after the output.

SAS Output for Example 10.1

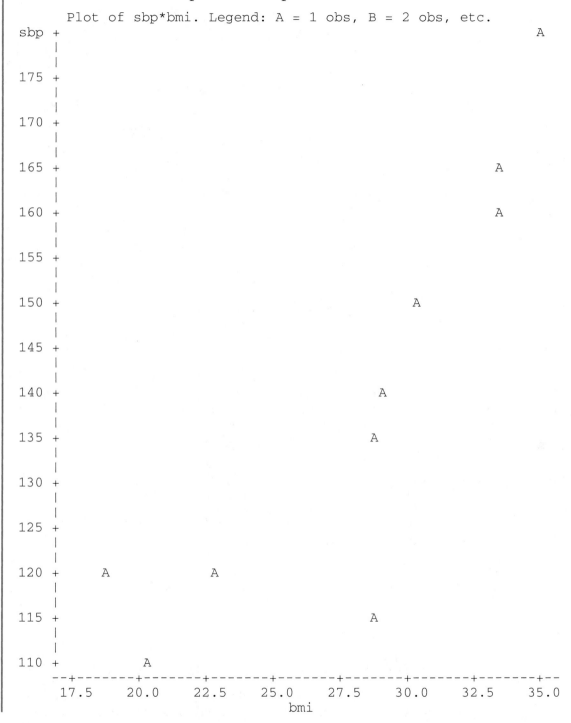

Plot of sbp*bmi. Legend: A = 1 obs, B = 2 obs, etc.

```
                         The CORR Procedure
                  2  Variables:  bmi  sbp

                     Covariance Matrix, DF = 9
                             bmi                sbp
            bmi       31.8521111         115.2166667
            sbp      115.2166667         563.6111111

                       Simple Statistics
Variable   N      Mean    Std Dev        Sum     Minimum     Maximum
bmi       10   27.29000    5.64377   272.90000   18.40000    34.70000
sbp       10  139.50000   23.74050       1395  110.00000   180.00000

            Pearson Correlation Coefficients, N = 10
                 Prob > |r| under H0: Rho = 0

                         bmi            sbp
            bmi      1.00000        0.85992
                                    0.0014
            sbp      0.85992        1.00000
                     0.0014
```

```
                       The REG Procedure
                       Model: MODEL1
                  Dependent Variable: sbp

                      Analysis of Variance
                          Sum of           Mean
Source            DF     Squares         Square    F Value    Pr > F
Model              1  3750.89494     3750.89494      22.71    0.0014
Error              8  1321.60506      165.20063
Corrected Total    9  5072.50000

            Root MSE            12.85304   R-Square   0.7395
            Dependent Mean     139.50000   Adj R-Sq   0.7069
            Coeff Var            9.21365

                      Parameter Estimates
                    Parameter        Standard
Variable    DF       Estimate           Error    t Value    Pr > |t|
Intercept    1       40.78558        21.11158       1.93      0.0895
bmi          1        3.61724         0.75913       4.76      0.0014
```

Interpretation of SAS Output for Example 10.1

SAS can produce high-level graphical displays; however, these require some programming to implement. The Plot procedure is easy to use and generates a scatter diagram for two continuous variables. SAS displays points in the scatter diagram with letters of the alphabet. The letter "A" appears when there is a single point in a particular location, "B" indicates that two points fall in the same location, and so on. Although the scatter diagram is somewhat crude, it clearly conveys the nature of the relationship between the two variables under investigation. In Example 10.1 there is a strong positive association between BMI and SBP.

The SAS Corr procedure generates a correlation analysis. In this example, we requested that SAS produce a Covariance Matrix, which appears at the top of the correlation output. The Covariance Matrix contains covariances between the variables in the rows and columns of the matrix. The covariance between a variable and itself is identical to the variance of that variable. For example, the Cov(SBP,SBP) = Var(SBP) = 563.61, the Cov(SBP,BMI) = Cov(BMI,SBP) = 115.22, and Cov(BMI,BMI) = Var(BMI) = 31.85. By default, the Corr procedure also generates summary statistics for each variable. The last section of the output contains the Correlation Matrix. The Correlation Matrix contains sample correlations (r) between variables in the rows and columns of the matrix. Underneath each sample correlation is a two-sided p value for testing H_0: $\rho = 0$ against H_1: $\rho \neq 0$. The correlation between a variable and itself is always 1.00 (and no test is performed). Of interest are the correlations between distinct variables. Specifically, Corr(SBP, BMI) = Corr(BMI,SBP) = 0.85992, and the correlation between SBP and BMI is significant at $p = 0.0014 < 0.05$.

The SAS Reg procedure generates an ANOVA table used to test whether the regression is significant or not, and then provides estimates of the parameters of the regression equation (e.g., estimates of the intercept and slope). The ANOVA table is set up exactly like the ANOVA tables we used in analysis of variance applications to test the equality of k treatment means. In regression analysis, the total sum of squares represents the variation in the dependent variable Y. The total variation is partitioned into two components, labeled Model and Error. The Model sum of squares is also referred to as the regression sums of squares and reflects the variation in Y accounted for by the regression equation. The Error sum of squares is also referred to as the residual sum of squares and reflects the variation in Y not accounted for by the regression. The p value (0.0014) is used to test whether the regression is significant. When there is a single independent variable (i.e., simple linear regression), this p value is identical to the p value in the correlation analysis and also identical to the p value used to test if the slope is significant. The next part of the regression output includes estimates of the regression parameters. The estimate of the intercept is 40.79 and the estimate of the slope is 3.62. SAS produces two-sided p values to test if the intercept and slope are significantly different from zero. It is usually not of interest to test if the intercept is significantly different from zero, but it is of interest to test if the slope is different from zero. Here, $p = 0.0014$, indicating that the slope (reflecting the change in SBP associated with a 1-unit change in BMI) is significant. ■

If a regression is run and it has been established that the regression is significant, we often want to quantify how much variation in the dependent variable is "explained" by the independent variable (or set of independent variables, in the case of multiple regression analysis, described in Section 10.3).

The *coefficient of determination,* denoted R^2, "R-squared," is the ratio of the regression (or Model) sums of squares to the total sums of squares (from the ANOVA table).

$$R^2 = \frac{SS_{model}}{SS_{total}} = \frac{SS_{regression}}{SS_{total}} \tag{10.7}$$

where $0 \le R^2 \le 1$

Higher values of R^2 imply that more variation in the dependent variable is "explained" by the independent variable(s).

In Example 10.1:

$$R^2 = \frac{3750.89}{5072.50} = 0.7395$$

Thus, 73.9% of the variation in systolic blood pressures is explained by body mass index. Because this example is artificial, the value of R^2 is very high. In practice, values of $R^2 > 0.1$ can be clinically significant.

EXAMPLE 10.2   Regression Analysis of Number of Hours of Exercise on Systolic Blood Pressure (continued)

We now estimate the simple linear regression equation for the data given in Example 10.2. The dependent variable (Y) is systolic blood pressure, the independent variable (X) is the number of hours of exercise per week, and the correlation was estimated at -0.75.

We first estimate the slope by substituting the appropriate statistics into (10.5):

$$\hat{\beta}_1 = r\sqrt{\frac{\text{Var}(Y)}{\text{Var}(X)}} = -0.75\sqrt{\frac{563.611}{7.73}} = -6.38$$

Now, substituting, we compute the Y-intercept:

$$\hat{\beta}_0 = \overline{Y} - \hat{\beta}_1\overline{X} = 139.5 - (-6.38)(3.2) = 159.9$$

The simple linear regression equation for Example 10.2 is

$$\hat{Y} = 159.9 - 6.38X$$

The Y-intercept is 159.9, which is the expected systolic blood pressure (Y) when the number of hours of exercise (X) is zero. In this example, the Y-intercept is meaningful because it is possible to observe $X = 0$ (i.e., no exercise). The estimated slope, $\hat{\beta}_1 = -6.38$, indicates that each additional hour of exercise per week (a 1-unit increase) is associated with a decrease of 6.38 units in systolic blood pressure. If we compare two males 50 years of age, the first exercising 5 hours in a typical week and the second exercising 4 hours in a typical week, we would expect the first subject's systolic blood pressure to be 6.38 units lower than the second subject's.

We can also use the equation to generate estimates of values of systolic blood pressure (Y) for a person who exercises a specific number of hours per week (X). For example, suppose we wish to estimate the systolic blood pressure of a male aged 50 who exercises 3 hours per week. Using the simple linear regression equation:

$$\hat{Y} = 159.9 - 6.38(3) = 159.9 - 19.14 = 140.76 \qquad ■$$

SAS EXAMPLE 10.2 Correlation and Regression Analysis of Number of Hours of Exercise on Systolic Blood Pressure Using SAS

The following output was generated using SAS Proc Corr and SAS Proc Reg, which perform, respectively, correlation and regression analyses. A brief interpretation appears after the output.

SAS Output for Example 10.2

```
                    The CORR Procedure
                 2 Variables: exercise sbp

                      Simple Statistics
Variable   N      Mean     Std Dev       Sum    Minimum     Maximum
exercise  10    3.20000    2.78089  32.00000          0    10.00000
sbp       10  139.50000   23.74050      1395  110.00000   180.00000

          Pearson Correlation Coefficients, N = 10
             Prob > |r| under H0: Rho=0

                        exercise          sbp
            exercise     1.00000     -0.74725
                                      0.0130

            sbp         -0.74725      1.00000
                         0.0130

                 The REG Procedure
                   Model: MODEL1
              Dependent Variable: sbp

                 Analysis of Variance
                       Sum of          Mean
Source           DF    Squares        Square    F Value    Pr > F
Model             1  2832.41379    2832.41379      10.12    0.0130
Error             8  2240.08621     280.01078
Corrected Total   9  5072.50000
```

```
Root MSE              16.73352  R-Square  0.5584
Dependent Mean       139.50000  Adj R-Sq  0.5032
Coeff Var             11.99536
```

```
                      Parameter Estimates
                  Parameter      Standard
Variable    DF     Estimate         Error    t Value    Pr > |t|
Intercept    1    159.91379       8.31854      19.22     <.0001
exercise     1     -6.37931       2.00578      -3.18     0.0130
```

Interpretation of SAS Output for Example 10.2

The SAS Corr procedure generates a correlation analysis. In this example, we did not request that SAS produce a Covariance Matrix (as we had requested in Example 10.1). The first part of the output contains summary statistics for each variable. The next section of the output includes the Correlation Matrix. The Correlation Matrix contains sample correlations (r) between variables showing in the rows and columns of the matrix. Underneath each sample correlation is a two-sided p value for testing H_0: $\rho = 0$ against H_1: $\rho \neq 0$. The correlation between a variable and itself is always 1.00. Of interest are the correlations between distinct variables. Specifically, Corr(SBP,EXERCISE) = Corr(EXERCISE,SBP) = -0.74725, and this correlation between SBP and EXERCISE is significant at $p = 0.0130 < 0.05$.

The SAS Reg procedure generates an ANOVA table used to test whether the regression is significant or not and then provides estimates of the parameters of the regression equation (e.g., estimates of the intercept and slope). In regression analysis, the total sums of squares represents the variation in the dependent variable Y. The total variation is partitioned into two components, labeled Model and Error. The Model sums of squares is also referred to as the regression sums of squares and reflects the variation in Y accounted for by the regression equation. The Error sums of squares is also referred to as the residual sums of squares and reflects the variation in Y that is not accounted for by the regression. The p value (0.0130) is used to test whether the regression is significant. When there is a single independent variable (i.e., simple linear regression), this p value is identical to the p value in the correlation analysis and also identical to the p value used to test if the slope is significant. The next part of the regression output includes estimates of the regression parameters. The estimate of the intercept is 159.91 and the estimate of the slope is -6.38. SAS produces two-sided p values to test if the intercept and slope are significantly different from zero. It is usually not of interest to test if the intercept is significantly different from zero, but it is of interest to test if the slope is different from zero. Here, $p = 0.0130$, indicating that the slope (reflecting the change in SBP associated with a 1-unit change in the number of hours of exercise per week) is significant. In this example, $R^2 = 0.56$, suggesting that 56% of the variation in systolic blood pressures is explained by the number of hours of exercise per week. ■

10.3 Multiple Regression Analysis

In Examples 10.1 and 10.2, we found that body mass index and the number of hours of exercise per week were significantly associated with systolic blood pressure (when considered separately). There are many characteristics that might be related to systolic blood pressure, such as age, gender, diet, family history of hypertension, race, smoking status, and so on. In many applications, we wish to assess the extent to which a set of candidate variables are related to a particular dependent variable and investigate their relative importance.

In multiple regression analysis, we consider applications involving a single continuous dependent variable, Y, and multiple independent variables, denoted X_1, X_2, and so on. The form of the multiple linear regression equation is

$$Y = \beta_0 + \beta_1 X_1 + \beta_2 X_2 + \cdots + \beta_p X_p + \varepsilon \tag{10.8}$$

where Y is the dependent variable
  X_1 to X_p are the independent variables
  β_0 is the intercept (i.e., the value of Y when $X_1 = X_2 = \cdots = X_p = 0$)
  β_i $(i = 1, \ldots, p)$ are the slope coefficients, also called the
    regression parameters (i.e., the expected change in Y relative to
    a 1-unit change in X_i)
  ε is the random error

NOTE: The dependent variable, Y, is a continuous variable. The independent variables, X_i, can be continuous variables or dichotomous (sometimes called *indicator*) variables. In the following example we consider both continuous independent variables (e.g., age) and indicator variables (e.g., smoking status, where 1 = smoker, 0 = nonsmoker).

The formulas used to estimate the intercept and slope coefficients are computationally complex. We therefore restrict our attention to interpreting a multiple regression analysis performed in SAS. Interested readers can refer to *Applied Regression Analysis and Other Multivariable Methods*, 2nd ed. by Kleinbaum, D. G., Kupper, L. L., and Muller, K. E. (1988), PWS-Kent Publishing Company, Boston, MA, for a more complete discussion of multiple regression analysis. Here we present only a single example as a means of illustrating the conceptual framework of the application in a basic sense. Consider the following example.

SAS EXAMPLE 10.3 **Multiple Linear Regression Analysis for Systolic Blood Pressure**
We used SAS to analyze a large data set ($n = 5078$). The dependent variable is systolic blood pressure (SBP) and we consider three independent variables: AGE (a continuous variable, measured in years), MALE (an indicator variable, coded 1 for males and 0 for females), and SMOKER (an indicator

variable, coded 1 for smokers and 0 for nonsmokers). In the following we estimated a multiple regression equation using SAS. A brief interpretation appears after the output.

SAS Output for Example 10.3

The REG Procedure
Model: MODEL1
Dependent Variable: SBP

Analysis of Variance

Source	DF	Sum of Squares	Mean Square	F Value	Pr > F
Model	3	446821	148940	493.03	<.0001
Error	5074	1532804	302.08991		
Corrected Total	5077	1979625			

Root MSE	17.38073	R-Square	0.2257	
Dependent Mean	129.60339	Adj R-Sq	0.2253	
Coeff Var	13.41071			

Parameter Estimates

Variable	DF	Parameter Estimate	Standard Error	t Value	DF	Pr > |t|
Intercept	1	90.57147	1.12386	80.59	1	<.0001
age	1	0.77398	0.02094	36.96	1	<.0001
male	1	4.39800	0.49022	8.97	1	<.0001
smoker	1	-1.84484	0.50366	-3.66	1	0.0003

Interpretation of SAS Output for Example 10.3

The SAS Reg procedure generates an ANOVA table used to test whether the regression is significant or not and then provides estimates of the regression parameters (e.g., estimates of the intercept and regression slopes). The ANOVA table is set up exactly like the ANOVA tables we used in analysis of variance applications to test the equality of k treatment means. In multiple regression analysis, the total sums of squares represents the variation in the dependent variable Y. The total variation is partitioned into two components, labeled Model and Error. The Model sums of squares, also referred to as the regression sums of squares, reflects the variation in Y accounted for by the regression equation. In multiple regression analysis, the ANOVA table is used to perform a global test—in particular, to test if the collection of variables is significant. The p value (<0.0001) is used to test whether the set of independent variables considered is significant—here, whether AGE, MALE, and SMOKER, considered simultaneously, are significant in explaining variation in SBP.

The next part of the regression output includes estimates of the regression parameters. The estimate of the intercept is 90.57 and is the expected systolic blood pressure if all of the independent variables are zero, that is, if AGE = 0 (which is unreasonable), MALE = 0 (which is the code for female), and SMOKER = 0 (which is the code for a nonsmoker). The estimate of the regression parameter associated with AGE is 0.77. A 1-year increase in AGE is associated with a 0.77-unit increase in SBP, holding the other variables in the equation constant. The estimate of the regression parameter associated with MALE is 4.40. Because MALE is an indicator variable, the effect is interpreted as follows. On average, males (MALE = 1) have SBPs 4.40 units higher than females (MALE = 0), holding the other variables constant. Smoking status is interpreted in a similar fashion. The estimate of the regression parameter associated with SMOKER is −1.84. On average, smokers (SMOKER = 1) have SBPs 1.84 units lower than nonsmokers (SMOKER = 0), holding the other variables constant. We should be cautious interpreting this effect of smoking, as smoking status may be related to other behaviors and smoking, per se, may not be associated with lower systolic blood pressure. Each of these parameter estimates is adjusted for the other independent variables in the model. They reflect the impact of each independent variable on SBP after considering the other variables.

SAS produces two-sided p values to test if the intercept and regression coefficients are significantly different from zero. Again, it is usually not of interest to test if the intercept is significantly different from zero, but it is of interest to test if the other coefficients are different from zero. Here, $p < 0.001$ for each independent variable, indicating that each of the variables is highly significant. In this example, the sample size is very large, which accounts, in part, for the highly significant results.

It is often of interest to determine the relative importance of variables in a multiple regression analysis. This is determined by the magnitude of the t statistics (or associated p values). One cannot determine relative importance based on the magnitude of the parameter estimates, as the estimates are affected by the scales on which variables are measured. In this example, age is the most important (most significant) independent variable, followed by gender and then smoking status. These three variables account for, or explain, 22.6% of the variation in the dependent variable SBP. ▪

10.4 Logistic Regression Analysis

In logistic regression analysis, we consider applications involving a single dependent variable, Y, which is dichotomous (i.e., Success versus Failure). The technique is discussed in detail in Chapter 11; here we present a brief overview for comparison to linear regression analysis. Suppose in Example 10.3 the dependent variable was not systolic blood pressure (a continuous variable), but instead diagnosis of hypertension (hypertensive or not, a dichotomous variable). Logistic regression applications can involve one or several independent variables, denoted X_1, X_2, and so on. The form of the logistic regression equation is

$$\ln\left\{\frac{p}{(1-p)}\right\} = \beta_0 + \beta_1 X_1 + \beta_2 X_2 + \cdots + \beta_p X_p + \varepsilon \qquad (10.9)$$

where Y is the dichotomous dependent variable (success or failure)
$\quad p$ is the proportion of successes
$\quad X_1$ to X_p are the independent variables
$\quad \beta_0$ is the intercept
$\quad \beta_i$ $(i = 1, \ldots, p)$ are the slope coefficients, also called the regression parameters
$\quad \varepsilon$ is the random error

NOTES: The left-hand side of the equation displays the dependent or outcome variable in a specific form. The quantity $p/(1-p)$ is the "odds" of possessing the event. The expression $\ln\{p/(1-p)\}$ is called the "log odds" or the "logit" of Y. The slope coefficients, β_i $(i = 1, \ldots, p)$, reflect the change in the logit or log odds of Y relative to a 1-unit change in X_i.

The dependent variable, Y, is a dichotomous variable. The independent variables, X_i, can be continuous variables or dichotomous (sometimes called *indicator*) variables. In Example 10.4 we consider both continuous independent variables (e.g., age) and indicator variables (e.g., smoking status, where $1 = $ smoker, $0 = $ nonsmoker).

The formulas to estimate the intercept and slope coefficients are computationally complex. We again restrict our attention to interpreting a logistic regression analysis performed in SAS. Consider the following example.

SAS Example 10.4 Multiple Logistic Regression Analysis for Hypertension

We used SAS to analyze a large data set ($n = 5078$). The dependent variable is a dichotomous variable (HTN, coded 1 for patients classified as hypertensive and 0 otherwise) and we consider three independent variables: AGE (a continuous variable, measured in years), MALE (an indicator variable, coded 1 for males and 0 for females), and SMOKER (an indicator variable, coded 1 for smokers and 0 for nonsmokers). In the following we estimated a logistic regression equation using SAS. A brief interpretation appears after the output.

SAS Output for Example 10.4

```
                    The LOGISTIC Procedure
                      Response Profile
        Ordered                          Total
        Value          HTN            Frequency
          1             1                 349
          2             0                4729
```

```
                  Probability modeled is HTN=1.
                    Model Convergence Status
             Convergence criterion (GCONV=1E-8) satisfied.
                      Model Fit Statistics
                         Intercept       Intercept and
           Criterion       Only            Covariates
           AIC            2544.410          2246.882
           SC             2550.943          2273.013
           -2 Log L       2542.410          2238.882

              Testing Global Null Hypothesis: BETA=0
           Test              Chi-Square   DF    Pr > ChiSq
           Likelihood Ratio    303.5282    3      <.0001
           Score               291.3059    3      <.0001
           Wald                259.4685    3      <.0001
                      The LOGISTIC Procedure

            Analysis of Maximum Likelihood Estimates
                               Standard        Wald
Parameter    DF    Estimate     Error     Chi-Square   Pr > ChiSq
Intercept    1     -7.1813      0.3371      453.9220      <.0001
age          1      0.0654      0.00534     149.6551      <.0001
male         1      1.2628      0.1254      101.4887      <.0001
smoker       1      0.9520      0.1187       64.3556      <.0001

                     Odds Ratio Estimates
                        Point          95% Wald
            Effect    Estimate    Confidence Limits
             age        1.068      1.056      1.079
             male       3.535      2.765      4.520
             smoker     2.591      2.053      3.269

    Association of Predicted Probabilities and Observed Responses
         Percent Concordant     77.0    Somers' D    0.550
         Percent Discordant     22.1    Gamma        0.555
         Percent Tied            0.9    Tau-a        0.070
         Pairs              1650421     c            0.775
```

Interpretation of SAS Output for Example 10.4

The SAS Logistic procedure first displays the numbers of subjects classified in each of the two response categories. In this example, 349 subjects are hypertensive and 4729 are not. SAS then indicates that the model is set up to predict the probability that a person is hypertensive (HTN = 1). The next sections of the output are entitled "Model Fit Statistics" and "Testing Global Null Hypothesis: BETA = 0." The statistics are used for testing whether the collection of variables considered are significant and for comparing models. In this application, we considered three independent variables. The statistic labeled "−2 log L" for model containing intercept only (2542.410) represents a measure of the overall variation in the dependent variable. This quantity is similar to the total sum of squares in a multiple regression analysis. The Chi-Square for Covariates (303.528) is computed by taking the difference between −2 log L for model containing intercept only and −2 log L for model with intercept and covariates. This statistic is used to perform a global test of significance of the set of independent variables considered. In Example 10.4, $p < 0.0001$, so AGE, MALE, and SMOKER, considered simultaneously, are significant in explaining variation in hypertensive status.

The next part of the regression output includes estimates of the regression parameters. The estimate of the intercept is −7.1813. The estimate of the regression parameter associated with AGE is 0.0654. A 1-year increase in AGE is associated with a 0.0654-unit increase in the logit or log odds of Y. SAS also produces odds ratios for each independent variable (computed by $\exp(\beta_i)$). The odds ratio for AGE is estimated as 1.068. The odds of having hypertension are 1.068 times higher with every additional year of age, holding the other variables constant. The regression parameter associated with MALE is 1.2628. Because MALE is an indicator variable, the effect is interpreted as follows. On average, the logit or log odds of Y is 1.2628 times higher for males (MALE = 1) as compared to females (MALE = 0). Smoking status is interpreted in a similar fashion. The estimate of the regression parameter associated with SMOKER is 0.9520. On average, smokers (SMOKER = 1) are more likely to be hypertensive than nonsmokers. The odds ratio for MALE is 3.535, so males are 3.535 times more likely to have hypertension than females (odds of being hypertensive are 3.535 times higher for males as compared to females), holding the other variables constant. The odds ratio for SMOKER is 2.591, so smokers are more likely to have hypertension than nonsmokers (odds of being hypertensive are 2.591 times higher for smokers as compared to nonsmokers), holding the other variables constant. Each of these parameter estimates and estimates of odds ratios is adjusted for the other independent variables in the model. They reflect the impact of each independent variable on hypertensive status after considering the other variables in the model.

SAS produces two-sided p values to test if the intercept and regression coefficients are significantly different from zero. Again, it is usually not of interest to test if the intercept is significantly different from zero, but it is of interest to test if the other coefficients are different from zero. Here, $p < 0.0001$ for each independent variable, indicating that each of the variables is highly significant. In this example, the sample size is very large, which accounts, in part, for the highly significant results.

It is often of interest to determine the relative importance of variables in a multiple logistic regression analysis. This is determined by the magnitude of the Wald chi-square statistics (or associated p values). One cannot determine relative importance based on the magnitude of the parameter estimates, as the estimates are affected by the scales on which variables are measured. In this example, age is the most important (most significant) independent variable, followed by gender and then smoking status. It is not possible to compute R^2 (10.7) for a logistic regression to describe how much variation in the dependent variable is explained by the independent variables. In logistic regression, we use the c statistic for a similar purpose. The c statistic in Example 10.4 is 0.775 and represents the extent to which the actual values of the dependent variable and the predicted values (generated by the estimated model) agree. Values exceeding 0.7 are generally considered adequate.

A more detailed discussion of logistic regression analysis is contained in Chapter 11. ■

10.5 Key Formulas

Application	Notation/Formula	Description
Estimate sample correlation coefficient	$r = \dfrac{\text{Cov}(X, Y)}{\sqrt{\text{Var}(X)\text{Var}(Y)}}$	quantifies the nature and extent of linear association between X and Y
Test H_0: $\rho = 0$	$t = r\sqrt{\dfrac{n-2}{1-r^2}}, \quad \text{df} = n - 2$	test for significant correlation between X and Y
Simple linear regression equation	$Y = \beta_0 + \beta_1 X + \varepsilon$	see (10.5) for least squares estimates of regression parameters
Estimate R^2	$R^2 = \dfrac{SS_{\text{model}}}{SS_{\text{total}}} = \dfrac{SS_{\text{regression}}}{SS_{\text{total}}}$	proportion of variation in dependent variable explained by independent variable(s)
Multiple linear regression equation	$Y = \beta_0 + \beta_1 X_1 + \beta_2 X_2 + \cdots + \beta_p X_p + \varepsilon$	continuous dependent variable Y, continuous or dichotomous independent variables X_i
Multiple logistic regression equation	$\ln\left\{\dfrac{p}{(1-p)}\right\}$ $= \beta_0 + \beta_1 X_1 + \beta_2 X_2 + \cdots + \beta_p X_p + \varepsilon$	dichotomous dependent variable Y, p is the proportion of successes, continuous or dichotomous independent variables X_i (see Chapter 11)

10.6 Statistical Computing

Following are the SAS programs used to generate a scatter diagram and to estimate correlations between variables, a simple linear regression equation, a multiple regression equation, and a logistic regression equation. The SAS procedures and brief descriptions of their use are noted in the header to each example. Notes are provided to the right of the SAS programs (in blue) for orientation purposes and are not part of the programs. In addition, the blank lines in the programs are solely to accommodate the notes. Blank lines and spaces can be used throughout SAS programs to enhance readability. A summary of the SAS procedures used in the examples is provided at the end of this section.

SAS EXAMPLE 10.1 Scatter Diagram, Correlation, and Simple Linear Regression Analysis
Assess the Relationship Between Body Mass Index and Systolic Blood Pressure (Example 10.1)

Suppose we are interested in the relationship between body mass index (computed as the ratio of weight in kilograms to height in meters squared) and systolic blood pressure in males 50 years of age. A random sample of 10 males 50 years of age is selected and their weights, heights, and systolic blood pressures are measured. Their weights and heights are transformed into body mass index scores and are given in the following table. In this analysis, the independent (or predictor) variable is body mass index, and the dependent (or outcome) variable is systolic blood pressure. Generate a scatter diagram to assess the relationship between body mass index and systolic blood pressure using SAS. In addition, estimate the correlation between body mass index and systolic blood pressure and the regression equation relating body mass index to systolic blood pressure using SAS.

X = Body Mass Index	Y = Systolic Blood Pressure
18.4	120
20.1	110
22.4	120
25.9	135
26.5	140
28.9	115
30.1	150
32.9	165
33.0	160
34.7	180

Program Code

```
options ps=62 ls=80;
```
Formats the output page to 62 lines in length and 80 columns in width

```
data in;
```
Beginning of Data Step

```
  input bmi sbp;
```
Inputs two variables **bmi** and **sbp** (both continuous variables).

```
cards;
```
Beginning of Raw Data section.

```
18.4 120
```
actual observations (value of **bmi** and

```
20.1 110
```
    sbp on each line)

```
22.4 120
25.9 135
26.5 140
28.9 115
30.1 150
32.9 165
33.0 160
34.7 180
run;
```

```
proc plot;
```
Procedure call. Proc Plot generates a scatter diagram for two continuous variables.

```
  plot sbp*bmi;
```
Specification of analytic variables. First variable specified is plotted on the vertical axis (*y*).

```
run;
```
End of procedure section.

```
proc corr cov;
```
Procedure call. Proc Corr estimates correlation coefficients and tests their significance. The Cov option produces a covariance matrix.

```
  var bmi sbp;
```
Specification of analytic variables.

```
run;
```
End of procedure section.

```
proc reg;
```
Procedure call. Proc Reg estimates regression parameters and tests significance.

```
  model sbp=bmi;
```
Specification of regression model. The format is dependent (**sbp**) = independent variable(s) (**bmi**).

```
run;
```
End of procedure section. ▪

SAS EXAMPLE 10.2 Correlation and Simple Linear Regression Analysis

Assess the Relationship Between Number of Hours of Exercise per Week and Systolic Blood Pressure (Example 10.2)

Consider the application described in Example 10.1. Suppose for the same sample, we also recorded the number of hours of vigorous exercise in a typical week. We wish to investigate the relationship between the number of hours of exercise per week and systolic blood pressure in males 50 years of age. In this analysis, the independent (or predictor) variable is number of hours of exercise per week and the dependent (or response) variable is systolic blood pressure. Estimate the correlation between the number of hours of exercise and systolic blood pressure and the regression equation relating the number of hours of exercise to systolic blood pressure using SAS.

X = Number of Hours of Exercise per Week	Y = Systolic Blood Pressure
4	120
10	110
2	120
3	135
3	140
5	115
1	150
2	165
2	160
0	180

Program Code

```
options ps=62 ls=80;

data in;
  input exercise sbp;



cards;
4 120
10 110
2 120
3 135
3 140
5 115
1 150
2 165
2 160
0 180
run;
```

Formats the output page to 62 lines in length and 80 columns in width
Beginning of Data Step
Inputs two variables *exercise* and *sbp* (both continuous variables).

Beginning of Raw Data section.
actual observations (value of *exercise* and *sbp* on each line)

```
proc corr;
```
Procedure call. Proc Corr estimates correlation coefficients and tests their significance.

```
 var exercise sbp;
```
Specification of analytic variables.

```
run;
```
End of procedure section.

```
proc reg;
```
Procedure call. Proc Reg estimates regression parameters and tests their significance.

```
  model sbp=exercise;
```
Specification of regression model. The format is dependent variable *(sbp)* = independent variable(s) *(exercise)*.

```
run;
```
End of procedure section. ▪

SAS EXAMPLE 10.3 Multiple Regression Analysis

Assess the Effects of Age, Gender, and Smoking Status, Considered Simultaneously, on Systolic Blood Pressure (Example 10.3)

In the following we use SAS to analyze a large data set ($n = 5078$). The dependent variable is systolic blood pressure (SBP) and we consider three independent variables: AGE (a continuous variable, measured in years), MALE (an indicator variable, coded 1 for males and 0 for females), and SMOKER (an indicator variable, coded 1 for smokers and 0 for nonsmokers). We estimated a multiple regression equation using SAS—the data are abbreviated.

Program Code

```
options ps=62 ls=80;
```
Formats the output page to 62 lines in length and 80 columns in width

```
data in;
  input sbp age male smoker;
```
Beginning of Data Step
Inputs four variables *sbp, age, male,* and *smoker.*

```
cards;
120 55 1 0
110 40 0 0
.
.
.
140 35 1 1
run;
```
Beginning of Raw Data section.
actual observations (value of *sbp, age, male,* and *smoker* on each line)

```
proc reg;
```
Procedure call. Proc Reg estimates regression parameters and tests their significance.

```
  model sbp=age male smoker;
```
Specification of regression model. The format is dependent variable *(sbp)* = independent variable(s) *(age, male,* and *smoker)*.

```
run;
```
End of procedure section. ▪

SAS EXAMPLE 10.4 Logistic Regression Analysis

Assess the Effects of Age, Gender, and Smoking Status, Considered Simultaneously, on Hypertensive Status (Example 10.4)

In the following we use SAS to analyze a large data set ($n = 5078$). The dependent variable is a dichotomous variable (HTN, coded 1 for patients classified as hypertensive and 0 otherwise) and we consider three independent variables: AGE (a continuous variable, measured in years), MALE (an indicator variable, coded 1 for males and 0 for females), and SMOKER (an indicator variable, coded 1 for smokers and 0 for nonsmokers). We estimated a logistic regression equation using SAS—the data are abbreviated.

Program Code

`options ps=62 ls=80;`	Formats the output page to 62 lines in length and 80 columns in width
`data in;`	Beginning of Data Step
`  input htn age male smoker;`	Inputs four variables *htn, age, male,* and *smoker*.
`cards;`	Beginning of Raw Data section.
`0 55 1 0`	actual observations (value of *htn,*
`0 40 0 0`	*age, male,* and *smoker* on each line)
`.`	
`.`	
`1 35 1 1`	
`run;`	
`proc logistic descending;`	Procedure call. Proc Logistic estimates logistic regression parameters and tests their significance. The descending option indicates that the value 1 is the outcome of interest (the value 0 is the comparison).
`  model htn=age male smoker;`	Specification of regression model. The format is dependent variable *(htn)* = independent variable(s) *(age, male,* and *smoker).*
`run;`	End of procedure section. ■

10.6.1 Summary of SAS Procedures

The SAS procedures for correlation and regression analysis are summarized in the following table. Specific options can be requested in these procedures to produce specific results. The options are shown in italics. Users should refer to the examples in this section for complete descriptions of the procedures and specific options.

Procedure	Sample Procedure Call	Description
proc plot	proc plot; plot y*x;	Generates a scatter diagram for two continuous variables with Y on the vertical axis and X on the horizontal axis.
proc corr	proc corr *cov*; var y x1 x2 x3 x4;	Produces a correlation matrix for specified variables. Matrix includes estimates of correlations between variables and associated significance tests. Cov option generates a covariance matrix.
proc reg	proc reg; model y = x1 x2 x3 x4;	Estimates a linear regression model relating dependent variable Y to independent variables X_1, X_2, X_3, X_4.
proc logistic	proc logistic *descending*; model y = x1 x2 x3 x4;	Estimates a logistic regression model relating dichotomous dependent variable Y to independent variables X_1, X_2, X_3, X_4. The descending option considers the response with the higher numerical code the outcome of interest (e.g., 1 = outcome of interest, 0 = comparison).

10.7 Analysis of Framingham Heart Study Data

The Framingham data set includes data collected from the original cohort. Participants contributed up to three examination cycles of data. Here we analyze data collected in the first examination cycle (called the period = 1 examination) and use SAS Proc Corr to investigate correlations between body mass index (BMI), number of cigarettes smoked per day, and systolic blood pressure. We then develop a simple linear regression model relating systolic blood pressure to BMI and a multiple linear regression model relating systolic blood pressure to BMI, age, gender, number of cigarettes smoked per day, and whether participants are on antihypertensive medications or not. For this analysis we exclude persons who are underweight (BMI < 18.5) because analysis has shown that there may be a different relationship between very low BMI and other risk factors. We also create an indicator variable for male gender. The SAS code to create the analytic data set along with the procedure calls is shown next.

Framingham Data Analysis—SAS Code

```
data fhs;
 set in.frmgham;
 if period=1;

if bmi<18.5 then delete;

if sex=1 then male=1;
else if sex=2 then male=0;
run;
```

```
proc corr data=fhs;
 var sysbp bmi cigpday;
run;

proc reg data=fhs;
 model sysbp=bmi;
run;

proc reg data=fhs;
 model sysbp=bmi age male bpmeds cigpday;
run;
```

Framingham Data Analysis—SAS Output

The CORR Procedure

3   Variables: SYSBP   BMI   CIGPDAY

Simple Statistics

Variable	N	Mean	Std Dev	Sum	Minimum	Maximum
SYSBP	4358	133.06436	22.36517	579895	83.50000	295.00000
BMI	4358	25.95387	4.01749	113107	18.52000	56.80000
CIGPDAY	4326	8.96163	11.95879	38768	0	70.00000

Simple Statistics

Variable	Label
SYSBP	Systolic BP mmHg
BMI	Body Mass Index (kr/(M*M)
CIGPDAY	Cigarettes per day

Pearson Correlation Coefficients
Prob > |r| under H0: Rho=0
Number of Observations

	SYSBP	BMI	CIGPDAY
SYSBP Systolic BP mmHg	1.00000 4358	0.32353 <.0001 4358	-0.09910 <.0001 4326
BMI Body Mass Index (kr/(M*M)	0.32353 <.0001 4358	1.00000 4358	-0.09824 <.0001 4326
CIGPDAY Cigarettes per day	-0.09910 <.0001 4326	-0.09824 <.0001 4326	1.00000 4326

The first part of the correlation output contains summary statistics on the three variables. SAS then produces the correlations between pairs of variables. For each pair, SAS produces the correlation, a p value to test if the correlation is significantly different from zero, and then the number of subjects available (i.e., the number of subjects who have complete data on both variables). What is the nature of the correlation between BMI and systolic blood pressure? Is it in the expected direction? Is it statistically significant? What is the nature of the correlation between the number of cigarettes smoked per day and systolic blood pressure? Is it in the expected direction? Is it statistically significant?

The following is a simple linear regression analysis relating BMI to systolic blood pressure. What is the estimate of the slope? How is the slope interpreted? What is the expected systolic blood pressure for a person with BMI = 25?

```
              The REG Procedure
               Model: MODEL1
   Dependent Variable: SYSBP Systolic BP mmHg
```

Analysis of Variance

Source	DF	Sum of Squares	Mean Square	F Value	Pr > F
Model	1	228114	228114	509.24	<.0001
Error	4356	1951261	447.94797		
Corrected Total	4357	2179375			

Root MSE	21.16478	R-Square	0.1047
Dependent Mean	133.06436	Adj R-Sq	0.1045
Coeff Var	15.90567		

Parameter Estimates

Variable	Label	DF	Parameter Estimate	Standard Error	t Value
Intercept	Intercept	1	86.32004	2.09608	41.18
BMI	Body Mass Index (kr/(M*M)	1	1.80105	0.07981	22.57

Parameter Estimates

Variable	Label	DF	Pr > |t|
Intercept	Intercept	1	<.0001
BMI	Body Mass Index (kr/(M*M)	1	<.0001

The following is a multiple linear regression analysis relating age, gender, BMI, antihypertensive treatment, and the number of cigarettes smoked per day to systolic blood pressure. Is the set of predictors statistically significant? How much variation in systolic blood pressure is explained by this set of predictors? Assess the slope coefficients—are they all in the expected direction? What is the relative importance of the risk factors? Do men or women have higher systolic blood pressures, adjusting for age, BMI, number of cigarettes smoked, and antihypertensive medication?

```
                    The REG Procedure
                     Model: MODEL1
          Dependent Variable: SYSBP Systolic BP mmHg

                    Analysis of Variance
                        Sum of          Mean
Source              DF  Squares         Square    F Value   Pr > F
Model                5   573039         114608     316.14   <.0001
Error             4260  1544325      362.51768
Corrected Total   4265  2117364


        Root MSE              19.03990   R-Square    0.2706
        Dependent Mean       132.99672   Adj R-Sq    0.2698
        Coeff Var             14.31607

                    Parameter Estimates
                                      Parameter   Standard
Variable    Label                 DF  Estimate      Error   t Value

Intercept   Intercept              1   51.24242    2.50258    20.48
AGE         Age (years) at examination  1   0.87136    0.03481    25.03
MALE                               1   -2.38027    0.62292    -3.82
BMI         Body Mass Index (kr/(M*M))  1   1.47203    0.07407    19.87
BPMEDS      Anti-hypertensive meds Y/N  1  23.80633    1.65356    14.40
CIGPDAY     Cigarettes per day     1    0.03689    0.02633     1.40

                    Parameter Estimates

     Variable     Label                              DF    Pr > |t|
     Intercept    Intercept                           1     <.0001
     AGE          Age (years) at examination          1     <.0001
     MALE                                             1     0.0001
     BMI          Body Mass Index (kr/(M*M))          1     <.0001
     BPMEDS       Anti-hypertensive meds Y/N          1     <.0001
     CIGPDAY      Cigarettes per day                  1     0.1613
```

10.8 Problems

1. Data were collected from a random sample of eight patients currently undergoing treatment for hypertension. Each subject reported the average number of cigarettes smoked per day and each was assigned a numerical value reflecting their risk of cardiovascular disease (CVD). The risk assessments were computed by physicians and based on blood pressure, cholesterol level, and exercise status. The risk assessments ranged from 0 to 100, with higher values indicating increased risk.

Number of Cigarettes:	0	2	6	8	12	0	2	20
Risk of CVD:	12	20	50	68	75	8	10	80

   a. Compute the correlation between number of cigarettes and risk of CVD.

   b. Based on this sample, is there evidence of a significant correlation between number of cigarettes and risk of CVD?

   c. Compute the equation of the line that best fits the data to predict risk of CVD from the number of cigarettes (i.e., risk of CVD = dependent variable).

2. Consider the following data reflecting lengths of stay in the hospital (recorded in days) and the total charge (in $000s) for six patients undergoing a minor surgical procedure:

Length of Stay:	5	7	9	10	12	15
Total Charge:	6	5	7.2	8	9.4	7.9

   Consider length of stay as the independent and total charge as the dependent variable.

   a. Describe the relationship between length of stay and total charge using a scatter diagram.

   b. Compute the sample correlation coefficient.

   c. Compute the regression equation.

   d. Estimate the total charge for an individual who stays 11 days in the hospital.

   e. Suppose we compare two patients and one stays 3 days longer in the hospital than the other. What is the expected difference in total charges between these patients?

3. Consider a study assessing the quality of life (QOL) of patients with arthritis. The following figure describes the relationship between QOL and the duration of arthritis, measured in years.

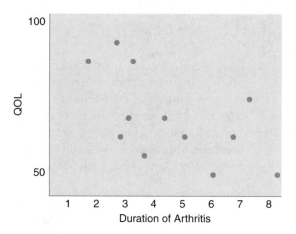

The mean QOL score is 75.8, with a standard deviation of 6.3. The mean duration of arthritis is 4 years, with a standard deviation of 0.7 years. The covariance between QOL and duration is −2.9.

a. What is the nature of the relationship between duration of arthritis and QOL based on the figure? Describe briefly.

b. Estimate the sample correlation coefficient.

c. Compute the equation of the line that best describes the relationship between duration of arthritis and QOL.

d. Interpret the estimated slope.

4. Consider the study described in Problem 3. Suppose that we investigate the relationship between age and the severity of arthritis (measured on a scale from 0 to 100, with higher scores indicative of more severe disease). The following summary statistics are available on the total sample ($n = 175$):

Characteristic	Mean	SD
Age	57.7	3.8
Severity score	59.8	19.5

a. The point estimate for the correlation between age and severity score is 0.36 ($p = 0.0001$). Is there evidence of a significant correlation between age and severity score among all patients?

b. Estimate the equation of the line that best describes the relationship between age and severity score (consider severity score as the dependent variable).

c. What is the expected severity score for an individual 60 years of age?

5. An analysis is performed to investigate the relationship between cardiovascular exercise and percent body fat. A sample of 50 subjects is

involved, and the number of hours of cardiovascular exercise over the previous week and percent body fat are measured on each subject. Summary statistics are shown here:

Variable	n	Mean	SD
# Hours of cardiovascular exercise	50	3.7	1.9
Percent body fat	50	28.5	5.8

a. Suppose the sample correlation coefficient is $r = -0.48$. Is there evidence of a significant correlation? Run the appropriate test at $\alpha = 0.05$.

b. Suppose that the investigation involved 20 men and 30 women and separate correlation analyses were performed that produced the following results: $r_{men} = -0.17$ ($p = 0.4748$), $r_{women} = -0.59$ ($p = 0.0006$). Describe the relationships between exercise and percent body fat in men and women.

c. Summary statistics for the 20 men and 30 women are

Variable	Men Mean (SD)	Women Mean (SD)
# Hours of cardiovascular exercise	4.7 (2.2)	2.5 (1.4)
Percent body fat	20.1 (3.1)	31.6 (8.1)

Estimate the regression equations relating the number of hours of exercise to percent body fat for men and women (separately). (Let Y = percent body fat.)

d. What is the expected percent body fat for a male who exercises 2 hours per week?

e. What is the expected percent body fat for a female who exercises 2 hours per week?

6. An investigation is performed to understand the relationship between age and satisfaction with medical care. Each individual in the investigation is asked to rate satisfaction with medical care on a scale of 0 to 100, with 100 denoting complete satisfaction. Other data, including sociodemographic characteristics (age, gender, race) are also recorded on each individual. A total of 200 individuals are involved in the investigation, and the following results were obtained: $r = 0.43$ ($p = 0.0001$), satisfaction $= 45.2 + 0.9$ age.

a. Based on the correlation, are older or younger individuals more likely to be more satisfied with medical care? (Be brief.)

b. Is the correlation between age and satisfaction significant? Justify.

c. Estimate the satisfaction rating of a 50-year-old individual.

7. Anecdotal evidence suggests that there is an inverse relationship between alcohol consumption and medication adherence (measured as percent of prescribed doses taken). A study is run involving $n = 50$ subjects, and data are measured on each subject reflecting the number of alcoholic drinks consumed in the week prior to study enrollment. The sample correlation coefficient between the number of alcoholic drinks consumed and medication adherence is -0.47.

a. Is there evidence of a significant correlation between the number of alcoholic drinks consumed and medication adherence? Run the appropriate test at $\alpha = 0.05$.

b. Using the following information, estimate the equation of the line that best describes the relationship between the number of alcoholic drinks consumed and medication adherence. The mean number of alcoholic drinks consumed (per week) is 14, with a standard deviation of 4.5. The mean medication adherence is 78.5, with a standard deviation of 13.5. Consider medication adherence as the dependent variable.

c. What is the expected medication adherence for a person who does not drink alchohol?

8. A study was conducted on a random sample of 15 undergraduates to assess whether there is a relationship between stress and GPA. Stress was measured on a scale of 0 to 100, with higher scores indicative of more stress. Descriptive statistics were generated in SAS and are given here.

```
                  Correlation Analysis
              2 'VAR' Variables: STRESS GPA
                 Covariance Matrix DF = 14
```

	STRESS	GPA
STRESS	1075.666667	-11.328571
GPA	-11.328571	0.226857

```
                    Simple Statistics
```

Variable	N	Mean	Std Dev	Sum	Minimum	Maximum
STRESS	15	54.66667	32.79736	820.00000	5.00000	100.00000
GPA	15	2.76000	0.47630	41.40000	2.00000	3.90000

a. Compute the sample correlation coefficient.

b. Is there evidence of a significant correlation between stress and GPA? Run the appropriate test at $\alpha = 0.05$.

c. Compute the equation of the line that best describes the relationship between stress and GPA. Assume that GPA is the dependent variable.

d. What is the expected GPA for a person with a stress score of 80?

SAS Problems

Use SAS to solve the following problems.

1. Data were collected from a random sample of eight patients currently undergoing treatment for hypertension. Each subject reported the average number of cigarettes smoked per day and each was assigned a numerical value reflecting risk of cardiovascular disease (CVD). The risk assessments were computed by physicians and based on blood pressure, cholesterol level, and exercise status. The risk assessments ranged from 0 to 100, with higher values indicating increased risk.

Number of Cigarettes:	0	2	6	8	12	0	2	20
Risk of CVD:	12	20	50	68	75	8	10	80

Use SAS to generate a scatter diagram, to estimate the correlation between variables, and to test if the correlation is significant. In addition, estimate a simple linear regression equation and compute R^2.

2. Consider the following data reflecting lengths of stay in the hospital (recorded in days) and the total charge (in $000s) for six patients undergoing a minor surgical procedure:

Length of Stay:	5	7	9	10	12	15
Total Charge:	6	5	7.2	8	9.4	7.9

Consider length of stay as the independent and total charge as the dependent variable and use SAS to generate a scatter diagram, to estimate the correlation between variables, and to test if the correlation is significant. In addition, estimate a simple linear regression equation and compute R^2.

Descriptive Statistics
(Ch. 2)

Probability
(Ch. 3)

Sampling Distributions
(Ch. 4)

Statistical Inference
(Chapters 5–13)

OUTCOME VARIABLE	GROUPING VARIABLE(S)/ PREDICTOR(S)	ANALYSIS	CHAPTER(S)
Continuous	—	Estimate μ; Compare μ to Known, Historical Value	5/12
Continuous	Dichotomous (2 groups)	Compare Independent Means (Estimate/Test $(\mu_1 - \mu_2)$) or the Mean Difference(μ_d)	6/12
Continuous	Discrete (> 2 groups)	Test the Equality of k Means using Analysis of Variance ($\mu_1 = \mu_2 = \cdots = \mu_k$)	9/12
Continuous	Continuous	Estimate Correlation or Determine Regression Equation	10/12
Continuous	Several Continuous or Dichotomous	Multiple Linear Regression Analysis	10
Dichotomous	—	Estimate p; Compare p to Known, Historical Value	7
Dichotomous	Dichotomous (2 groups)	Compare Independent Proportions (Estimate/Test $(p_1 - p_2)$)	7/8
Dichotomous	Discrete (>2 groups)	Test the Equality of k Proportions (Chi-Square Test)	7
Dichotomous	Several Continuous or Dichotomous	Multiple Logistic Regression Analysis	11
Discrete	Discrete	Compare Distributions Among k Populations (Chi-Square Test)	7
Time to Event	Several Continuous or Dichotomous	Survival Analysis	13

II

Logistic Regression Analysis

In Chapter 8 we introduced several effect measures used to compare risks in two populations. We presented estimation and statistical inference procedures for crude comparisons and a technique for estimation and statistical inference adjusting for one categorical confounder variable. In Chapter 10 we presented simple linear regression techniques used to estimate the linear equation describing the relationship between two continuous variables. We also introduced multiple linear regression techniques, which are used when the continuous outcome variable is linearly related to a set of independent variables.

In this chapter we describe logistic regression analysis, which can be used to model the effect of one or several independent variables on the risk of a dichotomous outcome. In Section 11.1 we introduce the logistic model, and in Section 11.2 we focus on statistical inference for situations with a single independent variable. In Section 11.3 we extend our discussion to the multiple logistic model, in which a single dichotomous outcome variable is modeled as a function of a set of independent variables. We briefly describe the use of the receiver operating characteristic (ROC) curve as a measure of the goodness-of-fit of logistic regression models in Section 11.4, and we summarize key formulas and present statistical computing applications in Sections 11.5 and 11.6, respectively. In Section 11.7 we use data from the Framingham Heart Study to illustrate the applications presented here.

As in linear regression, our goal is to estimate the regression coefficients in a model, given a sample of (X, Y) pairs. In the case of logistic regression, the Xs can be continuous or dichotomous, but the Ys are generally coded as 0 (for those who do not have the event) or 1 (for those who do have the event).

11.1 The Logistic Model

The simple logistic model is based on a linear relationship between the natural logarithm (ln) of the odds of an event and a continuous independent variable. The form of this relationship is as follows:

$$L = \ln(o) = \ln\left(\frac{p}{1-p}\right) = \beta_0 + \beta_1 X + \varepsilon \qquad (11.1)$$

where Y is the dichotomous event of interest, coded 0/1 for
   failure/success
   p is the proportion of successes
   o is the odds of the event
   L is the ln(odds of event)
   X is the independent variable
   β_0 is the intercept
   β_1 is the regression coefficient
   ε is the random error

The relationship between X and L is linear:

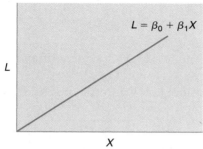

$$L = \beta_0 + \beta_1 X$$

Recall from Chapter 8 that if p is the probability of the event, then the odds of the event are

$$\text{Odds} = o = \frac{p}{(1 - p)}$$

We defined $L = \ln(\text{odds of event } Y)$, sometimes called the "log odds" or "logit" of Y. We can write L in terms of p, Probability$(Y = 1)$, as follows:

$$L = \ln(o) = \ln\left(\frac{p}{(1 - p)}\right)$$

We can then use the laws of exponents and logs and some algebra to express p (the proportion of successes or risk of the event) in terms of L:

$$e^L = o$$
$$e^L = \frac{p}{(1 - p)}$$
$$p = e^L(1 - p)$$
$$p = e^L - pe^L$$
$$p + pe^L = e^L$$
$$p(1 + e^L) = e^L$$
$$p = \frac{e^L}{(1 + e^L)}$$

This is called the logistic function and its graph is as follows. Notice that p, the probability of the event, increases from 0 to 1.

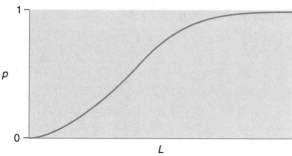

Logistic function: $p = \dfrac{e^L}{(1 + e^L)}$ (11.2)

11.2 Statistical Inference for Simple Logistic Regression

The logistic regression model (11.1) may be written in terms of p, the risk of event Y, using equations (11.1) and (11.2). We assume that L is a linear function of X, and we substitute (11.1) into the logistic function (11.2).

$$p = \frac{e^{\beta_0 + \beta_1 X + \varepsilon}}{(1 + e^{\beta_0 + \beta_1 X + \varepsilon})} \tag{11.3}$$

As in linear regression, our goal is to estimate the regression coefficients β_0 and β_1, given a sample of (X, Y) pairs. Here, as in the linear regression framework, X can be continuous or dichotomous. However, the Y is not continuous as in linear regression; rather, it is coded 0 (for those who do not have the event) or 1 (for those who do have the event). Unfortunately, the techniques used in linear regression to estimate the regression coefficients cannot be applied to the logistic regression case. The iterative maximum likelihood estimation process used to estimate coefficients is computationally complex and is beyond the scope of this book. Here we restrict our attention to interpretation of the logistic model estimated by SAS.

SAS Example 11.1
Logistic Regression Relating White Blood Count to Coronary Abnormalities

Recall the clinical trial described in Example 8.1 in which patients with Kawasaki syndrome were treated with gamma globulin (GG = 1) or with the standard treatment (GG = 0) and the primary outcome was the development of coronary abnormalities (CA). A possible explanation for the effect of treatment with gamma globulin on the development of CA is that it reduces the elevated white blood count (WBC) early in the course of the disease and that WBC is related to CA. This example addresses the relationship between WBC and CA. The data set contains a total of $n = 168$ observations; each participant has a continuous value for WBC and a dichotomous value (0 or 1) for the outcome, CA. The following is a subset of the output generated by SAS Proc Logistic. A brief interpretation follows the output.

SAS Output for Example 11.1

```
                    The LOGISTIC Procedure
                      Model Information
          Data Set                        WORK.ONE
          Response Variable               CA
          Number of Response Levels       2
          Number of Observations          156
          Model                           binary logit
          Optimization Technique          Fisher's scoring
```

                              Response Profile
            Ordered   Total
            Value   CA   Frequency
                1   1   25
                2   0   131

                   Probability modeled is CA=1.

NOTE: 12 observations were deleted due to missing values for the
response or explanatory variables.

            The LOGISTIC Procedure

        Analysis of Maximum Likelihood Estimates

                            Standard   Wald
Parameter   DF   Estimate   Error   Chi-Square   Pr > ChiSq

Intercept   1   -3.3848   0.5838   33.6120   <.0001
WBC   1   0.1253   0.0362   11.9835   0.0005

Interpretation of SAS Output for Example 11.1

We coded the variable, CA, to represent the development of CA and specified that the event of interest is CA $= 1$. SAS reports this under "Probability modeled is CA $= 1$." Among the 168 patients, 12 were missing the WBC measure and SAS notes that 12 observations were deleted due to missing values. Thus the analysis is based on $n = 156$. The results of the logistic regression estimation procedure are given under "Analysis of Maximum Likelihood Estimates." The estimated coefficients are in the column labeled "Estimate" and are $\hat{\beta}_0 = -3.385$ and $\hat{\beta}_1 = 0.125$.

SAS also provides the p value associated with the test $H_0: \beta_1 = 0$ in the column labeled "Pr > ChiSq." In this case, $p < 0.0005$, so we reject H_0 and conclude that there is significant evidence that WBC is linearly related to the log odds of CA. ■

Now that the coefficients have been estimated, we can estimate the risk, $\hat{p}$, of the event Y, given a specific value of X, by substituting estimates of β_0 and β_1 into equation (11.3) as follows:

$$\hat{p} = \frac{e^{\hat{\beta}_0 + \hat{\beta}_1 X}}{(1 + e^{\hat{\beta}_0 + \hat{\beta}_1 X})} \qquad (11.4)$$

Consider a person with $X = x_a$. We estimate the risk of Y given $X = x_a$ as

$$\hat{p}_a = \frac{e^{\hat{\beta}_0 + \hat{\beta}_1 x_a}}{1 + e^{\hat{\beta}_0 + \hat{\beta}_1 x_a}}$$

For example, in SAS Example 11.1,

$$\hat{p} = \frac{e^{-3.385 + 0.125(\text{WBC})}}{(1 + e^{-3.385 + 0.125(\text{WBC})})}$$

The risk of CA for a patient with WBC = 10 is

$$\hat{p} = \frac{e^{-3.385 + 0.125(10)}}{(1 + e^{-3.385 + 0.125(10)})}$$

$$= \frac{e^{-2.135}}{(1 + e^{-2.135})}$$

$$= 0.106$$

We can also use the estimate of β_1 to estimate the odds of the event for specified values of X and the odds ratio comparing the odds of the event for people with different values of X. Recall that the logistic model is based on the linear relationship between X and the log odds: $\ln(o) = L = \beta_0 + \beta_1 X$. Thus, the estimated odds are

$$\hat{o} = e^{\hat{L}} = e^{\hat{\beta}_0 + \hat{\beta}_1 X} \tag{11.5}$$

Consider two people with different values of X, one with $X = x_a$ and one with $X = x_b$. The odds of the event given $X = x_a$ are $\hat{o}_a = e^{\hat{\beta}_0 + \hat{\beta}_1 X_a}$ and, similarly, for $X = x_b$, the odds are $\hat{o}_b = e^{\hat{\beta}_0 + \hat{\beta}_1 X_b}$. For example, the odds of CA given WBC are

$$\hat{o} = e^{\hat{L}} = e^{\hat{\beta}_0 + \hat{\beta}_1 X} = e^{-3.385 + 0.125(\text{WBC})}$$

For a patient with WBC = 10, the odds of CA are

$$\hat{o} = e^{-3.385 + 0.125(10)} = e^{-2.135} = 0.118$$

The odds ratio comparing the odds of an event for two people, one with $X = x_a$ and one with $X = x_b$, is

$$\hat{\text{OR}} = \frac{e^{\hat{\beta}_0 + \hat{\beta}_1 x_a}}{e^{\hat{\beta}_0 + \hat{\beta}_1 x_b}}$$

Recall some of the properties of exponents:

$$e^{A+B} = e^A e^B$$
$$e^{A-B} = e^A / e^B \tag{11.6}$$
$$e^{AB} = (e^A)^B$$

The estimate of the odds ratio,

$$\hat{OR} = \frac{e^{\hat{\beta}_0 + \hat{\beta}_1 x_a}}{e^{\hat{\beta}_0 + \hat{\beta}_1 x_b}}$$

can be written as

$$\hat{OR} = e^{(\hat{\beta}_0 + \hat{\beta}_1 x_a) - (\hat{\beta}_0 + \hat{\beta}_1 x_b)}$$

which is equivalent to

$$\hat{OR} = e^{\hat{\beta}_1 (x_a - x_b)}$$

The odds ratio comparing the odds of the event for a person $X = x_a$ compared to one with $X = x_b$ is

$$\hat{OR} = e^{\hat{\beta}_1 (x_a - x_b)} \qquad (11.7)$$

In SAS Example 11.1, the estimate of the odds ratio comparing a person with WBC = 15 to a person with WBC = 10 is

$$\hat{OR} = e^{\hat{\beta}_1 (x_a - x_b)} = e^{0.1253(15-10)} = e^{0.1253(5)} = e^{0.6265} = 1.87$$

Thus the odds of CA is 1.87 times higher for every 5-unit increase in WBC. Notice that this depends only on the difference in the X values $(x_a - x_b)$, not on their actual values. Thus the estimated odds ratio comparing one person with WBC = 10 to another person with WBC = 5 is also equal to 1.87, as is the odds ratio comparing any two people whose WBC values are 5 units apart.

The formula for a 95% confidence interval around the odds ratio is

$$\left(e^{(\hat{\beta}_1 - 1.96 s.e.(\hat{\beta}_1))(x_a - x_b)}, \ e^{(\hat{\beta}_1 + 1.96 s.e.(\hat{\beta}_1))(x_a - x_b)} \right) \qquad (11.8)$$

where $s.e.(\hat{\beta}_1)$ is the standard error of $(\hat{\beta}_1)$. The calculation of $s.e.(\hat{\beta}_1)$ is beyond the scope of this book, but it is produced by SAS in the following sections. In SAS Example 11.1, the standard error of $\hat{\beta}_1$, $s.e.(\hat{\beta}_1) = 0.0362$. Thus, using (11.8), the 95% confidence interval for the odds ratio comparing the odds of CA in children with Kawasaki syndrome whose WBC are 5 units apart is as follows:

$$\left(e^{(\hat{\beta}_1 - 1.96 s.e.(\hat{\beta}_1))(x_a - x_b)}, \ e^{(\hat{\beta}_1 + 1.96 s.e.(\hat{\beta}_1))(x_a - x_b)} \right)$$

$$= \left(e^{(0.1253 - 1.96(0.0362))(5)}, \ e^{(0.1253 + 1.96(0.0362))(5)} \right)$$

$$= \left(e^{(0.2717)}, \ e^{(0.9813)} \right)$$

$$= (1.31, 2.67)$$

Thus, we conclude that the odds ratio comparing the odds of CA in one child with Kawasaki syndrome whose WBC is 5 units higher than another child is 1.87 and that the 95% confidence interval for the odds ratio is (1.31, 2.67).

In SAS Example 11.1, the odds ratio comparing two people with WBC values 1 unit apart is $\hat{OR} = e^{\hat{\beta}_1} = e^{0.125} = 1.13$. The 95% confidence interval is

$$\left(e^{(\hat{\beta}_1 - 1.96 s.e.(\hat{\beta}_1))(x_a - x_b)}, e^{(\hat{\beta}_1 + 1.96 s.e.(\hat{\beta}_1))(x_a - x_b)}\right)$$
$$= \left(e^{(0.1253 - 1.96(0.0362))(1)}, e^{(0.1253 + 1.96(0.0362))(1)}\right)$$
$$= \left(e^{(0.0543)}, e^{(0.1963)}\right)$$
$$= (1.056, 1.217)$$

SAS automatically provides the odds ratio (and the associated 95% confidence interval) comparing the odds of the event for two people with X values 1 unit apart $(x_a - x_b = 1)$. This is $\hat{OR} = e^{\hat{\beta}_1(x_a - x_b)} = e^{\hat{\beta}_1(1)} = e^{\hat{\beta}_1}$. The label is simply "Odds Ratio Estimates," but it is important to remember that this is the estimate of the odds ratio comparing people exactly 1 unit apart.

SAS Output for Example 11.1 (Continued)

```
              The LOGISTIC Procedure

        Analysis of Maximum Likelihood Estimates
```

			Standard	Wald	
Parameter	DF	Estimate	Error	Chi-Square	Pr > ChiSq
Intercept	1	-3.3848	0.5838	33.6120	<.0001
WBC	1	0.1253	0.0362	11.9835	0.0005

```
                  Odds Ratio Estimates
```

	Point	95% Wald	
Effect	Estimate	Confidence Limits	
WBC	1.133	1.056	1.217

In clinical trials, X is usually a dichotomous variable representing membership in one treatment group or another. For example, $X = 1$ could denote the treatment group, and $X = 0$ could denote the control group. Suppose that X is dichotomous (0 or 1) and we use (11.7) to estimate the odds ratio. The estimate of the odds ratio comparing those with $X = 1$ to those with $X = 0$ reduces to

$$\hat{OR} = e^{\hat{\beta}_1(1-0)} = e^{\hat{\beta}_1(1)} = e^{\hat{\beta}_1}$$

In this situation, where X is dichotomous, the null hypothesis, $H_0: \beta_1 = 0$, can be phrased, equivalently, as $H_0: e^{\beta} = 1$, or $H_0: OR = 1$.

SAS EXAMPLE 11.2 **Logistic Regression Relating Treatment to Coronary Abnormalities**

We now use SAS Proc Logistic to estimate the probabilities and odds of the development of coronary abnormalities (CA) in children with Kawasaki syndrome treated with gamma globulin (GG = 1) as compared to the standard treatment (GG = 0).

SAS Output for Example 11.2

```
                    Model Information

Data Set                          WORK.ONE
Response Variable                 CA
Number of Response Levels         2
Number of Observations            167
Model                             binary logit
Optimization Technique            Fisher's scoring
```

```
                   Response Profile

        Ordered                        Total
         Value            CA         Frequency

           1              1              26
           2              0             141
```

```
          Probability modeled is CA=1.
```

NOTE: 1 observation was deleted due to missing values for the response or explanatory variables.

```
                The LOGISTIC Procedure

      Analysis of Maximum Likelihood Estimates
```

Parameter	DF	Estimate	Standard Error	Wald Chi-Square	Pr > ChiSq
Intercept	1	-1.0986	0.2520	19.0094	<.0001
gg	1	-1.6487	0.5257	9.8370	0.0017

```
                Odds Ratio Estimates
```

Effect	Point Estimate	95% Wald Confidence Limits	
gg	0.192	0.069	0.539

Interpretation of SAS Output for Example 11.2

The estimated coefficients are in the column labeled "Estimate" and are $\hat{\beta}_0 = -1.099$ and $\hat{\beta}_1 = -1.649$. The p value associated with the test $H_0: \beta_1 = 0$ is $p = 0.0017$, so we reject H_0 and conclude that there is significant evidence that treatment (gamma globulin versus standard care) is associated with the development of CA.

The odds ratio comparing children with Kawasaki syndrome treated with gamma globulin (GG $= 1$) to those treated with the standard treatment (GG $= 0$) with respect to the development of coronary abnormalities (CA) is given here as 0.19, with 95% confidence interval $(0.069, 0.539)$. Note that the p value associated with the test $H_0: \beta_1 = 0$ is also the p value associated with the test $H_0: OR = 1$. Here, this p value is $p = 0.0017$, so we reject H_0 and conclude that there is significant evidence that the odds ratio is not equal to 1. ■

Recall that we used these data in Example 8.1 and estimated the odds ratio using equation (8.6). This estimate will be identical to that obtained using equation (11.8). In fact, we obtained the same result as we do here: $\hat{OR} = 0.19$. In Chapter 8, we used the formula given in Table 8.2 to estimate the 95% confidence interval for the odds ratio; in this chapter we provide an estimate of the confidence interval that assumes the logistic model. The difference in estimated confidence intervals will be very slight. Using Table 8.2, we estimated the 95% confidence interval as $(0.069, 0.539)$, which, in this case, is identical to the estimate provided by SAS.

SAS does not automatically provide an estimate of the relative risk as it does the odds ratio. To estimate the relative risk, RR, comparing a patient with $X = x_a$ to one with $X = x_b$, we must first estimate the individual probabilities using (11.4).

$$\hat{RR} = \hat{p}_a / \hat{p}_b = \frac{\left(\dfrac{e^{\hat{\beta}_0 + \hat{\beta}_1 x_a}}{1 + e^{\hat{\beta}_0 + \hat{\beta}_1 x_a}} \right)}{\left(\dfrac{e^{\hat{\beta}_0 + \hat{\beta}_1 x_b}}{1 + e^{\hat{\beta}_0 + \hat{\beta}_1 x_b}} \right)} \tag{11.9}$$

In SAS Example 11.1, the estimated regression coefficients were $\hat{\beta}_0 = -3.385$ and $\hat{\beta}_1 = 0.125$, and the estimated risk of CA given WBC was $\hat{p} = \dfrac{e^{-3.385 + 0.125(\text{WBC})}}{(1 + e^{-3.385 + 0.125(\text{WBC})})}$. We estimated the risk of CA for a patient with WBC $= 10$ to be $\hat{p}_{10} = 0.106$. Similarly, the estimate of the risk of CA for a patient with WBC $= 5$ was $\hat{p}_5 = 0.060$. Thus, the estimated relative risk of CA for a patient with WBC $= 10$ as compared to a patient with WBC $= 5$ is $\hat{RR} = (\hat{p}_{10} / \hat{p}_5) = (0.106/0.060) = 1.77$.

Remember that the odds ratio comparing a patient with $X = x_a$ to one with $X = x_b$ depends only on the difference $(x_a - x_b)$, not on the actual values

x_a and x_b. The relative risk comparing those with $X = x_a$ to those with $X = x_b$, however, does depend on the actual values x_a and x_b.

In SAS Example 11.1, the estimated risk of CA for a patient with WBC = 15 is $\hat{p}_{15} = 0.181$. Therefore, the relative risk of CA comparing patients with WBC = 15 to those with WBC = 10 is $\hat{RR} = (\hat{p}_{15}/\hat{p}_{10}) = (0.181/0.106) = 1.71$. An increase of 5 units in WBC is associated with a relative risk $\hat{RR} = 1.77$ if the WBC values are 5 and 10, but is associated with a relative risk $\hat{RR} = 1.71$ if the WBC values are 10 and 15.

Now consider the situation in SAS Example 11.2 in which the comparison is of patients in two treatment groups, of those with $X = 1$ to those with $X = 0$. We can modify (11.9) slightly to estimate the relative risk of the event in patients with $X = 1$ as compared to those with $X = 0$.

$$\hat{RR} = \hat{p}_1/\hat{p}_0 = \frac{\left(\dfrac{e^{\hat{\beta}_0 + \hat{\beta}_1}}{1 + e^{\hat{\beta}_0 + \hat{\beta}_1}}\right)}{\left(\dfrac{e^{\hat{\beta}_0}}{1 + e^{\hat{\beta}_0}}\right)} \tag{11.10}$$

In SAS Example 11.2, the relative risk of CA for patients treated with gamma globulin as compared to those on the standard treatment is estimated as follows from the SAS output.

$$\hat{RR} = \hat{p}_1/\hat{p}_0 = \frac{\left(\dfrac{e^{\hat{\beta}_0 + \hat{\beta}_1}}{1 + e^{\hat{\beta}_0 + \hat{\beta}_1}}\right)}{\left(\dfrac{e^{\hat{\beta}_0}}{1 + e^{\hat{\beta}_0}}\right)} = \frac{\left(\dfrac{e^{-1.099 - 1.649}}{1 + e^{-1.099 - 1.649}}\right)}{\left(\dfrac{e^{-1.099}}{1 + e^{-1.099}}\right)}$$

$$= (0.060/0.250) = 0.24$$

The estimated relative risk of CA for patients treated with gamma globulin as compared to those on the standard treatment is $\hat{RR} = 0.24$. This may be interpreted as the risk of CA among patients treated with gamma globulin is less than a quarter the risk for those on the standard treatment. Recall that this is the estimate we obtained in Chapter 8 using (8.5).

As discussed in Chapter 8, the odds ratio may be used to estimate the relative risk if the prevalence of the disease is low (the overall risk is small). Here we see that it is much easier to obtain the estimate of the odds ratio from the results of a logistic regression than it is to obtain an estimate of the relative risk.

11.3 Multiple Logistic Regression

As in the case of linear regression, logistic regression techniques may be generalized to models with more than one independent variable. The methods used to estimate coefficients, standard errors, and p values are beyond the scope of this book, but we present some examples of multiple logistic regression in this section using SAS. The general multiple logistic regression model is

$$p = \frac{e^{\beta_0 + \beta_1 X_1 + \beta_2 X_2 + \cdots + \beta_k X_k + \varepsilon}}{(1 + e^{\beta_0 + \beta_1 X_1 + \beta_2 X_2 + \cdots + \beta_k X_k + \varepsilon})} \tag{11.11}$$

where p is the probability of the event and $X_1, X_2, \ldots, X_k$ are independent variables. The independent variables may be dichotomous or continuous.

SAS EXAMPLE 11.3 **Multiple Logistic Regression Relating White Blood Count and Treatment to Coronary Abnormalities**

Suppose we want to use the sample described in SAS Example 11.2 to assess the effect of treatment with gamma globulin on CA but we are concerned that white blood count is a potential confounder. Although patients were randomized to one of the two treatment groups, it turns out that the mean WBC in patients treated with gamma globulin was significantly lower than the mean WBC in patients on the standard treatment. Thus, the investigators were concerned that the patients treated with gamma globulin were less sick than those on the standard treatment and that the observed impact of the gamma globulin treatment might not be real. The following output is a multiple logistic regression in which the outcome is the dichotomous variable CA and the two independent variables are the dichotomous treatment variable, GG, and the continuous variable, WBC.

SAS Output for Example 11.3

```
                  The LOGISTIC Procedure

          Analysis of Maximum Likelihood Estimates


                               Standard          Wald
Parameter     DF    Estimate     Error      Chi-Square      Pr > ChiSq

Intercept      1     -2.6352     0.6562       16.1283         <.0001
gg             1     -1.3152     0.5496        5.7272          0.0167
WBC            1      0.1048     0.0383        7.4781          0.0062


                  Odds Ratio Estimates

                      Point              95% Wald
          Effect     Estimate        Confidence Limits

          gg          0.268        0.091          0.788
          WBC         1.111        1.030          1.197
```

Interpretation of SAS Output for Example 11.3

In SAS Example 11.2, we estimated the *crude* odds ratio comparing patients treated with gamma globulin to patients on standard treatment. The estimated OR is $\hat{OR} = 0.19$ with $p = 0.0017$. The odds ratio estimated here is the odds ratio comparing patients treated with gamma globulin to patients on standard treatment, *adjusted* for white blood count. The estimated OR is $\hat{OR} = 0.27$ with $p = 0.0167$. As suspected, part of the observed effect of gamma globulin may have been due to the lower initial WBC in patients treated with gamma globulin. Thus, we see the estimated OR is closer to 1 after adjustment for WBC. The crude estimate indicates that gamma globulin reduces the odds by just over 80% (the crude $\hat{OR} = 0.19$), while the adjusted $\hat{OR}$ of 0.27 indicates that gamma globulin reduces the odds by 73%. Although the adjustment appears to reduce the impact of gamma globulin, the effect is still quite important and the hypothesis of no effect is rejected with $p = 0.0167$.

■

Another application of multiple logistic regression is in model prediction. The multiple logistic regression model can be used to predict probabilities of an event for subjects with specific characteristics (i.e., values of the independent variables included in the model).

SAS EXAMPLE 11.4 Multiple Logistic Regression Relating White Blood Count and Hemoglobin to Coronary Abnormalities

Suppose we want to use the sample described in SAS Example 11.1 to predict the probability of developing CA as a function of both white blood count and hemoglobin. The following output is based on a multiple logistic regression in which the outcome is the dichotomous variable, CA, and the two independent variables are the continuous variables, WBC and HEM.

SAS Output for Example 11.4

```
              The LOGISTIC Procedure


                 Model Information

     Data Set                    WORK.ONE
     Response Variable           CA
     Number of Response Levels   2
     Number of Observations      151
     Model                       binary logit
     Optimization Technique      Fisher's scoring
```

```
                      Response Profile

        Ordered                        Total
         Value              CA        Frequency

            1               1             25
            2               0            126


           Probability modeled is CA=1.
```

NOTE: 17 observations were deleted due to missing values for the response or explanatory variables.

```
              Model Convergence Status

     Convergence criterion (GCONV=1E-8) satisfied.

               Model Fit Statistics
                                       Intercept
                       Intercept          and
      Criterion          Only         Covariates

      AIC               137.532         124.420
      SC                140.549         133.472
      -2 Log L          135.532         118.420


       Testing Global Null Hypothesis: BETA=0

  Test                Chi-Square       DF      Pr > ChiSq

  Likelihood Ratio     17.1120          2        0.0002
  Score                19.1517          2        <.0001
  Wald                 14.1978          2        0.0008
              The LOGISTIC Procedure


        Analysis of Maximum Likelihood Estimates

                          Standard         Wald
  Parameter   DF  Estimate   Error    Chi-Square   Pr > ChiSq

  Intercept    1    1.2859   2.2545      0.3253       0.5684
  HEM          1   -0.4272   0.2078      4.2294       0.0397
  WBC          1    0.1135   0.0376      9.0948       0.0026
```

```
                   Odds Ratio Estimates

                     Point            95% Wald
       Effect      Estimate       Confidence Limits

       HEM          0.652        0.434        0.980
       WBC          1.120        1.041        1.206
```

```
Association of Predicted Probabilities and Observed Responses

      Percent Concordant      75.0      Somers' D    0.504
      Percent Discordant      24.6      Gamma        0.506
      Percent Tied             0.5      Tau-a        0.140
      Pairs                   3150      c            0.752
```

Interpretation of SAS Output for Example 11.4

We first consider the overall null hypothesis that all the β coefficients in (11.11) are zero. SAS provides three tests of this hypothesis; we recommend the *likelihood ratio test*. In this example, the likelihood ratio test statistic $\chi^2 = 17.11$, with 2 df and the associated $p < 0.0002$. Therefore, we reject H_0 and conclude that at least one of the regression coefficients is not equal to zero. We then look at the individual tests H_0: $\beta_1 = 0$ and H_0: $\beta_2 = 0$, where β_1 and β_2 are the coefficients associated with HEM and WBC, respectively. We reject H_0: $\beta_1 = 0$, with $p = 0.0397$, and conclude that hemoglobin is significantly associated with the development of CA, even after adjusting for white blood count. We also reject H_0: $\beta_2 = 0$, with $p = 0.0026$, and conclude that white blood count is significantly associated with the development of CA, even after adjusting for hemoglobin. In fact, we conclude that hemoglobin and white blood count each make an independent contribution to the risk of CA. ■

We now turn to the prediction of the risk of CA given hemoglobin and white blood count. In SAS Example 11.1, we used logistic regression to estimate the probability of CA given WBC. The estimated probability was

$$\hat{p} = \frac{e^{-3.385+0.125(\text{WBC})}}{(1 + e^{-3.385+0.125(\text{WBC})})}$$

We used this to estimate the risk of CA for a patient with WBC = 10 and obtained the estimate $\hat{p} = 0.106$.

The output for SAS Example 11.4 allows us to estimate the probability of CA given both white blood count and hemoglobin.

$$\hat{p} = \frac{e^{1.286-0.427(\text{HEM})+0.114(\text{WBC})}}{(1 + e^{1.286-0.427(\text{HEM})+0.114(\text{WBC})})}$$

Now, suppose WBC $= 10$ and HEM $= 10$. We estimate the risk of CA by substituting these values for WBC and HEM into the equation.

$$\hat{p} = \frac{e^{1.286-0.427(10)+0.114(10)}}{(1 + e^{1.286-0.427(10)+0.114(10)})}$$

$$= \frac{e^{-1.844}}{(1 + e^{-1.844})}$$

$$= 0.137$$

Similarly, we can estimate the risk of CA given WBC $= 10$ and HEM $= 12$, and obtain the estimated risk $\hat{p} = 0.063$. Notice that as HEM increases, the risk of CA decreases. This can be seen from the negative sign of the coefficient, β_1, associated with HEM. In contrast, β_2 is positive, indicating that the risk of CA increases as WBC increases.

The odds ratio describing the impact of a 1-unit increase in the independent variable on the risk of the event, adjusting for the other independent variables, is automatically displayed. In this example, an increase of 1 unit in HEM is associated with a 35% decrease in the odds of CA ($\hat{OR} = 0.65$). An increase of 1 unit in WBC is associated with a 12% increase in the odds of CA ($\hat{OR} = 1.12$).

11.4 Area under the ROC Curve

We use maximum likelihood estimation to estimate the coefficients in logistic regression models as opposed to the least squares estimation used in linear regression models. One application of logistic regression analysis is the comparison of risks between groups. We use the estimated coefficients to estimate odds ratios and relative risks. Another application of logistic regression is in prediction. The estimated coefficients may be used to estimate or predict probabilities of the event for subjects with specified characteristics (i.e., values on the independent variables included in the model). In this application, we may also want to assess the extent to which the set of independent variables is associated with the event.

It is relatively easy to compare pairs of observed and predicted values in the linear regression framework, as they are both measured on the same scale. For example, pairs $(Y, \hat{Y})$ can be compared by simple subtraction, $(Y - \hat{Y})$ or by squaring the difference, $(Y - \hat{Y})^2$. A summary over the Y values in the sample can be made by summing the squares of the differences (called the error sums of squares, denoted SSE) or by calculating $R^2 = \sum(\hat{Y} - \bar{Y})^2 / \sum(Y - \bar{Y})^2$. In linear regression modeling, R^2 may be interpreted as the percent of variability in Y that can be explained by the set of independent variables. This gives an idea of how well the selected model fits or how close the predicted values ($\hat{Y}$) are to the observed values, Y.

Unfortunately, this statistic has no obvious interpretation in the logistic regression setting and depends on the overall event prevalence. Consider pairs of observed and predicted values, $(Y, \hat{p})$. The Y values are all equal to either 0 or 1 and the $\hat{p}$ values are all between 0 and 1. Clearly, if $\hat{p}$ is equal to or close to 1 for an individual who had the event ($Y = 1$), we would say that the model was successful in predicting that individual's risk. However, what if $\hat{p} = 0.8$ or 0.25? Suppose that the model only yields the following predicted probabilities: 0, 0.06, 0.25, and 1.00. Suppose also that all the individuals who had the event had $\hat{p} = 0.25$ and those who did not have the event had $\hat{p} = 0.06$. The model seems to distinguish those who did and did not have the event even though the individual $\hat{p}$ values are not particularly close to the observed Y values. We would like a summary measure of the goodness-of-fit that takes into account whether the model distinguishes those who do and do not have the outcome.

Recall in SAS Example 11.2 that the risk of CA in patients treated with gamma globulin was estimated using logistic regression to be $\hat{p} = 0.25$ and the estimated risk of CA in patients on the standard treatment was $\hat{p} = 0.06$. Suppose we decide to classify individuals in terms of predicting event status by using a cutoff in predicted probability. We predict that individuals with $\hat{p} > 0.06$ will have the event and assign them the predicted value $\hat{Y} = 1$. We predict that individuals with $\hat{p} \leq 0.06$ will not have the event and assign them the predicted value $\hat{Y} = 0$.

Five of the 26 patients who did develop CA ($Y = 1$) had been treated with gamma globulin and have $\hat{p} = 0.06$; the other 21 were on standard treatment and have $\hat{p} = 0.25$. Thus, if we apply a cutoff of 0.06 and decide that patients with $\hat{p} > 0.06$ are predicted to have CA, then we have correctly predicted event status in 21 of the 26 patients who actually did have the event. This is called the *true positive rate,* or *sensitivity*. In this example, the sensitivity is $21/26 = 0.808$.

Conversely, 78 of the 141 patients who did not develop CA ($Y = 0$) had been on standard treatment and have $\hat{p} = 0.25$. If we again apply a cutoff of 0.06 (assigning patients with $\hat{p} > 0.06$ to have $\hat{Y} = 1$, and those with $\hat{p} \leq 0.06$ to have $\hat{Y} = 0$), then we have correctly predicted event status in 78 of the 141 patients who actually did not have the event. This is called the *true negative rate,* or *specificity*. In this example, the specificity is $78/141 = 0.553$.

Now consider the cutoff of 0 (individuals with $\hat{p} > 0$ will be assigned $\hat{Y} = 1$). Since all the $\hat{p}$ values are either $= 0.06$ or 0.25, all the patients have $\hat{p} > 0$ and thus all the patients will be assigned $\hat{Y} = 1$. The sensitivity $= 26/26 = 1$, and specificity $= 0/141 = 0$. The other extreme is the situation in which the cutoff is 1, all the patients are assigned $\hat{Y} = 0$, the sensitivity is 0, and the specificity is 1. Any other cutoffs will produce the same sensitivity-specificity combination as one of these three cutoffs. For example, any cutoff between 0 and 0.06 will have sensitivity and specificity 1 and 0, respectively. Cutoffs between 0.06 and 0.25 will have sensitivity and specificity 0.808 and 0.553, respectively, and cutoffs between 0.25 and 1 will have sensitivity and specificity 0 and 1, respectively.

Models with only one dichotomous independent variable produce only two possible predicted probabilities: individuals with $X = 0$ have predicted probability $\hat{p}_0$, and those with $X = 1$ have predicted probability $\hat{p}_1$. For example, the only possible predicted values in the example were $\hat{p}_0 = 0.25$ and $\hat{p}_1 = 0.06$. As a result, there is only one cutoff (in addition to cutoffs of 0 and 1) that produces distinct values of sensitivity and specificity. Models with one continuous independent variable or more than one independent variable all have more possible values for $\hat{p}$, and there may be many cutoffs that produce distinct values of sensitivity and specificity.

A measure of goodness-of-fit often used to evaluate the fit of a logistic regression model is based on the simultaneous measure of sensitivity and specificity for all possible cutoff points. First, we calculate sensitivity and specificity pairs for each possible cutoff point and plot sensitivity on the y axis by $(1 - \text{specificity})$ on the x axis. This curve is called the *receiver operating characteristic* (ROC) curve, and the area under the curve, the "*area under the ROC curve,*" is a summary of the sensitivities and specificities over all possible cutoff points. It ranges from 0.5 to 1.0, with larger values indicative of better fit. The following figure shows an application in which the area under the ROC curve (the area to the right under the curve) is equal to 0.68.

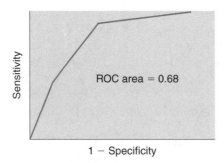

The following figure shows an application in which the area under the ROC curve is 0.5. When the area under the ROC curve is 0.5, the logistic model is said to classify events and nonevents simply by chance or at random.

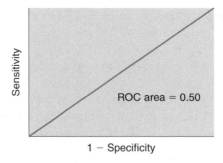

SAS EXAMPLE 11.5 **Area Under the ROC Curve**

Proc Logistic produces the "c" statistic automatically in the output of a logistic regression analysis; this is equal to the area under the ROC curve. We can compare the models used in SAS Examples 11.2, 11.3, and 11.4, each of which was used to predict the development of CA. Recall that the independent variables in SAS Examples 11.2, 11.3, and 11.4 were {gamma globulin alone}, {gamma globulin plus white blood count}, and {white blood count plus hemoglobin}, respectively.

SAS Output for Example 11.2 (Continued)

```
                    The LOGISTIC Procedure

         Analysis of Maximum Likelihood Estimates
```

Parameter	DF	Estimate	Standard Error	Wald Chi-Square	Pr > ChiSq
Intercept	1	-1.0986	0.2520	19.0094	<.0001
gg	1	-1.6487	0.5257	9.8370	0.0017

```
   Association of Predicted Probabilities and Observed Responses
```

Percent Concordant	44.7	Somers' D	0.361
Percent Discordant	8.6	Gamma	0.677
Percent Tied	46.7	Tau-a	0.095
Pairs	3666	c	0.680

SAS Output for Example 11.3 (Continued)

```
                    The LOGISTIC Procedure

         Analysis of Maximum Likelihood Estimates
```

Parameter	DF	Estimate	Standard Error	Wald Chi-Square	Pr > ChiSq
Intercept	1	-2.6352	0.6562	16.1283	<.0001
gg	1	-1.3152	0.5496	5.7272	0.0167
WBC	1	0.1048	0.0383	7.4781	0.0062

```
                         Odds Ratio Estimates


                      Point                 95% Wald
           Effect    Estimate           Confidence Limits


            gg         0.268           0.091         0.788
            WBC        1.111           1.030         1.197


   Association of Predicted Probabilities and Observed Responses


            Percent Concordant    75.5    Somers' D    0.518
            Percent Discordant    23.8    Gamma        0.521
            Percent Tied           0.7    Tau-a        0.140
            Pairs                 3275    c            0.759
```

SAS Output for Example 11.4 (Continued)

```
                       The LOGISTIC Procedure


             Analysis of Maximum Likelihood Estimates


                             Standard            Wald
   Parameter    DF    Estimate      Error    Chi-Square    Pr > ChiSq


   Intercept     1      1.2859     2.2545        0.3253        0.5684
   HEM           1     -0.4272     0.2078        4.2294        0.0397
   WBC           1      0.1135     0.0376        9.0948        0.0026


   Association of Predicted Probabilities and Observed Responses


            Percent Concordant    75.0    Somers' D    0.504
            Percent Discordant    24.6    Gamma        0.506
            Percent Tied           0.5    Tau-a        0.140
            Pairs                 3150    c            0.752
```

Interpretation of SAS Output for Example 11.4 (Continued)

The logistic regression model used to predict the development of CA based on treatment with gamma globulin alone (SAS Example 11.2) has $c = 0.680$. The addition of white blood count to the model increases c to 0.759 (SAS Example 11.3). Thus the model that includes white blood count fits better, as measured by c, the area under the ROC curve.

This model in SAS Example 11.4, which predicts the development of CA based on the combination of white blood count and hemoglobin, has $c = 0.752$. ■

11.5 Key Formulas

Application	Notation/Formula	Description
Logistic function	$L = \beta_0 + \beta_1 X + \varepsilon$ $p = \dfrac{e^L}{(1 + e^L)}$	$L = \ln(\text{odds of event } Y)$ Y is the dichotomous event of interest X is the independent variable β_0 is the intercept β_1 is the coefficient ε is the random error
Logistic regression model	$p = \dfrac{e^{\beta_0 + \beta_1 X + \varepsilon}}{(1 + e^{\beta_0 + \beta_1 X + \varepsilon})}$	
Estimated logistic regression equation	$\hat{p} = \dfrac{e^{\hat{\beta}_0 + \hat{\beta}_1 X}}{(1 + e^{\hat{\beta}_0 + \hat{\beta}_1 X})}$	
Estimate of the odds given X	$\hat{o} = e^{\hat{\beta}_0 + \hat{\beta}_1 X}$	
Estimate of the odds ratio	$\hat{OR} = e^{\hat{\beta}_1 (x_a - x_b)}$	compares odds of the event for those with $X = x_a$ to those with $X = x_b$
95% confidence interval for OR	$\left(e^{(\hat{\beta}_1 - 1.96 s.e.(\hat{\beta}_1))(x_a - x_b)}, \right.$ $\left. e^{(\hat{\beta}_1 + 1.96 s.e.(\hat{\beta}_1))(x_a - x_b)} \right)$	
Estimate of the OR in the special case where $x_a = 1$ and $x_b = 0$	$\hat{OR} = e^{\hat{\beta}_1 (1 - 0)} = e^{\hat{\beta}_1 (1)} = e^{\hat{\beta}_1}$	
Estimate of the relative risk	$\hat{RR} = \hat{p}_a / \hat{p}_b = \dfrac{\left(\dfrac{e^{\hat{\beta}_0 + \hat{\beta}_1 x_a}}{1 + e^{\hat{\beta}_0 + \hat{\beta}_1 x_a}} \right)}{\left(\dfrac{e^{\hat{\beta}_0 + \hat{\beta}_1 x_b}}{1 + e^{\hat{\beta}_0 + \hat{\beta}_1 x_b}} \right)}$	compares the risks of the event for those with $X = x_a$ to those with $X = x_b$
Estimate of the RR in the special case where $x_a = 1$ and $x_b = 0$	$\hat{RR} = \hat{p}_1 / \hat{p}_0 = \dfrac{\left(\dfrac{e^{\hat{\beta}_0 + \hat{\beta}_1}}{1 + e^{\hat{\beta}_0 + \hat{\beta}_1}} \right)}{\left(\dfrac{e^{\hat{\beta}_0}}{1 + e^{\hat{\beta}_0}} \right)}$	

11.6 Statistical Computing

Following are the SAS programs used to perform logistic regression analysis. The SAS procedures and brief descriptions are noted in the header to each example. Notes are provided to the right of the SAS programs (in blue) for orientation purposes and are not part of the programs. In addition, the blank lines in the programs are solely to accommodate the notes. Blank lines and spaces can be used throughout SAS programs to enhance readability. A summary of the SAS procedures used in the examples is provided at the end of this section.

SAS EXAMPLE 11.1 **Estimate Relative Risk and Odds Ratios in Logistic Regression Model**

Estimate Relative Risk and Odds Ratio of CA Comparing Patients with Kawasaki Syndrome Who Have Different White Blood Counts (Example 11.1)

A trial of gamma globulin in the treatment of children with Kawasaki syndrome randomized approximately half of the patients to receive gamma globulin plus aspirin; the other half received the standard treatment of aspirin. The outcome of interest was the development of coronary abnormalities (CA) within 7 weeks of treatment. A possible explanation for the effect of treatment with gamma globulin on CA is that it reduces the elevated white blood count (WBC) early in the course of the disease and that WBC is related to CA. Use SAS to estimate a simple logistic regression model relating WBC to the development of CA.

Program Code

```
options ps=62 ls=80;
```
Formats the output page to 62 lines in length and 80 columns in width

```
data in;
   input gg wbc ca;
```
Beginning of Data Step

Inputs three variables, **gg** (noGG or GG), **wbc** (white blood count), and **ca** (CA or noCA)

```
cards;
0 9.4 0
0 20.4 1
0 8.7 0
.
1 5.7 0
run;
proc logistic descending;
 model ca=wbc;
```
Beginning of Raw Data section.

actual observations (value of **group**, **wbc**, and **event** on each line)

Procedure call. Proc logistic provides estimates of the coefficients in a logistic regression model. The procedure is set up for outcomes with values of 1 or 2 instead of 0 and 1. The option "descending" ensures that outcomes of 1 are counted as events and outcomes of 0 are counted as nonevents. Proc logistic also provides estimates of the odds ratio and 95% confidence limits comparing the odds of the event (ca) for two people with a difference of 1 unit (in wbc).

```
run;
```
End of procedure section ■

SAS EXAMPLE 11.2 **Estimate Relative Risk and Odds Ratios in Logistic Regression Model**

Estimate Relative Risk and Odds Ratio of CA Comparing GG to no GG (Example 11.2)

A trial of gamma globulin in the treatment of children with Kawasaki syndrome randomized approximately half of the patients to receive gamma globulin plus aspirin; the other half received the standard treatment of aspirin. The outcome of interest was the development of coronary abnormalities (CA) within 7 weeks of treatment. Use SAS to estimate the effect of treatment on development of CA.

Program Code

`options ps=62 ls=80;`	Formats the output page to 62 lines in length and 80 columns in width
`data in;`	Beginning of Data Step
`  input gg wbc ca;`	Inputs three variables, **gg** (noGG or GG), **wbc** (white blood count), and **ca** (CA or noCA)
`cards;`	Beginning of Raw Data section.
`0 9.4 0`	actual observations (value of **group, wbc,** and **event** on each line)
`0 20.4 1`	
`0 8.7 0`	
`.`	
`.`	
`.`	
`1 5.7 0`	
`run;`	
`proc logistic descending;`	Procedure call. Proc logistic provides estimates of the coefficients in a logistic regression model. The procedure is set up for outcomes with values of 1 or 2 instead of 0 and 1. The option "descending" ensures that outcomes of 1 are counted as events and outcomes of 0 are counted as nonevents. Proc logistic also provides estimates of the odds ratio and 95% confidence limits comparing the odds of the event (ca) in those with gg=1 as compared to those with gg=0
`  model ca=gg;`	
`run;`	End of procedure section ■

SAS EXAMPLE 11.3 **Estimate Adjusted Relative Risk and Odds Ratios in Logistic Regression Model**

Estimate the Odds Ratio Comparing Patients on Gamma Globulin to Those on Standard Treatment with Respect to CA, Adjusted for WBC (Example 11.3)

Suppose we want to use the sample described in SAS Example 11.2 to assess the effect of treatment with gamma globulin on CA but we are concerned that white blood count is a potential confounder. Use SAS to estimate a multiple logistic regression model where the outcome is the dichotomous variable, CA, and the two independent variables are the dichotomous grouping variable, GG, and the continuous variable, WBC.

Program Code

`options ps=62 ls=80;`	Formats the output page to 62 lines in length and 80 columns in width
`data in;`	Beginning of Data Step
`  input gg wbc ca;`	Inputs three variables, **gg** (noGG or GG), **wbc** (white blood count), and **ca** (CA or noCA)
`cards;`	Beginning of Raw Data section.
`0 9.4 0`	actual observations (value of **group, wbc,** and **event** on each line)
`0 20.4 1`	
`0 8.7 0`	
`.`	
`.`	
`.`	
`1 5.7 0`	
`run;`	
`proc logistic descending;`	Procedure call. Proc logistic provides estimates of
`  model ca=gg wbc;`	the coefficients in a logistic regression model. Proc Logistic also provides estimates of the odds ratio and 95% confidence limits comparing the odds of the event (ca) in those with gg=1 as compared to those with gg=0, adjusted for wbc.
`run;`	End of procedure section ■

SAS Example 11.4 Estimate a Multiple Logistic Regression Model

Perform a Multiple Logistic Regression, Predicting the Risk of Developing CA Given Both Hemoglobin and White Blood Count (Example 11.4)

Suppose we want to use the sample described in SAS Example 11.2 to predict the probability of developing CA as a function of both white blood count and hemoglobin. Use SAS to estimate a multiple logistic regression model where the outcome is the dichotomous variable, CA, and the two independent variables are the continuous variables WBC and HEM.

Program Code

`options ps=62 ls=80;`	Formats the output page to 62 lines in length and 80 columns in width
`data in;`	Beginning of Data Step
`  input ca wbc hem;`	Inputs three variables, **ca** (CA or noCA) **wbc** (white blood count), and **hem** (hemoglobin).
`cards;`	Beginning of Raw Data section.
`0 9.4 11.0`	actual observations (value of **group**, **wbc**, and
`1 20.4 10.1`	**event** on each line)
`0 8.7 10.0`	
`.`	
`.`	
`0 5.7 10.2`	
`run;`	
`proc logistic descending;`	Procedure call. Proc logistic provides estimates of
`  model ca=wbc hem;`	the coefficients in a multiple logistic regression model.
`run;`	End of procedure section. ■

11.6.1 Summary of SAS Procedures

The SAS procedure for logistic regression analysis is summarized in the following table. Specific options can be requested in this procedure to produce specific results. These options are shown in italics. Users should refer to the examples in this section for complete descriptions of the procedure and specific options.

Procedure	Sample Procedure Call	Description

proc logistic	proc logistic *descending*;
	    model y = x1 x2 x3 x4*/rl*;
	    units x1 = d;
		Estimates a logistic regression model relating dichotomous dependent variable Y to independent variables X_1, X_2, X_3, X_4. The descending option considers the response with the higher numerical code the outcome of interest (e.g., 1 = outcome of interest, 0 = comparison). The units option produces CI for the odds ratios relative to a difference of d units (e.g., $d = 5$).

11.7 Analysis of Framingham Heart Study Data

The Framingham data set includes data collected from the original cohort. Participants contributed up to three examination cycles of data. Here we analyze data collected in the first examination cycle (called the period = 1 examination) and use SAS Proc Logistic to investigate predictors of incident hypertension. The candidate predictor variables are age, gender, and body mass index. For this analysis, we exclude persons who are underweight (BMI < 18.5) because analysis has shown that there may be a different relationship between very low BMI and other risk factors. We also create an indicator variable for male gender and two variables to represent BMI categories. The two BMI variables represent overweight and obese classifications. When they are included in the model together, we can estimate the effect of each of these BMI categories compared to normal-weight persons (called the referent group) on incident hypertension. The SAS code to create the analytic data set along with the procedure calls is shown next.

Framingham Data Analysis—SAS Code

```
data fhs;
  set in.frmgham;
  if period=1;

if bmi<18.5 then delete;

if sex=1 then male=1;
else if sex=2 then male=0;

if 25.0 le bmi lt 30.0 then overwt=1; else overwt=0;
if bmi ge 30.0 then obese=1; else obese=0;
run;

proc logistic descending data=fhs;
  model hyperten=age male bmi;
run;

proc logistic descending data=fhs;
  model hyperten=age male bmi;
  units age=10 bmi=5;
run;

proc logistic descending data=fhs;
  model hyperten=age male overwt obese;
  units age=10;
run;
```

Framingham Data Analysis—SAS Output

The LOGISTIC Procedure
Model Information

Data Set   WORK.FHS
Response Variable   HYPERTEN   Incident Hypertension
Number of Response Levels   2
Number of Observations   4358
Model   binary logit
Optimization Technique   Fisher's scoring

Response Profile

Ordered Value	HYPERTEN	Total Frequency
1	1	3210
2	0	1148

Probability modeled is HYPERTEN=1.

Model Convergence Status
Convergence criterion (GCONV=1E-8) satisfied.

Model Fit Statistics

Criterion	Intercept Only	Intercept and Covariates
AIC	5027.711	4606.015
SC	5034.091	4631.534
-2 Log L	5025.711	4598.015

Testing Global Null Hypothesis: BETA=0

Test	Chi-Square	DF	Pr > ChiSq
Likelihood Ratio	427.6957	3	<.0001
Score	378.3059	3	<.0001
Wald	350.6524	3	<.0001

The LOGISTIC Procedure

Analysis of Maximum Likelihood Estimates

Parameter	DF	Estimate	Standard Error	Wald Chi-Square	Pr > ChiSq
Intercept	1	-5.4069	0.3419	250.0608	<.0001
AGE	1	0.0519	0.00440	138.6171	<.0001
MALE	1	-0.2496	0.0738	11.4306	0.0007
BMI	1	0.1582	0.0113	196.4913	<.0001

```
                    Odds Ratio Estimates
                 Point              95% Wald
       Effect   Estimate       Confidence Limits
       AGE        1.053         1.044        1.062
       MALE       0.779         0.674        0.900
       BMI        1.171         1.146        1.198
```

Association of Predicted Probabilities and Observed Responses

```
    Percent Concordant     69.7   Somers' D  0.397
    Percent Discordant     29.9   Gamma      0.399
    Percent Tied            0.4   Tau-a      0.154
    Pairs              3685080    c          0.699
```

Wald Confidence Interval for Adjusted Odds Ratios

```
    Effect        Unit    Estimate    95% Confidence Limits
    AGE         1.0000      1.053       1.044        1.062
    MALE        1.0000      0.779       0.674        0.900
    BMI         1.0000      1.171       1.146        1.198
```

Is the model significant? How significant is each predictor? Is the effect of each predictor in the expected direction? The odds ratios are often reported as the measures of effect. SAS produces odds ratios relative to a 1-unit increase in each predictor. For example, each additional year of age is associated with a 5% increase in risk of hypertension (estimated $OR = 1.053$). Each additional increase of 1 unit of BMI is associated with a 17% increase in risk of hypertension. In the following logistic regression model, SAS produces the estimates of the OR and associated 95% confidence intervals relative to a 10-year increase in age and a 5-unit increase in BMI. These unit increases are specified by the user (see the preceding SAS program).

```
                    The LOGISTIC Procedure
                      Model Information
Data Set                   WORK.FHS
Response Variable          HYPERTEN          Incident Hypertension
Number of Response Levels  2
Number of Observations     4358
Model                      binary logit
Optimization Technique     Fisher's scoring
```

```
                    Response Profile
          Ordered                      Total
          Value      HYPERTEN      Frequency
            1           1              3210
            2           0              1148
```

Probability modeled is HYPERTEN=1.

```
             Model Convergence Status
      Convergence criterion (GCONV=1E-8) satisfied.
```

```
              Model Fit Statistics
                                        Intercept
                         Intercept         and
          Criterion        Only        Covariates
          AIC             5027.711       4606.015
          SC              5034.091       4631.534
          -2 Log L        5025.711       4598.015
```

```
          Testing Global Null Hypothesis: BETA=
     Test                 Chi-Square       DF       Pr > ChiSq
     Likelihood Ratio      427.6957         3         <.0001
     Score                 378.3059         3         <.0001
     Wald                  350.6524         3         <.0001
```

The LOGISTIC Procedure

Analysis of Maximum Likelihood Estimates

Parameter	DF	Estimate	Standard Error	Wald Chi-Square	Pr > ChiSq
Intercept	1	-5.4069	0.3419	250.0608	<.0001
AGE	1	0.0519	0.00440	138.6171	<.0001
MALE	1	-0.2496	0.0738	11.4306	0.0007
BMI	1	0.1582	0.0113	196.4913	<.0001

Odds Ratio Estimates

Effect	Point Estimate	95% Wald Confidence Limits	
AGE	1.053	1.044	1.062
MALE	0.779	0.674	0.900
BMI	1.171	1.146	1.198

```
Association of Predicted Probabilities and Observed Responses
      Percent Concordant         69.7     Somers' D    0.397
      Percent Discordant         29.9     Gamma        0.399
      Percent Tied                0.4     Tau-a        0.154
      Pairs                   3685080     c            0.699
```

```
          Wald Confidence Interval for Adjusted Odds Ratios

   Effect      Unit     Estimate    95% Confidence Limits
   AGE      10.0000       1.680        1.541       1.831
   BMI       5.0000       2.206        1.975       2.464
```

The odds ratios above can be interpreted as follows. Ten years of aging is associated with an increase in risk of hypertension of 68%. An increase of 5 units of BMI is associated with more than a twofold increase in risk of hypertension. In the next model, we consider BMI categories; the two indicator variables reflect overweight and obese categories, and the referent category is normal weight. An interpretation of the effect of BMI is given after the output.

```
                    The LOGISTIC Procedure
                       Model Information
Data Set                    WORK.FHS
Response Variable           HYPERTEN          Incident Hypertension
Number of Response Levels   2
Number of Observations      4358
Model                       binary logit
Optimization Technique      Fisher's scoring
```

```
                      Response Profile
                 Ordered                     Total
                   Value    HYPERTEN      Frequency
                       1           1           3210
                       2           0           1148
```

```
            Probability modeled is HYPERTEN=1.

                  Model Convergence Status
         Convergence criterion (GCONV=1E-8) satisfied.
```

Model Fit Statistics

Criterion	Intercept Only	Intercept and Covariates
AIC	5027.711	4634.628
SC	5034.091	4666.527
-2 Log L	5025.711	4624.628

Testing Global Null Hypothesis: BETA=0

Test	Chi-Square	DF	Pr > ChiSq
Likelihood Ratio	401.0833	4	<.0001
Score	367.6993	4	<.0001
Wald	330.8209	4	<.0001

The LOGISTIC Procedure
Analysis of Maximum Likelihood Estimates

Parameter	DF	Estimate	Standard Error	Wald Chi-Square	Pr > ChiSq
Intercept	1	-1.9005	0.2163	77.1786	<.0001
AGE	1	0.0528	0.00440	143.6058	<.0001
MALE	1	-0.2189	0.0738	8.7967	0.0030
OVERWT	1	0.7001	0.0770	82.5950	<.0001
OBESE	1	1.7752	0.1599	123.2305	<.0001

Odds Ratio Estimates

Effect	Point Estimate	95% Wald Confidence Limits	
AGE	1.054	1.045	1.063
MALE	0.803	0.695	0.928
OVERWT	2.014	1.732	2.342
OBESE	5.901	4.313	8.073

Association of Predicted Probabilities and Observed Responses

Percent Concordant	68.8	Somers' D	0.385
Percent Discordant	30.3	Gamma	0.388
Percent Tied	0.9	Tau-a	0.149
Pairs	3685080	c	0.692

```
         Wald Confidence Interval for Adjusted Odds Ratios
      Effect          Unit      Estimate      95% Confidence Limits
      AGE          10.0000        1.695        1.555        1.847
```

Persons who are overweight are 2.014 times more likely to develop hypertension as compared to persons of normal weight; persons who are obese are 5.901 times more likely to develop hypertension as compared to normal-weight persons, adjusting for age and gender.

It is sometimes of interest to compare different models in terms of their fit, and SAS produces a number of statistics for this purpose. One such statistic is the AIC statistic, a smaller value of which indicates better fit. Which model is better, the one with continuous BMI or the one with BMI categories?

11.8 Problems

1. A placebo-controlled clinical trial of a new treatment for chronic pain was performed on 200 patients in a pain clinic: 100 patients were given the new treatment and 100 were given a placebo. A pain questionnaire was administered to all 200 patients 1 week later. Based on the questionnaire, patients were classified as having moderate to severe pain (pain = 1) or mild or no pain (pain = 0).

   a. Use the following output from a logistic regression to estimate the probability of pain in patients on placebo (newtreat = 0).
   b. Use the following output from a logistic regression to estimate the probability of pain in patients on the new treatment (newtreat = 1).
   c. Estimate the relative risk of pain comparing those on the new treatment to those on placebo.
   d. Compare this to the estimated odds ratio provided in the output.
   e. Does the odds ratio provide a reasonable estimate of the relative risk?

```
                  The LOGISTIC Procedure

              Analysis of Maximum Likelihood Estimates

                              Standard         Wald
   Parameter     DF    Estimate    Error    Chi-Square    Pr > ChiSq

   Intercept      1     -1.2657   0.2414      27.4888        <.0001
   newtreat       1     -0.8251   0.4005       4.2435        0.0394
```

```
                    Odds Ratio Estimates

                  Point              95% Wald
      Effect     Estimate       Confidence Limits

    newtreat      0.438         0.200       0.961
```

In Problems 2–4, suppose we have 30-day follow-up data on 350 ischemic stroke patients and want to investigate whether the risk of recurrent stroke and/or death (RD) depends on the type of stroke. You classify patients according to initial stroke type—having a cerebral embolism (CE = 1) or not (CE = 0)—and perform a series of logistic regression analyses.

2. First, consider the crude relationship between CE and RD. The following output is produced by SAS.

```
                The LOGISTIC Procedure

      Analysis of Maximum Likelihood Estimates

                           Standard       Wald
Parameter    DF   Estimate   Error   Chi-Square   Pr > ChiSq

Intercept    1    -2.8034   0.5149     29.6390      <.0001
ce           1     1.8651   0.6479      8.2874      0.0040

                 Odds Ratio Estimates

                  Point              95% Wald
      Effect     Estimate       Confidence Limits

       ce        6.457         1.814       22.986
```

a. Use the output to estimate the odds ratio, provide a 95% confidence interval, and test H_0: $OR = 1$.
b. What is the risk of RD for CE patients?
c. What is the risk of RD for non-CE patients?
d. Is the risk of RD the same for CE and non-CE patients? Estimate the relative risk of RD.

3. Next, consider the relationship between age and RD. The following output is produced by SAS.

The LOGISTIC Procedure

Analysis of Maximum Likelihood Estimates

Parameter	DF	Estimate	Standard Error	Wald Chi-Square	Pr > ChiSq
Intercept	1	-8.0798	8.1749	0.9769	0.3230
age	1	0.0869	0.1149	0.5714	0.4497

Odds Ratio Estimates

Effect	Point Estimate	95% Wald Confidence Limits	
age	1.091	0.871	1.366

a. What is the odds ratio comparing the odds of RD in patients whose ages are 1 year apart? Provide a 95% confidence interval and test $H_0: OR = 1$.

b. What is the risk of RD for a patient aged 65?

c. What is the risk of RD for a patient aged 66?

d. Estimate the relative risk of RD comparing patients who are 65 and 66.

e. How does this compare with the odds ratio given in the output?

f. What is the risk of RD for a patient aged 75?

g. Estimate the relative risk of RD comparing patients who are 75 to those who are 65.

4. Finally, consider the relationship between CE and RD, adjusted for age.

The LOGISTIC Procedure

Analysis of Maximum Likelihood Estimates

Parameter	DF	Estimate	Standard Error	Wald Chi-Square	Pr > ChiSq
Intercept	1	-15.3151	9.4983	2.5999	0.1069
ce	1	2.0684	0.6822	9.1919	0.0024
age	1	0.1751	0.1318	1.7654	0.1840

Odds Ratio Estimates

Effect	Point Estimate	95% Wald Confidence Limits	
ce	7.912	2.078	30.129
age	1.191	0.920	1.542

a. Use the output to estimate the odds ratio comparing the odds of RD in patients with CE to those without CE, adjusted for age. Provide a 95% confidence interval and test $H_0: OR = 1$.

b. How do these adjusted results differ from the crude results in Problem 2a?

5. A clinical trial was designed to investigate the effect of a new treatment on the course of disease among pediatric patients with a particular infectious disease. Fifty boys and fifty girls were randomized to receive the new treatment (newtreat = 1) or the standard treatment (newtreat = 0), and the outcome was hospitalization (hospital = 1 or 0).

a. What is the odds ratio comparing patients treated with the new treatment to patients receiving standard treatment with respect to the odds of hospitalization? Test $H_0: OR = 1$.

The LOGISTIC Procedure

Analysis of Maximum Likelihood Estimates

Parameter	DF	Estimate	Standard Error	Wald Chi-Square	Pr > ChiSq
Intercept	1	-0.5754	0.2083	7.6273	0.0057
newtreat	1	-0.9410	0.3334	7.9660	0.0048

Odds Ratio Estimates

Effect	Point Estimate	95% Wald Confidence Limits	
newtreat	0.390	0.203	0.750

b. What is the odds ratio comparing boys treated with the new treatment to boys receiving standard treatment with respect to the odds of hospitalization? Test $H_0: OR = 1$.

```
-------------------- gender=boy --------------------
```

Analysis of Maximum Likelihood Estimates

Parameter	DF	Estimate	Standard Error	Wald Chi-Square	Pr > ChiSq
Intercept	1	-0.0800	0.2831	0.0800	0.7774
newtreat	1	-1.0726	0.4356	6.0626	0.0138

Odds Ratio Estimates

Effect	Point Estimate	95% Wald Confidence Limits	
newtreat	0.342	0.146	0.803

c. What is the odds ratio comparing girls treated with the new treatment to girls receiving standard treatment with respect to the odds of hospitalization? Test $H_0: OR = 1$.

```
-------------------- gender=girl --------------------
```
Analysis of Maximum Likelihood Estimates

Parameter	DF	Estimate	Standard Error	Wald Chi-Square	Pr > ChiSq
Intercept	1	-1.1527	0.3311	12.1175	0.0005
newtreat	1	-0.8397	0.5468	2.3581	0.1246

Odds Ratio Estimates

Effect	Point Estimate	95% Wald Confidence Limits	
newtreat	0.432	0.148	1.261

d. Does the new treatment have the same effect on girls and boys?

SAS Problems

Use SAS to solve the following problems.

1. A clinical trial was conducted to compare a new therapy for HIV to a standard therapy with respect to CD4 cell counts and incident opportunistic infections. A total of 30 subjects with diagnosed HIV were enrolled in the trial, randomized to receive standard or new therapy, and studied for 3 months. The following are data on age and OI (presence of opportunistic infection during the study period), coded 0 or 1 for absent or present, respectively.

Standard Therapy Group

Age	23	34	25	29	33	35	31	26	29	20	31	37	28	45	49
OI	0	1	0	0	0	1	1	1	1	1	1	1	0	1	1

New Therapy Group

Age	30	32	40	37	38	28	47	49	50	43	41	38	37	36	34
OI	0	0	0	0	0	0	0	1	1	0	1	1	0	0	0

    a. Use logistic regression to estimate the odds ratio comparing the therapies with respect to risk of opportunistic infection.

    b. Provide a 95% confidence interval for the odds ratio.

    c. Test the appropriate hypothesis.

    d. Use your output to provide estimates of the risk of opportunistic infection in each of the therapy groups.

    e. Estimate the relative risk of opportunistic infection comparing the new therapy to the standard therapy.

    f. Perform a separate logistic regression to estimate the odds ratio comparing the effect of therapy on the risk of opportunistic infection, adjusted for age. Test the appropriate hypothesis.

Descriptive Statistics
(Ch. 2)

Probability
(Ch. 3)

Sampling Distributions
(Ch. 4)

Statistical Inference
(Chapters 5–13)

OUTCOME VARIABLE	GROUPING VARIABLE(S)/ PREDICTOR(S)	ANALYSIS	CHAPTER(S)
Continuous	—	Estimate μ, Compare μ to Known, Historical Value	5/12
Continuous	Dichotomous (2 groups)	Compare Independent Means (Estimate/Test $(\mu_1 - \mu_2)$) or the Mean Difference (μ_d)	6/12
Continuous	Discrete (>2 groups)	Test the Equality of k Means Using Analysis of Variance $(\mu_1 = \mu_2 = \cdots = \mu_k)$	9/12
Continuous	Continuous	Estimate Correlation or Determine Regression Equation	10/12
Continuous	Several Continuous or Dichotomous	Multiple Linear Regression Analysis	10
Dichotomous	—	Estimate p; Compare p to Known, Historical Value	7
Dichotomous	Dichotomous (2 groups)	Compare Independent Proportions (Estimate/Test $(p_1 - p_2)$)	7/8
Dichotomous	Discrete (>2 groups)	Test the Equality of k Proportions (Chi-Square Test)	7
Dichotomous	Several Continuous or Dichotomous	Multiple Logistic Regression Analysis	11
Discrete	Discrete	Compare Distributions Among k Populations (Chi-Square Test)	7
Time to Event	Several Continuous or Dichotomous	Survival Analysis	13

Nonparametric Tests

In Chapters 5, 6, and 9 we presented techniques for tests of hypothesis concerning one, two (independent and dependent, matched or paired), and more than two population means. Valid application of these tests and procedures requires specific assumptions. These tests assume that the characteristic under investigation is approximately normally distributed or rely on large samples for the application of the Central Limit Theorem. If the samples are small and the analytic variable is clearly not normally distributed, we violate the assumptions for valid application of the procedures. In such cases, alternative methods are required. The techniques we presented in Chapters 5, 6, and 9 are called *parametric* procedures, as they are based on assumptions regarding the distributional form of the analytic variable (e.g., that the analytic variable follows a normal distribution). Methods that do not require such assumptions are called *nonparametric* procedures. Here we present four nonparametric procedures analogous to the two dependent samples test, the two independent samples test, the $k > 2$ independent samples test, and correlation analysis. The hypotheses tested with nonparametric procedures are more general than hypotheses we tested with parametric procedures. For example, the parametric test for two independent means is of the form $H_0: \mu_1 = \mu_2$. The nonparametric analog tests for equality of medians in the two distributions.

In Sections 12.1 and 12.2 we present nonparametric tests for two dependent samples. In Section 12.3 we present a technique for the two independent samples test. In Section 12.4 we present a technique for the $k > 2$ independent samples test, and in Section 12.5 we present a nonparametric approach to correlation analysis. We summarize key formulas in Section 12.6, and in Section 12.7 we display the statistical computing programs used to generate the nonparametric analyses presented in this chapter. In Section 12.8 we use data from the Framingham Heart Study to illustrate the applications presented here.

12.1 The Sign Test (Two Dependent Samples Test)

Recall the two dependent (also called matched or paired) samples procedures we presented in Chapter 6. In the examples we discussed, a single random sample of subjects was selected from the population of interest, and two measurements were taken on each subject. Because the measurements were dependent, we calculated differences between measurements and performed a test on the mean difference (e.g., $H_0: \mu_d = 0$). When the sample is small and/or if the differences are not normally distributed, we consider a nonparametric test. One popular nonparametric test for two dependent samples is the Sign test. The null hypothesis in the Sign test is that the median difference is zero. We illustrate the test using an example.

EXAMPLE 12.1

Test for Mean Difference in Systolic Blood Pressures Using Sign Test

Suppose we are investigating a new drug hypothesized to lower systolic blood pressure. We select a random sample of subjects and record their resting systolic blood pressure at baseline (i.e., prior to the investigation). Each subject is then given the new drug, and resting systolic blood pressure is measured again after 4 weeks of drug treatment. The data are shown in the following table. Based on the data, does it appear that the new drug lowers systolic blood pressure?

Subject Identification Number	Baseline Systolic Blood Pressure	Post-Treatment Systolic Blood Pressure
1	166	138
2	135	120
3	189	176
4	180	180
5	156	160
6	142	150
7	176	152
8	156	140
9	164	160
10	142	130

In the Sign test, as we did with the parametric test in Chapter 6, we focus on differences between measurements. In this application, we are investigating whether there is a significant reduction in systolic blood pressure following drug treatment. In the following table we display the differences between systolic blood pressures (baseline–post-treatment blood pressures). In the right-most column of the table, we record the sign of the difference (+ or −, or 0 if there is no difference).

Subject Identification Number	Baseline Systolic Blood Pressure	Post-Treatment Systolic Blood Pressure	Difference (Baseline–Post-Treatment)	Sign of Difference
1	166	138	28	+
2	135	120	15	+
3	189	176	13	+
4	180	180	0	0
5	156	160	−4	−
6	142	150	−8	−
7	176	152	24	+
8	156	140	16	+
9	164	160	4	+
10	142	130	12	+

The Sign test is based on the binomial distribution. There are a total of 10 difference scores (one for each subject in the sample). If the drug has no effect on systolic blood pressure, we expect the post-treatment systolic blood pressures to be the same as the baseline pressures. Because there are many factors that affect blood pressure, expecting identical results is unrealistic. If the drug has no effect, we expect some subjects' post-treatment systolic blood pressures to be higher than their baseline pressures (negative differences) and some subjects' post-treatment systolic blood pressures to be lower than their baseline pressures (positive differences). If the drug has no effect, we expect these differences to be small, and with respect to the signs of the differences, we expect to see about five + and five − signs (or 50% in each direction).

In Example 12.1 we observed seven (out of ten) + signs. Is a total of seven + signs indicative of a drug effect (here a + sign indicates a difference in the hypothesized direction, i.e., a lowering of systolic blood pressure)? The exact significance of the test is determined from the binomial distribution (discussed in Chapter 3). The null hypothesis is that, for each person, the probability of a + sign is 0.5. We therefore consider a binomial distribution with $n = 10$ (because there are 10 subjects, or trials) and $p = 0.5$. The following values are found in Table B.1 in the Appendix.

Number of Successes, x	$P(X = x \mid n = 10, p = 0.5)$
0	0.0010
1	0.0098
2	0.0439
3	0.1172
4	0.2051
5	0.2461
6	0.2051
7	0.1172
8	0.0439
9	0.0098
10	0.0010

The question of interest is, *How likely is it to observe as many as seven or more successes out of ten when the probability of success is 0.5?* The p value for the test statistic (seven successes or seven + signs out of ten) is the probability of observing seven or more successes. (Recall the p value is defined as the probability of observing a test statistic as more extreme.) In Example 12.1, the p value is $P(X \geq 7) = P(X = 7) + P(X = 8) + P(X = 9) + P(X = 10) = 0.1172 + 0.0439 + 0.0098 + 0.0010 = 0.1719$. We reject H_0 if the p value is small (i.e., 0.05 or less). In Example 12.1, we do not reject H_0. We do not have significant evidence, $p = 0.1719$, to show that the drug lowers systolic blood pressure. ■

There are several options for handling observations in which the difference is zero. We can ignore the zero (i.e., not assign a sign). Another alternative is to count a zero difference as $0.5+$ and $0.5-$. Had we used this strategy, we would have counted $7.5+$ signs in Example 12.1. The p value for the test would then have been $P(X \geq 7.5) = P(X = 8) + P(X = 9) + P(X = 10) = 0.0439 + 0.0098 + 0.0010 = 0.0547$, which is marginally significant.

Note that we performed a one-sided test with the alternative hypothesis that the drug lowers systolic blood pressure. The p value corresponding to a two-sided test with the alternative hypothesis that the drug either raises or lowers systolic blood pressure is simply twice the one-sided p value. In our example, the two-sided p value is $P(X \geq 7 \text{ or } X \leq 3) = 2P(X \geq 7) = 2(0.1719) = 0.3438$.

The analysis of the data presented in Example 12.1 was based solely on the signs ($+$ or $-$) of the differences in blood pressures from baseline to post-treatment. In the parametric test, we also computed differences but focused on the magnitude of those differences. Recall, our test statistic was based on $\overline{X}_d$, the mean of the difference scores. In the Sign test we do not capture the magnitude of the differences, only the direction. For example, if a particular patient has a baseline systolic blood pressure of 150 and a post-treatment blood pressure of 149, we count that observation as evidence in favor of the alternative hypothesis (lowering of blood pressure) because the difference is positive ($+$). Suppose a second patient has a baseline systolic blood pressure of 150 and a post-treatment blood pressure of 129, we again count this observation as evidence in favor of the alternative hypothesis. In the Sign test, there is no distinction between these subjects, when in fact the second has a much more substantial reduction in blood pressure. Because we are basing the Sign test on limited information (only the sign of the difference in measurements), this nonparametric test often has lower power than competing procedures. A second nonparametric procedure for two dependent samples is described in the next section. This procedure incorporates the magnitude of the differences in measurements.

12.2 The Wilcoxon Signed-Rank Test (Two Dependent Samples Test)

The Wilcoxon Signed-Rank test is a second nonparametric alternative for two dependent matched or paired samples. To perform the Wilcoxon Signed-Rank test we compute differences in measurements as we did to perform the Sign test. However, here we also take into account the relative magnitude of these differences. The null hypothesis is that the median difference is zero. We illustrate the test using the data we presented in Example 12.1.

EXAMPLE 12.2

Test for Mean Difference in Systolic Blood Pressures Using Signed-Rank Test

Consider the data presented in Example 12.1. Again, the question of interest is whether there is significant evidence that the new drug lowers systolic blood pressure. The data are shown here along with the difference scores computed by subtracting the post-treatment systolic blood pressures from the baseline pressures.

Subject Identification Number	Baseline Systolic Blood Pressure	Post-Treatment Systolic Blood Pressure	Difference (Baseline–Post-Treatment)
1	166	138	28
2	135	120	15
3	189	176	13
4	180	180	0
5	156	160	−4
6	142	150	−8
7	176	152	24
8	156	140	16
9	164	160	4
10	142	130	12

In the Wilcoxon Signed-Rank test we assign ranks to the *absolute values* of the difference scores (see the third column of the following table). We assign a 1 to the smallest absolute difference, 2 to the next smallest, and so on, up to *n*. For now, we will ignore the case with no difference, although there are alternative methods for handling these cases. If there are ties in the absolute values of the differences, then the mean rank is assigned to both. For example, subjects 5 and 9 had absolute differences of 4 units. These are the first and second smallest differences, so we assign a rank of 1.5 to each. The next (third) smallest absolute difference is measured in subject 6, who is assigned a rank of 3. The ranking continues until all nonzero differences are ranked. Once the ranks are assigned to the absolute differences, we then reattach the signs of the differences (+ or −) to the ranks (see the right-most column of the table).

Subject Identification Number	Difference (Baseline– Post-Treatment)	Ranks of Absolute Values of Differences	Signed Ranks
1	28	9	+9
2	15	6	+6
3	13	5	+5
4	0	—	—
5	−4	1.5	−1.5
6	−8	3	−3
7	24	8	+8
8	16	7	+7
9	4	1.5	+1.5
10	12	4	+4

If the drug has no effect on systolic blood pressures, we expect about half of the post-treatment systolic blood pressures to be higher than the baseline pressures and half of the post-treatment pressures to be lower than the base-line pressures. In addition, we expect the magnitudes of the increases to be about the same as the magnitudes of the decreases. In terms of the signed ranks, if the drug has no effect, we expect the sum of the positive ranks to be approximately equal to the sum of the negative ranks.

The test statistic in the Wilcoxon Signed-Rank test is the sum of the positive ranks, called T. If all the differences are negative, then all the signs will be negative, so all the signed ranks will be negative and T will equal 0. The other extreme is all positive differences, which will yield all positive signed ranks and T will therefore be the sum $1 + 2 + 3 + 4 + \cdots + n = n(n + 1)/2$ (where n represents the number of ranked observations, i.e., the number of observations with nonzero differences). Thus, the sum of positive ranks, T, ranges from 0 to $n(n + 1)/2$. The median value is $n(n + 1)/4$. In Example 12.2, the observed sum of the positive ranks is $T = 40.5$. The maximum possible value of the sum of positive ranks is $9(9 + 1)/2 = 45$, and the median is $9(9 + 1)/4 = 22.5$. The observed test statistic, $T = 40.5$, is very close to the theoretical maximum (i.e., it falls in the tail of the distribution). To draw a conclusion in the test, we compare the observed test statistic to an appropriate critical value. Because the observed test statistic is large, we would expect a very small p value and therefore would likely reject H_0. There are tables of critical values for this test, but one can use the normal approximation to the distribution instead, which is what we illustrate here. SAS runs the Wilcoxon Signed-Rank test and produces a p value based on a normal approximation to the distribution of T. Specifically, SAS converts the sum of positive ranks, T, to a Z score, and then produces a p value from the standard normal distribution. The Z score (12.1) is computed by subtracting the median, $n(n+1)/4$, and dividing by the standard error, which is $\sqrt{\dfrac{n(n+1)(2n+1)}{24}}$.

$$Z = \frac{T - \dfrac{n(n+1)}{4}}{\sqrt{\dfrac{n(n+1)(2n+1)}{24}}} \tag{12.1}$$

In Example 12.2,

$$Z = \frac{40.5 - \dfrac{9(9+1)}{4}}{\sqrt{\dfrac{9(9+1)(2(9)+1)}{24}}} = \frac{18}{8.44} = 2.13$$

The one sided p value is $P(Z \geq 2.13) = 1 - 0.9834 = 0.0166$. Based on the Wilcoxon Signed-Rank test, $p = 0.0166 < 0.05$; we therefore reject H_0: The medians are equal in favor of the alternative, H_1: The median systolic blood pressure is lower post-treatment as compared to baseline. ■

SAS EXAMPLE 12.2 Test for Mean Difference in Systolic Blood Pressures Using Signed-Rank Test in SAS

SAS performs the Wilcoxon Signed-Rank test in its Proc Univariate. We used Proc Univariate in Chapter 2 to generate summary statistics for continuous variables. The following output was generated by SAS on the difference variable (i.e., differences between baseline and post-treatment systolic blood pressures). A brief interpretation appears after the output.

SAS Output for Example 12.2

```
                    The UNIVARIATE Procedure
                      Variable:  diff
                         Moments
N                          10    Sum Weights              10
Mean                       10    Sum Observations        100
Std Deviation       11.785113    Variance          138.888889
Skewness           -0.0712764    Kurtosis          -0.9438912
Uncorrected SS           2250    Corrected SS           1250
Coeff Variation     117.85113    Std Error Mean     3.72677996

             Basic Statistical Measures
      Location                      Variability
  Mean      10.00000    Std Deviation         11.78511
  Median    12.50000    Variance             138.88889
  Mode         .        Range                 36.00000
                        Interquartile Range   16.00000
```

```
                     Tests for Location: Mu0=0
          Test              -Statistic-         -----p Value------
          Student's t    t  2.683282       Pr > |t|     0.0251
          Sign           M       2.5       Pr >= |M|    0.1797
          Signed Rank    S        18       Pr >= |S|    0.0313

                     Tests for Normality
Test                         --Statistic---       -----p Value------
Shapiro-Wilk                 W     0.962454       Pr < W        0.8135
Kolmogorov-Smirnov           D     0.167379       Pr > D       >0.1500
Cramer-von Mises             W-Sq  0.035094       Pr > W-Sq    >0.2500
Anderson-Darling             A-Sq  0.209068       Pr > A-Sq    >0.2500

                     Quantiles (Definition 5)
                      Quantile          Estimate
                      100% Max            28.0
                      99%                 28.0
                      95%                 28.0
                      90%                 26.0
                      75% Q3              16.0
                      50% Median          12.5
                      25% Q1               0.0

                     The UNIVARIATE Procedure
                         Variable:  diff
                     Quantiles (Definition 5)
                      Quantile          Estimate
                      10%                 -6.0
                      5%                  -8.0
                      1%                  -8.0
                      0% Min              -8.0

                     Extreme Observations
          ----Lowest----              ----Highest---
          Value        Obs            Value        Obs
            -8          6               13          3
            -4          5               15          2
             0          4               16          8
             4          9               24          7
            12         10               28          1
```

Interpretation of SAS Output for Example 12.2

The output for the Wilcoxon Signed-Rank test is in the third section of the output. SAS produces the numerator of the test statistic in (12.1) and gives

"Signed Rank S = 18," which is computed by subtracting the median = 22.5 from $T = 40.5$ (S = $40.5 - 22.5 = 18$). SAS then gives a two-sided p value "Pr >= |S| = 0.0313," which is based on the normal approximation. (SAS applies a correction factor to formula (12.1)—see also SAS Example 12.3.) If a one-sided test is desired (as is the case in Example 12.2), we divide $p = 0.0313/2 = 0.016$. Based on the observed p value, $p = 0.016$ (<0.05), we reject H_0: The medians are equal in favor of the alternative, H_1: The median systolic blood pressure is lower post-treatment as compared to baseline.

The Signed-Rank test uses more information in the data than the Sign test in that it incorporates the *relative* magnitude of the values through ranks. Note that the Signed-Rank test rejected the two-sided null hypothesis, with $p = 0.0313$, whereas the Sign test did not reject the same null hypothesis, with $p = 0.3438$. Ranks are particularly useful in the presence of outliers. However, tests based on ranks do not capture the *absolute* magnitude of the differences (in the two dependent samples case) as compared to parametric tests such as the two dependent samples t test.

In the next section of the output (just below the results of the Signed-Rank test), SAS produces several tests for normality. The null hypothesis is H_0: Differences follow a normal distribution. We use the Shapiro–Wilk test, with test statistic $W = 0.962454$ and p value $= 0.8135$. We would not reject H_0 based on the observed p value; therefore, we do not have significant evidence to show that the data do not follow a normal distribution. This test can be useful in assessing whether a parametric test (which assumes that the analytic variable follows a normal distribution) or a nonparametric test should be applied. In this example, we could use the parametric test, as there is no evidence that the differences are not normally distributed. ■

12.3 The Wilcoxon Rank Sum Test (Two Independent Samples Test)

We now present a nonparametric test for two independent samples. In these applications, we have two independent populations defined based on a specific attribute of the subjects under study (e.g., gender) or based on the study design (e.g., some patients are assigned to receive Drug A while others receive Drug B). The parametric procedure tests the equality of population means and assumes large samples ($n_i > 30$, $i = 1, 2$) or, if the sample sizes are small, that the analytic variable under investigation is approximately normally distributed. If these assumptions are not met, a nonparametric test might be appropriate. A popular nonparametric test comparing two independent samples is the Wilcoxon Rank Sum test, in which the null hypothesis is the equality of medians. This test is equivalent to the nonparametric Mann–Whitney U test. We illustrate the test using an example.

EXAMPLE 12.3

Test for Difference in Mean Health Status Between Treatments Using Wilcoxon Rank Sum Test

Suppose we wish to compare two competing treatments for the management of adult-onset diabetes. A total of eight subjects agree to participate in the investigation and are randomly assigned to one of the competing treatments. After following the prescribed treatment regimen for 6 weeks, we assess the patients' self-reported health status. Patients are asked to rate their current health on an ordinal scale from 0 to 20, with higher scores reflecting better health. To anchor the scale, the following instruction is provided: Subjects who feel that their health is poor should report a score of 0, and subjects who feel that their health is excellent should report a score of 20. The test of interest is a two-sided test (i.e., Are self-reported health status scores different between treatment groups?). The data are

Self-Reported Health Status

Treatment 1	Treatment 2
0	8
7	10
11	12
16	15

In this application we have small samples and the analytic variable is ordinal. Therefore, we consider a nonparametric test. To perform the Wilcoxon Rank Sum test, we pool the data from the two groups and order the values from lowest to highest (i.e., from lowest self-reported health status to highest). We then assign ranks to the values in increasing order. We assign a 1 to the lowest value, 2 to the next lowest, and so on, up to $N = n_1 + n_2$. When there are ties, the mean ranks are assigned (similar to the procedure we followed with the Signed-Rank test). The following table contains the ranks of the values:

Ranks

Treatment 1	Treatment 2
1	3
2	4
5	6
8	7

If there is no difference between treatments, we expect to see some low ranks and some high ranks in each group. If there is a difference between treatments, we expect to see clustering of lower ranks in one group and higher ranks in the other.

The Wilcoxon Rank Sum test statistic is S, the smaller of the sums of the ranks in the groups. The sum of all ranks is $N(N + 1)/2$. For Example 12.3, the sum of all ranks is $8(8 + 1)/2 = 36$. If there is no difference between treatments, we expect the sums of the ranks to be about 18 in each group. In the most extreme situation (for $N = 8$), the four smallest values would fall in one group, producing ranks in that group of 1, 2, 3, and 4, and the smaller sum of ranks would be equal to $S = 10$ ($S = 1 + 2 + 3 + 4$). In Example 12.3, the sum of ranks in Treatment 1 is 16 and the sum of ranks in Treatment 2 is 20. The smaller sum is $S = 16$. Is that significantly different from the expected sum (if there were no difference between treatments) of 18?

To draw a conclusion in the test, we compare the observed sum to an appropriate critical value. Again, we will use SAS to run the test. SAS runs the Wilcoxon Rank Sum test and produces a p value based on a normal approximation. Specifically, SAS converts the smaller sum of ranks, S, to a Z score and then produces a p value from the standard normal distribution. The Z score (12.2) is computed by subtracting the mean, $n_1(n_1 + n_2 + 1)/2$, and dividing by the standard error, which is $\sqrt{\dfrac{n_1 n_2(n_1 + n_2 + 1)}{12}}$.

$$Z = \frac{S - \dfrac{n_1(n_1 + n_2 + 1)}{2}}{\sqrt{\dfrac{n_1 n_2(n_1 + n_2 + 1)}{12}}} \tag{12.2}$$

In Example 12.3,

$$Z = \frac{16 - \dfrac{4(4 + 4 + 1)}{2}}{\sqrt{\dfrac{(4)4(4 + 4 + 1)}{12}}} = \frac{-2}{3.46} = -0.58$$

The one-sided p value is $P(Z \le -0.58) = 0.2810$. The two-sided p value $= 2(0.2810) = 0.5620$. Based on the Wilcoxon Rank Sum test, we would not reject H_0. We do not have significant evidence to show a difference between treatments with respect to self-reported health status because $p = 0.5620 > 0.05$. ■

SAS EXAMPLE 12.3 Test for Difference in Mean Health Status Between Treatments Using Wilcoxon Rank Sum Test in SAS

SAS performs the Wilcoxon Rank Sum test in its Proc Npar1way. The following output was generated by SAS. A brief interpretation appears after the output.

SAS Output for Example 12.3

```
                The NPAR1WAY Procedure
   Wilcoxon Scores (Rank Sums) for Variable hlthstat
                Classified by Variable trt
```

trt	N	Sum of Scores	Expected Under H0	Std Dev Under H0	Mean Score
1	4	16.0	18.0	3.464102	4.0
2	4	20.0	18.0	3.464102	5.0

```
                Wilcoxon Two-Sample Test
           Statistic                  16.0000
           Normal Approximation
           Z                          -0.4330
           One-Sided Pr < Z            0.3325
           Two-Sided Pr > |Z|          0.6650


           t Approximation
           One-Sided Pr < Z            0.3390
           Two-Sided Pr > |Z|          0.6780


      Z includes a continuity correction of 0.5.


                   Kruskal-Wallis Test
           Chi-Square                  0.3333
           DF                               1
           Pr > Chi-Square             0.5637
```

Interpretation of SAS Output for Example 12.3

For the Wilcoxon Rank Sum test, SAS produces the sum of the ranks (labeled "Sum of Scores") in each treatment group. SAS then gives the expected sums and the standard deviation (assuming no difference between treatments) and then the mean rank in each group. SAS then shows the sum of the smaller ranks ($S = 16$) and the corresponding Z score. It applies a correction of ½ to formula (12.2) in the standardization as follows:

$$Z = \frac{S - \dfrac{n_1(n_1 + n_2 + 1)}{2} + \dfrac{1}{2}}{\sqrt{\dfrac{n_1 n_2(n_1 + n_2 + 1)}{12}}}$$

For Example 12.3, SAS computes $Z = -0.4330$ (with the correction) and gives a two-sided p value of 0.6650. Based on the Wilcoxon Rank Sum test, we would not reject H_0 because $p = 0.6650 > 0.05$. We do not have significant evidence to show a difference between treatments with respect to self-reported health status.

The Rank Sum test provides a comparison of the relative magnitude of values in the two groups but does not use the actual observed values—as does the two independent samples (t or Z) test. Just below the results of the Rank Sum test, SAS provides an approximate p value for the two independent samples t test for means (Proc Ttest in SAS). For Example 12.3, SAS gives "t Approximation Pr > $|Z| = 0.6780$." Had we run a two independent samples test for equality of means we would not have rejected $H_0: \mu_1 = \mu_2$ because $p = 0.6780 > 0.05$. Note that this is an approximation to the t test. ■

12.4 The Kruskal–Wallis Test (k Independent Samples Test)

We now present a nonparametric procedure for the k independent samples test. This procedure is the nonparametric analog to analysis of variance (ANOVA), discussed in Chapter 9. A popular nonparametric test for the k independent samples test is the Kruskal–Wallis test. Like the Wilcoxon Signed-Rank and Rank Sum tests, the Kruskal–Wallis test is based on assigning ranks to the observed values and then comparing the observed sums of ranks to what would be expected if there were no difference among groups. The computations involved in the Kruskal–Wallis test are similar to those in the Wilcoxon Rank Sum test but are complicated by the fact that there are more groups involved. We illustrate the application of the Kruskal–Wallis test using SAS for the sample data in the next example.

EXAMPLE 12.4

Test for Difference in Mean Symptom Scores Between Treatments Using Kruskal–Wallis Test

Suppose we wish to compare four treatments for seasonal allergies. A total of 20 subjects agree to participate in the investigation and are randomly assigned to one of the four competing treatments. After following the prescribed treatment regimen for 2 weeks, we assess the subjects' status. The outcome variable in this application is an index score based on three distinct symptoms. At the end of treatment, subjects are asked if they are currently experiencing any (or all) of the following symptoms: scratchy throat, itchy eyes, runny nose. Each subject responds "yes" or "no" to each of the three symptoms. The index score is the sum of affirmative responses. The range of scores is 0 (no symptoms) to 3 (all three symptoms). The data are shown in the following table:

Treatment 1	Treatment 2	Treatment 3	Treatment 4
0	2	0	2
0	2	0	3
1	3	1	2
2	3	1	3
3	2	1	3

Is there is a significant difference among the treatments with respect to symptom scores? Here again, we have small samples and an analytic variable with limited response options. Therefore, we consider a nonparametric test. ■

SAS Example 12.4 Test for Difference in Mean Symptom Scores Between Treatments Using Kruskal–Wallis Test in SAS

SAS performs the Kruskal–Wallis test in its Proc Npar1way. The following output was generated by SAS. A brief interpretation appears after the output.

SAS Output for Example 12.4

```
                    The NPAR1WAY Procedure
        Wilcoxon Scores (Rank Sums) for Variable symptoms
                  Classified by Variable trt
                 Sum of      Expected      Std Dev       Mean
   trt    N      Scores      Under H0      Under H0      Score
   -------------------------------------------------------------
    1     5       40.50        52.50       11.062026      8.10
    2     5       69.50        52.50       11.062026     13.90
    3     5       24.50        52.50       11.062026      4.90
    4     5       75.50        52.50       11.062026     15.10

             Average scores were used for ties.
                     Kruskal-Wallis Test
                 Chi-Square            10.7013
                 DF                          3
                 Pr > Chi-Square       0.0135
```

Interpretation of SAS Output for Example 12.4

For the Kruskal–Wallis test, SAS produces the sum of the ranks (labeled "Sum of Scores") in each treatment group. SAS then gives the expected sums and the standard deviation (assuming no difference among treatments) and then the mean rank in each group. It then gives a test statistic—in this case, a chi-square statistic—and a corresponding p value. For Example 12.4, SAS computes $\chi^2 = 10.7013$, which has 3 degrees of freedom (number of groups $- 1$), and a *p* value $= 0.0135$. Based on the Kruskal–Wallis test, we reject H_0 because $p = 0.0135 < 0.05$. We have significant evidence of a difference among the treatments with respect to median symptom scores. ■

12.5 Spearman Correlation (Correlation Between Variables)

In Chapter 10 we presented a formula for the sample correlation coefficient, r $(r = \text{Cov}(X, Y)/\sqrt{\text{Var}(X)\text{Var}(Y)})$. The correlation coefficient quantifies the nature and strength of the linear association between X and Y. Extreme values can have a substantial impact on the value of the sample correlation coefficient. When data are subject to extremes, an alternative measure of correlation between variables is based on ranks. The correlation based on ranks is called the Spearman correlation. We illustrate the computation of the Spearman correlation using an example.

EXAMPLE 12.5

Spearman Correlation Between Number of Cigarettes Smoked and Aerobic Exercise

Suppose we wish to assess the relationship between the number of cigarettes smoked per day and the number of hours of aerobic exercise per week. A total of 12 subjects agree to participate in the investigation, and we measure the typical number of cigarettes smoked per day and the number of hours of exercise in a typical week on each subject. The data are shown next.

Raw Scores

$X = $ Number of Cigarettes per Day	$Y = $ Number of Hours of Exercise per Week
20	0
0	0
20	1
10	2
5	3
4	5
3	5
5	6
0	3
0	4
0	7
0	8

The first step in the analysis is to replace the raw X and Y scores with ranks. The ranks are assigned for each variable, considered separately, as shown in the following table. In this example, there are several instances of ties in raw scores. When scores are tied, the mean rank is applied to each of the tied values.

	Ranks	
Rx = Rank of Number of Cigarettes per Day		*Ry = Rank of Number of Hours of Exercise per Week*
11.5		1.5
3		1.5
11.5		3
10		4
8.5		5.5
7		8.5
6		8.5
8.5		10
3		5.5
3		7
3		11
3		12

The Spearman correlation is computed using the same formula we used in Chapter 10, based on the ranks. In the formula Rx and Ry are the ranks of X and Y, respectively:

$$r_s = \frac{\text{Cov}(Rx,\ Ry)}{\sqrt{\text{Var}(Rx)\text{Var}(Ry)}} \tag{12.3}$$

We now calculate the components for r_s.

We first compute the variance of Rx: $\text{Var}(Rx) = \sum(Rx - \overline{Rx})^2/(n-1)$. The following table summarizes the computations:

Rx = Rank of Number of Cigarettes per Day	(Rx − $\overline{Rx}$)	(Rx − $\overline{Rx}$)²
11.5	5.0	25.00
3	−3.5	12.25
11.5	5.0	25.00
10	3.5	12.25
8.5	2.0	4.00
7	0.5	0.25
6	−0.5	0.25
8.5	2.0	4.00
3	−3.5	12.25
3	−3.5	12.25
3	−3.5	12.25
3	−3.5	12.25
$\overline{Rx} = 6.5$	0	132.0

So, $\text{Var}(Rx) = \sum (Rx - \overline{Rx})^2/(n-1) = 132.0/11 = 12.0$.

We now compute the variance of Ry: $\text{Var}(Ry) = \sum (Ry - \overline{Ry})^2/(n-1)$. The following table summarizes the computations:

Ry = Rank of Number of Hours of Exercise per Week	(Ry − $\overline{Ry}$)	(Ry − $\overline{Ry}$)²
1.5	−5.0	25.00
1.5	−5.0	25.00
3	−3.5	12.25
4	−2.5	6.25
5.5	−1.0	1.00
8.5	2.0	4.00
8.5	2.0	4.00
10	3.5	12.25
5.5	−1.0	1.00
7	0.5	0.25
11	4.5	20.25
12	5.5	30.25
$\overline{Ry} = 6.5$	0	141.50

So, $\text{Var}(Ry) = \sum(Ry - \overline{Ry})^2/(n-1) = 141.50/11 = 12.86$.

Finally, we compute the covariance of Rx and Ry: $\text{Cov}(Rx, Ry) = \sum(Rx - \overline{Rx})(Ry - \overline{Ry})/(n-1)$. The following table summarizes the computations:

$(Rx - \overline{Rx})$	$(Ry - \overline{Ry})$	$(Rx - \overline{Rx})(Ry - \overline{Ry})$
5.0	−5.0	−25.00
−3.5	−5.0	17.50
5.0	−3.5	−17.50
3.5	−2.5	−8.75
2.0	−1.0	−2.00
0.5	2.0	1.00
−0.5	2.0	−1.00
2.0	3.5	7.00
−3.5	−1.0	3.50
−3.5	0.5	−1.75
−3.5	4.5	−15.75
−3.5	5.5	−19.25
		−62.0

So, $\text{Cov}(Rx, Ry) = \sum(Rx - \overline{Rx})(Ry - \overline{Ry})/(n-1) = -62.0/11 = -5.64$.
Substituting into (12.3):

$$r_s = \frac{\text{Cov}(Rx, Ry)}{\sqrt{\text{Var}(Rx)\text{Var}(Ry)}} = \frac{-5.64}{\sqrt{(12.05)(12.86)}} = -0.453$$

Based on the sign and the magnitude of r_s, there is a moderately strong, inverse association between the number of cigarettes smoked per day and the number of hours of aerobic exercise per week. ▪

SAS Example 12.5 Spearman Correlation Between Number of Cigarettes Smoked and Aerobic Exercise Using SAS

Users can request that SAS compute a correlation based on ranks by specifying the Spearman option in Proc Corr. Before computing the rank correlation, we generated a scatter diagram using Proc Plot. The following output was generated by SAS. A brief interpretation appears after the output.

SAS Output for Example 12.5

```
          Plot of exercise*cigarett.   Legend: A = 1 obs, B = 2 obs, etc.
exercise |
         |
     8 +  A
         |
         |
         |
         |
     7 +  A
         |
         |
         |
         |
     6 +                   A
         |
         |
         |
         |
     5 +             A
         |
         |
         |
         |
     4 +  A          A
         |
         |
         |
         |
     3 +  A          A
         |
         |
         |
         |
     2 +                        A
         |
         |
         |
         |
     1 +                                                    A
         |
         |
         |
         |
     0 +  A                                                 A
         |
         ---+------------+------------+------------+------------+--
            0            5           10           15           20
                                 cigarett
```

```
                    The CORR Procedure
              2 Variables:    cigarett exercise
                    Simple Statistics
Variable   N      Mean    Std Dev     Median    Minimum    Maximum
cigarett   12   5.58333   7.39113    3.50000          0   20.00000
exercise   12   3.66667   2.64002    3.50000          0    8.00000

           Spearman Correlation Coefficients, N = 12
                 Prob > |r| under H0: Rho=0
                             cigarett         exercise
              cigarett       1.00000          -0.45366
                                               0.1385
              exercise      -0.45366           1.00000
                              0.1385
```

Interpretation of SAS Output for Example 12.5

The scatterplot shows the inverse relationship between the number of hours of exercise per week and the number of cigarettes smoked per day. The Corr Procedure first generates summary statistics on the raw scores of the variables specified. For example, the mean number of hours of exercise per week in the sample is 3.67 and the mean number of cigarettes smoked per day is 5.58. Because we requested a correlation based on ranks (Spearman correlation), SAS makes this note in the header to the correlation matrix. The Spearman correlation is estimated at -0.45 and the two-sided p value $= 0.1385$. The correlation is not significantly different from zero, $p = 0.1385 > 0.05$. ▪

12.6 Key Formulas

Application	Notation/Formula	Description
Two dependent samples test	$Z = \dfrac{T - \dfrac{n(n+1)}{4}}{\sqrt{\dfrac{n(n+1)(2n+1)}{24}}},$ where T = sum of positive ranks	normal approximation for Signed-Rank test
Two independent samples test	$Z = \dfrac{S - \dfrac{n_1(n_1+n_2+1)}{2}}{\sqrt{\dfrac{n_1 n_2(n_1+n_2+1)}{12}}},$ where S = smaller sum of ranks	normal approximation for Rank Sum test
Rank correlation	$r_s = \dfrac{\text{Cov}(Rx, Ry)}{\sqrt{\text{Var}(Rx)\text{Var}(Ry)}}$	Spearman correlation based on ranks

12.6.1 Guidelines for Determining When to Use a Nonparametric Procedure

There are no "rules" for determining when it is appropriate to use a nonparametric procedure instead of a parametric one. Substantial work has been done to show the robustness of many parametric procedures in the presence of violations of the normality assumption. The following are guidelines that may be useful in determining when it might be appropriate to employ a nonparameteric test:

- When substantive knowledge of the analytic variable suggests nonnormality. Some variables clearly do not follow a normal distribution (e.g., ordinal variables with few response options); analyses focused on these variables are candidates for nonparametric procedures.

- Observed nonnormality in the sample data. Visual inspection of the distribution of a variable might suggest that the variable is highly skewed. Comparison of measures of central tendency, such as the mean and median, which are equal in symmetric distributions, can help determine the extent of skewness. Situations in which the standard deviation is much larger than the mean also suggest nonnormality. A formal test for normality (such as one of the tests available in Proc Univariate) might indicate that a variable does not follow a normal distribution. (These tests have been shown to be very sensitive and might suggest nonnormality when in fact the data are not substantially different from normal, so they should be interpreted with caution.) In all of these cases, nonparametric procedures might be appropriate.

12.7 Statistical Computing

Following are the SAS programs used to conduct the nonparametric tests for two dependent samples, two independent samples, and k independent samples. Also included is the SAS program used to estimate a correlation based on ranks. The SAS procedures and brief descriptions of their use are noted in the header to each example. Notes are provided to the right of the SAS programs (in blue) for orientation purposes and are not part of the programs. In addition, blank lines in the programs are solely to accommodate the notes. Blank lines and spaces can be used throughout SAS programs to enhance readability. A summary of the SAS procedures used in the examples is provided at the end of this section.

SAS EXAMPLE 12.2 **Wilcoxon Signed-Rank Test: Two Dependent Samples**
Test to Determine If New Drug Lowers Systolic Blood Pressure (Example 12.2)

Suppose we are investigating a new drug hypothesized to lower systolic blood pressure. We select a random sample of subjects and record their resting systolic blood pressure at baseline (i.e., prior to the investigation). Each subject is then given the new drug, and resting systolic blood pressure is again measured after 4 weeks of drug treatment. The data are shown in the following table. Based on the data, does it appear that the new drug lowers systolic blood pressure? Run a Wilcoxon Signed-Rank test using SAS.

Subject Identification Number	Baseline Systolic Blood Pressure	Post-Treatment Systolic Blood Pressure
1	166	138
2	135	120
3	189	176
4	180	180
5	156	160
6	142	150
7	176	152
8	156	140
9	164	160
10	142	130

Program Code

`options ps=62 ls=80;`	Formats the output page to 62 lines in length and 80 columns in width
`data in;`	Beginning of Data Step
`   input baseline post_trt;`	Inputs two variables, **baseline** and **post_trt**, for each subject.
`diff=baseline-post_trt;`	Compute the difference between blood pressures (**diff**).
`cards;`	Beginning of Raw Data section.
`166 138`	actual observations
`135 120`	
`189 176`	
`180 180`	
`156 160`	
`142 150`	
`176 152`	
`156 140`	
`164 160`	
`142 130`	
`run;`	
`proc univariate normal;`	Procedure call. Proc Univariate generates summary statistics and the Wilcoxon Signed-Rank test. The normal option requests a test for normality.
`   var diff;`	Specification of analytic variable (**diff**).
`run;`	End of procedure section. ■

SAS EXAMPLE 12.3 Wilcoxon Rank Sum Test: Two Independent Samples

Test for Difference in Treatments with Respect to Health Status (Example 12.3)

We wish to compare two competing treatments for the management of adult-onset diabetes. A total of eight subjects agree to participate in the investigation and are randomly assigned to one of the competing treatments. After subjects follow the prescribed treatment regimen for 6 weeks, we assess their self-reported health status. Subjects are asked to rate their current health using an ordinal scale with the following response options: Excellent, Very Good, Good, Fair, or Poor. Investigators often assign numerical values to ordinal responses and analyze these values as if they were continuous. The assignment of numerical values can be made in a variety of ways. Suppose we assign increasing numerical values to the response options as follows: Poor = 0, Fair = 5, Good = 10, Very Good = 15, and Excellent = 20. Here, higher values indicate better self-reported health status. Are self-reported health status scores different between treatment groups? Run a Wilcoxon Signed-Rank test using SAS.

Self-Reported Health Status

Treatment 1	Treatment 2
0	8
7	10
11	12
16	15

Program Code

```
options ps=62 ls=80;          Formats the output page to 62 lines in length and
                                 80 columns in width
data in;                      Beginning of Data Step
  input trt hlthstat;         Inputs two variables, trt and hlthstat, for each subject.
cards;                        Beginning of Raw Data section.
1 0                           actual observations
1 7
1 11
1 16
2 8
2 10
2 12
2 15
run;
```

```
proc npar1way wilcoxon;
```
Procedure call. Proc Npar1way runs nonparametric tests. We request a Wilcoxon Rank Sum test by specifying the wilcoxon option.

```
    class trt;
```
Specification of grouping variable (**trt**).

```
    var hlthstat;
```
Specification of analytic variable (**hlthstat**).

```
run;
```
End of procedure section.   ■

SAS EXAMPLE 12.4 Kruskal–Wallis Test: *k* Independent Samples

Test for Difference Among Treatments with Respect to Symptom Scores (Example 12.4)

Suppose we wish to compare four treatments for seasonal allergies. A total of 20 subjects agree to participate in the investigation and are randomly assigned to one of the four competing treatments. After subjects follow the prescribed treatment regimen for 2 weeks, we assess their status. The outcome variable in this application is an index score based on three distinct symptoms. At the end of treatment, subjects are asked if they are currently experiencing any (or all) of the following symptoms: scratchy throat, itchy eyes, runny nose. Each subject responds "yes" or "no" to each of the three symptoms. This index score is the sum of affirmative responses. The range of scores is 0 (no symptoms) to 3 (all three symptoms). Is there a significant difference among the treatments with respect to symptom scores? Run a Kruskal–Wallis test using SAS.

Treatment 1	*Treatment 2*	*Treatment 3*	*Treatment 4*
0	2	0	2
0	2	0	3
1	3	1	2
2	3	1	3
3	2	1	3

Program Code

```
options ps=62 ls=80;
```
Formats the output page to 62 lines in length and 80 columns in width

```
data in;
  input trt symptoms;
```
Beginning of Data Step

Inputs two variables, **trt** and **symptoms,** for each subject.

```
cards;                          Beginning of Raw Data section.
1 0                             actual observations
1 0
1 1
1 2
1 3
2 2
2 2
2 3
2 3
2 2
3 0
3 0
3 1
3 1
3 1
4 2
4 3
4 2
4 3
4 3
run;

proc npar1way wilcoxon;         Procedure call. Proc Npar1way runs nonparametric
                                tests. When there are more than two comparison
                                groups, the Kruskal–Wallis test is run.
  class trt;                    Specification of grouping variable (trt).
  var symptoms;                 Specification of analytic variable
                                (symptoms).
run;                            End of procedure section.              ■
```

SAS EXAMPLE 12.5 **Spearman (Rank) Correlation**

Assess Relationship Between Number of Cigarettes Smoked per Day and Number of Hours of Exercise per Week (Example 12.5)

Suppose we wish to assess the relationship between the number of cigarettes smoked per day and the number of hours of aerobic exercise per week. A total of 12 subjects agree to participate in the investigation, and we measure the average number of cigarettes smoked per day and the number of hours of exercise in a typical week on each subject. Estimate the Spearman correlation using SAS.

Raw Scores	
X = Number of Cigarettes per Day	Y = Number of Hours of Exercise per Week
20	0
0	0
20	1
10	2
5	3
4	5
3	5
5	6
0	3
0	4
0	7
0	8

Program Code

```
options ps=62 ls=80;
```
Formats the output page to 62 lines in length and 80 columns in width

```
data in;
  input cigarett exercise;
```
Beginning of Data Step

Inputs two variables, **cigarett** and **exercise**, for each subject.

```
cards;
```
Beginning of Raw Data section.

```
20 0
```
actual observations

```
0 0
20 1
10 2
5 3
4 5
3 5
5 6
0 3
0 4
0 7
0 8
run;
```

`proc plot;`	Procedure call. Proc Plot generates a scatter diagram.
`  plot exercise*cigarett;`	Specification of analytic variables; **exercise** is plotted on the vertical axis and **cigarett** on the horizontal.
`run;`	End of procedure section.
`proc corr spearman;`	Procedure call. Proc Corr runs a correlation analysis. The Spearman option requests that the correlation be based on ranks.
`  var exercise cigarett;`	Specification of analytic variables.
`run;`	End of procedure section. ■

12.7.1 Summary of SAS Procedures

The SAS procedures for nonparametric analysis are outlined in the following table. Specific options can be requested in these procedures to generate specific tests or analyses. The options are shown in italics. Users should refer to the examples in this section for complete descriptions of the procedure and specific options.

Procedure	Sample Procedure Call	Description
proc univariate	proc univariate *normal;* var diff;	Runs a Wilcoxon Signed-Rank test. The normal option produces a test for normality. The analytic variable is the difference between measures in dependent samples.
proc npar1way	proc npar1way *wilcoxon;* class group; var x;	Runs nonparametric tests for equality of medians among independent groups. The wilcoxon option requests the Wilcoxon (ranks-based) test. When the group variable is dichotomous, a Wilcoxon Rank Sum test is run; when the group variable has $k > 2$ levels a Kruskal–Wallis test is run.
proc corr	proc corr *spearman;* var x y;	Runs a correlation analysis. The Spearman option requests that the correlation be based on ranks.

12.8 Analysis of Framingham Heart Study Data

The Framingham data set includes data collected from the original cohort. Participants contributed up to three examination cycles of data. First we analyze data collected in the first examination cycle (called the period = 1 examination) and use SAS Proc Npar1way to test for differences in medians between two and more than two independent groups. We then construct a data set that contains risk factors measured at period 1 and then again at period 2 (approximately 6 years later) and evaluate changes in risk factors over time using the nonparametric procedures for two dependent samples implemented through Proc Univariate.

Framingham Data Analysis—SAS Code

```
data fhs;
   set in.frmgham;
   if period=1;
run;


proc sort data=fhs;
   by sex;
run;


proc univariate normal data=fhs;
   var glucose;
   by sex;
   where diabetes=1;
run;


proc npar2way wilcoxon data=fhs;
   class sex;
   var glucose;
   where diabetes=1;
run;


proc ttest data=fhs;
   class sex;
   var glucose;
   where diabetes=1;
run;
```

Framingham Data Analysis—SAS Output

---------------------------- SEX=M ----------------------------

The UNIVARIATE Procedure
Variable: GLUCOSE (Casual Glucose mg/dL)

Moments

N	57	Sum Weights	57
Mean	168.824561	Sum Observations	9623
Std Deviation	78.748724	Variance	6201.36153
Skewness	0.98724434	Kurtosis	0.50420161
Uncorrected SS	1971875	Corrected SS	347276.246
Coeff Variation	46.6453005	Std Error Mean	10.4305233

Basic Statistical Measures

Location		Variability	
Mean	168.8246	Std Deviation	78.74872
Median	150.0000	Variance	6201
Mode	120.0000	Range	321.00000
		Interquartile Range	99.00000

Tests for Location: Mu0=0

Test	-Statistic-		-----p Value------	
Student's t	t	16.18563	Pr > |t|	<.0001
Sign	M	28.5	Pr >= |M|	<.0001
Signed Rank	S	826.5	Pr >= |S|	<.0001

Tests for Normality

Test	--Statistic---		-----p Value------	
Shapiro-Wilk	W	0.91455	Pr < W	0.0007
Kolmogorov-Smirnov	D	0.123195	Pr > D	0.0301
Cramer-von Mises	W-Sq	0.196426	Pr > W-Sq	0.0056
Anderson-Darling	A-Sq	1.325553	Pr > A-Sq	<0.0050

Quantiles (Definition 5)

Quantile	Estimate
100% Max	394
99%	394
95%	332
90%	292
75% Q3	207

```
----------------------------- SEX=M -----------------------------
```

The UNIVARIATE Procedure
Variable: GLUCOSE (Casual Glucose mg/dL)

Quantiles (Definition 5)

Quantile	Estimate
50% Median	150
25% Q1	108
10%	80
5%	75
1%	73
0% Min	73

Extreme Observations

----Lowest----		----Highest---	
Value	Obs	Value	Obs
73	46	297	23
73	3	325	17
75	28	332	51
78	15	370	54
80	58	394	39

Missing Values

Missing Value	Count	----Percent Of-----	
		All Obs	Missing Obs
.	2	3.39	100.00

```
----------------------------- SEX=F -----------------------------
```

The UNIVARIATE Procedure
Variable: GLUCOSE (Casual Glucose mg/dL)

Moments

N	59	Sum Weights	59
Mean	169.915254	Sum Observations	10025
Std Deviation	91.0915205	Variance	8297.66511
Skewness	0.92774764	Kurtosis	0.00307055
Uncorrected SS	2184665	Corrected SS	481264.576
Coeff Variation	53.6099722	Std Error Mean	11.8591059

```
                    Basic Statistical Measures
          Location                    Variability
    Mean      169.9153     Std Deviation         91.09152
    Median    137.0000     Variance                  8298
    Mode      107.0000     Range              347.00000
                           Interquartile Range 128.00000
```

NOTE: The mode displayed is the smallest of 3 modes with a count of 2.

```
                   Tests for Location: Mu0=0
          Test            -Statistic-     -----p Value------
          Student's t     t  14.32783     Pr > |t|    <.0001
          Sign            M      29.5     Pr >= |M|   <.0001
          Signed Rank     S       885     Pr >= |S|   <.0001
```

```
                      Tests for Normality
      Test                 --Statistic---     -----p Value------
      Shapiro-Wilk         W     0.905032     Pr < W       0.0002
      Kolmogorov-Smirnov   D     0.171072     Pr > D      <0.0100
      Cramer-von Mises     W-Sq  0.333946     Pr > W-Sq   <0.0050
      Anderson-Darling     A-Sq  1.910492     Pr > A-Sq   <0.0050
```

```
                 Quantiles (Definition 5)
                 Quantile        Estimate
                 100% Max            394
                 99%                 394
                 95%                 368
```

------------------------------ SEX=F ------------------------------

```
                The UNIVARIATE Procedure
        Variable:  GLUCOSE   (Casual Glucose mg/dL)

                 Quantiles (Definition 5)
                 Quantile        Estimate
                 90%                 320
                 75% Q3              235
                 50% Median          137
                 25% Q1              107
                 10%                  78
                 5%                   57
                 1%                   47
                 0% Min               47
```

Extreme Observations

----Lowest----		----Highest---	
Value	Obs	Value	Obs
47	27	348	55
55	2	366	19
57	43	368	54
63	18	386	30
66	21	394	38

Missing Values

Missing		----Percent Of-----	
			Missing
Value	Count	All Obs	Obs
.	3	4.84	100.00

The NPAR1WAY Procedure

Wilcoxon Scores (Rank Sums) for Variable GLUCOSE
Classified by Variable SEX

SEX	N	Sum of Scores	Expected Under H0	Std Dev Under H0	Mean Score
M	57	3388.50	3334.50	181.070365	59.447368
F	59	3397.50	3451.50	181.070365	57.584746

Average scores were used for ties.

Wilcoxon Two-Sample Test

Statistic	3388.5000

Normal Approximation
Z	0.2955
One-Sided Pr > Z	0.3838
Two-Sided Pr > |Z|	0.7676

t Approximation
One-Sided Pr > Z	0.3841
Two-Sided Pr > |Z|	0.7682

Z includes a continuity correction of 0.5.

```
              Kruskal-Wallis Test

        Chi-Square                0.0889
        DF                             1
        Pr > Chi-Square           0.7655

              The TTEST Procedure

                  Statistics
```

| | | | Lower CL | | Upper CL | Lower CL | |
Variable	SEX	N	Mean	Mean	Mean	Std Dev	Std Dev
GLUCOSE	M	57	147.93	168.82	189.72	66.484	78.749
GLUCOSE	F	59	146.18	169.92	193.65	77.112	91.092
GLUCOSE	Diff (1-2)		-32.46	-1.091	30.275	75.477	85.252

```
                  Statistics
```

| | | Upper CL | | | |
Variable	SEX	Std Dev	Std Err	Minimum	Maximum
GLUCOSE	M	96.605	10.431	73	394
GLUCOSE	F	111.31	11.859	47	394
GLUCOSE	Diff (1-2)	97.959	15.833		

```
                   T-Tests
```

Variable	Method	Variances	DF	t Value	Pr > |t|
GLUCOSE	Pooled	Equal	114	-0.07	0.9452
GLUCOSE	Satterthwaite	Unequal	113	-0.07	0.9451

```
             Equality of Variances
```

Variable	Method	Num DF	Den DF	F Value	Pr > F
GLUCOSE	Folded F	58	56	1.34	0.2758

We used SAS to first generate descriptive statistics on GLUCOSE levels in male and female participants with diabetes. Is the distribution of glucose levels approximately normal for men and women? We then ran a Wilcoxon Rank Sum test to compare median glucose levels for male and female diabetics. Are the medians similar or different? For comparison purposes, we also ran a *t* test comparing mean glucose levels. How do the nonparametric and parametric tests compare? Which test is appropriate here?

Framingham Data Analysis—SAS Code

```
data fhs1;
   set in.frmgham;
   if period=1;

glucose1=glucose;
keep randid sex glucose1 diabetes;
run;

data fhs2;
   set in.frmgham;
   if period=2;

glucose2=glucose;
keep randid glucose2;
run;

proc sort data=fhs1; by randid; run;
proc sort data=fhs2; by randid; run;

data fhs12;
   merge fhs1(in=a)  fhs2(in=b);
   by randid;
   if a and b;

diffglucose=glucose2-glucose1;
run;

proc univariate normal data=fhs12;
   var diffglucose;
run;
```

We created two data sets, called fhs1 and fhs2, which contain the glucose data measured at periods 1 and 2, respectively. We then merged them into a single data set so that the period 1 and period 2 values are together (one SAS record per participant). In the merged data set, we created a difference score by subtracting the glucose measured at period 1 from the glucose measured at period 2. The difference reflects how much the glucose level changed over time (specifically, how it increased from period 1 to period 2). In the following, we used SAS Proc Univariate to test for differences in glucose levels over time among participants with diabetes. Here we perform a nonparametric test.

Framingham Data Analysis—SAS Output

The UNIVARIATE Procedure
Variable: diffglucose

Moments

N	71	Sum Weights	71
Mean	-11.492958	Sum Observations	-816
Std Deviation	64.8745764	Variance	4208.71066
Skewness	-0.7233025	Kurtosis	0.68490171
Uncorrected SS	303988	Corrected SS	294609.746
Coeff Variation	-564.47242	Std Error Mean	7.69919574

Basic Statistical Measures

Location		Variability	
Mean	-11.4930	Std Deviation	64.87458
Median	4.0000	Variance	4209
Mode	5.0000	Range	315.00000
		Interquartile Range	86.00000

Tests for Location: Mu0=0

Test		-Statistic-	------p Value-----	
Student's t	t	-1.49275	Pr > |t|	0.1400
Sign	M	1.5	Pr >= |M|	0.8126
Signed Rank	S	-169.5	Pr >= |S|	0.3350

Tests for Normality

Test		---Statistic---	------p Value-----	
Shapiro-Wilk	W	0.965272	Pr < W	0.0466
Kolmogorov-Smirnov	D	0.109515	Pr > D	0.0343
Cramer-von Mises	W-Sq	0.087716	Pr > W-Sq	0.1657
Anderson-Darling	A-Sq	0.574876	Pr > A-Sq	0.1356

Quantiles (Definition 5)

Quantile	Estimate
100% Max	105
99%	105
95%	84
90%	66
75% Q3	34
50% Median	4
25% Q1	-52

How much did glucose change over time? Is the distribution of change over time approximately normal? Is there a significant difference in median glucose over time? Is there a significant difference in mean glucose over time?

12.9 Problems

1. A study is conducted comparing two competing medications for asthma. Sixteen subjects are involved in the investigation. The data shown reflect asthma symptom scores for patients randomly assigned to each treatment. Higher scores are indicative of worse asthma symptoms. Test the null hypothesis that there is no difference in asthma symptoms between medications. Run the Wilcoxon Rank Sum test using the normal approximation at a 5% level of significance.

Treatment A:	55	60	80	65	72	78	68	71
Treatment B:	80	82	86	89	76	81	90	76

2. We wish to evaluate a program designed to improve quality of life in older patients with coronary heart disease. Quality of life is measured on a scale of 0–100, with higher scores indicative of better quality of life. Quality-of-life measures are taken on each subject at baseline and then again after participating in the program. Based on the following data, is there evidence that the program significantly improves quality of life? Run the Wilcoxon Signed-Rank test using the normal approximation at a 5% level of significance.

Baseline:	80	55	63	76	88	45	65	77
Post-Program:	85	50	65	78	82	55	68	90

3. Data were collected from a random sample of eight patients currently undergoing treatment for hypertension. Each subject reported the average number of cigarettes smoked per day and was assigned a numerical value reflecting risk of cardiovascular disease (CVD). The risk assessments were computed by physicians and based on blood pressure, cholesterol level, and exercise status. The risk assessments ranged from 0 to 100, higher values indicating increased risk. Compute the Spearman correlation coefficient.

Number of Cigarettes:	0	2	6	8	12	0	2	20
Risk of CVD:	12	20	50	68	75	8	10	80

4. Suppose we are interested in whether there is a difference between the mean numbers of sick days taken by men and women in a local company. The numbers of sick days taken by men and women are shown here. Use the data to test the null hypothesis that there is no difference in sick days between men and women. Run the Wilcoxon Rank Sum test using the normal approximation at a 5% level of significance.

Men:	5	10	2	0	6	4	5	15
Women:	8	9	3	5	0	4	15	

5. A nutritionist is investigating the effects of a rigorous walking program on systolic blood pressure (SBP) in patients with mild hypertension.

Subjects who agree to participate have their systolic blood pressure measured at the start of the study and then after completing the 6-week walking program. The data are

Subject	Starting SBP	Ending SBP
1	140	128
2	130	125
3	150	140
4	160	162
5	135	137
6	128	130
7	142	135
8	151	140

Based on the data, is there evidence that SBP is significantly reduced by the walking program? Run the Wilcoxon Signed-Rank test using the normal approximation at a 5% level of significance.

6. An antismoking campaign is being evaluated prior to its implementation in high schools across the state. A pilot study involving six volunteers who smoke is conducted. Each volunteer reports the number of cigarettes smoked the day before enrolling in the study. Each is then subjected to the antismoking campaign, which involves educational material, support groups, formal programs designed to reduce or quit smoking, etc. After 4 weeks, each volunteer again reports the number of cigarettes smoked the day before. Based on the following pilot data, does it appear that the program is effective?

At Enrollment:	21	15	8	6	12	20
After Campaign:	12	10	10	6	10	20

Run the Wilcoxon Signed-Rank test using the normal approximation at a 5% level of significance.

SAS Problems

Use SAS to solve the following problems.

1. A study is conducted comparing two competing medications for asthma. Sixteen subjects are involved in the investigation. The data shown reflect asthma symptom scores for patients randomly assigned to each treatment. Higher scores are indicative of worse asthma symptoms. Test the null hypothesis that there is no difference in asthma symptoms between medications.

Treatment A:	55	60	80	65	72	78	68	71
Treatment B:	80	82	86	89	76	81	90	76

Use SAS Proc Univariate to run the Wilcoxon Signed-Rank test. Include the option to perform a test for normality.

2. We wish to evaluate a program designed to improve quality of life in older patients with coronary heart disease. Quality of life is measured on a scale of 0–100, with higher scores indicative of better quality of life. Quality-of-life measures are taken on each subject at baseline and then again after participating in the program. Based on the following data, is there evidence that the program significantly improves quality of life?

Baseline:	80	55	63	76	88	45	65	77
Post-Program:	85	50	65	78	82	55	68	90

Use SAS to perform a Wilcoxon Rank Sum test. Perform the test at a 5% level of significance.

3. A pharmaceutical company is interested in the effectiveness of a new preparation designed to relieve arthritis pain. Three variations of the compound have been prepared for investigation, which differ according to the proportion of the active ingredients: T15 contains 15% active ingredients, T40 contains 40% active ingredients, and T50 contains 50% active ingredients. A sample of 20 patients is selected to participate in a study comparing the three variations of the compound. A control compound, which is currently available over the counter, is also included in the investigation. Patients are randomly assigned to one of the four treatments (control, T15, T40, T50) and the time (in minutes) until pain relief is recorded on each subject. The data are

Control:	12	15	18	16	20
T15:	20	21	22	19	20
T40:	17	16	19	15	19
T50:	14	13	12	14	11

Use SAS to perform a Kruskal–Wallis test. Perform the test at a 5% level of significance.

4. Data were collected from a random sample of eight patients currently undergoing treatment for hypertension. Each subject reported the average number of cigarettes smoked per day and each was assigned a numerical value reflecting risk of cardiovascular disease (CVD). The risk assessments were computed by physicians and based on blood pressure, cholesterol level, and exercise status. The risk assessments ranged from 0 to 100, higher values indicating increased risk.

Number of Cigarettes:	0	2	6	8	12	0	2	20
Risk of CVD:	12	20	50	68	75	8	10	80

Use SAS to estimate the Spearman correlation coefficient. Test if the correlation is significant using a 5% level of significance.

Descriptive Statistics
(Ch. 2)

Probability
(Ch. 3)

Sampling Distributions
(Ch. 4)

Statistical Inference
(Chapters 5–13)

OUTCOME VARIABLE	GROUPING VARIABLE(S)/ PREDICTOR(S)	ANALYSIS	CHAPTER(S)
Continuous	—	Estimate μ; Compare μ to Known, Historical Value	5/12
Continuous	Dichotomous (2 groups)	Compare Independent Means (Estimate/Test $(\mu_1 - \mu_2)$) or the Mean Difference (μ_d)	6/12
Continuous	Discrete (>2 groups)	Test the Equality of k Means using Analysis of Variance ($\mu_1 = \mu_2 = \cdots = \mu_k$)	9/12
Continuous	Continuous	Estimate Correlation or Determine Regression Equation	10/12
Continuous	Several Continuous or Dichotomous	Multiple Linear Regression Analysis	10
Dichotomous	—	Estimate p; Compare p to Known, Historical Value	7
Dichotomous	Dichotomous (2 groups)	Compare Independent Proportions (Estimate/Test $(p_1 - p_2)$)	7/8
Dichotomous	Discrete (>2 groups)	Test the Equality of k Proportions (Chi-Square Test)	7
Dichotomous	Several Continuous or Dichotomous	Multiple Logistic Regression Analysis	11
Discrete	Discrete	Compare Distributions Among k Populations (Chi-Square Test)	7
Time to Event	Several Continuous or Dichotomous	Survival Analysis	13

Introduction to Survival Analysis

13

In Chapter 6 we presented estimation and statistical inference procedures for crude or unadjusted comparisons of two groups with respect to a continuous outcome variable. When the outcome variable in a prospective study or a clinical trial (i.e., a study in which the outcome is measured subsequent to treatment allocation) is continuous, the effect measure is the difference in means. The test of the effect of treatment is carried out using a two independent sample *t* test. To adjust for other variables when the outcome is continuous, multiple regression analysis can be used, as described in Chapter 10.

In Chapter 8 we presented estimation and statistical inference procedures for crude or unadjusted comparisons of two proportions and a technique for estimation and statistical inference adjusting for one categorical confounder. In Chapter 11 we presented logistic regression, which can be used to model the effect of independent variables on the risk of a dichotomous outcome. When the outcome variable in a prospective study or a clinical trial is dichotomous, the effect measure is the relative risk and the crude test of the effect of treatment is carried out using a chi-square test. If there are confounders to consider, the adjusted effect is assessed using the Cochran–Mantel–Haenszel chi-square statistic or with logistic regression analysis.

In this section we give a brief overview of survival analysis, which is used to compare groups with respect to a dichotomous outcome, as is logistic regression, but survival analysis techniques also takes into account different lengths of follow-up. Survival analysis techniques are used to estimate risks using all available data and are also used to estimate survival over time and to compare groups with respect to time to an event. Survival analysis techniques are beyond the scope of this book, but we do present a brief heuristic explanation of the theory.

13.1 Incomplete Follow-Up

In Chapters 8 and 11, we described procedures to compare proportions or risks between comparison groups. Some studies involve a long follow-up or observation period during which time each participant is measured for the occurrence or nonoccurrence of the event of interest. When the follow-up or observation period is long, subjects may withdraw from the study before the end of the observation period. Some might move or simply decide that they no longer wish to participate. These subjects are often termed "lost to follow-up." In addition, subjects may die during the study period. Subjects who are lost to follow-up or who die during the study period do not contribute to the overall counts of outcomes measured at the end of the study period, but their data should be included as much as possible in all analyses.

EXAMPLE 13.1   **Risk of Stroke**

Consider a very small study of five subjects in which the outcome of interest is stroke. Suppose that the study period is 5 years; that is, subjects are classified at the end of 5 years according to whether or not they had a stroke. Now suppose that among the five subjects in our study, two have strokes (after 2 and 3 years, respectively), one completes the 5-year period free of stroke, one subject dies after 1 year, and one subject drops out of the study after 4 years.

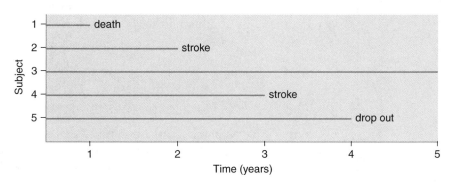

Based on these data, what can we say about the 5-year risk of stroke? We could restrict our analysis to those whose 5-year data are available, excluding the subject who died during the study period and the subject who dropped out. This would yield a 5-year risk estimate of $2/3 = 67\%$. However, the person who dropped out survived for 4 of the 5 years free of stroke, and the person who died survived for 1 year free of stroke. Thus, our estimate is probably too high. In fact, if we estimated 1-year, 2-year, 3-year, and 4-year risks of stroke using this method, we would obtain estimates of 0% (0/5), 25% (1/4), 50% (2/4), and 50% (2/4), respectively. Then, with no additional strokes, our 5-year estimate jumps to 67%! ■

EXAMPLE 13.2   **Risk of Stroke Using Conditional Probabilities**

Suppose we sample a group of $n = 100$ individuals from a population and follow them for 2 years. The goal is to estimate the probability of surviving for 2 years. Twenty people die each year, and we estimate the 2-year survival probability as 0.60 (60/100).

Suppose that we then sample another group of $n = 100$ from the same population and follow them for 1 year. Seventy-five people survive. How can we use this additional information to better estimate the 2-year survival probability?

Recall the *conditional probability rule* from Chapter 3,

$$P(B \mid A) = \frac{P(A \text{ and } B)}{P(A)}$$

We can rewrite this as $P(A \text{ and } B) = P(A)P(B \mid A)$ and apply it to the example. Let A = survive year 1 and B = survive year 2. Then, using sample 1 only,

$$P(A) = P(\text{survive year 1}) = 80/100 = 0.80$$
$$P(B \mid A) = P(\text{survive year 2 given survival through year 1}) = 60/80 = 0.75$$
$$P(A \text{ and } B) = P(\text{survive both years}) = P(A)P(B \mid A) = (0.80)(0.75) = 0.60$$

Note that this is the correct survival probability.

Now we can use the combined data from the two samples to estimate year 1 survival by combining the two samples:

$$P(A) = P(\text{survive year 1 (both samples combined)})$$
$$= \frac{(80 + 75)}{(100 + 100)} = 155/200 = 0.775$$

$$P(A \text{ and } B) = P(\text{survive both years}) = P(A)P(B \mid A) = (0.775)(0.75) = 0.58$$

Notice that $P(A)$ is based on the combined sample and $P(B \mid A)$ is based on sample 1 only. Thus our estimate of 2-year survival is 0.60 if we only use sample 1 but is 0.58 if we include all available data. Survival analysis techniques allow us to do this. ■

13.2 Time to Event

Another application of survival analysis techniques is based on the date at which events occur, in addition to they simply do or do not occur.

EXAMPLE 13.3

Risk of Stroke Using Time to Event Data

Consider another very small study of five subjects in which the outcome of interest is stroke, also with a 5-year study period. Suppose that in this study we have complete data, and that two of the five participants had strokes during the study period. The estimated 5-year risk of stroke is thus $2/5 = 40\%$. Consider the following two graphs:

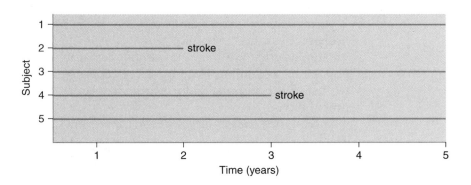

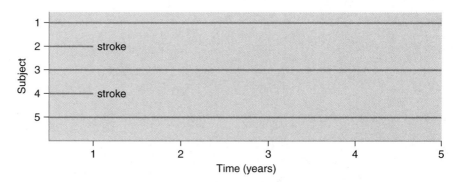

In both cases, there are two strokes among the five subjects. However, the strokes in the first graph occur much later than those in the second graph. Using a simple proportion to estimate risk yields does not allow us to distinguish these two scenarios. Survival analysis techniques take into account time to event in addition to the occurrence of the event.   ■

13.3 Survival Analysis Techniques

Survival analysis methods are used when the time to event is important and/or when there are withdrawals or losses to follow-up. The probability of survival through a specified time is calculated using (a) the probability of survival through the previous interval and (b) the conditional probability of surviving the current interval given survival through the previous interval.

The Kaplan–Meier method creates a new interval each time there is an event; the life table method uses prespecified intervals, such as months or years. Cox regression analysis allows the inclusion of potential confounders in survival analysis models. We will not present these techniques here, but we encourage the reader to explore them in more advanced texts.

Introduction to Statistical Computing Using SAS

A.1 Introduction to SAS

A *SAS program* is a list of commands in the SAS language. We use the SAS editor to create SAS programs and then execute them using the *submit* command. The components of a SAS program are outlined in the following sections. When a SAS program is executed, a *log* and *output* are produced. The program, log, and output, described in detail here, can be printed and/or saved in files on your computer (or computer account).

SAS can be implemented on a variety of operating platforms (e.g., personal computer, mainframe). The operating platform does not dramatically change the structure and operation of the program. SAS can also be implemented in several operational modes, including a mode in which the user enters SAS program code to perform a specific application as well as a menu-driven mode in which the user selects the desired statistical application from a list of options by pointing and clicking. In the following, we illustrate the use of SAS in the former mode, in all cases specifying the SAS code to perform specific analyses.

A.1.1 Components of a SAS Program

Each basic SAS program is made up of three components:

- The **Data Step,** which contains variable names and labels
- The **Raw Data Section,** or the actual observations considered for analysis

- The **Procedure Section,** which contains the call statements for the appropriate statistical analysis (e.g., summary statistics, two-sample *t* test, analysis of variance); several procedures can be invoked in this section to perform the desired analyses

The following sample SAS program displays each component. The parts of the program presented in boldface represent those that are specific to each application and can be changed, as appropriate, by the user. The notes to the right of the SAS program (in blue) are for orientation and are not part of the program. In addition, the blank lines are there only to accommodate the notes. In practice, the SAS program will be single-spaced. Indentation is used solely for readability and does not influence processing.

Sample SAS Program	Notes
`data one;`	Beginning of the Data Step. SAS allows the user to choose a data set name (e.g., **one**).
`  input x;`	Specification of variables. Here we consider a single variable, which we name **x**.
`cards;`	Beginning of the Raw Data section.
`5`	Actual observations—one observation (**x**) per subject.
`6`	
`12`	
`run;`	End of the Raw Data section.
`proc print;`	SAS Procedure that prints the data.
`  var x;`	Specification of the variable to be printed (**x**).
`run;`	End of the Procedure call.
`proc means;`	SAS Procedure to generate summary statistics (e.g., mean, standard deviation).
`  var x;`	Specification of the analytic variable (**x**).
`run;`	End of the Procedure call.

The data set called **one** is a temporary SAS data set and exists only for the duration of the program execution. It does not exist after the SAS session is over and it will not be saved. It is, however, possible to create permanent SAS data sets. At this point, our focus is on small, temporary data sets.

The statements that make up a SAS program are generally called the SAS code. The three components of SAS programs will be discussed in detail in subsequent sections.

The sample SAS program shown produces the following output. The notes to the right of the SAS output (in blue) are for orientation and are not part of the output. In addition, the blank lines are there only to accommodate the notes. In practice, the SAS output will be single-spaced.

Sample SAS Output	Notes

```
OBS  X
 1   5
 2   6
 3   12
```

Output from the Print Procedure:
SAS prints a column labeled "OBS"
   denoting the observation number, followed by each
   value of the variable **x**.

```
Analysis Variable: X
```

Output from the Means Procedure:
Summary statistics (See Chapter 3 for a complete
   discussion).

N	Mean	Std Dev	Minimum	Maximum
3	7.666	3.7859	5.0000	12.0000

A.1.2 Overview of the SAS System

SAS uses three windows to display information:

- The **PROGRAM,** or **PGM** Window (also called the Editor or Advanced Editor Window), which contains the SAS program code
- The **LOG** Window, which displays the SAS execution statements, notes, error messages, etc.
- The **OUTPUT,** or **OUT** Window, which displays the results of the statistical analysis

Throughout this book we present SAS programs and corresponding output. We do not illustrate the contents of the LOG Window. The information in the LOG Window is very useful, primarily to verify that the data are read correctly by SAS and that the procedure statements are entered and interpreted appropriately.

To view the contents of any window (i.e., PGM, LOG, or OUTPUT), simply click on any part of that window. The box in the top corner of each window (i.e., PGM, LOG, or OUTPUT) is used to maximize or minimize the window. Clicking on the box will enlarge or shrink the current window display.

Program creation, editing, and execution may be simplified by using the command line option within SAS. To move from one window to another, click on any part of that window or type <u>PGM</u>, <u>LOG</u>, or <u>OUTPUT</u> in the command line at the top of any window, followed by [enter] to move to the PROGRAM, LOG, and OUTPUT windows, respectively.

The following table displays some of the more commonly used SAS commands, a description of their functions, and the windows in which the commands are used. All commands are issued in the command line. Either the arrow keys or the [Home] key, which moves the cursor directly to the command line, can be used to move to the command line.

On some platforms, SAS also displays a toolbox window where commands can be issued to perform specific functions. Commands that are issued at the command line (see table) can also be issued in the appropriate space in the toolbox window. The toolbox also has some built-in functions that appear as buttons in the toolbox window (e.g., submit, cut, copy, paste). Specific functions can be performed using the buttons.

SAS COMMAND	Function	Windows
submit	executes SAS program	PGM
recall	recalls last program executed into the PGM window	PGM
file filename (ext[1])	saves the contents of window into the file named "filename"	PGM (use ext=SAS) LOG (use ext=LOG) OUTPUT (use ext=LST)
inc filename	includes a copy of the file named "filename.sas" in the PGM window	PGM
pgm	moves to the PROGRAM window	LOG OUTPUT
log	moves to the LOG window	PGM OUTPUT
out	moves to the OUTPUT window	PGM LOG
help	displays online help	PGM LOG OUTPUT
clear	clears the contents of the window	PGM LOG OUTPUT
bye	exits SAS	PGM LOG OUTPUT

[1] The extension (ext) is a three-character code used to distinguish between program files (extension 'SAS'), log files (extension 'LOG'), and output files (extension 'LST'). SAS *automatically* appends these extensions when you save code in a particular window.

A.1.3 Creating and Executing a SAS Program

The following is a step-by-step guide to creating and executing a SAS program. The guide can be applied in any statistical computing application presented in this book.

1. Start SAS (Type SAS).
2. Click on the PGM window (or enter PGM from LOG or OUTPUT command lines).
3. Enter the program code (statements) by typing directly into the SAS Program window. The SAS Program editor has line numbers in the left-hand margin. The program code is typed to the right of the line numbers. The column directly adjacent to the line numbers is reserved and cannot be used to enter data or program code. The arrow keys can be used to move around in the SAS Program editor to edit the program. Typing SAS commands into the SAS editor creates a *temporary* SAS program file, which you will save as a *permanent* program file after checking that it works correctly.
4. Once the program code is complete, execute the program. Move the cursor to the command line in the PGM window (use arrow keys or [Home] key). Enter <u>submit</u>.
5. Check the LOG for error messages and for documentation of desired procedure.
   If there are errors (in which case no output appears in the OUTPUT Window) or if the program did not execute correctly, do the following:

   a. Return to the PGM window (click on PGM window or type PGM followed by [enter] in command line of LOG window). Recall the SAS program by typing <u>recall</u> followed by [enter] in the command line of the PGM window.

   b. Edit the program code to fix the errors and/or modify the code to perform the desired analysis. Use the arrow keys to move around within the program window.

   c. Go to the LOG window (click on LOG window or enter LOG in command line of the PGM window). At the command line, type <u>clear</u> followed by [enter] to clear the contents of the window for the next program execution.

   d. Go to the OUTPUT window (click on OUTPUT window or enter OUT in command line of current window). At the command line, type <u>clear</u> followed by [enter] to clear the contents of the window for the next program execution.

  e. Return to the PGM window (click on PGM window or type PGM followed by [enter] in command line of OUTPUT window).

  f. Return to step 4 and continue the process.

If there are no errors, then continue to step 6.

6. Save the SAS PROGRAM code, the LOG, and the SAS OUTPUT into files by performing the following steps:

  a. Return to the PGM window (click on PGM window or enter PGM in command line of current window).

   Recall the SAS program by typing <u>recall</u> followed by [enter] in the command line.

   At the command line, type <u>file 'filename'</u> followed by [enter], where filename is the selected name. The '.sas' extension is automatically appended to the filename.

  b. Go to the LOG window (click on LOG window or enter LOG in command line of current window).

   At the command line, type <u>file 'filename'</u> followed by [enter]. The '.log' extension is automatically appended to the filename.

  c. Go to the OUTPUT window (click on OUTPUT window or enter OUTPUT in the command line of current window). At the command line, type <u>file 'filename'</u> followed by [enter]. The '.lst' extension is automatically appended to the filename.

  ** You have now saved three files on your computer or computer account called filename.sas, filename.log, and filename.lst.

7. Exit SAS by typing **bye** followed by [enter] in the command line of the current window.

8. Print hard copies of the three files (program, log, and output) using the appropriate print command on your system. For example, <u>print filename.pgm [enter]</u>, <u>print filename.log [enter]</u>, and <u>print filename.out [enter]</u>

9. Log off or sign off the system using the appropriate logout command.

A.2 The Data Step

The Data Step, the first component of a SAS program, contains (among other items) the variable names, specifications, and labels. We will illustrate a variety of programming statements that can be used in the Data Step.

  SAS, like many computing languages, offers a variety of options for inputting and manipulating data. SAS can read and analyze both numeric and alphanumeric, or character, data. We present several different techniques for inputting and manipulating data in the following Data Step. Sample data sets are provided to illustrate each technique.

A.2.1 Inputting Data: Types of Data

In SAS, data are classified as numeric, alphanumeric (or character), or date variables (e.g., 01/15/1998). Numeric data include numbers, such as 48, 4.8, and 0.48. SAS expects to read and analyze numeric data and no special considerations are necessary to input numeric data. Alphanumeric data, or character data, include letters and symbols (and possibly numbers). These include data elements, such as male, female, M, or F. To input character data, you must inform SAS that the data are character as opposed to numeric. To do this, simply follow the variable name in the input statement with a "$" symbol, which denotes a character variable. The following data set includes age (in years) and gender (denoted M or F) recorded on each of four subjects:

    20 F
    32 M
    24 F
    31 M

The following SAS code will correctly input the data:

```
data one;
  input age gender $;
cards;
20 F
32 M
24 F
31 M
run;
proc print;
run;
```

The following output is produced by the Print procedure. Since no variables are specified (e.g., var age), SAS prints the values of all of the variables in the data set. Note that SAS inserts a column labeled "OBS," indicating the observation numbers:

OBS	AGE	GENDER
1	20	F
2	32	M
3	24	F
4	31	M

A.2.2 Inputting Data: Types of Input

Two of the more common techniques for inputting data are (1) list, or free, format, and (2) column, or fixed, format. (A third technique, formatted data, is described in Section A.2.4. "Advanced Data Input and Data Manipulation"). To illustrate the different techniques, consider the following data set, which includes four variables (listed in the columns below) recorded on five subjects (listed on each row):

```
10      12      11      1.4
22      12      99      4.3
15      12      64      3.1
12       5       3      8.9
99      45      66      0.5
```

A. LIST, OR FREE, FORMAT

As long as the data are separated by at least one space, the following SAS code may be used to input the data. The data do not have to be lined up in columns, only separated by one or more spaces on each line. Notice that four variables, called x1, x2, x3 and x4, are input.

```
data one;
  input x1 x2 x3 x4;
cards;
10        12      11      1.4
22        12      99      4.3
15        12      64      3.1
12         5       3      8.9
99        45      66      0.5
run;
proc print;
run;
```

B. COLUMN, OR FIXED, FORMAT

If the data set is rectangular (i.e., the observations are lined up in columns and rows), then the following SAS code is used to input the data, where the numbers following each variable name denote the columns in which the actual data elements are located in the Raw Data section.

```
data one;
  input x1 1-2 x2 4-5 x3 7-8 x4 10-12;
cards;
10 12 11 1.4
22 12 99 4.3
```

```
15 12 64 3.1
12  5  3 8.9
99 45 66 0.5
run;
proc print;
run;
```

Both programs shown produce the following output:

OBS	X1	X2	X3	X4
1	10	12	11	1.4
2	22	12	99	4.3
3	15	12	64	3.1
4	12	5	3	8.9
5	99	45	66	0.5

A.2.3 Missing Values

Every effort should always be made to ensure that data are complete and accurate. However, even with intense effort on the part of investigators, it is not always possible to obtain values for every variable under study for each subject.

For example, suppose a study is conducted in which seven variables are measured on each of five subjects: height (in inches), weight (in pounds), systolic blood pressure (in mmHg), diastolic blood pressure (in mmHg), gender (M or F), age (in years), and race (Asian, Black, Hispanic, Other, White). Suppose that one subject does not have his weight measured and another does not report her race. We would not want to eliminate these subjects altogether since there are some variables collected on these subjects that are nonmissing and can be analyzed.

SAS can input and maintain missing data values. Missing values of numeric variables (e.g., sbp) are denoted by a period "."; missing values of alphanumeric, or character, variables are denoted by a blank space " ". The following data set includes seven variables measured on five subjects and includes two missing values. Notice that the missing values are denoted with a period or a blank space (without quotation marks) for the numeric and alphanumeric variables, respectively.

```
63  145  135  70  F  32  black
67  132  120  60  F  31  hispanic
60  117  130  80  F  33
68  187  140  60  M  26  white
72  145   .   55  M  29  white
```

The following SAS code inputs seven variables using column formats:

```
data one;
  input height 1-2 weight 4-6 sbp 8-10 dbp 12-13
        gender $ 15 age 17-18 race $ 20-27;
cards;
63  145  135  70  F  32  black
67  132  120  60  F  31  hispanic
60  117  130  80  F  33
68  187  140  60  M  26  white
72  145   .   55  M  29  white
run;
proc print;
run;
```

The following output is produced by the Print procedure.

OBS	HEIGHT	WEIGHT	SBP	DBP	GENDER	AGE	RACE
1	63	145	135	70	F	32	black
2	67	132	120	60	F	31	hispanic
3	60	117	130	80	F	33	
4	68	187	140	60	M	26	white
5	72	.	145	55	M	29	white

A.2.4 Advanced Data Input and Data Manipulation*

*This section contains advanced statistical computing techniques. It is designed for more advanced users and can be omitted without disrupting the progression of material.

This section covers inputting formatted data, the use of line pointers, and the drop and keep functions.

INPUTTING FORMATTED DATA

If a data set is rectangular, the following SAS code may be used to input the data, where the "@" symbol "points to" the starting column location of each variable and the number following each variable name denotes the variable format (e.g., 2.0 denotes a variable of length 2 with no decimal places, 3.1 denotes a variable of length 3 with 1 decimal place). *Note:* 2. can be used in place of 2.0.

```
data one;
  input @1 x1 2.0 @4 x2 2.0 @7 x3 2.0 @10 x4 3.1;
```

Notice that x1, x2, and x3 all have the same format (2.0). The following SAS code is equivalent:

```
data one;
  input (x1 x2 x3) (2. 2. 2.) @10 x4 3.1;
```

The following is also equivalent, where the "+1" moves the column pointer to the right 1 column after inputting x1, x2, and x3:

```
data one;
  input (x1-x3) (2. +1) @10 x4 3.1;
```

SAS allows the user to mix input options (e.g., formatted input and list, or free, format):

```
data one;
  input (x1-x3) (2. +1) x4;
```

INPUTTING DATA: LINE POINTERS

In addition to the column pointers, just illustrated, SAS allows the user to manipulate the rows, or lines, in a data set. For example, suppose we record total serum cholesterol values on a sample of five subjects at three different points in time. The data set consists of three lines (rows) per person, with exam 1 values for total cholesterol on the first line, exam 2 values on the second line, and exam 3 values on the third line. The data follow; the first entry on each line (row) is the exam number (i.e., 1, 2, or 3), the second entry is the subject identifier (sequential numbers), and the third entry is the observed total cholesterol value. A "." denotes a missing value.

1 1 180	exam number=1, subject identifier=1, cholesterol=180
2 1 160	exam number=2, subject identifier=1, cholesterol=160
3 1 177	exam number=3, subject identifier=1, cholesterol=177
1 2 240	exam number=1, subject identifier=2, cholesterol=240
2 2 220	
3 2 146	
1 3 .	
2 3 230	
3 3 210	
1 4 170	
2 4 200	
3 4 240	
1 5 180	
2 5 .	
3 5 190	exam number=3, subject identifier=5, cholesterol=190

The data are input as follows. The "#" symbol denotes the line number in the data set; the numbers following the variable names denote the columns in which the data are stored. Notice that the exam number (i.e., 1, 2, 3), stored in column 1 on each line, is ignored (i.e., not input).

```
data one;
  input #1 id 3 tc1 5-7
        #2 tc2 5-7
        #3 tc3 5-7;
cards;
1 1 180
2 1 160
3 1 177
1 2 240
2 2 220
3 2 146
1 3  .
2 3 230
3 3 210
1 4 170
2 4 200
3 4 240
1 5 180
2 5  .
3 5 190
run;
proc print;
run;
```

The Print procedure generates the following output:

OBS	ID	TC1	TC2	TC3
1	1	180	160	177
2	2	240	220	146
3	3	.	230	210
4	4	170	200	240
5	5	180	.	190

THE DROP FUNCTION

Suppose (for the same data set) we are only interested in the first two exams. The "drop" function can be used to remove variables from a data set. Note that the data are never deleted or removed permanently—a *variable* is simply omitted from a SAS data set:

```
data one;
  input #1 id 3 tc1 5-7
        #2 tc2 5-7
        #3 tc3 5-7;
  drop tc3;
cards;
1 1 180
2 1 160
3 1 177
1 2 240
2 2 220
3 2 146
1 3 .
2 3 230
3 3 210
1 4 170
2 4 200
3 4 240
1 5 180
2 5 .
3 5 190
run;
proc print;
run;
```

Or equivalently:

```
data one;
  input #1 id 3 tc1 5-7
        #2 tc2 5-7
        #3;
cards;
1 1 180
2 1 160
3 1 177
1 2 240
2 2 220
3 2 146
1 3 .
2 3 230
3 3 210
1 4 170
```

```
2 4 200
3 4 240
1 5 180
2 5 .
3 5 190
run;
proc print;
run;
```

NOTE: The #3 indicates that there are three lines of data per subject and is necessary so that the correct observations are read from the appropriate lines in the data set.

Both programs shown produce the following output:

OBS	ID	TC1	TC2
1	1	180	160
2	2	240	220
3	3	.	230
4	4	170	200
5	5	180	.

Suppose we only want the second exam:

```
data one;
  input #1 id 3 tc1 5-7
        #2 tc2 5-7
        #3;
  drop tc1;
cards;
1 1 180
2 1 160
3 1 177
1 2 240
2 2 220
3 2 146
1 3 .
2 3 230
3 3 210
1 4 170
2 4 200
3 4 240
1 5 180
2 5 .
3 5 190
```

```
run;
proc print;
run;
```

The following program is equivalent and illustrates the use of line holders. The "@" symbol at the end of an input line holds the pointer at that line until the next input statement.

```
data one;
  input exam 1 @;
  if exam eq 2 then input id 3 tc2 5-7;
cards;
1 1 180
2 1 160
3 1 177
1 2 240
2 2 220
3 2 146
1 3 .
2 3 230
3 3 210
1 4 170
2 4 200
3 4 240
1 5 180
2 5 .
3 5 190
run;
proc print;
run;
```

Both programs shown produce the following output:

```
OBS       ID       TC1
1         1        160
2         2        220
3         3        230
4         4        200
5         5          .
```

Now, suppose that only lines with nonmissing data are in the data set. Again, suppose we are interested only in the exam 2 values. In this case, we no longer have exactly three lines per person. Instead, the data is as follows. Notice that subject 3 has only two lines of data.

```
1 1 180
2 1 160
3 1 177
1 2 240
2 2 220
3 2 146
2 3 230
3 3 210
1 4 170
2 4 200
3 4 240
1 5 180
3 5 190
```

To input only the second exam value from a data set that does not contain three lines per person, we must use the line holder:

```
data one;
  input exam 1 @;
  if exam eq 2 then input id 3 tc2 5-7;
cards;
1 1 180
2 1 160
3 1 177
1 2 240
2 2 220
3 2 146
2 3 230
3 3 210
1 4 170
2 4 200
3 4 240
1 5 180
3 5 190
run;
proc print;
run;
```

The program shown above produces the following output:

OBS	ID	TC2
1	1	160
2	2	220

```
3          3          230
4          4          200
5          5            .
```

Suppose we record data (a continuous variable) on 10 subjects and want descriptive statistics on that variable. The data could be entered in 10 rows, and SAS Proc Means (see Chapter 2) could be used to generate descriptive statistics:

```
data one;              Data set name=one.
  input id x;          Two variables are input—id and x.
cards;                 Beginning of Raw Data Section.
1 23                   id=1, x=23
2 43                   id=2, x=43
3 66
4 31
5 28
6 73
7 92
8 19
9 33
10 55
run;
proc means;            Procedure call to generate summary
                          statistics.
 var x;                Specification of analytic variable x.
run;
```

The same can be achieved using line holders:

```
data one;
  input id x @@;

cards;
1 23 2 43 3 66 4 31 5 28
6 73 7 92 8 19 9 33 10 55
run;

proc means;
 var x;
run;
```

NOTE: "@@" holds the line until all data have been read and then moves to the next line.

The following program is equivalent, only the id variable is created in the program instead of input:

```
data one;
  input x @@;                Input variable x.
  id=id+1;                   Create the variable id by adding 1 for each subject.

cards;
23 43 66 31 28 73 92 19 33 55
run;

proc means;
  var x;
run;
```

The three programs shown produce the following output:

```
Analysis Variable: X

N       Mean      Std Dev    Minimum    Maximum
------------------------------------------------
10      46.300    25.2901    19.0000    92.0000
------------------------------------------------
```

APPENDIX B

Statistical Tables

B.1 Statistical Tables

Table B.1 *Probabilities of the Binomial Distribution*

Table entries represent $P(X = x)$ for n trials and $P(\text{success}) = p$, e.g., $P(X = 4 \mid n = 10, p = 0.2) = 0.0881$.

n	x	.10	.20	.30	.40	p .50	.60	.70	.80	.90
2	0	0.8100	0.6400	0.4900	0.3600	0.2500	0.1600	0.0900	0.0400	0.0100
	1	0.1800	0.3200	0.4200	0.4800	0.5000	0.4800	0.4200	0.3200	0.1800
	2	0.0100	0.0400	0.0900	0.1600	0.2500	0.3600	0.4900	0.6400	0.8100
3	0	0.7290	0.5120	0.3430	0.2160	0.1250	0.0640	0.0270	0.0080	0.0010
	1	0.2430	0.3840	0.4410	0.4320	0.3750	0.2880	0.1890	0.0960	0.0270
	2	0.0270	0.0960	0.1890	0.2880	0.3750	0.4320	0.4410	0.3840	0.2430
	3	0.0010	0.0080	0.0270	0.0640	0.1250	0.2160	0.3430	0.5120	0.7290
4	0	0.6561	0.4096	0.2401	0.1296	0.0625	0.0256	0.0081	0.0016	0.0001
	1	0.2916	0.4096	0.4116	0.3456	0.2500	0.1536	0.0756	0.0256	0.0036
	2	0.0486	0.1536	0.2646	0.3456	0.3750	0.3456	0.2646	0.1536	0.0486
	3	0.0036	0.0256	0.0756	0.1536	0.2500	0.3456	0.4116	0.4096	0.2916
	4	0.0001	0.0016	0.0081	0.0256	0.0625	0.1296	0.2401	0.4096	0.6561
5	0	0.5905	0.3277	0.1681	0.0778	0.0313	0.0102	0.0024	0.0003	0.0000
	1	0.3280	0.4096	0.3601	0.2592	0.1562	0.0768	0.0284	0.0064	0.0005

n	x	.10	.20	.30	.40	.50	.60	.70	.80	.90
	2	0.0729	0.2048	0.3087	0.3456	0.3125	0.2304	0.1323	0.0512	0.0081
	3	0.0081	0.0512	0.1323	0.2304	0.3125	0.3456	0.3087	0.2048	0.0729
	4	0.0005	0.0064	0.0283	0.0768	0.1563	0.2592	0.3601	0.4096	0.3281
	5	0.0000	0.0003	0.0024	0.0102	0.0313	0.0778	0.1681	0.3277	0.5905
6	0	0.5314	0.2621	0.1176	0.0467	0.0156	0.0041	0.0007	0.0001	0.0000
	1	0.3543	0.3932	0.3025	0.1866	0.0938	0.0369	0.0102	0.0015	0.0001
	2	0.0984	0.2458	0.3241	0.3110	0.2344	0.1382	0.0595	0.0154	0.0012
	3	0.0146	0.0819	0.1852	0.2765	0.3125	0.2765	0.1852	0.0819	0.0146
	4	0.0012	0.0154	0.0595	0.1382	0.2344	0.3110	0.3241	0.2458	0.0984
	5	0.0001	0.0015	0.0102	0.0369	0.0938	0.1866	0.3025	0.3932	0.3543
	6	0.0000	0.0001	0.0007	0.0041	0.0156	0.0467	0.1176	0.2621	0.5314
7	0	0.4783	0.2097	0.0824	0.0280	0.0078	0.0016	0.0002	0.0000	0.0000
	1	0.3720	0.3670	0.2471	0.1306	0.0547	0.0172	0.0036	0.0004	0.0000
	2	0.1240	0.2753	0.3177	0.2613	0.1641	0.0774	0.0250	0.0043	0.0002
	3	0.0230	0.1147	0.2269	0.2903	0.2734	0.1935	0.0972	0.0287	0.0026
	4	0.0026	0.0287	0.0972	0.1935	0.2734	0.2903	0.2269	0.1147	0.0230
	5	0.0002	0.0043	0.0250	0.0774	0.1641	0.2613	0.3177	0.2753	0.1240
	6	0.0000	0.0004	0.0036	0.0172	0.0547	0.1306	0.2471	0.3670	0.3720
	7	0.0000	0.0000	0.0002	0.0016	0.0078	0.0280	0.0824	0.2097	0.4783
8	0	0.4305	0.1678	0.0576	0.0168	0.0039	0.0007	0.0001	0.0000	0.0000
	1	0.3826	0.3355	0.1977	0.0896	0.0313	0.0079	0.0012	0.0001	0.0000
	2	0.1488	0.2936	0.2965	0.2090	0.1094	0.0413	0.0100	0.0011	0.0000
	3	0.0331	0.1468	0.2541	0.2787	0.2188	0.1239	0.0467	0.0092	0.0004
	4	0.0046	0.0459	0.1361	0.2322	0.2734	0.2322	0.1361	0.0459	0.0046
	5	0.0004	0.0092	0.0467	0.1239	0.2188	0.2787	0.2541	0.1468	0.0331
	6	0.0000	0.0011	0.0100	0.0413	0.1094	0.2090	0.2965	0.2936	0.1488
	7	0.0000	0.0001	0.0012	0.0079	0.0313	0.0896	0.1977	0.3355	0.3826
	8	0.0000	0.0000	0.0001	0.0007	0.0039	0.0168	0.0576	0.1678	0.4305
9	0	0.3874	0.1342	0.0404	0.0101	0.0020	0.0003	0.0000	0.0000	0.0000
	1	0.3874	0.3020	0.1556	0.0605	0.0176	0.0035	0.0004	0.0000	0.0000
	2	0.1722	0.3020	0.2668	0.1612	0.0703	0.0212	0.0039	0.0003	0.0000
	3	0.0446	0.1762	0.2668	0.2508	0.1641	0.0743	0.0210	0.0028	0.0001
	4	0.0074	0.0661	0.1715	0.2508	0.2461	0.1672	0.0735	0.0165	0.0008
	5	0.0008	0.0165	0.0735	0.1672	0.2461	0.2508	0.1715	0.0661	0.0074
	6	0.0001	0.0028	0.0210	0.0743	0.1641	0.2508	0.2668	0.1762	0.0446
	7	0.0000	0.0003	0.0039	0.0212	0.0703	0.1612	0.2668	0.3020	0.1722
	8	0.0000	0.0000	0.0004	0.0035	0.0176	0.0605	0.1556	0.3020	0.3874
	9	0.0000	0.0000	0.0000	0.0003	0.0020	0.0101	0.0404	0.1342	0.3874

p

		p								
n	x	.10	.20	.30	.40	.50	.60	.70	.80	.90
10	0	0.3487	0.1074	0.0282	0.0060	0.0010	0.0001	0.0000	0.0000	0.0000
	1	0.3874	0.2684	0.1211	0.0403	0.0098	0.0016	0.0001	0.0000	0.0000
	2	0.1937	0.3020	0.2335	0.1209	0.0439	0.0106	0.0014	0.0001	0.0000
	3	0.0574	0.2013	0.2668	0.2150	0.1172	0.0425	0.0090	0.0008	0.0000
	4	0.0112	0.0881	0.2001	0.2508	0.2051	0.1115	0.0368	0.0055	0.0001
	5	0.0015	0.0264	0.1029	0.2007	0.2461	0.2007	0.1029	0.0264	0.0015
	6	0.0001	0.0055	0.0368	0.1115	0.2051	0.2508	0.2001	0.0881	0.0112
	7	0.0000	0.0008	0.0090	0.0425	0.1172	0.2150	0.2668	0.2013	0.0574
	8	0.0000	0.0001	0.0014	0.0106	0.0439	0.1209	0.2335	0.3020	0.1937
	9	0.0000	0.0000	0.0001	0.0016	0.0098	0.0403	0.1211	0.2684	0.3874
	10	0.0000	0.0000	0.0000	0.0001	0.0010	0.0060	0.0282	0.1074	0.3487
11	0	0.3138	0.0859	0.0198	0.0036	0.0005	0.0000	0.0000	0.0000	0.0000
	1	0.3835	0.2362	0.0932	0.0266	0.0054	0.0007	0.0000	0.0000	0.0000
	2	0.2131	0.2953	0.1998	0.0887	0.0269	0.0052	0.0005	0.0000	0.0000
	3	0.0710	0.2215	0.2568	0.1774	0.0806	0.0234	0.0037	0.0002	0.0000
	4	0.0158	0.1107	0.2201	0.2365	0.1611	0.0701	0.0173	0.0017	0.0000
	5	0.0025	0.0388	0.1321	0.2207	0.2256	0.1471	0.0566	0.0097	0.0003
	6	0.0003	0.0097	0.0566	0.1471	0.2256	0.2207	0.1321	0.0388	0.0025
	7	0.0000	0.0017	0.0173	0.0701	0.1611	0.2365	0.2201	0.1107	0.0158
	8	0.0000	0.0002	0.0037	0.0234	0.0806	0.1774	0.2568	0.2215	0.0710
	9	0.0000	0.0000	0.0005	0.0052	0.0269	0.0887	0.1998	0.2953	0.2131
	10	0.0000	0.0000	0.0000	0.0007	0.0054	0.0266	0.0932	0.2362	0.3835
	11	0.0000	0.0000	0.0000	0.0000	0.0005	0.0036	0.0198	0.0859	0.3138
12	0	0.2824	0.0687	0.0138	0.0022	0.0002	0.0000	0.0000	0.0000	0.0000
	1	0.3766	0.2062	0.0712	0.0174	0.0029	0.0003	0.0000	0.0000	0.0000
	2	0.2301	0.2835	0.1678	0.0639	0.0161	0.0025	0.0002	0.0000	0.0000
	3	0.0852	0.2362	0.2397	0.1419	0.0537	0.0125	0.0015	0.0001	0.0000
	4	0.0213	0.1329	0.2311	0.2128	0.1208	0.0420	0.0078	0.0005	0.0000
	5	0.0038	0.0532	0.1585	0.2270	0.1934	0.1009	0.0291	0.0033	0.0000
	6	0.0005	0.0155	0.0792	0.1766	0.2256	0.1766	0.0792	0.0155	0.0005
	7	0.0000	0.0033	0.0291	0.1009	0.1934	0.2270	0.1585	0.0532	0.0038
	8	0.0000	0.0005	0.0078	0.0420	0.1208	0.2128	0.2311	0.1329	0.0213
	9	0.0000	0.0001	0.0015	0.0125	0.0537	0.1419	0.2397	0.2362	0.0852
	10	0.0000	0.0000	0.0002	0.0025	0.0161	0.0639	0.1678	0.2835	0.2301
	11	0.0000	0.0000	0.0000	0.0003	0.0029	0.0174	0.0712	0.2062	0.3766
	12	0.0000	0.0000	0.0000	0.0000	0.0002	0.0022	0.0138	0.0687	0.2824
13	0	0.2542	0.0550	0.0097	0.0013	0.0001	0.0000	0.0000	0.0000	0.0000
	1	0.3672	0.1787	0.0540	0.0113	0.0016	0.0001	0.0000	0.0000	0.0000

						p				
n	x	.10	.20	.30	.40	.50	.60	.70	.80	.90
	2	0.2448	0.2680	0.1388	0.0453	0.0095	0.0012	0.0001	0.0000	0.0000
	3	0.0997	0.2457	0.2181	0.1107	0.0349	0.0065	0.0006	0.0000	0.0000
	4	0.0277	0.1535	0.2337	0.1845	0.0873	0.0243	0.0034	0.0001	0.0000
	5	0.0055	0.0691	0.1803	0.2214	0.1571	0.0656	0.0142	0.0011	0.0000
	6	0.0008	0.0230	0.1030	0.1968	0.2095	0.1312	0.0442	0.0058	0.0001
	7	0.0001	0.0058	0.0442	0.1312	0.2095	0.1968	0.1030	0.0230	0.0008
	8	0.0000	0.0011	0.0142	0.0656	0.1571	0.2214	0.1803	0.0691	0.0055
	9	0.0000	0.0001	0.0034	0.0243	0.0873	0.1845	0.2337	0.1535	0.0277
	10	0.0000	0.0000	0.0006	0.0065	0.0349	0.1107	0.2181	0.2457	0.0997
	11	0.0000	0.0000	0.0001	0.0012	0.0095	0.0453	0.1388	0.2680	0.2448
	12	0.0000	0.0000	0.0000	0.0001	0.0016	0.0113	0.0540	0.1787	0.3672
	13	0.0000	0.0000	0.0000	0.0000	0.0001	0.0013	0.0097	0.0550	0.2542
14	0	0.2288	0.0440	0.0068	0.0008	0.0001	0.0000	0.0000	0.0000	0.0000
	1	0.3559	0.1539	0.0407	0.0073	0.0009	0.0001	0.0000	0.0000	0.0000
	2	0.2570	0.2501	0.1134	0.0317	0.0056	0.0005	0.0000	0.0000	0.0000
	3	0.1142	0.2501	0.1943	0.0845	0.0222	0.0033	0.0002	0.0000	0.0000
	4	0.0349	0.1720	0.2290	0.1549	0.0611	0.0136	0.0014	0.0000	0.0000
	5	0.0078	0.0860	0.1963	0.2066	0.1222	0.0408	0.0066	0.0003	0.0000
	6	0.0013	0.0322	0.1262	0.2066	0.1833	0.0918	0.0232	0.0020	0.0000
	7	0.0002	0.0092	0.0618	0.1574	0.2095	0.1574	0.0618	0.0092	0.0002
	8	0.0000	0.0020	0.0232	0.0918	0.1833	0.2066	0.1262	0.0322	0.0013
	9	0.0000	0.0003	0.0066	0.0408	0.1222	0.2066	0.1963	0.0860	0.0078
	10	0.0000	0.0000	0.0014	0.0136	0.0611	0.1549	0.2290	0.1720	0.0349
	11	0.0000	0.0000	0.0002	0.0033	0.0222	0.0845	0.1943	0.2501	0.1142
	12	0.0000	0.0000	0.0000	0.0005	0.0056	0.0317	0.1134	0.2501	0.2570
	13	0.0000	0.0000	0.0000	0.0001	0.0009	0.0073	0.0407	0.1539	0.3559
	14	0.0000	0.0000	0.0000	0.0000	0.0001	0.0008	0.0068	0.0440	0.2288
15	0	0.2059	0.0352	0.0047	0.0005	0.0000	0.0000	0.0000	0.0000	0.0000
	1	0.3432	0.1319	0.0305	0.0047	0.0005	0.0000	0.0000	0.0000	0.0000
	2	0.2669	0.2309	0.0916	0.0219	0.0032	0.0003	0.0000	0.0000	0.0000
	3	0.1285	0.2501	0.1700	0.0634	0.0139	0.0016	0.0001	0.0000	0.0000
	4	0.0428	0.1876	0.2186	0.1268	0.0417	0.0074	0.0006	0.0000	0.0000
	5	0.0105	0.1032	0.2061	0.1859	0.0916	0.0245	0.0030	0.0001	0.0000
	6	0.0019	0.0430	0.1472	0.2066	0.1527	0.0612	0.0116	0.0007	0.0000
	7	0.0003	0.0138	0.0811	0.1771	0.1964	0.1181	0.0348	0.0035	0.0000
	8	0.0000	0.0035	0.0348	0.1181	0.1964	0.1771	0.0811	0.0138	0.0003
	9	0.0000	0.0007	0.0116	0.0612	0.1527	0.2066	0.1472	0.0430	0.0019
	10	0.0000	0.0001	0.0030	0.0245	0.0916	0.1859	0.2061	0.1032	0.0105
	11	0.0000	0.0000	0.0006	0.0074	0.0417	0.1268	0.2186	0.1876	0.0428

n	x	p .10	.20	.30	.40	.50	.60	.70	.80	.90
	12	0.0000	0.0000	0.0001	0.0016	0.0139	0.0634	0.1700	0.2501	0.1285
	13	0.0000	0.0000	0.0000	0.0003	0.0032	0.0219	0.0916	0.2309	0.2669
	14	0.0000	0.0000	0.0000	0.0000	0.0005	0.0047	0.0305	0.1319	0.3432
	15	0.0000	0.0000	0.0000	0.0000	0.0000	0.0005	0.0047	0.0352	0.2059
16	0	0.1853	0.0281	0.0033	0.0003	0.0000	0.0000	0.0000	0.0000	0.0000
	1	0.3294	0.1126	0.0228	0.0030	0.0002	0.0000	0.0000	0.0000	0.0000
	2	0.2745	0.2111	0.0732	0.0150	0.0018	0.0001	0.0000	0.0000	0.0000
	3	0.1423	0.2463	0.1465	0.0468	0.0085	0.0008	0.0000	0.0000	0.0000
	4	0.0514	0.2001	0.2040	0.1014	0.0278	0.0040	0.0002	0.0000	0.0000
	5	0.0137	0.1201	0.2099	0.1623	0.0667	0.0142	0.0013	0.0000	0.0000
	6	0.0028	0.0550	0.1649	0.1983	0.1222	0.0392	0.0056	0.0002	0.0000
	7	0.0004	0.0197	0.1010	0.1889	0.1746	0.0840	0.0185	0.0012	0.0000
	8	0.0001	0.0055	0.0487	0.1417	0.1964	0.1417	0.0487	0.0055	0.0001
	9	0.0000	0.0012	0.0185	0.0840	0.1746	0.1889	0.1010	0.0197	0.0004
	10	0.0000	0.0002	0.0056	0.0392	0.1222	0.1983	0.1649	0.0550	0.0028
	11	0.0000	0.0000	0.0013	0.0142	0.0667	0.1623	0.2099	0.1201	0.0137
	12	0.0000	0.0000	0.0002	0.0040	0.0278	0.1014	0.2040	0.2001	0.0514
	13	0.0000	0.0000	0.0000	0.0008	0.0085	0.0468	0.1465	0.2463	0.1423
	14	0.0000	0.0000	0.0000	0.0001	0.0018	0.0150	0.0732	0.2111	0.2745
	15	0.0000	0.0000	0.0000	0.0000	0.0002	0.0030	0.0228	0.1126	0.3294
	16	0.0000	0.0000	0.0000	0.0000	0.0000	0.0003	0.0033	0.0281	0.1853
17	0	0.1668	0.0225	0.0023	0.0002	0.0000	0.0000	0.0000	0.0000	0.0000
	1	0.3150	0.0957	0.0169	0.0019	0.0001	0.0000	0.0000	0.0000	0.0000
	2	0.2800	0.1914	0.0581	0.0102	0.0010	0.0001	0.0000	0.0000	0.0000
	3	0.1556	0.2393	0.1245	0.0341	0.0052	0.0004	0.0000	0.0000	0.0000
	4	0.0605	0.2093	0.1868	0.0796	0.0182	0.0021	0.0001	0.0000	0.0000
	5	0.0175	0.1361	0.2081	0.1379	0.0472	0.0081	0.0006	0.0000	0.0000
	6	0.0039	0.0680	0.1784	0.1839	0.0944	0.0242	0.0026	0.0001	0.0000
	7	0.0007	0.0267	0.1201	0.1927	0.1484	0.0571	0.0095	0.0004	0.0000
	8	0.0001	0.0084	0.0644	0.1606	0.1855	0.1070	0.0276	0.0021	0.0000
	9	0.0000	0.0021	0.0276	0.1070	0.1855	0.1606	0.0644	0.0084	0.0001
	10	0.0000	0.0004	0.0095	0.0571	0.1484	0.1927	0.1201	0.0267	0.0007
	11	0.0000	0.0001	0.0026	0.0242	0.0944	0.1839	0.1784	0.0680	0.0039
	12	0.0000	0.0000	0.0006	0.0081	0.0472	0.1379	0.2081	0.1361	0.0175
	13	0.0000	0.0000	0.0001	0.0021	0.0182	0.0796	0.1868	0.2093	0.0605
	14	0.0000	0.0000	0.0000	0.0004	0.0052	0.0341	0.1245	0.2393	0.1556
	15	0.0000	0.0000	0.0000	0.0001	0.0010	0.0102	0.0581	0.1914	0.2800
	16	0.0000	0.0000	0.0000	0.0000	0.0001	0.0019	0.0169	0.0957	0.3150
	17	0.0000	0.0000	0.0000	0.0000	0.0000	0.0002	0.0023	0.0225	0.1668

n	x	.10	.20	.30	.40	p .50	.60	.70	.80	.90
18	0	0.1501	0.0180	0.0016	0.0001	0.0000	0.0000	0.0000	0.0000	0.0000
	1	0.3002	0.0811	0.0126	0.0012	0.0001	0.0000	0.0000	0.0000	0.0000
	2	0.2835	0.1723	0.0458	0.0069	0.0006	0.0000	0.0000	0.0000	0.0000
	3	0.1680	0.2297	0.1046	0.0246	0.0031	0.0002	0.0000	0.0000	0.0000
	4	0.0700	0.2153	0.1681	0.0614	0.0117	0.0011	0.0000	0.0000	0.0000
	5	0.0218	0.1507	0.2017	0.1146	0.0327	0.0045	0.0002	0.0000	0.0000
	6	0.0052	0.0816	0.1873	0.1655	0.0708	0.0145	0.0012	0.0000	0.0000
	7	0.0010	0.0350	0.1376	0.1892	0.1214	0.0374	0.0046	0.0001	0.0000
	8	0.0002	0.0120	0.0811	0.1734	0.1669	0.0771	0.0149	0.0008	0.0000
	9	0.0000	0.0033	0.0386	0.1284	0.1855	0.1284	0.0386	0.0033	0.0000
	10	0.0000	0.0008	0.0149	0.0771	0.1669	0.1734	0.0811	0.0120	0.0002
	11	0.0000	0.0001	0.0046	0.0374	0.1214	0.1892	0.1376	0.0350	0.0010
	12	0.0000	0.0000	0.0012	0.0145	0.0708	0.1655	0.1873	0.0816	0.0052
	13	0.0000	0.0000	0.0002	0.0045	0.0327	0.1146	0.2017	0.1507	0.0218
	14	0.0000	0.0000	0.0000	0.0011	0.0117	0.0614	0.1681	0.2153	0.0700
	15	0.0000	0.0000	0.0000	0.0002	0.0031	0.0246	0.1046	0.2297	0.1680
	16	0.0000	0.0000	0.0000	0.0000	0.0006	0.0069	0.0458	0.1723	0.2835
	17	0.0000	0.0000	0.0000	0.0000	0.0001	0.0012	0.0126	0.0811	0.3002
	18	0.0000	0.0000	0.0000	0.0000	0.0000	0.0001	0.0016	0.0180	0.1501
19	0	0.1351	0.0144	0.0011	0.0001	0.0000	0.0000	0.0000	0.0000	0.0000
	1	0.2852	0.0685	0.0093	0.0008	0.0000	0.0000	0.0000	0.0000	0.0000
	2	0.2852	0.1540	0.0358	0.0046	0.0003	0.0000	0.0000	0.0000	0.0000
	3	0.1796	0.2182	0.0869	0.0175	0.0018	0.0001	0.0000	0.0000	0.0000
	4	0.0798	0.2182	0.1491	0.0467	0.0074	0.0005	0.0000	0.0000	0.0000
	5	0.0266	0.1636	0.1916	0.0933	0.0222	0.0024	0.0001	0.0000	0.0000
	6	0.0069	0.0955	0.1916	0.1451	0.0518	0.0085	0.0005	0.0000	0.0000
	7	0.0014	0.0443	0.1525	0.1797	0.0961	0.0237	0.0022	0.0000	0.0000
	8	0.0002	0.0166	0.0981	0.1797	0.1442	0.0532	0.0077	0.0003	0.0000
	9	0.0000	0.0051	0.0514	0.1464	0.1762	0.0976	0.0220	0.0013	0.0000
	10	0.0000	0.0013	0.0220	0.0976	0.1762	0.1464	0.0514	0.0051	0.0000
	11	0.0000	0.0003	0.0077	0.0532	0.1442	0.1797	0.0981	0.0166	0.0002
	12	0.0000	0.0000	0.0022	0.0237	0.0961	0.1797	0.1525	0.0443	0.0014
	13	0.0000	0.0000	0.0005	0.0085	0.0518	0.1451	0.1916	0.0955	0.0069
	14	0.0000	0.0000	0.0001	0.0024	0.0222	0.0933	0.1916	0.1636	0.0266
	15	0.0000	0.0000	0.0000	0.0005	0.0074	0.0467	0.1491	0.2182	0.0798
	16	0.0000	0.0000	0.0000	0.0001	0.0018	0.0175	0.0869	0.2182	0.1796
	17	0.0000	0.0000	0.0000	0.0000	0.0003	0.0046	0.0358	0.1540	0.2852
	18	0.0000	0.0000	0.0000	0.0000	0.0000	0.0008	0.0093	0.0685	0.2852
	19	0.0000	0.0000	0.0000	0.0000	0.0000	0.0001	0.0011	0.0144	0.1351

n	x					p				
		.10	.20	.30	.40	.50	.60	.70	.80	.90
20	0	0.1216	0.0115	0.0008	0.0000	0.0000	0.0000	0.0000	0.0000	0.0000
	1	0.2702	0.0576	0.0068	0.0005	0.0000	0.0000	0.0000	0.0000	0.0000
	2	0.2852	0.1369	0.0278	0.0031	0.0002	0.0000	0.0000	0.0000	0.0000
	3	0.1901	0.2054	0.0716	0.0123	0.0011	0.0000	0.0000	0.0000	0.0000
	4	0.0898	0.2182	0.1304	0.0350	0.0046	0.0003	0.0000	0.0000	0.0000
	5	0.0319	0.1746	0.1789	0.0746	0.0148	0.0013	0.0000	0.0000	0.0000
	6	0.0089	0.1091	0.1916	0.1244	0.0370	0.0049	0.0002	0.0000	0.0000
	7	0.0020	0.0545	0.1643	0.1659	0.0739	0.0146	0.0010	0.0000	0.0000
	8	0.0004	0.0222	0.1144	0.1797	0.1201	0.0355	0.0039	0.0001	0.0000
	9	0.0001	0.0074	0.0654	0.1597	0.1602	0.0710	0.0120	0.0005	0.0000
	10	0.0000	0.0020	0.0308	0.1171	0.1762	0.1171	0.0308	0.0020	0.0000
	11	0.0000	0.0005	0.0120	0.0710	0.1602	0.1597	0.0654	0.0074	0.0001
	12	0.0000	0.0001	0.0039	0.0355	0.1201	0.1797	0.1144	0.0222	0.0004
	13	0.0000	0.0000	0.0010	0.0146	0.0739	0.1659	0.1643	0.0545	0.0020
	14	0.0000	0.0000	0.0002	0.0049	0.0370	0.1244	0.1916	0.1091	0.0089
	15	0.0000	0.0000	0.0000	0.0013	0.0148	0.0746	0.1789	0.1746	0.0319
	16	0.0000	0.0000	0.0000	0.0003	0.0046	0.0350	0.1304	0.2182	0.0898
	17	0.0000	0.0000	0.0000	0.0000	0.0011	0.0123	0.0716	0.2054	0.1901
	18	0.0000	0.0000	0.0000	0.0000	0.0002	0.0031	0.0278	0.1369	0.2852
	19	0.0000	0.0000	0.0000	0.0000	0.0000	0.0005	0.0068	0.0576	0.2702
	20	0.0000	0.0000	0.0000	0.0000	0.0000	0.0000	0.0008	0.0115	0.1216
21	0	0.1094	0.0092	0.0006	0.0000	0.0000	0.0000	0.0000	0.0000	0.0000
	1	0.2553	0.0484	0.0050	0.0003	0.0000	0.0000	0.0000	0.0000	0.0000
	2	0.2837	0.1211	0.0215	0.0020	0.0001	0.0000	0.0000	0.0000	0.0000
	3	0.1996	0.1917	0.0585	0.0086	0.0006	0.0000	0.0000	0.0000	0.0000
	4	0.0998	0.2156	0.1128	0.0259	0.0029	0.0001	0.0000	0.0000	0.0000
	5	0.0377	0.1833	0.1643	0.0588	0.0097	0.0007	0.0000	0.0000	0.0000
	6	0.0112	0.1222	0.1878	0.1045	0.0259	0.0027	0.0001	0.0000	0.0000
	7	0.0027	0.0655	0.1725	0.1493	0.0554	0.0087	0.0005	0.0000	0.0000
	8	0.0005	0.0286	0.1294	0.1742	0.0970	0.0229	0.0019	0.0000	0.0000
	9	0.0001	0.0103	0.0801	0.1677	0.1402	0.0497	0.0063	0.0002	0.0000
	10	0.0000	0.0031	0.0412	0.1342	0.1682	0.0895	0.0176	0.0008	0.0000
	11	0.0000	0.0008	0.0176	0.0895	0.1682	0.1342	0.0412	0.0031	0.0000
	12	0.0000	0.0002	0.0063	0.0497	0.1402	0.1677	0.0801	0.0103	0.0001
	13	0.0000	0.0000	0.0019	0.0229	0.0970	0.1742	0.1294	0.0286	0.0005
	14	0.0000	0.0000	0.0005	0.0087	0.0554	0.1493	0.1725	0.0655	0.0027
	15	0.0000	0.0000	0.0001	0.0027	0.0259	0.1045	0.1878	0.1222	0.0112
	16	0.0000	0.0000	0.0000	0.0007	0.0097	0.0588	0.1643	0.1833	0.0377
	17	0.0000	0.0000	0.0000	0.0001	0.0029	0.0259	0.1128	0.2156	0.0998

n	x	.10	.20	.30	.40	p .50	.60	.70	.80	.90
	18	0.0000	0.0000	0.0000	0.0000	0.0006	0.0086	0.0585	0.1917	0.1996
	19	0.0000	0.0000	0.0000	0.0000	0.0001	0.0020	0.0215	0.1211	0.2837
	20	0.0000	0.0000	0.0000	0.0000	0.0000	0.0003	0.0050	0.0484	0.2553
	21	0.0000	0.0000	0.0000	0.0000	0.0000	0.0000	0.0006	0.0092	0.1094
22	0	0.0985	0.0074	0.0004	0.0000	0.0000	0.0000	0.0000	0.0000	0.0000
	1	0.2407	0.0406	0.0037	0.0002	0.0000	0.0000	0.0000	0.0000	0.0000
	2	0.2808	0.1065	0.0166	0.0014	0.0001	0.0000	0.0000	0.0000	0.0000
	3	0.2080	0.1775	0.0474	0.0060	0.0004	0.0000	0.0000	0.0000	0.0000
	4	0.1098	0.2108	0.0965	0.0190	0.0017	0.0001	0.0000	0.0000	0.0000
	5	0.0439	0.1898	0.1489	0.0456	0.0063	0.0004	0.0000	0.0000	0.0000
	6	0.0138	0.1344	0.1808	0.0862	0.0178	0.0015	0.0000	0.0000	0.0000
	7	0.0035	0.0768	0.1771	0.1314	0.0407	0.0051	0.0002	0.0000	0.0000
	8	0.0007	0.0360	0.1423	0.1642	0.0762	0.0144	0.0009	0.0000	0.0000
	9	0.0001	0.0140	0.0949	0.1703	0.1186	0.0336	0.0032	0.0001	0.0000
	10	0.0000	0.0046	0.0529	0.1476	0.1542	0.0656	0.0097	0.0003	0.0000
	11	0.0000	0.0012	0.0247	0.1073	0.1682	0.1073	0.0247	0.0012	0.0000
	12	0.0000	0.0003	0.0097	0.0656	0.1542	0.1476	0.0529	0.0046	0.0000
	13	0.0000	0.0001	0.0032	0.0336	0.1186	0.1703	0.0949	0.0140	0.0001
	14	0.0000	0.0000	0.0009	0.0144	0.0762	0.1642	0.1423	0.0360	0.0007
	15	0.0000	0.0000	0.0002	0.0051	0.0407	0.1314	0.1771	0.0768	0.0035
	16	0.0000	0.0000	0.0000	0.0015	0.0178	0.0862	0.1808	0.1344	0.0138
	17	0.0000	0.0000	0.0000	0.0004	0.0063	0.0456	0.1489	0.1898	0.0439
	18	0.0000	0.0000	0.0000	0.0001	0.0017	0.0190	0.0965	0.2108	0.1098
	19	0.0000	0.0000	0.0000	0.0000	0.0004	0.0060	0.0474	0.1775	0.2080
	20	0.0000	0.0000	0.0000	0.0000	0.0001	0.0014	0.0166	0.1065	0.2808
	21	0.0000	0.0000	0.0000	0.0000	0.0000	0.0002	0.0037	0.0406	0.2407
	22	0.0000	0.0000	0.0000	0.0000	0.0000	0.0000	0.0004	0.0074	0.0985
23	0	0.0886	0.0059	0.0003	0.0000	0.0000	0.0000	0.0000	0.0000	0.0000
	1	0.2265	0.0339	0.0027	0.0001	0.0000	0.0000	0.0000	0.0000	0.0000
	2	0.2768	0.0933	0.0127	0.0009	0.0000	0.0000	0.0000	0.0000	0.0000
	3	0.2153	0.1633	0.0382	0.0041	0.0002	0.0000	0.0000	0.0000	0.0000
	4	0.1196	0.2042	0.0818	0.0138	0.0011	0.0000	0.0000	0.0000	0.0000
	5	0.0505	0.1940	0.1332	0.0350	0.0040	0.0002	0.0000	0.0000	0.0000
	6	0.0168	0.1455	0.1712	0.0700	0.0120	0.0008	0.0000	0.0000	0.0000
	7	0.0045	0.0883	0.1782	0.1133	0.0292	0.0029	0.0001	0.0000	0.0000
	8	0.0010	0.0442	0.1527	0.1511	0.0584	0.0088	0.0004	0.0000	0.0000
	9	0.0002	0.0184	0.1091	0.1679	0.0974	0.0221	0.0016	0.0000	0.0000
	10	0.0000	0.0064	0.0655	0.1567	0.1364	0.0464	0.0052	0.0001	0.0000
	11	0.0000	0.0019	0.0332	0.1234	0.1612	0.0823	0.0142	0.0005	0.0000

						p				
n	x	.10	.20	.30	.40	.50	.60	.70	.80	.90
	12	0.0000	0.0005	0.0142	0.0823	0.1612	0.1234	0.0332	0.0019	0.0000
	13	0.0000	0.0001	0.0052	0.0464	0.1364	0.1567	0.0655	0.0064	0.0000
	14	0.0000	0.0000	0.0016	0.0221	0.0974	0.1679	0.1091	0.0184	0.0002
	15	0.0000	0.0000	0.0004	0.0088	0.0584	0.1511	0.1527	0.0442	0.0010
	16	0.0000	0.0000	0.0001	0.0029	0.0292	0.1133	0.1782	0.0883	0.0045
	17	0.0000	0.0000	0.0000	0.0008	0.0120	0.0700	0.1712	0.1455	0.0168
	18	0.0000	0.0000	0.0000	0.0002	0.0040	0.0350	0.1332	0.1940	0.0505
	19	0.0000	0.0000	0.0000	0.0000	0.0011	0.0138	0.0818	0.2042	0.1196
	20	0.0000	0.0000	0.0000	0.0000	0.0002	0.0041	0.0382	0.1633	0.2153
	21	0.0000	0.0000	0.0000	0.0000	0.0000	0.0009	0.0127	0.0933	0.2768
	22	0.0000	0.0000	0.0000	0.0000	0.0000	0.0001	0.0027	0.0339	0.2265
	23	0.0000	0.0000	0.0000	0.0000	0.0000	0.0000	0.0003	0.0059	0.0886
24	0	0.0798	0.0047	0.0002	0.0000	0.0000	0.0000	0.0000	0.0000	0.0000
	1	0.2127	0.0283	0.0020	0.0001	0.0000	0.0000	0.0000	0.0000	0.0000
	2	0.2718	0.0815	0.0097	0.0006	0.0000	0.0000	0.0000	0.0000	0.0000
	3	0.2215	0.1493	0.0305	0.0028	0.0001	0.0000	0.0000	0.0000	0.0000
	4	0.1292	0.1960	0.0687	0.0099	0.0006	0.0000	0.0000	0.0000	0.0000
	5	0.0574	0.1960	0.1177	0.0265	0.0025	0.0001	0.0000	0.0000	0.0000
	6	0.0202	0.1552	0.1598	0.0560	0.0080	0.0004	0.0000	0.0000	0.0000
	7	0.0058	0.0998	0.1761	0.0960	0.0206	0.0017	0.0000	0.0000	0.0000
	8	0.0014	0.0530	0.1604	0.1360	0.0438	0.0053	0.0002	0.0000	0.0000
	9	0.0003	0.0236	0.1222	0.1612	0.0779	0.0141	0.0008	0.0000	0.0000
	10	0.0000	0.0088	0.0785	0.1612	0.1169	0.0318	0.0026	0.0000	0.0000
	11	0.0000	0.0028	0.0428	0.1367	0.1488	0.0608	0.0079	0.0002	0.0000
	12	0.0000	0.0008	0.0199	0.0988	0.1612	0.0988	0.0199	0.0008	0.0000
	13	0.0000	0.0002	0.0079	0.0608	0.1488	0.1367	0.0428	0.0028	0.0000
	14	0.0000	0.0000	0.0026	0.0318	0.1169	0.1612	0.0785	0.0088	0.0000
	15	0.0000	0.0000	0.0008	0.0141	0.0779	0.1612	0.1222	0.0236	0.0003
	16	0.0000	0.0000	0.0002	0.0053	0.0438	0.1360	0.1604	0.0530	0.0014
	17	0.0000	0.0000	0.0000	0.0017	0.0206	0.0960	0.1761	0.0998	0.0058
	18	0.0000	0.0000	0.0000	0.0004	0.0080	0.0560	0.1598	0.1552	0.0202
	19	0.0000	0.0000	0.0000	0.0001	0.0025	0.0265	0.1177	0.1960	0.0574
	20	0.0000	0.0000	0.0000	0.0000	0.0006	0.0099	0.0687	0.1960	0.1292
	21	0.0000	0.0000	0.0000	0.0000	0.0001	0.0028	0.0305	0.1493	0.2215
	22	0.0000	0.0000	0.0000	0.0000	0.0000	0.0006	0.0097	0.0815	0.2718
	23	0.0000	0.0000	0.0000	0.0000	0.0000	0.0001	0.0020	0.0283	0.2127
	24	0.0000	0.0000	0.0000	0.0000	0.0000	0.0000	0.0002	0.0047	0.0798
25	0	0.0718	0.0038	0.0001	0.0000	0.0000	0.0000	0.0000	0.0000	0.0000
	1	0.1994	0.0236	0.0014	0.0000	0.0000	0.0000	0.0000	0.0000	0.0000

		p							
n	x	.10	.20	.30	.40	.50	.60	.70	.80	.90
	2	0.2659	0.0708	0.0074	0.0004	0.0000	0.0000	0.0000	0.0000	0.0000
	3	0.2265	0.1358	0.0243	0.0019	0.0001	0.0000	0.0000	0.0000	0.0000
	4	0.1384	0.1867	0.0572	0.0071	0.0004	0.0000	0.0000	0.0000	0.0000
	5	0.0646	0.1960	0.1030	0.0199	0.0016	0.0000	0.0000	0.0000	0.0000
	6	0.0239	0.1633	0.1472	0.0442	0.0053	0.0002	0.0000	0.0000	0.0000
	7	0.0072	0.1108	0.1712	0.0800	0.0143	0.0009	0.0000	0.0000	0.0000
	8	0.0018	0.0623	0.1651	0.1200	0.0322	0.0031	0.0001	0.0000	0.0000
	9	0.0004	0.0294	0.1336	0.1511	0.0609	0.0088	0.0004	0.0000	0.0000
	10	0.0001	0.0118	0.0916	0.1612	0.0974	0.0212	0.0013	0.0000	0.0000
	11	0.0000	0.0040	0.0536	0.1465	0.1328	0.0434	0.0042	0.0001	0.0000
	12	0.0000	0.0012	0.0268	0.1140	0.1550	0.0760	0.0115	0.0003	0.0000
	13	0.0000	0.0003	0.0115	0.0760	0.1550	0.1140	0.0268	0.0012	0.0000
	14	0.0000	0.0001	0.0042	0.0434	0.1328	0.1465	0.0536	0.0040	0.0000
	15	0.0000	0.0000	0.0013	0.0212	0.0974	0.1612	0.0916	0.0118	0.0001
	16	0.0000	0.0000	0.0004	0.0088	0.0609	0.1511	0.1336	0.0294	0.0004
	17	0.0000	0.0000	0.0001	0.0031	0.0322	0.1200	0.1651	0.0623	0.0018
	18	0.0000	0.0000	0.0000	0.0009	0.0143	0.0800	0.1712	0.1108	0.0072
	19	0.0000	0.0000	0.0000	0.0002	0.0053	0.0442	0.1472	0.1633	0.0239
	20	0.0000	0.0000	0.0000	0.0000	0.0016	0.0199	0.1030	0.1960	0.0646
	21	0.0000	0.0000	0.0000	0.0000	0.0004	0.0071	0.0572	0.1867	0.1384
	22	0.0000	0.0000	0.0000	0.0000	0.0001	0.0019	0.0243	0.1358	0.2265
	23	0.0000	0.0000	0.0000	0.0000	0.0000	0.0004	0.0074	0.0708	0.2659
	24	0.0000	0.0000	0.0000	0.0000	0.0000	0.0000	0.0014	0.0236	0.1994
	25	0.0000	0.0000	0.0000	0.0000	0.0000	0.0000	0.0001	0.0038	0.0718

Table B.2 *Probabilities of the Standard Normal Distribution Z*

Table entries represent $P(Z < Z_i)$, e.g., $P(Z < -1.96) = 0.0250$, $P(Z < 1.96) = 0.9750$.

Z_i	.00	.01	.02	.03	.04	.05	.06	.07	.08	.09
-3.0	0.0013	0.0013	0.0013	0.0012	0.0012	0.0011	0.0011	0.0011	0.0010	0.0010
-2.9	0.0019	0.0018	0.0018	0.0017	0.0016	0.0016	0.0015	0.0015	0.0014	0.0014
-2.8	0.0026	0.0025	0.0024	0.0023	0.0023	0.0022	0.0021	0.0021	0.0020	0.0019
-2.7	0.0035	0.0034	0.0033	0.0032	0.0031	0.0030	0.0029	0.0028	0.0027	0.0026
-2.6	0.0047	0.0045	0.0044	0.0043	0.0041	0.0040	0.0039	0.0038	0.0037	0.0036
-2.5	0.0062	0.0060	0.0059	0.0057	0.0055	0.0054	0.0052	0.0051	0.0049	0.0048
-2.4	0.0082	0.0080	0.0078	0.0075	0.0073	0.0071	0.0069	0.0068	0.0066	0.0064
-2.3	0.0107	0.0104	0.0102	0.0099	0.0096	0.0094	0.0091	0.0089	0.0087	0.0084
-2.2	0.0139	0.0136	0.0132	0.0129	0.0125	0.0122	0.0119	0.0116	0.0113	0.0110
-2.1	0.0179	0.0174	0.0170	0.0166	0.0162	0.0158	0.0154	0.0150	0.0146	0.0143

Z_i	.00	.01	.02	.03	.04	.05	.06	.07	.08	.09
-2.0	0.0228	0.0222	0.0217	0.0212	0.0207	0.0202	0.0197	0.0192	0.0188	0.0183
-1.9	0.0287	0.0281	0.0274	0.0268	0.0262	0.0256	0.0250	0.0244	0.0239	0.0233
-1.8	0.0359	0.0351	0.0344	0.0336	0.0329	0.0322	0.0314	0.0307	0.0301	0.0294
-1.7	0.0446	0.0436	0.0427	0.0418	0.0409	0.0401	0.0392	0.0384	0.0375	0.0367
-1.6	0.0548	0.0537	0.0526	0.0516	0.0505	0.0495	0.0485	0.0475	0.0465	0.0455
-1.5	0.0668	0.0655	0.0643	0.0630	0.0618	0.0606	0.0594	0.0582	0.0571	0.0559
-1.4	0.0808	0.0793	0.0778	0.0764	0.0749	0.0735	0.0721	0.0708	0.0694	0.0681
-1.3	0.0968	0.0951	0.0934	0.0918	0.0901	0.0885	0.0869	0.0853	0.0838	0.0823
-1.2	0.1151	0.1131	0.1112	0.1093	0.1075	0.1056	0.1038	0.1020	0.1003	0.0985
-1.1	0.1357	0.1335	0.1314	0.1292	0.1271	0.1251	0.1230	0.1210	0.1190	0.1170
-1.0	0.1587	0.1562	0.1539	0.1515	0.1492	0.1469	0.1446	0.1423	0.1401	0.1379
-0.9	0.1841	0.1814	0.1788	0.1762	0.1736	0.1711	0.1685	0.1660	0.1635	0.1611
-0.8	0.2119	0.2090	0.2061	0.2033	0.2005	0.1977	0.1949	0.1922	0.1894	0.1867
-0.7	0.2420	0.2389	0.2358	0.2327	0.2296	0.2266	0.2236	0.2206	0.2177	0.2148
-0.6	0.2743	0.2709	0.2676	0.2643	0.2611	0.2578	0.2546	0.2514	0.2483	0.2451
-0.5	0.3085	0.3050	0.3015	0.2981	0.2946	0.2912	0.2877	0.2843	0.2810	0.2776
-0.4	0.3446	0.3409	0.3372	0.3336	0.3300	0.3264	0.3228	0.3192	0.3156	0.3121
-0.3	0.3821	0.3783	0.3745	0.3707	0.3669	0.3632	0.3594	0.3557	0.3520	0.3483
-0.2	0.4207	0.4168	0.4129	0.4090	0.4052	0.4013	0.3974	0.3936	0.3897	0.3859
-0.1	0.4602	0.4562	0.4522	0.4483	0.4443	0.4404	0.4364	0.4325	0.4286	0.4247
-0.0	0.5000	0.4960	0.4920	0.4880	0.4840	0.4801	0.4761	0.4721	0.4681	0.4641
0.0	0.5000	0.5040	0.5080	0.5120	0.5160	0.5199	0.5239	0.5279	0.5319	0.5359
0.1	0.5398	0.5438	0.5478	0.5517	0.5557	0.5596	0.5636	0.5675	0.5714	0.5753
0.2	0.5793	0.5832	0.5871	0.5910	0.5948	0.5987	0.6026	0.6064	0.6103	0.6141
0.3	0.6179	0.6217	0.6255	0.6293	0.6331	0.6368	0.6406	0.6443	0.6480	0.6517
0.4	0.6554	0.6591	0.6628	0.6664	0.6700	0.6736	0.6772	0.6808	0.6844	0.6879
0.5	0.6915	0.6950	0.6985	0.7019	0.7054	0.7088	0.7123	0.7157	0.7190	0.7224
0.6	0.7257	0.7291	0.7324	0.7357	0.7389	0.7422	0.7454	0.7486	0.7517	0.7549
0.7	0.7580	0.7611	0.7642	0.7673	0.7704	0.7734	0.7764	0.7794	0.7823	0.7852
0.8	0.7881	0.7910	0.7939	0.7967	0.7995	0.8023	0.8051	0.8078	0.8106	0.8133
0.9	0.8159	0.8186	0.8212	0.8238	0.8264	0.8289	0.8315	0.8340	0.8365	0.8389
1.0	0.8413	0.8438	0.8461	0.8485	0.8508	0.8531	0.8554	0.8577	0.8599	0.8621
1.1	0.8643	0.8665	0.8686	0.8708	0.8729	0.8749	0.8770	0.8790	0.8810	0.8830
1.2	0.8849	0.8869	0.8888	0.8907	0.8925	0.8944	0.8962	0.8980	0.8997	0.9015
1.3	0.9032	0.9049	0.9066	0.9082	0.9099	0.9115	0.9131	0.9147	0.9162	0.9177
1.4	0.9192	0.9207	0.9222	0.9236	0.9251	0.9265	0.9279	0.9292	0.9306	0.9319

z_i	.00	.01	.02	.03	.04	.05	.06	.07	.08	.09
1.5	0.9332	0.9345	0.9357	0.9370	0.9382	0.9394	0.9406	0.9418	0.9429	0.9441
1.6	0.9452	0.9463	0.9474	0.9484	0.9495	0.9505	0.9515	0.9525	0.9535	0.9545
1.7	0.9554	0.9564	0.9573	0.9582	0.9591	0.9599	0.9608	0.9616	0.9625	0.9633
1.8	0.9641	0.9649	0.9656	0.9664	0.9671	0.9678	0.9686	0.9693	0.9699	0.9706
1.9	0.9713	0.9719	0.9726	0.9732	0.9738	0.9744	0.9750	0.9756	0.9761	0.9767
2.0	0.9772	0.9778	0.9783	0.9788	0.9793	0.9798	0.9803	0.9808	0.9812	0.9817
2.1	0.9821	0.9826	0.9830	0.9834	0.9838	0.9842	0.9846	0.9850	0.9854	0.9857
2.2	0.9861	0.9864	0.9868	0.9871	0.9875	0.9878	0.9881	0.9884	0.9887	0.9890
2.3	0.9893	0.9896	0.9898	0.9901	0.9904	0.9906	0.9909	0.9911	0.9913	0.9916
2.4	0.9918	0.9920	0.9922	0.9925	0.9927	0.9929	0.9931	0.9932	0.9934	0.9936
2.5	0.9938	0.9940	0.9941	0.9943	0.9945	0.9946	0.9948	0.9949	0.9951	0.9952
2.6	0.9953	0.9955	0.9956	0.9957	0.9959	0.9960	0.9961	0.9962	0.9963	0.9964
2.7	0.9965	0.9966	0.9967	0.9968	0.9969	0.9970	0.9971	0.9972	0.9973	0.9974
2.8	0.9974	0.9975	0.9976	0.9977	0.9977	0.9978	0.9979	0.9979	0.9980	0.9981
2.9	0.9981	0.9982	0.9982	0.9983	0.9984	0.9984	0.9985	0.9985	0.9986	0.9986
3.0	0.9987	0.9987	0.9987	0.9988	0.9988	0.9989	0.9989	0.9989	0.9990	0.9990

Table B.2A *Z Values for Confidence Intervals*

Confidence Level	$Z_{1-\alpha/2}$	α (Total Tail Area)
99.99%	3.819	0.0001
99.9%	3.291	0.001
99%	2.576	0.01
95%	1.960	0.05
90%	1.645	0.10
80%	1.282	0.20

Table B.2B *Z Values for Tests of Hypothesis*

Lower-Tailed Tests	α	Z_α	Decision Rule
H_0: $\mu = \mu_0$	0.0001	-3.719	Reject H_0 if $Z \leq Z_\alpha$
H_1: $\mu < \mu_0$	0.001	-3.090	
	0.005	-2.576	
	0.010	-2.326	
	0.025	-1.960	
	0.050	-1.645	
	0.100	-1.282	

Table B.2B (continued)

Upper-Tailed Tests	α	$Z_{1-\alpha}$	Decision Rule
$H_0: \mu = \mu_0$	0.0001	3.719	Reject H_0 if $Z \geq Z_{1-\alpha}$
$H_1: \mu > \mu_0$	0.001	3.090	
	0.005	2.576	
	0.010	2.326	
	0.025	1.960	
	0.050	1.645	
	0.100	1.282	

Two-Tailed Tests	α	$Z_{1-\alpha/2}$	Decision Rule
$H_0: \mu = \mu_0$	0.0001	3.819	Reject H_0 if $Z \leq -Z_{1-\alpha/2}$
$H_1: \mu \neq \mu_0$	0.001	3.291	or if $Z \geq Z_{1-\alpha/2}$
	0.010	2.576	
	0.050	1.960	
	0.100	1.645	
	0.200	1.282	

Table B.3 Critical Values of the t Distribution

Table entries represent values from t distribution with upper-tail area equal to α, e.g., $P(t_{df} > t) = \alpha$, $P(t_6 > 1.943) = 0.05$.

Confidence Level CL		80%	90%	95%	98%	99%
Two Sided α	α_2	.20	.10	.05	.02	.01
One Sided α	α	.10	.05	.025	.01	.005
	df					
	1	3.078	6.314	12.71	31.82	63.66
	2	1.886	2.920	4.303	6.965	9.925
	3	1.638	2.353	3.182	4.541	5.841
	4	1.533	2.132	2.776	3.747	4.604
	5	1.476	2.015	2.571	3.365	4.032
	6	1.440	1.943	2.447	3.143	3.707
	7	1.415	1.895	2.365	2.998	3.499
	8	1.397	1.860	2.306	2.896	3.355
	9	1.383	1.833	2.262	2.821	3.250
	10	1.372	1.812	2.228	2.764	3.169

Confidence Level	CL	80%	90%	95%	98%	99%
Two Sided α	α_2	.20	.10	.05	.02	.01
One Sided α	α	.10	.05	.025	.01	.005
	df					
	11	1.363	1.796	2.201	2.718	3.106
	12	1.356	1.782	2.179	2.681	3.055
	13	1.350	1.771	2.160	2.650	3.012
	14	1.345	1.761	2.145	2.624	2.977
	15	1.341	1.753	2.131	2.602	2.947
	16	1.337	1.746	2.120	2.583	2.921
	17	1.333	1.740	2.110	2.567	2.898
	18	1.330	1.734	2.101	2.552	2.878
	19	1.328	1.729	2.093	2.539	2.861
	20	1.325	1.725	2.086	2.528	2.845
	21	1.323	1.721	2.080	2.518	2.831
	22	1.321	1.717	2.074	2.508	2.819
	23	1.319	1.714	2.069	2.500	2.807
	24	1.318	1.711	2.064	2.492	2.797
	25	1.316	1.708	2.060	2.485	2.787
	26	1.315	1.706	2.056	2.479	2.779
	27	1.314	1.703	2.052	2.473	2.771
	28	1.313	1.701	2.048	2.467	2.763
	29	1.311	1.699	2.045	2.462	2.756
	30	1.310	1.697	2.042	2.457	2.750
	31	1.309	1.696	2.040	2.453	2.744
	32	1.309	1.694	2.037	2.449	2.738
	33	1.308	1.692	2.035	2.445	2.733
	34	1.307	1.691	2.032	2.441	2.728
	35	1.306	1.690	2.030	2.438	2.724
	36	1.306	1.688	2.028	2.434	2.719
	37	1.305	1.687	2.026	2.431	2.715
	38	1.304	1.686	2.024	2.429	2.712
	39	1.304	1.685	2.023	2.426	2.708
	40	1.303	1.684	2.021	2.423	2.704
	41	1.303	1.683	2.020	2.421	2.701
	42	1.302	1.682	2.018	2.418	2.698
	43	1.302	1.681	2.017	2.416	2.695

Confidence Level	CL	80%	90%	95%	98%	99%
Two Sided α	α_2	.20	.10	.05	.02	.01
One Sided α	α	.10	.05	.025	.01	.005
	df					
	44	1.301	1.680	2.015	2.414	2.692
	45	1.301	1.679	2.014	2.412	2.690
	46	1.300	1.679	2.013	2.410	2.687
	47	1.300	1.678	2.012	2.408	2.685
	48	1.299	1.677	2.011	2.407	2.682
	49	1.299	1.677	2.010	2.405	2.680
	50	1.299	1.676	2.009	2.403	2.678
	51	1.298	1.675	2.008	2.402	2.676
	52	1.298	1.675	2.007	2.400	2.674
	53	1.298	1.674	2.006	2.399	2.672
	54	1.297	1.674	2.005	2.397	2.670
	55	1.297	1.673	2.004	2.396	2.668
	56	1.297	1.673	2.003	2.395	2.667
	57	1.297	1.672	2.002	2.394	2.665
	58	1.296	1.672	2.002	2.392	2.663
	59	1.296	1.671	2.001	2.391	2.662
	60	1.296	1.671	2.000	2.390	2.660
	61	1.296	1.670	2.000	2.389	2.659
	62	1.295	1.670	1.999	2.388	2.657
	63	1.295	1.669	1.998	2.387	2.656
	64	1.295	1.669	1.998	2.386	2.655
	65	1.295	1.669	1.997	2.385	2.654
	66	1.295	1.668	1.997	2.384	2.652
	67	1.294	1.668	1.996	2.383	2.651
	68	1.294	1.668	1.995	2.382	2.650
	69	1.294	1.667	1.995	2.382	2.649
	70	1.294	1.667	1.994	2.381	2.648
	71	1.294	1.667	1.994	2.380	2.647
	72	1.293	1.666	1.993	2.379	2.646
	73	1.293	1.666	1.993	2.379	2.645
	74	1.293	1.666	1.993	2.378	2.644
	75	1.293	1.665	1.992	2.377	2.643
	∞	1.282	1.645	1.960	2.326	2.576

Table B.4A *Critical Values of the F Distribution with Upper-Tail Area = 0.05*

$P(F > F_{df1,df2}) = 0.05$, e.g., $P(F_{3,20} > 3.10) = 0.05$

denominator df (df2)	\multicolumn numerator df (df1)														
	1	2	3	4	5	6	7	8	9	10	20	30	40	50
1	161.4	199.5	215.7	224.6	230.2	234.0	236.8	238.9	240.5	241.9	248.0	250.1	251.1	251.8
2	18.51	19.00	19.16	19.25	19.30	19.33	19.35	19.37	19.38	19.40	19.45	19.46	19.47	19.48
3	10.13	9.55	9.28	9.12	9.01	8.94	8.89	8.85	8.81	8.79	8.66	8.62	8.59	8.58
4	7.71	6.94	6.59	6.39	6.26	6.16	6.09	6.04	6.00	5.96	5.80	5.75	5.72	5.70
5	6.61	5.79	5.41	5.19	5.05	4.95	4.88	4.82	4.77	4.74	4.56	4.50	4.46	4.44
6	5.99	5.14	4.76	4.53	4.39	4.28	4.21	4.15	4.10	4.06	3.87	3.81	3.77	3.75
7	5.59	4.74	4.35	4.12	3.97	3.87	3.79	3.73	3.68	3.64	3.44	3.38	3.34	3.32
8	5.32	4.46	4.07	3.84	3.69	3.58	3.50	3.44	3.39	3.35	3.15	3.08	3.04	3.02
9	5.12	4.26	3.86	3.63	3.48	3.37	3.29	3.23	3.18	3.14	2.94	2.86	2.83	2.80
10	4.96	4.10	3.71	3.48	3.33	3.22	3.14	3.07	3.02	2.98	2.77	2.70	2.66	2.64
11	4.84	3.98	3.59	3.36	3.20	3.09	3.01	2.95	2.90	2.85	2.65	2.57	2.53	2.51
12	4.75	3.89	3.49	3.26	3.11	3.00	2.91	2.85	2.80	2.75	2.54	2.47	2.43	2.40
13	4.67	3.81	3.41	3.18	3.03	2.92	2.83	2.77	2.71	2.67	2.46	2.38	2.34	2.31
14	4.60	3.74	3.34	3.11	2.96	2.85	2.76	2.70	2.65	2.60	2.39	2.31	2.27	2.24
15	4.54	3.68	3.29	3.06	2.90	2.79	2.71	2.64	2.59	2.54	2.33	2.25	2.20	2.18
16	4.49	3.63	3.24	3.01	2.85	2.74	2.66	2.59	2.54	2.49	2.28	2.19	2.15	2.12
17	4.45	3.59	3.20	2.96	2.81	2.70	2.61	2.55	2.49	2.45	2.23	2.15	2.10	2.08
18	4.41	3.55	3.16	2.93	2.77	2.66	2.58	2.51	2.46	2.41	2.19	2.11	2.06	2.04
19	4.38	3.52	3.13	2.90	2.74	2.63	2.54	2.48	2.42	2.38	2.16	2.07	2.03	2.00
20	4.35	3.49	3.10	2.87	2.71	2.60	2.51	2.45	2.39	2.35	2.12	2.04	1.99	1.97
21	4.32	3.47	3.07	2.84	2.68	2.57	2.49	2.42	2.37	2.32	2.10	2.01	1.96	1.94
22	4.30	3.44	3.05	2.82	2.66	2.55	2.46	2.40	2.34	2.30	2.07	1.98	1.94	1.91
23	4.28	3.42	3.03	2.80	2.64	2.53	2.44	2.37	2.32	2.27	2.05	1.96	1.91	1.88
24	4.26	3.40	3.01	2.78	2.62	2.51	2.42	2.36	2.30	2.25	2.03	1.94	1.89	1.86
25	4.24	3.39	2.99	2.76	2.60	2.49	2.40	2.34	2.28	2.24	2.01	1.92	1.87	1.84
26	4.23	3.37	2.98	2.74	2.59	2.47	2.39	2.32	2.27	2.22	1.99	1.90	1.85	1.82
27	4.21	3.35	2.96	2.73	2.57	2.46	2.37	2.31	2.25	2.20	1.97	1.88	1.84	1.81
28	4.20	3.34	2.95	2.71	2.56	2.45	2.36	2.29	2.24	2.19	1.96	1.87	1.82	1.79
29	4.18	3.33	2.93	2.70	2.55	2.43	2.35	2.28	2.22	2.18	1.94	1.85	1.81	1.77
30	4.17	3.32	2.92	2.69	2.53	2.42	2.33	2.27	2.21	2.16	1.93	1.84	1.79	1.76
31	4.16	3.30	2.91	2.68	2.52	2.41	2.32	2.25	2.20	2.15	1.92	1.83	1.78	1.75
32	4.15	3.29	2.90	2.67	2.51	2.40	2.31	2.24	2.19	2.14	1.91	1.82	1.77	1.74
33	4.14	3.28	2.89	2.66	2.50	2.39	2.30	2.23	2.18	2.13	1.90	1.81	1.76	1.72
34	4.13	3.28	2.88	2.65	2.49	2.38	2.29	2.23	2.17	2.12	1.89	1.80	1.75	1.71
35	4.12	3.27	2.87	2.64	2.49	2.37	2.29	2.22	2.16	2.11	1.88	1.79	1.74	1.70
36	4.11	3.26	2.87	2.63	2.48	2.36	2.28	2.21	2.15	2.11	1.87	1.78	1.73	1.69
37	4.11	3.25	2.86	2.63	2.47	2.36	2.27	2.20	2.14	2.10	1.86	1.77	1.72	1.68
38	4.10	3.24	2.85	2.62	2.46	2.35	2.26	2.19	2.14	2.09	1.85	1.76	1.71	1.68
39	4.09	3.24	2.85	2.61	2.46	2.34	2.26	2.19	2.13	2.08	1.85	1.75	1.70	1.67
40	4.08	3.23	2.84	2.61	2.45	2.34	2.25	2.18	2.12	2.08	1.84	1.74	1.69	1.66
41	4.08	3.23	2.83	2.60	2.44	2.33	2.24	2.17	2.12	2.07	1.83	1.74	1.69	1.65
42	4.07	3.22	2.83	2.59	2.44	2.32	2.24	2.17	2.11	2.06	1.83	1.73	1.68	1.65
43	4.07	3.21	2.82	2.59	2.43	2.32	2.23	2.16	2.11	2.06	1.82	1.72	1.67	1.64
44	4.06	3.21	2.82	2.58	2.43	2.31	2.23	2.16	2.10	2.05	1.81	1.72	1.67	1.63
45	4.06	3.20	2.81	2.58	2.42	2.31	2.22	2.15	2.10	2.05	1.81	1.71	1.66	1.63

denominator df (df2)	1	2	3	4	5	6	numerator df (df1) 7	8	9	10	20	30	40	50
46	4.05	3.20	2.81	2.57	2.42	2.30	2.22	2.15	2.09	2.04	1.80	1.71	1.65	1.62
47	4.05	3.20	2.80	2.57	2.41	2.30	2.21	2.14	2.09	2.04	1.80	1.70	1.65	1.61
48	4.04	3.19	2.80	2.57	2.41	2.29	2.21	2.14	2.08	2.03	1.79	1.70	1.64	1.61
49	4.04	3.19	2.79	2.56	2.40	2.29	2.20	2.13	2.08	2.03	1.79	1.69	1.64	1.60
50	4.03	3.18	2.79	2.56	2.40	2.29	2.20	2.13	2.07	2.03	1.78	1.69	1.63	1.60
75	3.97	3.12	2.73	2.49	2.34	2.22	2.13	2.06	2.01	1.96	1.71	1.61	1.55	1.52
100	3.94	3.09	2.70	2.46	2.31	2.19	2.10	2.03	1.97	1.93	1.68	1.57	1.52	1.48
125	3.92	3.07	2.68	2.44	2.29	2.17	2.08	2.01	1.96	1.91	1.66	1.55	1.49	1.45
150	3.90	3.06	2.66	2.43	2.27	2.16	2.07	2.00	1.94	1.89	1.64	1.54	1.48	1.44
175	3.90	3.05	2.66	2.42	2.27	2.15	2.06	1.99	1.93	1.89	1.63	1.52	1.46	1.42
200	3.89	3.04	2.65	2.42	2.26	2.14	2.06	1.98	1.93	1.88	1.62	1.52	1.46	1.41

Table B.4B *Critical Values of the F Distribution with Upper Tail Area = 0.025*

$P(F > F_{df1,df2}) = 0.025$, e.g., $P(F_{3,20} > 3.86) = 0.025$

denominator df (df2)	1	2	3	4	5	6	numerator df (df1) 7	8	9	10	20	30	40	50
1	647.8	799.5	864.2	899.6	921.8	937.1	948.2	956.7	963.3	968.6	993.1	1001	1006	1008
2	38.51	39.00	39.17	39.25	39.30	39.33	39.36	39.37	39.39	39.40	39.45	39.46	39.47	39.48
3	17.44	16.04	15.44	15.10	14.88	14.73	14.62	14.54	14.47	14.42	14.17	14.08	14.04	14.01
4	12.22	10.65	9.98	9.60	9.36	9.20	9.07	8.98	8.90	8.84	8.56	8.46	8.41	8.38
5	10.01	8.43	7.76	7.39	7.15	6.98	6.85	6.76	6.68	6.62	6.33	6.23	6.18	6.14
6	8.81	7.26	6.60	6.23	5.99	5.82	5.70	5.60	5.52	5.46	5.17	5.07	5.01	4.98
7	8.07	6.54	5.89	5.52	5.29	5.12	4.99	4.90	4.82	4.76	4.47	4.36	4.31	4.28
8	7.57	6.06	5.42	5.05	4.82	4.65	4.53	4.43	4.36	4.30	4.00	3.89	3.84	3.81
9	7.21	5.71	5.08	4.72	4.48	4.32	4.20	4.10	4.03	3.96	3.67	3.56	3.51	3.47
10	6.94	5.46	4.83	4.47	4.24	4.07	3.95	3.85	3.78	3.72	3.42	3.31	3.26	3.22
11	6.72	5.26	4.63	4.28	4.04	3.88	3.76	3.66	3.59	3.53	3.23	3.12	3.06	3.03
12	6.55	5.10	4.47	4.12	3.89	3.73	3.61	3.51	3.44	3.37	3.07	2.96	2.91	2.87
13	6.41	4.97	4.35	4.00	3.77	3.60	3.48	3.39	3.31	3.25	2.95	2.84	2.78	2.74
14	6.30	4.86	4.24	3.89	3.66	3.50	3.38	3.29	3.21	3.15	2.84	2.73	2.67	2.64
15	6.20	4.77	4.15	3.80	3.58	3.41	3.29	3.20	3.12	3.06	2.76	2.64	2.59	2.55
16	6.12	4.69	4.08	3.73	3.50	3.34	3.22	3.12	3.05	2.99	2.68	2.57	2.51	2.47
17	6.04	4.62	4.01	3.66	3.44	3.28	3.16	3.06	2.98	2.92	2.62	2.50	2.44	2.41
18	5.98	4.56	3.95	3.61	3.38	3.22	3.10	3.01	2.93	2.87	2.56	2.44	2.38	2.35
19	5.92	4.51	3.90	3.56	3.33	3.17	3.05	2.96	2.88	2.82	2.51	2.39	2.33	2.30
20	5.87	4.46	3.86	3.51	3.29	3.13	3.01	2.91	2.84	2.77	2.46	2.35	2.29	2.25
21	5.83	4.42	3.82	3.48	3.25	3.09	2.97	2.87	2.80	2.73	2.42	2.31	2.25	2.21
22	5.79	4.38	3.78	3.44	3.22	3.05	2.93	2.84	2.76	2.70	2.39	2.27	2.21	2.17
23	5.75	4.35	3.75	3.41	3.18	3.02	2.90	2.81	2.73	2.67	2.36	2.24	2.18	2.14
24	5.72	4.32	3.72	3.38	3.15	2.99	2.87	2.78	2.70	2.64	2.33	2.21	2.15	2.11
25	5.69	4.29	3.69	3.35	3.13	2.97	2.85	2.75	2.68	2.61	2.30	2.18	2.12	2.08
26	5.66	4.27	3.67	3.33	3.10	2.94	2.82	2.73	2.65	2.59	2.28	2.16	2.09	2.05
27	5.63	4.24	3.65	3.31	3.08	2.92	2.80	2.71	2.63	2.57	2.25	2.13	2.07	2.03
28	5.61	4.22	3.63	3.29	3.06	2.90	2.78	2.69	2.61	2.55	2.23	2.11	2.05	2.01
29	5.59	4.20	3.61	3.27	3.04	2.88	2.76	2.67	2.59	2.53	2.21	2.09	2.03	1.99
30	5.57	4.18	3.59	3.25	3.03	2.87	2.75	2.65	2.57	2.51	2.20	2.07	2.01	1.97

denominator						numerator df (df1)								
df (df2)	1	2	3	4	5	6	7	8	9	10	20	30	40	50
31	5.55	4.16	3.57	3.23	3.01	2.85	2.73	2.64	2.56	2.50	2.18	2.06	1.99	1.95
32	5.53	4.15	3.56	3.22	3.00	2.84	2.71	2.62	2.54	2.48	2.16	2.04	1.98	1.93
33	5.51	4.13	3.54	3.20	2.98	2.82	2.70	2.61	2.53	2.47	2.15	2.03	1.96	1.92
34	5.50	4.12	3.53	3.19	2.97	2.81	2.69	2.59	2.52	2.45	2.13	2.01	1.95	1.90
35	5.48	4.11	3.52	3.18	2.96	2.80	2.68	2.58	2.50	2.44	2.12	2.00	1.93	1.89
36	5.47	4.09	3.50	3.17	2.94	2.78	2.66	2.57	2.49	2.43	2.11	1.99	1.92	1.88
37	5.46	4.08	3.49	3.16	2.93	2.77	2.65	2.56	2.48	2.42	2.10	1.97	1.91	1.87
38	5.45	4.07	3.48	3.15	2.92	2.76	2.64	2.55	2.47	2.41	2.09	1.96	1.90	1.85
39	5.43	4.06	3.47	3.14	2.91	2.75	2.63	2.54	2.46	2.40	2.08	1.95	1.89	1.84
40	5.42	4.05	3.46	3.13	2.90	2.74	2.62	2.53	2.45	2.39	2.07	1.94	1.88	1.83
41	5.41	4.04	3.45	3.12	2.89	2.74	2.62	2.52	2.44	2.38	2.06	1.93	1.87	1.82
42	5.40	4.03	3.45	3.11	2.89	2.73	2.61	2.51	2.43	2.37	2.05	1.92	1.86	1.81
43	5.39	4.02	3.44	3.10	2.88	2.72	2.60	2.50	2.43	2.36	2.04	1.92	1.85	1.80
44	5.39	4.02	3.43	3.09	2.87	2.71	2.59	2.50	2.42	2.36	2.03	1.91	1.84	1.80
45	5.38	4.01	3.42	3.09	2.86	2.70	2.58	2.49	2.41	2.35	2.03	1.90	1.83	1.79
46	5.37	4.00	3.42	3.08	2.86	2.70	2.58	2.48	2.41	2.34	2.02	1.89	1.82	1.78
47	5.36	3.99	3.41	3.07	2.85	2.69	2.57	2.48	2.40	2.33	2.01	1.89	1.82	1.77
48	5.35	3.99	3.40	3.07	2.84	2.69	2.56	2.47	2.39	2.33	2.01	1.88	1.81	1.77
49	5.35	3.98	3.40	3.06	2.84	2.68	2.56	2.46	2.39	2.32	2.00	1.87	1.80	1.76
50	5.34	3.97	3.39	3.05	2.83	2.67	2.55	2.46	2.38	2.32	1.99	1.87	1.80	1.75
75	5.23	3.88	3.30	2.96	2.74	2.58	2.46	2.37	2.29	2.22	1.90	1.76	1.69	1.65
100	5.18	3.83	3.25	2.92	2.70	2.54	2.42	2.32	2.24	2.18	1.85	1.71	1.64	1.59
125	5.15	3.80	3.22	2.89	2.67	2.51	2.39	2.30	2.22	2.15	1.82	1.68	1.61	1.56
150	5.13	3.78	3.20	2.87	2.65	2.49	2.37	2.28	2.20	2.13	1.80	1.67	1.59	1.54
175	5.11	3.77	3.19	2.86	2.64	2.48	2.36	2.27	2.19	2.12	1.79	1.65	1.57	1.52
200	5.10	3.76	3.18	2.85	2.63	2.47	2.35	2.26	2.18	2.11	1.78	1.64	1.56	1.51

Table B.5 *Critical Values of the χ^2 Distribution*

Table entries represent values from χ^2 distribution with upper-tail area equal to α.
$P(\chi^2 > \chi^2_{df}) = \alpha$, e.g., $P(\chi^2_3 > 7.81) = 0.05$

df	α				
	.10	.05	.025	.01	.005
1	2.71	3.84	5.02	6.63	7.88
2	4.61	5.99	7.38	9.21	10.60
3	6.25	7.81	9.35	11.34	12.84
4	7.78	9.49	11.14	13.28	14.86
5	9.24	11.07	12.83	15.09	16.75
6	10.64	12.59	14.45	16.81	18.55
7	12.02	14.07	16.01	18.48	20.28
8	13.36	15.51	17.53	20.09	21.95

df	.10	.05	α .025	.01	.005
9	14.68	16.92	19.02	21.67	23.59
10	15.99	18.31	20.48	23.21	25.19
11	17.28	19.68	21.92	24.72	26.76
12	18.55	21.03	23.34	26.22	28.30
13	19.81	22.36	24.74	27.69	29.82
14	21.06	23.68	26.12	29.14	31.32
15	22.31	25.00	27.49	30.58	32.80
16	23.54	26.30	28.85	32.00	34.27
17	24.77	27.59	30.19	33.41	35.72
18	25.99	28.87	31.53	34.81	37.16
19	27.20	30.14	32.85	36.19	38.58
20	28.41	31.41	34.17	37.57	40.00
21	29.62	32.67	35.48	38.93	41.40
22	30.81	33.92	36.78	40.29	42.80
23	32.01	35.17	38.08	41.64	44.18
24	33.20	36.42	39.36	42.98	45.56
25	34.38	37.65	40.65	44.31	46.93
26	35.56	38.89	41.92	45.64	48.29
27	36.74	40.11	43.19	46.96	49.64
28	37.92	41.34	44.46	48.28	50.99
29	39.09	42.56	45.72	49.59	52.34
30	40.26	43.77	46.98	50.89	53.67
40	51.81	55.76	59.34	63.69	66.77
50	63.17	67.50	71.42	76.15	79.49
60	74.40	79.08	83.30	88.38	91.95
70	85.53	90.53	95.02	100.4	104.2
80	96.58	101.9	106.6	112.3	116.3
90	107.6	113.1	118.1	124.1	128.3
100	118.5	124.3	129.6	135.8	140.2

Table B.6 *Critical Values of the Studentized Range Distribution, $\alpha = 0.05$*

df_{error} (df_{within})	\multicolumn{8}{c}{k = Number of treatments (groups) being compared}
	3	*4*	*5*	*6*	*7*	*8*	*9*	*10*
5	4.60	5.22	5.67	6.03	6.33	6.58	6.80	6.99
6	4.34	4.90	5.30	5.63	5.90	6.12	6.32	6.49
7	4.16	4.68	5.06	5.36	5.61	5.82	6.00	6.16

	k = *Number of treatments (groups) being compared*							
df_{error} (df_{within})	3	4	5	6	7	8	9	10
8	4.04	4.53	4.89	5.17	5.40	5.60	5.77	5.92
9	3.95	4.41	4.76	5.02	5.24	5.43	5.59	5.74
10	3.88	4.33	4.65	4.91	5.12	5.30	5.46	5.60
11	3.82	4.26	4.57	4.82	5.03	5.20	5.35	5.49
12	3.77	4.20	4.51	4.75	4.95	5.12	5.27	5.39
13	3.73	4.15	4.45	4.69	4.88	5.05	5.19	5.32
14	3.70	4.11	4.41	4.64	4.83	4.99	5.13	5.25
15	3.67	4.08	4.37	4.59	4.78	4.94	5.08	5.20
16	3.65	4.05	4.33	4.56	4.74	4.90	5.03	5.15
17	3.63	4.02	4.30	4.52	4.70	4.86	4.99	5.11
18	3.61	4.00	4.28	4.49	4.67	4.82	4.96	5.07
19	3.59	3.98	4.25	4.47	4.65	4.79	4.92	5.04
20	3.58	3.96	4.23	4.45	4.62	4.77	4.90	5.01
30	3.49	3.85	4.10	4.30	4.46	4.60	4.72	4.82
40	3.44	3.79	4.04	4.23	4.39	4.52	4.63	4.73
60	3.40	3.74	3.98	4.16	4.31	4.44	4.55	4.65
120	3.36	3.68	3.92	4.10	4.24	4.36	4.47	4.56
∞	3.31	3.63	3.86	4.03	4.17	4.29	4.39	4.47

B.2 SAS[1] Programs Used to Generate Table Entries

Table B.1 Probabilities of the Binomial Distribution

```
options ps=55 ls=80;
data in;
k=1;
file 'c:\sas\btable.out' noprint notitle;
do n=2 to 25 by 1;
x=0;
 do while (x le n);
 put +4 x 2.0 @@;
  do pi=0.1 to 0.9 by 0.10;
```

[1]SAS Institute Inc. *SAS® User's Guide: Basics, Version 6, Edition 8.* Cary, NC: SAS Institute Inc., 1985.

```
 if x=0 then p=probbnml(pi,n,x);
 else if x^=0 then p=probbnml(pi,n,x)-probbnml(pi,n,x-1);
 output;
 if mod(k,9) ^= 0 then put +2 p 6.4 @@;
 else if mod(k,9) = 0 then put +2 p 6.4 /;
 k+1;
 end;
 x+1;
 end;
end;
run;
```

Table B.2 Probabilities of the Standard Normal Distribution

```
options ps=55 ls=80;
data in;
k=1;
file 'c:\sas\ztable.out' noprint notitle;
do i=-3.09 to 3.09 by 0.01;
 p=probnorm(i);
 output;
 if mod(k,10) ^= 0 then put +2 p 6.4 @@;
 else if mod(k,10) = 0 then put +2 p 6.4 /;
 k+1;
end;
run;
```

Table B.3 Critical Values of the *t* Distribution

```
options ps=55 ls=80;
data in;
k=1;
file 'c:\sas\ttable.out' noprint notitle;
do df=1 to 75 by 1;
 do a=0.10,0.05,0.025,0.01,0.005;
 t=abs(tinv(a,df));
 output;
 if mod(k,5) ^= 0 then put +2 t 5.3 @@;
 else if mod(k,5) = 0 then put +2 t 5.3 /;
 k+1;
 end;
end;
run;
```

Table B.4 Critical Values of the *F* Distribution

```
options ps=55 ls=80;
data in;
k=1;
file 'c:\sas\ftable.out' noprint notitle;
do df2=1 to 49,50 to 200 by 25;
 do df1=1 to 9 by 1,10 to 50 by 10;
 f=finv(0.95,df1,df2);
 output;
 if mod(k,14) ^= 0 then put +2 f 5.2 @@;
 else if mod(k,14) = 0 then put +2 f 5.2 /;
 k+1;
 end;
end;
run;
```

Table B.5 Critical Values of the χ^2 Distribution

```
options ps=55 ls=80;
data in;
k=1;
file 'c:\sas\ctable.out' noprint notitle;
do df=1 to 29,30 to 100 by 10;
 do a=0.10,0.05,0.025,0.01,0.005;
 c=cinv(1-a,df);
 output;
 if mod(k,5) ^= 0 then put +2 c 5.2 @@;
 else if mod(k,5) = 0 then put +2 c 5.2 /;
 k+1;
 end;
end;
run;
```

APPENDIX C

Framingham Heart Study Longitudinal Data Documentation

The authors wish to thank Paul Sorlie, Ph.D., Leader of Analytical Resources Scientific Group, and Sean Coady, M.A., Health Statistician at the National Heart, Lung, and Blood Institute for preparing the Framingham Heart Study Data and Documentation.

The Framingham Heart Study is a long-term prospective study of the etiology of cardiovascular disease among a population of adults living in the community of Framingham, Massachusetts. The Framingham Heart Study was a landmark study in epidemiology in that it was the first prospective study of cardiovascular disease and identified the concept of risk factors and their joint effects. Beginning in 1948, the study initially enrolled 5209 subjects. Participants have been examined biennially since the inception of the study, and all subjects are continuously followed through regular surveillance for cardiovascular outcomes. Clinic examination data has included cardiovascular disease risk factors and markers of disease, such as blood pressure, blood chemistry, lung function, smoking history, health behaviors, ECG tracings, echocardiography, and medication use. Through regular surveillance of area hospitals, participant contact, and death certificates, the Framingham Heart Study reviews and adjudicates events for the occurrence of angina pectoris, myocardial infarction, heart failure, and cerebrovascular disease.

The enclosed data set is a subset of the data collected as part of the Framingham study and includes laboratory, clinic, questionnaire, and adjudicated event data on 4434 participants. Participant clinic data were collected during three examination periods, approximately 6 years apart, from roughly 1956 to 1968. Each participant was followed for a total of

24 years for the outcome of the following events: angina pectoris, myocardial infarction, atherothrombotic infarction or cerebral hemorrhage (stroke), or death. (*Note:* **Although the enclosed data set contains Framingham data "as collected" by Framingham investigators, specific methods were employed to ensure an anonymous data set that protects patient confidentiality; therefore, this data set is inappropriate for publication purposes.**)

The data is provided in longitudinal form. Each participant has 1 to 3 observations depending on the number of exams the subject attended, and as a result there are 11,627 observations on the 4434 participants. Event data for each participant have been added without regard for prevalent disease status or when examination data were collected. For example, consider the following participant:

RANDID	age	SEX	time	period	prevchd	mi_fchd	timemifc
95148	52	2	0	1	0	1	3607
95148	58	2	2128	2	0	1	3607
95148	64	2	4192	3	1	1	3607

Participant 95148 entered the study (time = 0 or period = 1) free of prevalent coronary heart disease (prevchd = 0 at period = 1); however, during follow-up, an MI event occurred at day 3607 following the baseline examination. The MI occurred after the second exam the subject attended (period = 2 or time = 2128 days), but before the third attended exam (period = 3 or time = 4192 days). Since the event occurred prior to the third exam, the subject was prevalent for CHD (prevchd = 1) at the third examination. Note that the event data (mi_fchd, timemifc) covers the entire follow-up period and does not change according to exam.

The following characteristics or risk factor data are provided in the data set. Missing values in the data set are indicated by a period (.). In SAS, missing values are numerically the smallest possible values (for example, < 0 or < -99999999).

Variable	Description	Units	Range or Count
RANDID	Unique identification number for each participant		2448–9999312
SEX	Participant sex	1 = Men	$n = 5022$
		2 = Women	$n = 6605$
PERIOD	Examination cycle	1 = Period 1	$n = 4434$
		2 = Period 2	$n = 3930$
		3 = Period 3	$n = 3263$

Variable	Description	Units	Range or Count
TIME	Number of days since baseline exam		0–4854
AGE	Age at exam (years)		32–81
SYSBP	Systolic blood pressure (mean of last two of three measurements) (mmHg)		83.5–295
DIABP	Diastolic blood pressure (mean of last two of three measurements) (mmHg)		30–150
BPMEDS	Use of antihypertensive medication at exam	0 = Not currently used 1 = Current use	$n = 10090$ $n = 944$
CURSMOKE	Current cigarette smoking at exam	0 = Not current smoker 1 = Current smoker	$n = 6598$ $n = 5029$
CIGPDAY	Number of cigarettes smoked each day	0 = Not current smoker 1–90 cigarettes per day	
TOTCHOL	Serum total cholesterol (mg/dL)		107–696
HDLC	High-density lipoprotein cholesterol (mg/dL)	Available for period 3 only	10–189
LDLC	Low-density lipoprotein cholesterol (mg/dL)	Available for period 3 only	20–565
BMI	Body mass index, weight in kilograms/height meters squared		14.43–56.8
GLUCOSE	Casual serum glucose (mg/dL)		39–478
DIABETES	Diabetic according to criteria of first exam treated or first exam with casual glucose of 200 mg/dL or more	0 = Not a diabetic 1 = Diabetic	$n = 11097$ $n = 530$
HEARTRTE	Heart rate (ventricular rate) in beats/min		37–220
PREVAP	Prevalent angina pectoris at exam	0 = Free of disease 1 = Prevalent disease	$n = 11000$ $n = 627$
PREVCHD	Prevalent coronary heart disease defined as preexisting angina pectoris, myocardial infarction (hospitalized, silent or unrecognized), or coronary insufficiency (unstable angina)	0 = Free of disease 1 = Prevalent disease	$n = 10785$ $n = 842$
PREVMI	Prevalent myocardial infarction	0 = Free of disease 1 = Prevalent disease	$n = 11253$ $n = 374$
PREVSTRK	Prevalent stroke	0 = Free of disease 1 = Prevalent disease	$n = 11475$ $n = 152$
PREVHYP	Prevalent hypertensive. Subject was defined as hypertensive if treated or if second exam at which mean systolic was >= 140 mmHg or mean diastolic >=90 mmHg	0 = Free of disease 1 = Prevalent disease	$n = 6283$ $n = 5344$

For each participant the following event data is provided. For each type of event, '0' indicates the event did not occur during follow-up, and '1' indicates an event did occur during follow-up. Only the first event occurring during the interval of baseline (PERIOD = 1) to end of follow-up is provided:

Variable Name	Description
ANGINA	Angina pectoris
HOSPMI	Hospitalized myocardial infarction
MI_FCHD	Hospitalized myocardial infarction or fatal coronary heart disease
ANYCHD	Angina pectoris, myocardial infarction (hospitalized and silent or unrecognized), coronary insufficiency (unstable angina), or fatal coronary heart disease
STROKE	Atherothrombotic infarction, cerebral embolism, intracerebral hemorrhage, or subarachnoid hemorrhage or fatal cerebrovascular disease
CVD	Myocardial infarction (hospitalized and silent or unrecognized), fatal coronary heart disease, atherothrombotic infarction, cerebral embolism, intracerebral hemorrhage, or subarachnoid hemorrhage or fatal cerebrovascular disease
HYPERTEN	Hypertensive. Defined as the first exam treated for high blood pressure or second exam in which either systolic is ≥ 140 mmHg or diastolic ≥ 90 mmHg
DEATH	Death from any cause
TIMEAP	Number of days from baseline exam to first angina during the follow-up or number of days from baseline to censor date. Censor date may be end of follow-up, death, or last known contact date if subject is lost to follow-up
TIMEMI	Defined as above for the first HOSPMI event during follow-up
TIMEMIFC	Defined as above for the first MI_FCHD event during follow-up
TIMECHD	Defined as above for the first ANYCHD event during follow-up
TIMESTRK	Defined as above for the first STROKE event during follow-up
TIMECVD	Defined as above for the first CVD event during follow-up
TIMEHYP	Defined as above for the first HYPERTEN event during follow-up
TIMEDTH	Number of days from baseline exam to death if occurring during follow-up or number of days from baseline to censor date. Censor date may be end of follow-up or last known contact date if subject is lost to follow-up.

Note that defining hypertensive requires exam participation and bias can therefore occur. Subjects attending exams regularly have a greater opportunity to be defined as hypertensive. Subjects not attending exams would be assumed to be free of hypertension. Since hypertension is highly prevalent, this misclassification could potentially be large.

Defining Incident Events

Frequently, epidemiologists need to define the population at risk for some disease or event outcome, and individuals who have previously had an event need to be excluded from the analysis so that only new or first events are counted. Incidence or first event rates can be calculated using any of the three examinations as a baseline exam. The variables PREVAP, PREVMI, PREVCHD, PREVSTRK, and PREVHYP will define the population at risk for the outcome of interest. For example, assume we are interested in incident hospitalized myocardial infarction or fatal coronary heart disease. Consider again participant 95148 and participants 477082 and 1140225, whose data are given below.

RANDID	age	SEX	time	period	prevchd	mi_fchd	timemifc
95148	52	2	0	1	0	1	3607
95148	58	2	2128	2	0	1	3607
95148	64	2	4192	3	1	1	3607
477082	38	1	0	1	0	1	1718
477082	44	1	2119	2	1	1	1718
1140225	58	2	0	1	0	0	8766
1140225	64	2	2172	2	0	0	8766
1140225	69	2	4287	3	0	0	8766

Participants are often enrolled in an observational study without regard to past medical history. The study investigators will review the medical record to determine if the participant had any preexisting disease at the time of the first study examination. If preexisting disease is found, then the data for that subject will reflect prevalent disease at the first exam; however, the subject will continue to be followed for any new events. All participants, regardless of their prevalent disease status, will continue to be followed and events recorded until the study ends, the participant dies, or the participant cannot be contacted to ascertain status (lost to follow-up). For participants who enter the study free of disease, the incident events are used to determine prevalent disease status at later exams. For the three participants here, none entered the study with prevalent disease, and using period 1 as the baseline exam, the population at risk could be defined using code similiar to the following SAS code:

```
data work;
 set frmgham;
 if period=1 and prevchd=0;
run;
```

The data would appear as the following:

RANDID	age	SEX	time	period	prevchd	mi_fchd	timemifc
95148	52	2	0	1	0	1	3607
477082	38	1	0	1	0	1	1718
1140225	58	2	0	1	0	0	8766

The population at risk consists of all three participants (prevchd = 0), and follow-up time for the event of hospitalized MI or fatal CHD would be the time indicated under TIMEMIFC. The first two participants (95148 and 477082) would be regarded as having an incident event during follow-up.

Likewise, the second examination, or period = 2, could also be used as a baseline exam. The full data set can be subset to include only those at risk at the start of the second period. For example,

```
data work;
 set frmgham;
 if period=2 and prevchd=0;
run;
```

Since time to event is provided as days since the first visit, a new time variable would need to be created so that number of days under study extends from the second exam until the end of follow-up:

```
newtime=timemifc-time;
```

The revised data set that includes the population at risk beginning at period = 2 and extends until the end of follow-up would be

RANDID	age	SEX	time	period	prevchd	mi_fchd	timemifc	newtime
95148	58	2	2128	2	0	1	3607	1479
1140225	64	2	2172	2	0	0	8766	6594

The population at risk (those free of prevalent disease) now includes only participants 95148 and 1140225. The variable NEWTIME correctly reflects the number of days of follow-up from the second exam, or period = 2, until the first event or a censor point.

The same procedure can be used to define the third exam, or period = 3, as the baseline exam.

For more complex analyses, such as time-dependent analysis, or a counting process style of input, the user would have to subset the population to those free of disease at all exams, and event data would have to be modified to reflect when the event occurred relative to the examinations. Consider the following SAS code, which would modify the data set to a counting process style of input for an analysis on the Hospitalized MI-Fatal CHD endpoint. The variable NEWEVNT is modified from MI_FCHD so that the event indicator is '1' only once for each participant. The variables TIME and END-TIME define the interval the subject is at risk.

```
data analysis;
 set work;
 if prevchd=0;
run;

proc sort data=analysis;
 by randid descending period;
run;

data analysis;
 set analysis;
 by randid;

newevnt=mi_fchd;
 retain exmtime;
 if first.randid then do;
    endtime=timemifc; exmtime=time;
  end;
  else do;
    newevnt=0;
    endtime=exmtime;
    exmtime=time;
  end;
 run;

proc sort data=analysis;
 by randid period;
run;
```

The data would appear, for example, as follows for three participants:

RANDID	age	SEX	period	time	endtime	newevnt	mi_fchd	timemifc
11263	43	2	1	0	2178	0	1	5719
11263	49	2	2	2178	4351	0	1	5719
11263	55	2	3	4351	5719	1	1	5719
12629	63	2	1	0	8766	0	0	8766
9069458	42	2	1	0	4362	0	0	8766
9069458	54	2	3	4362	8766	0	0	8766

SAS PROC CONTENTS PROCEDURE ON FRAMINGHAM LONGITUDINAL DATASET

The CONTENTS Procedure

Data Set Name: WORK.FRMGHAM	Observations:	11627
Member Type: DATA	Variables:	38
Engine: V8	Indexes:	0
Created: 14:50 Tuesday, March 2, 2004	Observation Length:	304
Last Modified: 14:50 Tuesday, March 2, 2004	Deleted Observations:	0
Protection:	Compressed:	NO
Data Set Type:	Sorted:	NO
Label:		

-----Engine/Host Dependent Information-----

Data Set Page Size:	16384
Number of Data Set Pages:	220
First Data Page:	1
Max Obs per Page:	53
Obs in First Data Page:	35
Number of Data Set Repairs:	0
Release Created:	8.0202M0
Host Created:	WIN_PRO

The CONTENTS Procedure

-----Variables Ordered by Position-----

#	Variable	Type	Len	Label
1	SEX	Num	4	SEX
2	RANDID	Num	8	Random ID
3	totchol	Num	8	Serum Cholesterol mg/dL
4	age	Num	8	Age (years) at examination
5	sysbp	Num	8	Systolic BP mmHg
6	diabp	Num	8	Diastolic BP mmHg
7	cursmoke	Num	8	Current Cig Smoker Y/N
8	cigpday	Num	8	Cigarettes per day
9	bmi	Num	8	Body Mass Index (kg/(M*M)
10	diabetes	Num	8	Diabetic Y/N
11	bpmeds	Num	8	Anti-hypertensive meds Y/N
12	heartrte	Num	8	Ventricular Rate (beats/min)
13	glucose	Num	8	Casual Glucose mg/dL
14	prevchd	Num	8	Prevalent CHD (MI,AP,CI)
15	prevap	Num	8	Prevalent Angina
16	prevmi	Num	8	Prevalent MI (Hosp,Silent)
17	prevstrk	Num	8	Prevalent Stroke (Infarct,Hem)

18	prevhyp	Num	8	Prevalent Hypertension
19	time	Num	8	Days since Index Exam
20	period	Num	8	Examination cycle
21	hdlc	Num	8	HDL Cholesterol mg/dL
22	ldlc	Num	8	LDL Cholesterol mg/dL
23	death	Num	8	Death indicator
24	angina	Num	8	Incident Angina Pectoris
25	hospmi	Num	8	Incident Hospitalized MI
26	mi_fchd	Num	8	Incident Hosp MI-Fatal CHD
27	anychd	Num	8	Incident Hosp MI, AP, CI, Fatal CHD
28	stroke	Num	8	Incident Stroke Fatal/non-fatal
29	cvd	Num	8	Incident Hosp MI or Stroke, Fatal or Non
30	hyperten	Num	8	Incident Hypertension
31	timeap	Num	8	Days Baseline-Inc Angina
32	timemi	Num	8	Days Baseline-Inc Hosp MI
33	timemifc	Num	8	Days Baseline-Inc MI-Fatal CHD
34	timechd	Num	8	Days Baseline-Inc Any CHD
35	timestrk	Num	8	Days Baseline-Inc Stroke
36	timecvd	Num	8	Days Baseline-Inc CVD
37	timedth	Num	8	Days Baseline-Death
38	timehyp	Num	8	Days Baseline-Inc Hypertension

Distributions of selected variables by period and sex

Examination cycle 1

Means selected Risk factors	N	NMiss	Mean	Std	Min	P25	Median	P75	Max
Men									
Days since Index Exam	1944	0	0.00	0.00	0.00	0.00	0.00	0.00	0.00
Age (years) at examination	1944	0	49.79	8.72	33.00	42.00	49.00	57.00	69.00
Body Mass Index (kg/(M*M)	1939	5	26.17	3.41	15.54	23.97	26.08	28.32	40.38
Systolic BP mmHg	1944	0	131.74	19.44	83.50	118.00	129.00	141.50	235.00
Diastolic BP mmHg	1944	0	83.71	11.44	48.00	76.00	82.00	90.00	136.00
Serum Cholesterol mg/dL	1937	7	233.58	42.36	113.00	206.00	231.00	259.00	696.00
HDL Cholesterol mg/dL	0	1944	.	.	.	.	.	.	.
LDL Cholesterol mg/dL	0	1944	.	.	.	.	.	.	.
Casual Glucose mg/dL	1824	120	82.32	24.72	40.00	71.00	78.00	87.00	394.00
Cigarettes per day	1928	16	13.23	13.78	0.00	0.00	10.50	20.00	70.00
Ventricular Rate (beats/min)	1943	1	74.40	11.90	44.00	66.00	75.00	80.00	130.00
Women									
Days since Index Exam	2490	0	0.00	0.00	0.00	0.00	0.00	0.00	0.00
Age (years) at examination	2490	0	50.03	8.64	32.00	43.00	49.00	57.00	70.00
Body Mass Index (kg/(M*M)	2476	14	25.59	4.56	15.96	22.54	24.83	27.82	56.80
Systolic BP mmHg	2490	0	133.82	24.46	83.50	116.00	128.50	146.50	295.00
Diastolic BP mmHg	2490	0	82.60	12.50	50.00	74.00	81.00	89.00	142.50
Serum Cholesterol mg/dL	2445	45	239.68	46.22	107.00	206.00	237.00	269.00	600.00
HDL Cholesterol mg/dL	0	2490	.	.	.	.	.	.	.
LDL Cholesterol mg/dL	0	2490	.	.	.	.	.	.	.
Casual Glucose mg/dL	2213	277	82.07	24.14	40.00	72.00	78.00	86.00	394.00
Cigarettes per day	2474	16	5.65	8.96	0.00	0.00	0.00	10.00	50.00
Ventricular Rate (beats/min)	2490	0	77.06	12.15	46.00	69.00	75.00	85.00	143.00

Means selected Risk factors	N	NMiss	Mean	Std	Min	P25	Median	P75	Max
Examination cycle 2									
Men									
Days since Index Exam	1691	0	2173.67	72.44	1577.00	2142.00	2174.00	2205.00	2520.00
Age (years) at examination	1691	0	55.10	8.51	39.00	48.00	54.00	62.00	75.00
Body Mass Index (kg/(M*M)	1685	6	26.23	3.40	16.24	24.05	26.09	28.23	39.46
Systolic BP mmHg	1691	0	135.48	19.90	88.00	120.00	132.00	148.00	216.00
Diastolic BP mmHg	1691	0	84.61	10.91	53.00	78.00	84.00	91.00	124.00
Serum Cholesterol mg/dL	1666	25	241.82	42.14	115.00	214.00	240.00	266.00	614.00
HDL Cholesterol mg/dL	0	1691	.	.	.	.	.	.	.
LDL Cholesterol mg/dL	0	1691	.	.	.	.	.	.	.
Casual Glucose mg/dL	1518	173	82.24	23.31	40.00	70.00	77.00	88.00	362.00
Cigarettes per day	1682	9	12.23	15.04	0.00	0.00	2.00	20.00	90.00
Ventricular Rate (beats/min)	1691	0	75.92	12.66	42.00	68.00	75.00	83.00	130.00
Women									
Days since Index Exam	2239	0	2176.22	76.20	1633.00	2144.00	2175.00	2207.00	2765.00
Age (years) at examination	2239	0	55.66	8.56	39.00	48.00	55.00	62.00	76.00
Body Mass Index (kg/(M*M)	2229	10	25.65	4.58	15.33	22.54	24.88	27.85	56.80
Systolic BP mmHg	2239	0	138.06	24.30	88.00	121.00	134.00	151.00	282.00
Diastolic BP mmHg	2239	0	83.57	11.79	47.00	76.00	82.00	90.00	150.00
Serum Cholesterol mg/dL	2121	118	255.67	47.53	122.00	223.00	252.00	285.00	638.00
HDL Cholesterol mg/dL	0	2239	.	.	.	.	.	.	.
LDL Cholesterol mg/dL	0	2239	.	.	.	.	.	.	.
Casual Glucose mg/dL	1931	308	81.76	21.32	39.00	71.00	78.00	87.00	420.00
Cigarettes per day	2215	24	5.97	10.00	0.00	0.00	0.00	10.00	60.00
Ventricular Rate (beats/min)	2238	1	78.36	12.76	45.00	70.00	75.00	85.00	220.00
Examination cycle 3									
Men									
Days since Index Exam	1387	0	4353.75	97.74	3748.00	4312.00	4361.00	4403.00	4816.00
Age (years) at examination	1387	0	60.35	8.19	45.00	53.00	60.00	67.00	80.00
Body Mass Index (kg/(M*M)	1380	7	26.22	3.49	14.43	24.02	26.09	28.25	45.43
Systolic BP mmHg	1387	0	139.26	21.15	91.00	123.00	136.00	152.00	225.00
Diastolic BP mmHg	1387	0	82.55	11.29	30.00	75.00	81.50	90.00	123.00
Serum Cholesterol mg/dL	1312	75	225.74	41.13	130.00	198.00	222.00	252.00	413.00
HDL Cholesterol mg/dL	1304	83	43.71	13.30	10.00	35.00	42.00	51.00	138.00
LDL Cholesterol mg/dL	1304	83	170.55	44.66	34.00	140.00	167.50	199.00	376.00
Casual Glucose mg/dL	1163	224	91.17	28.99	49.00	77.00	85.00	97.00	423.00
Cigarettes per day	1380	7	8.70	13.51	0.00	0.00	0.00	20.00	80.00
Ventricular Rate (beats/min)	1387	0	75.88	12.73	43.00	66.00	75.00	85.00	150.00
Women									
Days since Index Exam	1876	0	4353.61	93.13	3919.00	4313.00	4362.00	4402.50	4854.00
Age (years) at examination	1876	0	60.87	8.37	44.00	54.00	60.00	67.00	81.00
Body Mass Index (kg/(M*M)	1866	10	25.65	4.45	14.53	22.59	24.80	27.94	56.80
Systolic BP mmHg	1876	0	140.92	24.14	86.00	123.00	138.00	156.00	267.00
Diastolic BP mmHg	1876	0	81.23	11.23	46.00	73.00	80.00	88.00	130.00
Serum Cholesterol mg/dL	1737	139	245.00	45.08	112.00	214.00	242.00	270.00	625.00
HDL Cholesterol mg/dL	1723	153	53.64	15.90	11.00	43.00	52.00	62.00	189.00
LDL Cholesterol mg/dL	1722	154	180.95	48.00	20.00	149.00	177.00	208.00	565.00
Casual Glucose mg/dL	1538	338	88.72	27.48	46.00	76.00	84.00	95.00	478.00
Cigarettes per day	1869	7	5.35	9.78	0.00	0.00	0.00	8.00	60.00
Ventricular Rate (beats/min)	1872	4	78.45	12.20	37.00	70.00	77.00	85.00	130.00

Examination cycle

	Cycle 1 Men		Cycle 1 Women		Cycle 2 Men		Cycle 2 Women		Cycle 3 Men		Cycle 3 Women	
	N	Percent	N	Percent	N	Percent	N	Percent	N	Percent	N	Percent
Total	1944	100.00	2490	100.00	1691	100.00	2239	100.00	1387	100.00	1876	100.00
Current Cig Smoker Y/N												
No	769	39.56	1484	59.60	811	47.96	1392	62.17	848	61.14	1294	68.98
Yes	1175	60.44	1006	40.40	880	52.04	847	37.83	539	38.86	582	31.02
Diabetic Y/N												
No	1885	96.97	2428	97.51	1617	95.62	2158	96.38	1267	91.35	1742	92.86
Yes	59	3.03	62	2.49	74	4.38	81	3.62	120	8.65	134	7.14
Anti-hypertensive meds Y/N												
Missing	22	1.13	39	1.57	37	2.19	49	2.19	189	13.63	257	13.70
No	1880	96.71	2349	94.34	1553	91.84	1920	85.75	1060	76.42	1328	70.79
Yes	42	2.16	102	4.10	101	5.97	270	12.06	138	9.95	291	15.51
Prevalent CHD (MI,AP,CI)												
No	1820	93.62	2420	97.19	1516	89.65	2126	94.95	1187	85.58	1716	91.47
Yes	124	6.38	70	2.81	175	10.35	113	5.05	200	14.42	160	8.53
Prevalent MI (Hosp,Silent)												
No	1874	96.40	2474	99.36	1588	93.91	2212	98.79	1272	91.71	1833	97.71
Yes	70	3.60	16	0.64	103	6.09	27	1.21	115	8.29	43	2.29
Prevalent Angina												
No	1852	95.27	2435	97.79	1564	92.49	2146	95.85	1254	90.41	1749	93.23
Yes	92	4.73	55	2.21	127	7.51	93	4.15	133	9.59	127	6.77
Prevalent Stroke (Infarct,Hem)												
No	1930	99.28	2472	99.28	1675	99.05	2204	98.44	1357	97.84	1837	97.92
Yes	14	0.72	18	0.72	16	0.95	35	1.56	30	2.16	39	2.08
Prevalent Hypertension												
No	1313	67.54	1691	67.91	841	49.73	1130	50.47	542	39.08	766	40.83
Yes	631	32.46	799	32.09	850	50.27	1109	49.53	845	60.92	1110	59.17

Event Counts by sex

Counts of Endpoints by Sex	SEX			
	Men		Women	
	N	Percent	N	Percent
All	1944	100.00	2490	100.00
Incident Hypertension				
No	540	27.78	642	25.78
Yes	1404	72.22	1848	74.22
Incident Angina Pectoris				
No	1561	80.30	2148	86.27
Yes	383	19.70	342	13.73
Incident Hospitalized MI				
No	1624	83.54	2356	94.62
Yes	320	16.46	134	5.38
Incident Hosp MI-Fatal CHD				
No	1453	74.74	2250	90.36
Yes	491	25.26	240	9.64
Incident Stroke Fatal/non-fatal				
No	1751	90.07	2268	91.08
Yes	193	9.93	222	8.92
Incident Hosp MI, AP, CI, Fatal CHD				
No	1234	63.48	1960	78.71
Yes	710	36.52	530	21.29
Incident Hosp MI or Stroke, Fatal or Non				
No	1258	64.71	2019	81.08
Yes	686	35.29	471	18.92
Death indicator				
No	1101	56.64	1783	71.61
Yes	843	43.36	707	28.39

Distributions of Time to Event by sex

Time to Event	N	NMiss	Mean	Std	Min	P25	Median	P75	Max
Men									
Days Baseline-Inc Hypertension	1944	0	3313	3391	0	0	2156	6491	8766
Days Baseline-Inc Angina	1944	0	6507	2929	0	4572	8486	8766	8766
Days Baseline-Inc Hosp MI	1944	0	6736	2771	0	5006	8766	8766	8766
Days Baseline-Inc MI-Fatal CHD	1944	0	6655	2816	0	4822	8743	8766	8766
Days Baseline-Inc Stroke	1944	0	7003	2509	0	5608	8766	8766	8766
Days Baseline-Inc Any CHD	1944	0	6156	3067	0	3853	7653	8766	8766
Days Baseline-Inc CVD	1944	0	6274	3015	0	4009	7895	8766	8766
Days Baseline-Death	1944	0	7194	2386	26	6047	8766	8766	8766
Women									
Days Baseline-Inc Hypertension	2490	0	3532	3496	0	0	2219	7340	8766
Days Baseline-Inc Angina	2490	0	7209	2559	0	6132	8766	8766	8766
Days Baseline-Inc Hosp MI	2490	0	7634	2154	0	7541	8766	8766	8766
Days Baseline-Inc MI-Fatal CHD	2490	0	7600	2197	0	7452	8766	8766	8766
Days Baseline-Inc Stroke	2490	0	7540	2262	0	7283	8766	8766	8766
Days Baseline-Inc Any CHD	2490	0	7065	2656	0	5618	8766	8766	8766
Days Baseline-Inc CVD	2490	0	7243	2549	0	6241	8766	8766	8766
Days Baseline-Death	2490	0	7749	2037	34	8016	8766	8766	8766

Age Specific Angina and Hospitalized MI-Fatal CHD Incidence Rates by Sex

		Angina				Hospitalized MI - Fatal CHD		
	N*	Person Years	Events	Rate/ 1,000PY	N*	Person Years	Events	Rate/ 1,000PY
Men								
35-44	649	3,053	12	3.9	644	3,037	8	2.6
45-54	1,278	9,587	52	5.4	1,269	9,498	67	7.1
55-64	1,646	12,241	135	11.0	1,629	12,274	154	12.5
65-74	1,115	7,488	78	10.4	1,125	7,623	117	15.3
75-84	416	2,165	13	6.0	432	2,210	43	19.5
85+	52	93	1	10.8	54	97	6	62.0
Women								
35-44	783	3,765	3	0.8	783	3,769	2	0.5
45-54	1,634	12,316	26	2.1	1,631	12,400	12	1.0
55-64	2,229	17,261	123	7.1	2,238	17,675	60	3.4
65-74	1,640	11,679	98	8.4	1,705	12,421	78	6.3
75-84	707	3,815	35	9.2	769	4,262	55	12.9
85+	106	287	2	7.0	121	316	7	22.1

N* - Number of persons contributing person years to that age group. Incidence rates are calculated using derived age at time of event.

Age Specific Stroke and Cardiovascular Disease (Fatal and Non-Fatal) Incidence Rates by Sex

		Stroke				Cardiovascular Disease (CVD)		
	N*	Person Years	Events	Rate/ 1,000PY	N*	Person Years	Events	Rate/ 1,000PY
Men								
35-44	655	3,082	1	0.3	643	3,010	13	4.3
45-54	1,313	9,921	14	1.4	1,260	9,353	95	10.2
55-64	1,743	13,293	42	3.2	1,588	11,769	202	17.2
65-74	1,256	8,471	74	8.7	1,058	6,920	185	26.7
75-84	477	2,402	44	18.3	378	1,839	75	40.8
85+	50	97	4	41.1	41	65	9	138.0
Women								
35-44	782	3,761	2	0.5	781	3,759	5	1.3
45-54	1,638	12,420	10	0.8	1,621	12,282	31	2.5
55-64	2,283	17,932	47	2.6	2,209	17,180	133	7.7
65-74	1,760	12,713	83	6.5	1,631	11,588	148	12.8
75-84	774	4,230	52	12.3	695	3,737	85	22.7
85+	124	322	10	31.0	103	264	15	56.8

N* - Number of persons contributing person years to that age group. Incidence rates are calculated using derived age at time of event.
For CVD endpoint, population at risk defined by PREVCHD=0 AND PREVSTRK=0

Index